T0312281

Numerical Methods for Partial Differential Equations

Finite Difference and Finite Volume Methods

Numerical Methods for Partial Differential Equations
Finite Difference and Finite Volume Methods

Sandip Mazumder
The Ohio State University

AMSTERDAM • BOSTON • HEIDELBERG • LONDON
NEW YORK • OXFORD • PARIS • SAN DIEGO
SAN FRANCISCO • SINGAPORE • SYDNEY • TOKYO

Academic Press is an imprint of Elsevier

Academic Press is an imprint of Elsevier
125, London Wall, EC2Y 5AS, UK
525 B Street, Suite 1800, San Diego, CA 92101-4495, USA
225 Wyman Street, Waltham, MA 02451, USA
The Boulevard, Langford Lane, Kidlington, Oxford OX5 1GB, UK

Notices
Knowledge and best practice in this field are constantly changing. As new research and experience
broaden our understanding, changes in research methods, professional practices, or medical treatment
may become necessary.

Practitioners and researchers must always rely on their own experience and knowledge in evaluating and
using any information, methods, compounds, or experiments described herein. In using such information
or methods they should be mindful of their own safety and the safety of others, including parties for whom
they have a professional responsibility.

To the fullest extent of the law, neither the Publisher nor the authors, contributors, or editors, assume any
liability for any injury and/or damage to persons or property as a matter of products liability, negligence
or otherwise, or from any use or operation of any methods, products, instructions, or ideas contained in
the material herein.

Library of Congress Cataloging-in-Publication Data
A catalog record for this book is available from the Library of Congress

British Library Cataloguing-in-Publication Data
A catalogue record for this book is available from the British Library

ISBN: 978-0-12-849894-1

For information on all Academic Press publications
visit our website at http://store.elsevier.com/

Working together
to grow libraries in
developing countries

www.elsevier.com • www.bookaid.org

To my teachers

Contents

Instructor website for this book is available in
http://textbooks.elsevier.com/web/product_details.aspx?isbn=9780128498941

About the Author

Sandip Mazumder was born in Calcutta, India. Following his bachelor's degree in Mechanical Engineering from the Indian Institute of Technology, Kharagpur, he started his graduate education in the autumn of 1991. In 1997, he graduated with a PhD in Mechanical Engineering from the Pennsylvania State University. After graduation, he joined CFD Research Corporation, where he was one of the architects and early developers of the commercial computational fluid dynamics code CFD-ACE +. In 2004, he joined the Ohio State University, where he presently teaches both graduate and undergraduate courses in heat and mass transfer, thermodynamics, numerical methods, and computational fluid dynamics. He is the author of over 50 journal publications, which have been cited more than 2000 times. Dr Mazumder is the recipient of several research and teaching awards, and is a Fellow of the American Society of Mechanical Engineers.

Preface

The desire to write this book has stemmed from the realization of the undeniable benefits of formalizing, organizing, and cataloging material that I have developed and, to some extent, collected, for more than a decade as part of an annual ritual – that of teaching an entry-level graduate course entitled "Intermediate Numerical Methods." Through this book, I also intend to pay tribute to many great teachers who taught me the fundamentals of numerical methods in graduate school, but have never had the time or inclination to put their deep knowledge of the subject in writing.

Tremendous advancement in computer hardware over the past two decades has resulted in the proliferation of scientific and engineering analysis, and powerful user-friendly tools that enable such analysis. Modeling and simulation has now permeated the industry and is slowly beginning to occupy a position of importance in the design cycle of many products. Students graduating today with a degree in engineering or applied sciences are almost mandated by their recruiters to have a working knowledge of the preferred analysis tool in their discipline.

At the heart of many engineering and scientific analyses is the solution of differential equations – both ordinary and partial differential equations (PDEs). The solution of the latter type of equation can be very challenging, depending on the type of equation, the number of independent variables, the boundary and initial conditions, and other factors. A variety of broadly applicable methods have been developed to this end. Among the deterministic methods for solving differential equations, the most popular ones are the finite element method, the finite difference method, and the finite volume method. Each method has its own *pros* and *cons*, and shines for a certain class of problems, for reasons that are deeply rooted in the mathematical foundation of the method. Although trends are slowly changing, the finite element method has been traditionally used for solving problems in solid mechanics, while the finite difference and finite volume methods have been traditionally used to solve problems involving fluid flow and heat transfer. These boundaries, though, are strictly defined by history, not by the underlying mathematics.

Although this book is supposedly a general book on numerical methods for solving PDEs, and therefore, quite rightly, is expected to cover all relevant topics, certain practical constraints, such as the size of the book, as well as assessment of what material exists in other texts, has prompted me to exclude many topics. For example, I have chosen to adhere to deterministic methods only, and stochastic methods for solving PDEs have been excluded. Perhaps, the most important topic that has been excluded is the finite element method. This important decision was prompted by the fact that the finite element method has a deep-rooted history, and several well-written, high-quality texts already exist on this topic. Thus, in this book, I have chosen to focus on the finite difference and finite volume methods. In my quest for finding a book that adequately covers all aspects of the finite difference and finite volume methods, I have found that either the coverage is incomplete or too primitive. For example, many books cover the finite difference method in great detail, with a passing reference to the finite volume method, almost as if these methods are synonymous, when, in fact, the two methods are fundamentally of different philosophies. Similarly, most existing books rarely go beyond a Cartesian stencil and, in some cases, coordinate transformations (body-fitted mesh). The all-important topic of finite volume formulation for an unstructured mesh, which is the foundation, for example, of modern computational fluid dynamics codes, is usually missing from most texts.

The finite difference and finite volume methods have their early roots in computational fluid dynamics and computational heat transfer areas. Most texts on computational fluid dynamics commence with a discussion of these two methods. However, it is uncommon to find ample coverage of the application of these two methods to elliptic PDEs. Such texts, for example, dwell at length on treatment of the time-derivative and explicit time-marching methods as applied to the Euler equation with little or no coverage of the implicit solution of the algebraic equations that might result from discretization of an elliptic operator. While linear algebra is a routine topic in applied mathematics texts, in my opinion, there is a dearth of textbook literature that teaches an entry-level graduate student how to solve a relatively large system of algebraic equations by iterative methods. It is true that with the advent of linear algebraic equation solver suites (libraries) such as PetSc, it is no longer necessary to develop solvers from the ground up. However, in many instances, use of these libraries is overkill, and a student would, arguably, benefit from developing a simple solver from the ground up in some situations. If nothing else, fundamental knowledge of how iterative sparse matrix solvers and preconditioners function can be of great value to a budding numerical analyst. After all, when all is said and done, the computer is set the task of solving a system of linear algebraic equations, no matter what the underlying method of discretization. Therefore, I consider solution of a system of linear algebraic equations by iterative methods an important and all-encompassing topic, and have devoted significant coverage to it.

The sequence of topics covered in this text is one that I have converged upon after numerous iterations of teaching the aforementioned course. Chapter 1 commences with an overview of PDEs, and the various popular methods for solving them. Chapter 2 covers the fundamentals of the finite difference method. It introduces the reader to the basic methodology of using Taylor series expansions to develop algebraic approximations to derivatives, estimation of discretization errors, and other related concepts. The presentation is easy enough for an entry-level graduate student in science or engineering – one who is equipped with a basic knowledge of calculus and algebra. All of the concepts developed in this chapter are general and also carry over to the finite volume method in later chapters. The step that immediately follows the derivation of the linear system of equations is their solution. Thus, Chapter 3 dwells on solution of the linear system with an emphasis on iterative solvers. The sequence followed here is one that increases in degree of complexity starting from point-wise (Jacobi) iterations to conjugate gradient solvers. A benchmark boundary value problem is considered to demonstrate the *pros* and *cons* of each solver. Chapter 4 covers the theory behind the stability and convergence of iterative solvers – the objective being to elucidate the reader on the "why" of each method that was covered in the preceding chapter. The discussion in the first half of Chapter 4, in fact, lays the foundation for the multigrid method, which is discussed in the second half of the same chapter. Chapter 5 switches gears and delves into parabolic and hyperbolic PDEs, with particular emphasis on the treatment of the time derivative. Both explicit and implicit methods are covered, and accompanied by formal stability analysis of each method. Chapter 6 introduces the reader to the finite volume method. Benchmark problems are explored to demonstrate the conservation properties of the finite volume method – an issue that is often overlooked in many numerical methods texts. Chapter 7 extends the idea of the finite volume method to control volumes of arbitrary shape (unstructured mesh). Mathematical formulations and algorithmic steps needed to develop an unstructured finite volume code from the ground up are systematically laid out in this chapter. The book could have ended after Chapter 7 since Chapters 1–7 take the reader all the way from discretization of the governing PDE to the solution of the resulting algebraic equations. However, in the final chapter (Chapter 8), I decided to include a few important topics that are encountered routinely by practicing numerical analysts for postprocessing of solutions and for treatment of nonlinearities: interpolation, numerical integration, and root finding of nonlinear equations. Chapter 8 also introduces the reader to the solution of coupled PDEs – a situation that is often encountered in practical problems. Clearly, Chapter 8 can be treated independently of the other chapters. As far as the other chapters are concerned, I have tried to present the material in a manner that allows the chapters to be grouped into modules. For example, Chapters 3 and 4 may be thought of as the "solution" module, and an instructor, who may be teaching a finite element course, may find this material useful and suitable for his or her students. One does not need to read through Chapters 1 and 2 in order to benefit from Chapters 3 and 4. Also, rather than focus on any particular programming language or paradigm, I have presented algorithms as bulleted lists, and in some cases, Matlab-style code snippets that can be easily understood even by readers who have little programming experience. A conscious effort has been made to emphasize the "ground up" approach to learning the concepts. Rather than choose examples that "advertise" methods, I have chosen example problems that are simple enough for the reader to replicate, while still demanding a decent level of understanding and aptitude in numerical methods and computer programming.

Throughout the process of writing this book, my endeavor has been to present methods and concepts that are applicable to canonical PDEs of the three types, namely elliptic, parabolic, and hyperbolic, rather than equations with a distinctive disciplinary flavor. For example, it is my opinion that detailed discussion of advection and the treatment of the advective flux are best suited to texts on computational fluid dynamics rather than a text such as this one. That being the case, wherever possible, I have attempted to make a connection between the canonical PDEs and equations commonly encountered in physics and engineering. Since my own background is in thermofluid sciences, perhaps, involuntarily, I have drawn more examples from this discipline than any other. For that, I take full responsibility.

This book is not intended for advanced numerical analysts. On the other hand, it is not intended for undergraduate students who are encountering numerical analysis for the very first time. It is intended for beginning graduate students in physical sciences and engineering disciplines, although advanced undergraduate students may find it equally useful. The material in this text is meant to serve as a prerequisite for students who might go on to take additional courses in computational mechanics, computational fluid dynamics, or computational electromagnetics. The notations, language, and technical jargon used in the book are those that can be easily understood by scientists and engineers who may not have completed graduate-level applied mathematics or computer science courses. Readers with a strong applied mathematics background may find this book either too slow or dilute in rigor. On the other hand, readers with an engineering or science background are likely to find both the pace and rigor optimum for their taste. It is my earnest desire to see this book serve as a lifelong companion to graduate students in science and engineering disciplines.

The greatest critic of the contents of this book and the style of presentation has been the graduate students in my classroom. Over the years, their feedback has not only led to improvement of the technical content, but, most importantly, also provided a solid perception of the logical sequence of topics, the depth at which they need to be covered, as well as the presentation style. I am deeply indebted to all my students for their constructive criticism. My parents have sacrificed a lot for my education. For that, and their unconditional love and support, I am eternally grateful. Finally, I would like to thank my wife, Srirupa, and my son, Abhik, for their endless love, support, and understanding during the process of writing this book. They have always been, and will always be, the joy of my life.

Sandip Mazumder
Columbus, Ohio
2015

List of Symbols

a_i	Coefficients in polynomial or spline $[-]$
a_i	Subdiagonal elements of tridiagonal matrix $[-]$
a_i	Link coefficients $[-]$
A	Area $[\text{m}^2]$
$A_{i,j}$	Elements of coefficient matrix $[-]$
A_k	Link coefficients $[-]$
$[A]$	Generic coefficient matrix $[-]$
A, B, C	Constants in general form of partial differential equation $[-]$
$[B]$	Iteration or amplification matrix $[-]$
c	Speed of wave $[\text{m/s}]$
c_i	Superdiagonal elements of tridiagonal matrix $[-]$
C_m	Amplitude of mth Fourier mode $[-]$
d_i	Diagonal elements of tridiagonal matrix $[-]$
$[D]$	Direction vector (CG and related methods) $[-]$
E, F, D, B, H	Diagonals in Stone's method $[-]$
G	Component of mass flux vector $[\text{kg/s/m}^2]$
I	Value of integral $[-]$
I_T	Integral evaluated using trapezoidal rule $[-]$
$[J]$	Jacobian matrix of forward transformation $[-]$
J	Determinant of Jacobian matrix of forward transformation $[-]$
J	Components of flux vector $[\text{s}^{-1}\text{m}^{-2}]$
$[J]$	Jacobian matrix in Newton's method $[-]$
$\mathbf{J}$	Flux vector $[\text{s}^{-1}\text{m}^{-2}]$
K	Total number of equations in linear system $[-]$
$\mathbf{l}$	Vector joining two adjacent cell centers $[\text{m}]$
L	Length of 1D computational domain $[\text{m}]$
$\mathcal{L}$	Lagrange basis function $[-]$
M	Number of nodes in y direction $[-]$
M_c	Number of cells in y direction $[-]$
$\hat{\mathbf{n}}$	Outward pointing unit surface normal vector $[-]$
N	Total number of nodes or number of nodes in x direction $[-]$
N_c	Total number of cells or number of cells in x direction $[-]$
$p(x)$	Polynomial function $[-]$
Pe	Peclet number
$[Q]$	Source or right-hand side vector in linear system $[-]$
r	Radial coordinate (cylindrical or spherical) $[\text{m}]$
r_o	Outer radius (cylindrical coordinates) $[\text{m}]$
R	Courant or CFL number $[-]$
$[R]$	Residual vector $[-]$
$R2$	L^2 Norm of residual vector $[-]$
$s_i(x)$	Local spline function $[-]$

S	Surface area of control volume or cell [m^2]
$S(x)$	Global spline function [$-$]
S_ϕ	Generic source term or production rate of ϕ [s^{-1}m^{-3}]
S_i	Source term or production rate of ϕ at node or cell or vertex i [s^{-1}m^{-3}]
t	Time [s]
$\hat{\mathbf{t}}$	Unit tangent vector [$-$]
Δt	Time step size [s]
u, v, w	Cartesian components of velocity vector [m/s]
$\mathbf{U}$	Velocity vector [m/s]
U_i	Components of contravariant velocity vector [m/s]
v	Test function [$-$]
V	Volume [m^3]
V_i	Volume of cell i [m^3]
V_Σ	Volume of entire computational domain [m^3]
w	Interpolation function (either cell to face or cell to vertex) [$-$]
w_i	Weights in Gaussian quadrature formula [$-$]
x_1, x_2, x_3	Cartesian coordinates [m]
x, y, z	Cartesian coordinates [m]
$\Delta x, \Delta y$	Grid spacing [m]

Greek

α	Stone's calibration or fudge factor [$-$]
α	Inertial damping factor [$-$]
α	Parameter in MSD and CG method [$-$]
α	Generic constant in parabolic PDE [m^2/s]
α, β, γ	Coefficients in expression for Robin boundary condition [$-$]
β	Parameter in CG method [$-$]
$[\beta]$	Cofactor matrix of Jacobian matrix of forward transformation [$-$]
β_{ij}	Elements of cofactor matrix [$-$]
δ	Distance between adjacent cell centers in direction of surface normal [m]
ε	Error [$-$]
ϕ	Generic dependent variable [$-$]
Φ	Vector of generic dependent variables [$-$]
$\phi', \Delta\phi$	Correction (or change) in ϕ between successive iterations [$-$]
ϕ_i	Value of ϕ at node i [$-$]
Γ	Generic transport (diffusion) coefficient [m^2/s]
η_i	Location of ith knot of spline [$-$]
κ	Condition number [$-$]
λ	Eigenvalue [$-$]
λ_m	Amplification factor for mth Fourier mode [$-$]
λ_{SR}	Spectral radius of convergence [$-$]
v	Kinematic viscosity [m^2/s]

θ	Azimuthal (cylindrical) or polar (spherical) angle [rad]
θ	Phase angle in Fourier series [rad]
ρ	Mass density [kg/m³]
ψ_i	Basis function [−]
ω	Linear relaxation factor [−]
ξ_1, ξ_2, ξ_3	Curvilinear coordinates [−]
ξ_m	Complex amplitude of mth Fourier mode [−]

Subscripts

B	At boundary
C	Coarse grid
f	At face
e, w, n, s	Directional face indices used for structured mesh
E, W, N, S	Directional cell or nodal indices used for structured mesh
F	Fine grid
i, j, k	Spatial node or cell index
n	Time index
nb	At neighboring cell
v	At vertex

Superscripts

E	Exact
(n)	Iteration counter
$*$	Previous iteration or complex conjugate
$'$	Change between iterations

Abbreviations and Acronyms

1D	One-dimensional
2D	Two-dimensional
3D	Three-dimensional
CG	Conjugate gradient
CGS	Conjugate gradient squared
FDM	Finite difference method
FEM	Finite element method
FVM	Finite volume method
AMG	Algebraic multigrid
GMG	Geometric multigrid
MG	Multigrid
ODE	Ordinary differential equation
PDE	Partial differential equation
TDMA	Tridiagonal matrix algorithm

Introduction to Numerical Methods for Solving Differential Equations

The immense power of mathematics is, arguably, best divulged by "crunching" numbers. While an equation or a formula can provide significant insight into a physical phenomenon, its depth, as written on paper, can only be appreciated by a limited few – ones that already have a fairly rigorous understanding of the phenomenon to begin with. The same equation or formula, however, when put to use to generate numbers, reveals significantly more. For example, the Navier–Stokes equations, which govern fluid flow, are not particularly appealing on paper except, perhaps, to a select few. However, their solution, when appropriately postprocessed and depicted in the form of line plots, field plots, and animations, can be eye-opening even to a middle-school student! In realization of the fact that the numbers generated out of sophisticated equations are far more revealing than the equations themselves, for more than a century, applied mathematicians have endeavored to find ways to rapidly generate numbers from equations. The desire to generate numbers has also been partly prompted by the fact that closed-form analytical solutions exist only for a limited few scenarios, and even those require number crunching or computing to some degree.

Although the history of computing can be traced back to Babylon, where the abacus was believed to have been invented around 2400 BC, it was not until the nineteenth century that the development of devices that could, according to the modern sense of the word, compute, came to be realized. While the industrial revolution created machines that made our everyday life easier, the nineteenth and twentieth century witnessed strong interest among mathematicians and scientists in building a machine that could crunch numbers or compute repeatedly and rapidly. The so-called Analytical Engine, proposed by Charles Babbage around 1835, is believed to be the first computer design capable of logic-based computing. Unfortunately, it was never built due to political and economic turn of events. In 1872, Sir William Thomson built an analog tide-predicting machine that could integrate differential equations. The Russian naval architect and mathematician, Alexei Krylov (1863–1945), also built a machine capable of integrating an ordinary differential equation in 1904. These early analog machines were based on mechanical principles and built using mechanical parts. As a result, they were slow. The Second World War stimulated renewed interest in computing both on the German and British sides. The Zuse Z3, designed by Conrad Zuse (1910–1995), was built by German engineers in 1941. It is believed to be the world's first programmable electromechanical computer. It was also around this time that the British cryptanalyst Alan Turing, known as the father of computer

Numerical Methods for Partial Differential Equations: Finite Difference and Finite Volume Methods
http://dx.doi.org/10.1016/B978-0-12-849894-1.00001-9

science and artificial intelligence, and brought to the limelight recently by *The Imitation Game*, built an electromechanical machine to decode the Enigma machine that was being used by the German military for their internal communication. Shortly after the war, Turing laid the theoretical foundation for the modern stored-program programmable computer – a machine that does not require any rewiring to execute a different set of instructions. This so-called Turing Machine later became the theoretical standard for computer design, and modern computer designs, upon satisfying a set of mandatory design requirements, are referred to as "Turing complete."

With the invention of the bipolar transistor in 1947, and integrated circuits in 1952, the world witnessed a meteoric rise in computer hardware technology and computing power. In 1965, in a famous statement [1], known today as the Moore's Law, Gordon E. Moore predicted that the number of transistors in an integrated circuit would approximately double every two years. Over the past four decades, the growth in Very Large Scale Integrated (VLSI) circuit technology has roughly followed Moore's law. The 62-core Xeon Phi processor, released by Intel Corporation in 2012, has 5 billion transistors, compared with 2,300 transistors in the Intel 4004 processor released in 1971. The number of millions of instructions per second (MIPS), an essential marker of processor speed, scales directly as the number of transistors per processor. The Intel Core i7, one of the most prevalent processors in modern (as of 2014) laptop computers, executes 10^5 MIPS. In contrast, the 1971 Intel 4004 processor executed 0.09 MIPS. Fortunately, the increase in processor speed has not come with a heavy penalty in cost. With the advent of modern manufacturing techniques and large-scale production, the cost of semiconductor devices has, instead, plummeted. The capital (hardware) cost needed to execute one Giga floating point operations (flops) today is roughly 10 cents, while it was about $30,000 in 1997. On a final note, the improvement in processor speed has, in parallel, been accompanied by a relentless drive toward miniaturization. Consequently, we not only have immense affordable computing power today, but also have it in the palm of our hand!

Over the past two decades, this dramatic increase in computing power and reduction in cost has stimulated significant research, development, and educational activities that focus on computing. The US National Academy of Engineering now recognizes finite element analysis and computational fluid dynamics, for example, as core disciplines. Most undergraduate curricula in engineering now have a portfolio of courses that cater to students' desire to learn to use computational tools in their own discipline. In mainstream industries, such as the automotive industry, the traditional methodology of "make and break" is rapidly being replaced by a paradigm in which analysis and analysis-based understanding of the working principles of a device, part, or process, is considered imperative. As a result, having a working knowledge of the popular simulation tools is rapidly becoming a necessity rather than a choice for the next-generation workforce. In the academic and research communities, the drastic increase in computing power has also prompted renewed interest in the development of algorithms that make use of the enhanced computing power to execute complex tasks in an efficient manner. Today, computers can simulate processes that span length scales all the way from picometers to light years, and timescales that span femtoseconds to decades. From collisions of black holes to the computation of atomic vibrational spectra to unsteady turbulence in atmospheric boundary layers to weather prediction, the possibilities for computer simulations to promote scientific discoveries, and make our everyday life better, is endless!

1.1 ROLE OF ANALYSIS

Prior to the advent of digital computers and computing technology, machines and devices were built by trial and error. The paradigm of trial-and-error or make-and-break continues to be used even today, albeit to a limited degree and in a fashion more informed by previous experience and knowhow. As a matter of fact, building something with the intention to prove or disprove that it works is the only conclusive way to establish its feasibility. Such an approach not only confirms the science behind the device but also its feasibility from an engineering standpoint, and in some cases, even from an economic standpoint. For example, a newly designed gas turbine blade not only has to be able to deliver a certain amount of power (the science), but also has to be able to be withstand thermal loads (the engineering). Despite the long history of success of the make-and-break paradigm, it fails to capitalize upon the full power of science. Returning to the previous example, a turbine blade that delivers the required power and also withstands heat loads is not necessarily the best design under the prescribed operating conditions. It is just one design that works! There may be others that are far more efficient (deliver more power) than the one that was built and tested. Of course, one could build and test a large number of blades of different designs to answer the question as to which one is the best. However, such an approach has two major drawbacks. First, it would require enormous time and resources. Second, such an approach may not still be able to exhaust the entire design space. In other words, there may have been potentially better designs that were left unexplored. This is where scientific or engineering analysis of the problem becomes useful.

During the industrial revolution, and for decades afterwards, the need for analysis of man-made devices was not critical. The so-called factor of safety in-built into the design of most devices was so large that they rarely failed. Most devices were judged by their ability or inability to perform a certain task, not necessarily by how efficiently the task was performed. One marveled at an automobile because of its ability to transport passengers from point A to point B. Metrics, such as miles per gallon, was not even remotely in the picture. With an exponential rise in the world's population, and dwindling natural resources, building efficient devices and conserving natural resources is now a critical need rather than a luxury. Improvement in the efficiency requires understanding of the functioning of a device or system at a deeper level.

Analysis refers to the use of certain physical and mathematical principles to establish the relationship between cause and effect as applied to the device or system in question. The causal relationship, often referred to as the mathematical model, may be in the form of a simple explicit equation or in the form of a complex set of partial differential equations (PDEs). It may be based on empirical correlations or fundamental laws such as the conservation of mass, momentum, energy or other relevant quantities. Irrespective of whether the mathematical model is fundamental physics based, or empirical, it enables us to ask "what if?" questions. What if the blade angle was changed by 2 degrees? What if the blade speed was altered by 2%? Such analysis enables us to explore the entire design space in a relatively short period of time and, hopefully, with little use of resources. It also helps eliminate designs that are not promising from a scientific standpoint, thereby narrowing down the potential designs that warrant further experimental study.

Broadly, the mathematical models used for analysis may be classified into two categories: empirical and fundamental physics based. Empirical models are typically developed using experimental data and extensive experience. A large number of experiments are generally conducted on the device or system under scrutiny, and the data is appropriately processed to generate one or more equations that describe a causal relationship. Examples of empirical models include the correlations used in the area of heat transfer to compute heat transfer coefficients as a function of the fluids' thermophysical properties and the flow speed, or the correlations used in the area of fluid mechanics to compute drag over regular-shaped objects. The advantage of empirical models is that they are usually very easy to understand and use. The disadvantage is that their range of applicability is always limited. For example, an empirical correlation developed to compute the heat transfer coefficient on a flat surface is invalid (may produce inaccurate results) if the surface is curved. In contrast, fundamental physics-based models usually have a much larger range of applicability. For example, the Fourier law of heat conduction is valid for the description of conduction heat transfer as long as the continuum assumption is valid. Likewise, the energy conservation equation (first law of thermodynamics) is always valid no matter what the media, the geometry, or operating conditions. The former is an example of a so-called phenomenological law, while the latter is an example of a fundamental law. Phenomenological laws are laws that are borne out of experimental observations coupled with some physical insights rather than fundamental axioms. Their range of applicability is usually somewhat limited compared with fundamental laws but still significantly larger than empirical correlations.

Creation of a model to adequately describe a certain physical phenomenon implies that we have a sufficiently deep understanding of the physical phenomenon, and its cause-effect relationship, to begin with. On the other hand, as mentioned earlier, one of the purposes of model-based analysis is to discover the causal relationship in a phenomenon or system. How do we reconcile with this paradox? Fortunately, if the model being used for analysis is based on fundamental laws, it can almost be used blindly. For example, if we believe that Newton's laws of motion are universal, we can apply them to model extremely complex phenomena without even understanding the phenomena precisely. In fact, in such a scenario, the model reveals the actual physics. The physics of shock-turbulence interactions behind a hypersonic aircraft may be extremely complex, but it is embedded in the Navier–Stokes equation that, at its heart, has the principles of conservation of mass, momentum, and energy. Thus, as long as the analysis tool uses fundamental physics-based laws, it can be relied upon to reveal the causal relationship of interest. Therein lies the beauty of fundamental physics-based laws, and realization of their immense power has led to their increased use in analysis in the modern era.

Despite the aforementioned advantage of using physics-based models, either phenomenological or fundamental, their use is still somewhat limited in routine day-to-day analysis. This is attributed to their complexity. Some of the most general fundamental laws, such as those of conservation of mass, momentum, energy, charge, and other quantities, are typically described by differential equations. The solution to such equations is far from trivial, and requires advanced methods. In most cases, their solutions cannot be obtained without resorting to a computer. It is due to this reason that the increase in the use of fundamental models in scientific or engineering analysis has gone hand in hand with the dramatic advancement of computer technology. Looked at from a slightly different perspective, with the advent of modern computing technology, it is now possible to use fundamental law-based models for the simulation of devices and systems.

While analysis helps establish the relationship between the various parameters that govern the system, and also explore a large part of the design space, it is not, and should not, be perceived as a substitute for experimental observations. No matter how sophisticated or fundamental a model, it always has some assumptions that make it deviate from the real situation. For example, a simulation of flow through a pipe may ignore microscopic defects on the walls of the pipe – the assumption being that these defects do not alter the flow. In some cases, the assumptions may be justifiable, while in other situations, they may not. To gain a proper understanding of the effect of the assumptions, one must conduct *validation* of the model. The concept of validation is discussed in the final section of this chapter.

A mathematical model, by definition, relates inputs to outputs – the aforementioned causal relationship. Even if the model has been validated and is perfectly suited for describing this relationship, it requires proper inputs to begin with. In order to match an observed phenomenon, the modeler must ascertain that the inputs closely match those in the experimental setup. However, this is easier said than done since the experiment itself is not one hundred percent repeatable. For example, the inlet flow rate in a pipe flow experiment may vary slightly from day to day or hour to hour or even over the duration of the experiment. The small errors in the inputs – referred to as *uncertainties* – may either amplify or deamplify when fed into the model. In other words, a 1% error in an input may either cause a 10% change in the output or a 0.1% change in the output. The quantification of changes in the output as a function of perturbations in the input is referred to as *uncertainty quantification*. In recent years, in realization of its importance in mimicking the physical setup, uncertainty analysis has evolved into a field of its own under the broader umbrella of numerical analysis.

The primary focus of this text, of course, is PDEs. As discussed earlier, PDEs usually arise out of physics-based models. Such models are capable of predicting the spatiotemporal evolution of the physical quantity of interest. For example, solution of the unsteady heat conduction equation, which is a PDE with four independent variables, can predict the spatiotemporal evolution of the temperature, which, in this particular case, is the dependent variable. Since the solution of the governing PDE can potentially predict the quantity of interest (temperature, in the preceding example) at any point in space and time, one can liken such solutions to virtual experiments, in which probes may be imagined to have been placed at numerous locations in space at will. While placing probes in the actual physical setup requires additional considerations, such as whether the probe will intrude on the physics, whether there is adequate space available for it, cost, and other practical constraints, placing probes (nodes in a numerical simulation) has no such constraints. Therefore, one of the roles that analysis can play is to complement experimental observations by filling in the gaps – gaps in space and/or time where probes could not be placed due to the aforementioned practical constraints. Broadly, analysis can fill gaps in our understanding of the system under study by generating a wealth of information (spatiotemporal data) that would otherwise be impossible to measure. This idea is best illustrated by an example.

Figure 1.1 shows a prototype catalytic combustor in which a methane–air mixture is injected at one end. The monolithic catalytic combustor has 13 channels, each with a cross-sectional dimension of roughly 1 mm, as shown in Fig. 1.1. The channels are separated by walls made of cordierite, which also makes up the outer cylindrical wall of the combustor. As the methane–air mixture flows through the monolith channels, it is oxidized to carbon dioxide and water vapor by interacting with the platinum catalyst that is coated on to the channel walls. For this particular application, one of the critical design questions is what the length of the combustor should be so that no unburnt carbon monoxide – one of the products of combustion – goes out through the outlet. In the traditional paradigm of build-and-test, this question can only be answered by building combustors of various lengths, and then testing each one. This

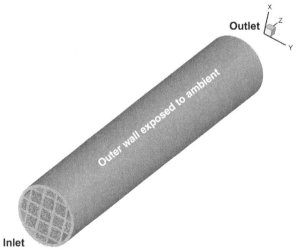

FIGURE 1.1 A Monolithic Catalytic Combustor Prototype with 13 Channels, Designed for Combustion of a Methane–Air Mixture on Platinum Catalyst

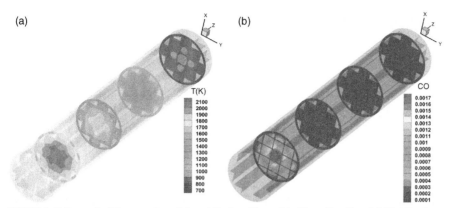

FIGURE 1.2 Computed Temperature (a) and Carbon Monoxide Mass Fraction (b) Distributions for the Test Case

This is depicted in Fig. 1.1, and described in the previous page.

is because devices that traditionally measure concentration of chemical species, such as mass spectrometers, cannot be placed inside the monolith channels and *in situ* measurements are difficult to perform in this particular case. Typically, such measurements can only be conducted *ex situ* at the outlet. In the analysis framework, however, if the governing PDEs, representing mass, momentum, and energy conservation, are solved accurately, one can determine the carbon monoxide concentration not only at the outlet of the combustor, but also at any point inside the combustor, as shown in Fig. 1.2. In this particular case, it is not even necessary to analyze different lengths because the downstream solution has negligible effect on the upstream solution. For example, from Fig. 1.2(b), it is evident that under the operating conditions used for the analysis, it would have been sufficient to use a combustor about half of the length considered in the study. The final implication of this finding is that about half of the platinum could have been saved, leading to significant cost reduction.

The preceding discussion sheds light on the various facets of scientific or engineering analysis. To summarize, analysis serves the following purpose:

1. It helps establish and understand the relationship between cause (input parameters) and effect (output), thereby providing clues on how a certain device or system works or why it fails under certain conditions.
2. For engineering applications, it enables exploration of parts of the design space that may be difficult to explore experimentally either because it is too time and resource-consuming, or due to other practical constraints. Such exhaustive exploration of the design space can potentially lead to better (optimal) designs or help filter out unworthy designs.
3. During the process of analysis of a complex system within which multiple phenomena may be occurring simultaneously, the effect of individual physical phenomena can often be isolated with relative ease. For example, the effect of natural convection on a flow pattern can be easily isolated by setting

the gravity vector to zero. In contrast, gravity is omnipresent, and cannot be turned off in a terrestrial experimental investigation.

4. Analysis that utilizes physics-based models can generate a large volume of spatio-temporal data that essentially mimic data measured by probes in an experiment. These data can fill critical gaps in our understanding of the working principle of the device or system under consideration, especially when used in conjunction with experimental data.

1.2 CLASSIFICATION OF PDEs

As discussed in the preceding section, at the heart of analysis that utilizes fundamental physics-based principles are differential equations. In the most general case, when the behavior of the system or device is sought as a function of both time and space, the governing equations are PDEs. The solution of a PDE – either by analytical or numerical means – is generally quite complex, and requires a deep understanding of the key attributes of the PDE. These attributes dictate the basic method of solution, how and where boundary and initial conditions must be applied, and what the general nature of the solution is expected to be. Therefore, in this section, we classify PDEs into broad canonical types, and also apply this classification to PDEs commonly encountered in engineering analysis.

We begin our classification by considering a PDE of the following general form:

$$A\frac{\partial^2 \phi}{\partial x^2} + B\frac{\partial^2 \phi}{\partial x \partial y} + C\frac{\partial^2 \phi}{\partial y^2} + ... = 0, \tag{1.1}$$

where x and y are so-called independent variables, while ϕ is the dependent variable. The coefficients A, B, and C are either real constants or real functions of the independent or dependent variables. If any of the three coefficients is a function of the dependent variable ϕ, the PDE becomes nonlinear. Otherwise, it is linear. It is worth pointing out that the distinction between linear and nonlinear PDEs is not related to whether the solution to the PDE is a linear or nonlinear function of x and y. As a matter of fact, most linear PDEs yield nonlinear solutions! What matters is whether the PDE has any nonlinearity in the dependent variable ϕ.

Eq. (1.1) is a PDE in two independent variables. In general, of course, PDEs can be in more than two independent variables, and such scenarios will be discussed in due course. For now, we will restrict ourselves to the bare minimum number of independent variables required to deem a differential equation a PDE. Depending on the values of the coefficients A, B, and C, PDEs are classified as follows:

If $B^2 - 4AC < 0$, then the PDE is *elliptic*.
If $B^2 - 4AC = 0$, then the PDE is *parabolic*.
If $B^2 - 4AC > 0$, then the PDE is *hyperbolic*.

Next, we examine a variety of PDEs, commonly encountered in science and engineering disciplines, and use the preceding criteria to identify its type. We begin with the *steady-state diffusion equation*, written in two independent variables as [2]

$$\frac{\partial^2 \phi}{\partial x^2} + \frac{\partial^2 \phi}{\partial y^2} = S_\phi, \tag{1.2}$$

where S_ϕ is the so-called source or source term that, in general, could be a function of either the dependent variables or the independent variables, or both, i.e., $S_\phi = S_\phi(x, y, \phi)$. If the source term is equal to zero, Eq. (1.2) is the so-called *Laplace equation*. If the source term is a function of the independent variables only, or a constant, i.e., $S_\phi = S_\phi(x, y)$, Eq. (1.2) reduces to the so-called *Poisson equation*. If the source term is a linear function of the dependent variable, i.e., $S_\phi = a\phi + b$, Eq. (1.2) is referred to as the *Helmholtz Equation*. The term "diffusion equation" stems from the fact that the differential operators shown in Eq. (1.2) usually arise out of modeling diffusion-like processes such as heat conduction, current conduction, molecular mass diffusion, and other similar phenomena, as is discussed in more detail in Chapters 6 and 7. Irrespective of the aforementioned three forms assumed by Eq. (1.2), comparing it with Eq. (1.1) yields $A = C = 1$, and $B = 0$, resulting in $B^2 - 4AC < 0$. Thus, the steady-state diffusion equation, which includes equations of the Laplace, Poisson, or Helmholtz type, is an elliptic PDE. An important characteristic of elliptic PDEs is that they require specification of boundary conditions on all surfaces that bound the domain of solution.

A variation of the PDE just considered is the *unsteady diffusion equation*, written in two independent variables as [2,3]

$$\frac{\partial \phi}{\partial t} = \Gamma \frac{\partial^2 \phi}{\partial x^2} + S_\phi, \tag{1.3}$$

where Γ is a positive constant. Comparing Eq. (1.3) with Eq. (1.1), we obtain $A = \Gamma$, and $B = C = 0$, which results in $B^2 - 4AC = 0$. Thus, the unsteady diffusion equation is parabolic. In Eq. (1.3), t denotes time, while x denotes space. Hence, Eq. (1.3) describes the transient evolution of the solution in one-dimensional space. The same equation can be extended to multiple spatial dimensions to write [2,3]

$$\frac{\partial \phi}{\partial t} = \Gamma \left(\frac{\partial^2 \phi}{\partial x^2} + \frac{\partial^2 \phi}{\partial y^2} \right) + S_\phi. \tag{1.4}$$

In this case, $A = C = \Gamma$, and $B = 0$, yielding $B^2 - 4AC < 0$. Thus, Eq. (1.4) is elliptic. The finding that the spatially one-dimensional unsteady diffusion equation is parabolic, while the spatially two-dimensional version is elliptic, is somewhat confusing. One must note that Eq. (1.4) actually has three independent variables: t, x, and y. The identification of its type depends on which of the two independent variables are considered in order to compare to the standard form in Eq. (1.1). If t and either x or y were chosen as the two independent variables, Eq. (1.4) is indeed parabolic. However, if x and y are chosen as the two independent variables, Eq. (1.4) is elliptic. To avoid such ambiguities, it is customary to often use a qualifier preceding the equation type. For example, Eq. (1.4) is traditionally referred to as parabolic in time and elliptic in space. Such qualifiers are warranted whenever PDEs with more than two independent variables are classified into the aforementioned three categories.

It is also common practice to ascribe the tags of elliptic, parabolic, or hyperbolic to differential operators used in vector calculus in lieu of ascribing them to the full PDE. For example, the Laplacian operator, $\nabla^2 \equiv (\partial^2/\partial x^2)+(\partial^2/\partial y^2)+(\partial^2/\partial z^2)$, is often referred to as an elliptic operator. The implication of such a practice is that if the Laplacian operator is in the PDE, then the PDE has elliptic characteristics. Specifically, such a PDE will require boundary conditions on all spatial boundaries of the computational domain to make the problem well-posed.

A common equation encountered in the analysis of fluid flow past solid surfaces is the so-called *boundary layer equation*. For a two-dimensional geometry, the steady-state boundary layer equation (the *x*-component of the momentum conservation equation) over a flat surface is written as [4]

$$u\frac{\partial u}{\partial x} + v\frac{\partial u}{\partial y} = \nu\frac{\partial^2 u}{\partial x^2}, \tag{1.5}$$

where u and v are *x*- and *y*-components of the velocity vector. It should suffice to note at this point that these two quantities could be either positive or negative. The quantity, ν, on the other hand, denotes the kinematic viscosity of the fluid, and is always positive. The dependent variable in Eq. (1.5) is u, and x and y are the two independent variables. Comparing Eq. (1.5) with Eq. (1.1), we obtain $A = \nu$, $B = C = 0$, and $B^2 - 4AC = 0$, which makes the equation parabolic. One of the critical attributes of a parabolic PDE is that conditions (either boundary or initial, depending on the independent variable) are not required at all boundaries (either in time or space) of the computational domain. For example, to make Eq. (1.5) well-posed, boundary conditions for u are needed only at $x = 0$ if u is positive. No boundary conditions are necessary at $x = L$, with L being the span of the computational domain in the x direction. Similarly, for Eq. (1.3), conditions are only necessary at $t = 0$ (initial conditions). No conditions are necessary at $t = \infty$. The specific boundary where conditions are not necessary can only be identified by considering the physics of the problem under consideration. Mathematically, it should suffice to state that one "end" does not require a condition. As opposed to an elliptic problem, wherein the information from the boundary propagates inward to determine the solution at any point within the computational domain, in a parabolic problem, information propagates only one way for one of the independent variables. In the specific case of Eq. (1.5), information only propagates downstream (along the flow direction).

The governing equation for viscous fluid flow is the well-known Navier–Stokes equation. Essentially, it represents the conservation of momentum on a differential fluid element. It is a vector equation. Under the assumption of constant viscosity and density, the x-component of the steady-state *Navier–Stokes equation* in 2D may be written as [4]

$$u\frac{\partial u}{\partial x}+v\frac{\partial u}{\partial y}=v\left(\frac{\partial^2 u}{\partial x^2}+\frac{\partial^2 u}{\partial y^2}\right)-\frac{1}{\rho}\frac{\partial p}{\partial x}+B_x, \tag{1.6}$$

where ρ is the density of the fluid, p is pressure, and B_x is the x-component of an external acceleration. The other symbols carry the same meanings as described for Eq. (1.5). Comparison of Eq. (1.6) with Eq. (1.1) reveals that $A = C = v$, $B = 0$, which implies that $B^2 - 4AC < 0$. Hence, the Navier–Stokes equation is elliptic. In contrast, the boundary layer equation [Eq. (1.5)], which is a simplified version of the Navier–Stokes equation under the so-called boundary layer approximation [2,4], was found to be parabolic. This difference is noteworthy because of its implications on the treatment of the boundary conditions in the two cases, as discussed earlier. Indeed, on account of this important difference, historically, the techniques used for solving the boundary layer equation have been quite different than the ones used for solving the Navier–Stokes equations. The unsteady form of the Navier–Stokes equation, of course, will have at least three independent variables. Hence, its classification faces the same ambiguities discussed earlier for Eq. (1.4). Like Eq. (1.4), it has dual nature: elliptic in space and either parabolic or hyperbolic in time, depending on the flow regime.

The parabolic nature in time of the Navier–Stokes equation becomes evident in the low flow speed regime (the so-called Stokes regime [4]), in which case, advection, as described by the left-hand side of Eq. (1.6), is negligible compared with diffusion, as described by the right-hand side of Eq. (1.6). In such a scenario, Eq. (1.6) essentially takes the same form as Eq. (1.4) with ϕ being replaced by u. This is the so-called *unsteady Stokes equation* [4]. It is parabolic in time and elliptic in space. The *incompressible Navier–Stokes equation*, as characterized by the flow regime in which the Mach number is low [4], also has the dual characteristics of being elliptic in space and parabolic in time.

The hyperbolic nature of the Navier–Stokes equation arises from the advection terms of the Navier–Stokes equation, and is best understood by considering its unsteady one-dimensional inviscid (without viscosity) counterpart – the so-called *Euler equation*, written as [4–6]

$$\frac{\partial u}{\partial t}+u\frac{\partial u}{\partial x}=0, \tag{1.7}$$

where the symbols carry the same meaning as before. Since Eq. (1.7) does not have any second derivative, it cannot be compared directly to the standard form of a PDE shown in Eq. (1.1). Therefore, we first differentiate Eq. (1.7) with respect to time, yielding

$$\frac{\partial^2 u}{\partial t^2}+u\frac{\partial^2 u}{\partial t\,\partial x}+\frac{\partial u}{\partial t}\frac{\partial u}{\partial x}=0. \tag{1.8}$$

Comparison of Eq. (1.8) to Eq. (1.1) yields $A = 1$, $B = u$, and $C = 0$. Therefore, $B^2 - 4AC = u^2$, which implies that the equation is hyperbolic since u^2 is always positive. When the flow is supersonic (Mach number greater than unity), the compressible Navier–Stokes equation is elliptic in space and hyperbolic in time [6]. At intermediate Mach numbers – the so-called transonic regime – the Navier–Stokes equation exhibits a mixture of all three PDE types: elliptic in space and a mixture of parabolic and hyperbolic in time. It is clear from the preceding discussion that the spatially multidimensional form of the Navier–Stokes equation is elliptic in space and either parabolic or hyperbolic in time or a mixture thereof. With increase in flow speed (Mach number), the nature of the PDE shifts gradually from parabolic in time to mixed parabolic–hyperbolic in time to purely hyperbolic. These distinctions generally necessitate special treatment of boundary conditions depending on the flow regime. For example, for supersonic flows, it is common practice to adopt the method of characteristics for application of boundary conditions [6,7] to the Navier–Stokes equation.

The so-called *wave equation* is another popular PDE used in a variety of disciplines. It is used for modeling vibrations of strings, membranes, beams and for solving other structural deformation problems. *Maxwell's equation* is also a wave equation that describes propagation of electromagnetic waves. The unsteady, spatially multidimensional classical wave equation is written as [6,8]

$$\frac{\partial^2 u}{\partial t^2} = c^2 \left(\frac{\partial^2 u}{\partial x^2} + \frac{\partial^2 u}{\partial y^2} \right). \tag{1.9}$$

Following preceding discussions, if x and y are considered as the two independent variables, then Eq. (1.9) is elliptic. If, however, t and x are considered as the two independent variables, then $A = 1$, $B = 0$, and $C = -c^2$, leading to $B^2 - 4AC > 0$. Thus, the classical wave equation is elliptic in space and hyperbolic in time, and is often referred to as the *hyperbolic wave equation*. Since Eq. (1.9) has a second derivative in time, it requires two conditions in time. As opposed to elliptic operators (second derivatives in space) that require boundary conditions on all bounding surfaces of the computational domain, the hyperbolic operator (second derivative in time) requires the two conditions at the same initial instance of time, i.e., at $t = 0$. No condition is required at large times (i.e., at $t \to \infty$). Physically, the solution at future time does not affect the solution at the current instant of time, while the solution at the current instant of time affects the solution at future instances of time. In other words, the propagation of information in one of the independent variables is only in one direction.

In summary, this section introduced various types of PDE and how they are classified into different types. Several examples from physics and engineering disciplines were chosen to highlight the classification procedure. The classification of PDEs into the three canonical types helps elucidate where and how boundary and initial conditions need to be applied, and subsequently, enables the development of appropriate numerical methods to solve each type of PDE, as we shall witness in the remainder of the text.

1.3 **OVERVIEW OF METHODS FOR SOLVING PDEs**

The solution of PDEs is quite challenging. The number of methods available to find closed-form analytical solutions to canonical PDEs is limited. These include separation of variables, superposition, product solution methods, Fourier transforms, Laplace transforms, and perturbation methods, among a few others. Even these methods are limited by constraints such as regular geometry, linearity of the equation, constant coefficients, and others. The imposition of these constraints severely curtails the range of applicability of analytical techniques for solving PDEs, rendering them almost irrelevant for problems of practical interest. In realization of this fact, applied mathematicians and scientists have endeavored to build machines that can solve differential equations by numerical means, as outlined in the brief history of computing presented at the beginning of this chapter.

The methods for numerical solution to PDEs can broadly be classified into two types: deterministic and stochastic. A *deterministic method* is one in which, for a given input to an equation, the output is always the same. The output does not depend on how many times one solves the equation, at what time of the day it is solved, or what computer it is solved on (disregarding precision errors, that may be slightly different on different computers). On the other hand, a *stochastic method* is based on statistical principles, and the output can be slightly different for the same input depending on how many times the calculation is performed, and other factors. In this case, by "slightly different," we mean within the statistical error bounds. The difference between these two approaches is best elucidated by a simple example. Let us consider a scenario in which a ball is released from a certain height, h_i, above the horizontal ground. Upon collision with the ground, the ball bounces back to a height h_o. Let us assume that based on experimental observations or other physical laws, we know that the ball always bounces back to half the height from which it is released. Following this information, we may construct the following deterministic equation: $h_0 = (1/2)h_i$. If this equation is used to compute the height of a bounced ball, it would always be one-half of the height of release. In other words, the equation (or method of calculation) has one hundred percent confidence built into it. Hence, it is termed a deterministic method. The stochastic viewpoint of the same problem would be quite different. In this viewpoint, one would argue that if n balls were made to bounce, by the laws of theoretical probability, $n/2$ balls would bounce to a height slightly above half the released height, and the remaining $n/2$ balls would bounce to a height slightly below the released height, such that in the end, when tallied, the mean height to which the balls bounce back to would be exactly half the height of release. Whether this exact result is recovered or not would depend on how many balls are bounced, i.e., the number of statistical samples used. The stochastic viewpoint may be implemented, for example, by drawing random numbers from a uniform deviate between 0 and 1. If the random number is greater than half, then the ball is made to bounce to a height greater than half the released height, while if it is less than half, it is made to bounce to a height less than half the released height. This viewpoint is more in keeping with experimental observations because in reality, it is unlikely for each ball to bounce back to exactly half the height from which it is released. In other words, statistical variability, as prevalent in any experimental measurement, is already built into the stochastic viewpoint.

The most widely used stochastic method for solving PDEs is the *Monte Carlo method* [9]. Despite the inherent statistical error in this method, as alluded to in the preceding paragraph, the method enjoys widespread popularity in a number of important disciplines due to some notable strengths. First, the Monte Carlo method is better suited to addressing complex physics. For example, the Monte Carlo method has witnessed tremendous success in the area of radiation transport [10] wherein material properties may be wavelength and direction dependent, making the physics of radiation transport quite challenging to address from a deterministic viewpoint. In contrast, extending Monte Carlo methods to account for the aforementioned dependencies in the material properties is quite straightforward. Second, the Monte Carlo method is amenable for use in the solution of high-dimensional PDEs. It has been used extensively for the solution of PDEs that have more than six independent variables, such as the radiative transfer equation [10,11], the Boltzmann Transport Equation for phonons [12–14], electrons [15,16] and rarefied gases [17], and for the transport equations in turbulent reacting flows [18,19]. Finally, algorithms based on the Monte Carlo method are generally strongly suited to vector implementations and parallel computing. An example is presented next to illustrate the effect of statistical sample size on the results generated by the Monte Carlo method. In this example, the radiative transfer equation is solved in an evacuated two-dimensional square cavity enclosed by gray diffuse walls having a reflectivity equal to 0.5, and with temperatures as shown in Fig. 1.3(a). The objective is to determine the radiative heat flux in each of the four walls. The deterministic computations were performed using the surface-to-surface radiation exchange equations [10]. Since the Monte Carlo method is stochastic, each calculation (or each case of the ensemble) yields a slightly different result

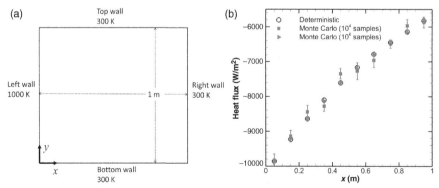

FIGURE 1.3 Example Illustrating the Solution of the Radiative Transfer Equation Using Deterministic and Stochastic (Monte Carlo) Methods

(a) Geometry and boundary conditions, and (b) results computed using the two methods.

since the random numbers generated and used are different. Therefore, in order to obtain the mean result, an ensemble average was performed. The standard deviation was then computed to assess the statistical errors in the solution. The procedure, just described, is identical to conducting experiments and processing experimental data, wherein an experiment is repeated many times and the mean value of the observed quantity and its standard deviation are ultimately computed. In Fig. 1.3(b), the standard deviation, σ, is depicted by an error bar corresponding to $\pm 3\sigma$. It is clear that as the number of statistical samples (photon bundles) used in the Monte Carlo calculations is increased, the error decreases. When 10^6 statistical samples are used, the results are almost identical to the results computed using the deterministic method. The error bars (statistical error) are so small that they are not visible in Fig. 3(b), although they have been plotted.

As shown in the preceding example, the downside of the Monte Carlo method is that the results generated by it are dependent on the statistical sample size. It is due to the presence of this inherent statistical error and the need to perform an ensemble of calculations to quantify it, that numerical analysts prefer deterministic methods over stochastic methods for solution of PDEs unless there is a specific reason not to do so. Over the past century, a number of powerful deterministic methods have been developed for the solution of PDEs. In the discussion to follow, a brief top-level overview of three of the most widely used methods is provided, along with cursory examination of a few other emerging methods.

1.3.1 FINITE DIFFERENCE METHOD

The finite difference method is based on the calculus of finite differences. It is a relatively straightforward method in which the governing PDE is satisfied at a set of prescribed interconnected points within the computational domain, referred to as *nodes*. The boundary conditions, in turn, are satisfied at a set of prescribed nodes located on the boundaries of the computational domain. The framework of interconnected nodes is referred to as a *grid* or *mesh*. Each derivative in the PDE is approximated using a difference approximation that is typically derived using Taylor series expansions. To illustrate the method, let us consider solution of the Poisson equation [Eq. (1.2)] in the computational domain shown in Fig. 1.4. Let us also assume that the value of the dependent variable, ϕ, is prescribed at the boundary, and is known at all boundary nodes.

The objective is to determine the value of ϕ at all nodes. The total number of nodes is denoted by N. This implies that for the specific nodal arrangement shown in Fig. 1.4, $N = 16$, and ϕ_{10} through ϕ_{16} needs to be determined. In order to determine these seven unknowns, seven equations are needed. In the finite difference method, these seven equations are formulated by satisfying the governing equation at the nodes 10–16, yielding

$$\left.\frac{\partial^2 \phi}{\partial x^2}\right|_i + \left.\frac{\partial^2 \phi}{\partial y^2}\right|_i = S_i \qquad \forall i = 10, 11, ..., 16. \tag{1.10}$$

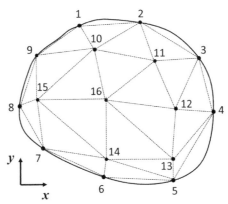

FIGURE 1.4 Schematic Representation of a 2D Computational Domain and Grid

The nodes to be used for the finite difference method are denoted by solid circles. Nodes 1–9 denote boundary nodes, while nodes 10–16 denote interior nodes.

Next, the second derivatives are approximated using the nodal values of ϕ. For example, one may write (approximate) the second derivatives with respect to x and y, respectively, at node 16 as

$$\frac{\partial^2 \phi}{\partial x^2}\bigg|_{16} \approx \sum_{k=1}^{N} A_{16,k}^x \phi_k, \tag{1.11a}$$

$$\frac{\partial^2 \phi}{\partial y^2}\bigg|_{16} \approx \sum_{k=1}^{N} A_{16,k}^y \phi_k, \tag{1.11b}$$

where $A_{16,k}^x$ and $A_{16,k}^y$ represent coefficients (or weights) for node 16 that expresses the two second derivatives as a linear combination of the nodal values of ϕ. These coefficients may be derived in a variety of ways: using Taylor series expansions, interpolation functions, and splines, among others. Generally, the nodes considered in the summation shown in Eq. (1.11) include a small subset of the total number of nodes – those being the immediate neighbors and the node in question itself. Thus, for node 16, only nodes 10–16 may be used in the summations shown in Eq. (1.11), implying that the coefficients would be zero for $k = 1, 2, ..., 10$. If the radius of influence is extended further and the neighbors of the neighbors are used, a more accurate approximation – so-called higher-order approximation – may be possible to derive. The exact mathematical procedure to derive the coefficients in the finite difference method, along with the errors incurred in the approximations, is described in Chapter 2. Upon substitution of Eq. (1.11) into Eq. (1.10) for node 16, we obtain

$$\sum_{k=1}^{N} A_{16,k}^x \phi_k + \sum_{k=1}^{N} A_{16,k}^y \phi_k = \sum_{k=1}^{N} A_{16,k} \phi_k = S_{16}, \tag{1.12}$$

where $A_{16,k} = A_{16,k}^x + A_{16,k}^y$. Equation (1.12) has $N = 16$ terms on the left-hand side, of which, depending on the radius of influence of each node, many are zeroes. Similar equations may be written for the other nodes to yield the following linear system of algebraic equations:

$$\sum_{k=1}^{N} A_{i,k}\phi_k = S_i \qquad \forall i = 1, 2, ..., N. \tag{1.13}$$

Its solution will yield the nodal value of ϕ for all the nodes. Since the solution to the PDE is obtained by satisfying it at all the interior nodes and the boundary conditions at all the boundary nodes, such a solution is referred to as the *strong form* solution, i.e., a solution in which the governing PDE is not modified whatsoever prior to its discretization and solution. Although the formulation outlined here is for a mesh in which the nodes are arranged in an arbitrary fashion, in practice, finding the coefficients of the linear system [as in Eq. (1.13)] for such a mesh is prohibitive for reasons discussed in Chapter 2. Nonetheless, the preceding discussion provides a brief overview of the finite difference method in the most general sense. Further details are presented in Chapters 2 and 5, and methods of solution of the linear system of algebraic equations [Eq. (1.13)] are presented in Chapters 3 and 4.

1.3.2 FINITE VOLUME METHOD

The finite volume method derives its name from the fact that in this method the governing PDE is satisfied over finite-sized control volumes, rather than at points. The first step in this method is to split the computational domain into a set of control volumes known as *cells*, as shown in Fig. 1.5. In general, these cells may be of arbitrary shape and size, although, traditionally, the cells are convex polygons (in 2D) or polyhedrons (in 3D), i.e., they are bounded by straight edges (in 2D) or planar surfaces (in 3D). As a result, if the bounding surface is curved, it is approximated by straight edges or planar faces, as is evident in Fig. 1.5. These bounding discrete surfaces are

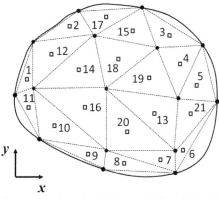

FIGURE 1.5 Schematic Representation of a 2D Ccomputational Domain and Mesh

The cell centers to be used for the finite volume method are denoted by hollow squares. The nodes, denoted by solid circles, are what are used in the finite difference method, and are the vertices of the cells.

known as cell faces or simply, *faces*. The vertices of the cells, on the other hand, are called *nodes*, and are, in fact, the same nodes that were used in the finite difference method. All information is stored at the geometric centroids of the cells, referred to as *cell centers*.

The derivation of the finite volume equations commences by integrating the governing PDE over the cells constituting the computational domain. In the case of the Poisson equation, this yields

$$\int_{V_i} \nabla^2 \phi \, dV = \int_{V_i} S_\phi \, dV,$$

(1.14)

where V_i is the volume of the i-th cell. The volume integral on the left-hand side of Eq. (1.14) can be simplified by writing the Laplacian, as $\nabla^2 \phi = \nabla \cdot (\nabla \phi)$, and by applying the Gauss divergence theorem, to yield

$$\int_{S_i} (\nabla \phi) \cdot \hat{\mathbf{n}} \, dA = \int_{V_i} S_\phi \, dV,$$

(1.15)

where S_i is the surface area of the surface bounding the cell i, and dA is a differential area on the surface with outward pointing unit surface normal $\hat{\mathbf{n}}$. For details on the Gauss divergence theorem, and its application to the finite volume procedure, the interested reader is referred to Chapter 7. The right-hand side of Eq. (1.15) can be simplified by applying the mean value theorem and by assuming that the mean value of S_ϕ over the volume V_i is the same as the value of S_ϕ evaluated at the centroid of the cell i. This simplification yields

$$\int_{S_i} (\nabla \phi) \cdot \hat{\mathbf{n}} \, dA = S_i V_i.$$

(1.16)

As mentioned earlier, the cells are usually enclosed by a set of discrete faces. Therefore, the integral on the left-hand side of Eq. (1.16) can be replaced by a summation over the faces bounding the cell i, to yield

$$\sum_{f=1}^{N_{f,i}} \int_{S_f} (\nabla \phi) \cdot \hat{\mathbf{n}} \, dA = \sum_{f=1}^{N_{f,i}} (\nabla \phi)_f \cdot \hat{\mathbf{n}}_f A_f = S_i V_i,$$

(1.17)

where the index "f" refers to faces of the cell, $N_{f,i}$ is the total number of faces of cell i, and A_f is the area of face f. The quantity, $(\nabla \phi)_f \cdot \hat{\mathbf{n}}_f$, represents the average value of the normal component of the gradient of ϕ at face f, i.e., $(\nabla \phi)_f \cdot \hat{\mathbf{n}}_f = \frac{1}{A_f} \int_{S_f} (\nabla \phi) \cdot \hat{\mathbf{n}} \, dA$.

As in the case of the finite difference method (Section 1.3.1) in which the second derivatives at the nodes were approximated using the nodal values of ϕ, in the case of the finite volume method, the normal gradients at the faces are approximated using the cell center values of ϕ. Once again, this task may be conducted using a variety of approaches. Traditionally, Taylor series expansions are used for this purpose. The end result of expressing the normal gradient of ϕ in terms of cell center values, followed by substitution into Eq. (1.17), is a set of discrete linear algebraic equations of the form

$$\sum_{i=1}^{N_C} A_{i,k}\phi_k = S_i V_i \qquad \forall i = 1, 2, ..., N_C,$$ (1.18)

where N_C is the total number of cells. The exact procedure to derive the coefficients, $A_{i,k}$, appearing in Eq. (1.18), is presented in Chapters 6 and 7 for structured and unstructured meshes, respectively. The procedure to apply boundary conditions is also presented in these chapters in considerable detail.

In contrast with the finite difference method, in the finite volume method, the governing PDE is not solved directly, as evident from the preceding discussion. Rather, it is first integrated over the control volume and then approximated and solved. Furthermore, since no cell center is located at the boundary, the boundary conditions cannot be satisfied directly. It is due to these reasons that the solution to the PDE, obtained using the finite volume method, is known as the *weak form* solution.

One of the notable properties of the finite volume method is its so-called conservation property. The rate of transfer of a quantity per unit area is known as its flux. For example, heat flux is the rate of transfer of heat energy per unit area, and has units of J/s/m^2 (or W/m^2) in the SI system. In the specific case where diffusion is responsible for the movement of the quantity under question, the flux of that quantity is proportional to the gradient of the potential that drives the movement of that quantity. For example, heat diffusion is caused by a temperature gradient, mass diffusion is caused by a concentration gradient, and charge diffusion is caused by an electric potential gradient. Many phenomenological laws (see Section 1.1 for an explanation) that describe diffusion follow this gradient diffusion paradigm. In general, such laws may be expressed mathematically as $\mathbf{J} = -\Gamma \nabla \phi$, where $\mathbf{J}$ is the flux (a vector), ϕ is the aforementioned driving potential, and Γ is the constant of proportionality. If $\Gamma = 1$, then $\mathbf{J} = -\nabla \phi$. Thus, Eq. (1.17) may be written as

$$\sum_{f=1}^{N_{f,i}} \mathbf{J}_f \bullet \hat{\mathbf{n}}_f A_f = -S_i V_i.$$ (1.19)

Noting that $\mathbf{J}_f \bullet \hat{\mathbf{n}}_f$ represents the rate of transport of the quantity in question, such as mass, energy, or charge, leaving the control volume through the face f per unit area, the left-hand side of Eq. (1.19) represents the net rate of the quantity in question that leaves the control volume. If S_i is interpreted as the destruction rate (or $-S_i$ as the production rate) per unit volume of the same quantity in question, then Eq. (1.19) physically states that the net rate of the quantity leaving the control volume is equal to the net rate of production of the same quantity within the control volume. In other words, Eq. (1.19) represents a fundamental conservation statement at steady state. Eq. (1.19) also illustrates that although canonical PDEs, such as the Poisson equation, do not obviously appear to have any relation to conservation laws, in most cases, they do stem from conservation principles and do have a physical interpretation.

The preceding discussion highlights the fact that the finite volume equation is, in essence, a flux balance equation if the governing PDE can be recast in divergence form. Hence, it may be concluded that the conservation property is inherent to the finite volume method. We shall see later in Chapters 6 and 7 that this important property differentiates it from the finite difference method, although the mathematical procedures used to derive the discrete equations in the two methods are similar. In light of this important property, the finite volume method's popularity has grown by leaps and bounds for disciplines in which the governing PDEs represent conservation laws. Examples include computational fluid dynamics [5], computational heat transfer [6], radiation transport [10], electrostatics [20,21], among others. While the finite difference method continues to be used in some of these areas, its popularity is rapidly fading in realization of the fact that the finite difference method does not guarantee conservation. Further discussion to this effect is presented in Section 6.7.

1.3.3 FINITE ELEMENT METHOD

The finite element method is based on the principle of variation of parameters, first proposed by Ritz and Galerkin [22]. Like the finite volume method, the finite element method also determines the *weak form solution* to the governing PDE. The method is comprised of two steps. The first step is to convert the strong form of the PDE to the weak form. The second step is to use finite-sized elements to discretize and solve the resulting weak form problem. The derivation of the weak form commences by multiplying the governing PDE with a so-called *test function*, and then integrating the whole equation over the entire computational domain. For the specific case of the Poisson equation considered in the preceding two sections, this yields

$$\int_{V_\Sigma} [\nabla^2\phi] v \, dV = \int_{V_\Sigma} S_\phi \, v \, dV, \tag{1.20}$$

where V_Σ is the volume of the entire computational domain, and v is the test function: $v = v(x, y)$ in 2D and $v = v(x, y, z)$ in 3D. The test function is a scalar function. Noting that $\nabla \cdot (v\nabla\phi) = v\nabla^2\phi + (\nabla v) \cdot (\nabla\phi)$, Eq. (1.20) may be written as

$$\int_{V_\Sigma} [\nabla \cdot (v\nabla\phi) - (\nabla v) \cdot (\nabla\phi)] \, dV = \int_{V_\Sigma} S_\phi v \, dV. \tag{1.21}$$

Next, applying the Gauss-divergence theorem to the first term of Eq. (1.21), we obtain

$$\int_{S_\Sigma} (v\nabla\phi) \cdot \hat{\mathbf{n}} \, dA - \int_{V_\Sigma} [(\nabla v) \cdot (\nabla\phi)] \, dV = \int_{V_\Sigma} S_\phi v \, dV, \tag{1.22}$$

where, as in Section (1.3.2), $\hat{\mathbf{n}}$ is the outward pointing surface normal on a differential area dA on the bounding surface of the control volume. The total surface area of the bounding surface is denoted by S_Σ. Equation (1.22) represents the generalized weak form of the Poisson equation. Any further simplification of this equation will require choice of the test function v, which, in turn, is dictated by the boundary conditions. For convenience, it is customary [22] to choose $v = 0$ on all boundaries where Dirichlet boundary conditions (i.e., value of the dependent variable ϕ is known) are prescribed. Until this point, the governing equation has simply been recast to the weak form. The finite element procedure is yet to be applied.

Prior to application of the finite element procedure, the computational domain is discretized into a set of convex elements. The elements could be of arbitrary polygonal shape in 2D or arbitrary polyhedrons in 3D. Historically, triangular elements in 2D and tetrahedral elements in 3D have been preferred since the mathematical formulations are easiest for these shapes. Also, the two aforementioned basic element shapes are, by definition, convex, thereby satisfying the most important criterion for the element shape. The elemental arrangement follows that shown in Fig. 1.4. The next step is to describe the solution and the test function in each element by a linear combination of basis functions:

$$\phi(x,y) = \sum_{i=1}^{N} a_i \psi_i(x,y), \tag{1.23a}$$

$$v(x,y) = \sum_{i=1}^{N} b_i \psi_i(x,y), \tag{1.23b}$$

where N is the number of nodes (same as the solid circles in Fig. 1.4), and ψ_i are the basis functions. These basis functions may be of arbitrary order, the minimum order being dictated by the order of the PDE being solved. These basis functions are linearly combined using undetermined coefficients a_i and b_i in Eq. (1.23). Substitution of Eq. (1.23) in Eq. (1.22), followed by simplification and rearrangement yields a linear system of algebraic equations of the form

$$\sum_{k=1}^{N} K_{i,k} a_k = F_i \qquad \forall i = 1, 2, ..., N, \tag{1.24}$$

which may be solved to obtain the undetermined coefficients a_i. Substitution of these coefficients back into Eq. (1.23) will yield the final solution for ϕ. The square matrix $K_{i,k}$ is known as the *stiffness matrix*, while the column matrix (vector) F_i is known as the *load vector* – the names having been derived from structural mechanics. For a multidimensional problem, the exact procedure for deriving the stiffness matrix and the load vector is beyond the scope of this text, and may be found in texts dedicated to the finite element method [22]. However, since the present section is the only section in this text dedicated to the finite element method, for the sake of completeness, we present here an example that demonstrates the finite element method for a 1D problem.

EXAMPLE 1.1

In this example, we consider the solution of the following ordinary differential equation in [0,1] using the Galerkin finite element method:

$$\frac{d^2\phi}{dx^2} = e^x,$$

subject to the following boundary conditions: $\phi(0) = 0$, and $\phi(1) = 1$. The analytical solution to this system is $\phi(x) = e^x + (2 - e)x - 1$. Next, we embark upon the numerical solution to the problem using the Galerkin finite element method.

For this particular case, Eq. (1.22) reduces to

$$\left[v\frac{d\phi}{dx}\right]_{x=1} - \left[v\frac{d\phi}{dx}\right]_{x=0} - \int_0^1 \frac{dv}{dx}\frac{d\phi}{dx}dx = \int_0^1 e^x v\,dx.$$

As discussed earlier, it is customary to set the test function equal to zero at Dirichlet boundaries. Thus, $v(0) = 0$ and $v(1) = 0$, and the weak form governing equation further reduces to

$$-\int_0^1 \frac{dv}{dx}\frac{d\phi}{dx}dx = \int_0^1 e^x v\,dx.$$

Next, we employ a linear combination of basis functions to express the test function and the solution, such that

$$v(x) = \sum_{j=1}^N b_j \psi_j(x),$$

and

$$\phi(x) = \sum_{j=1}^N a_j \psi_j(x),$$

where ψ_j are the basis functions to be chosen later. Substituting the previous series into the governing equations, we obtain

$$-\int_0^1 \sum_{j=1}^N \left(b_j \frac{d\psi_j}{dx}\right) \sum_{i=1}^N \left(a_i \frac{d\psi_i}{dx}\right) dx = \int_0^1 e^x \sum_{j=1}^N b_j \psi_j(x)\,dx.$$

Interchanging the order of integration and summations, we can rewrite the previous equation as

$$-\sum_{j=1}^N b_j \sum_{i=1}^N a_i \int_0^1 \left(\frac{d\psi_j}{dx}\right)\left(\frac{d\psi_i}{dx}\right) dx = \sum_{j=1}^N b_j \int_0^1 e^x \psi_j(x)\,dx,$$

which may be further rearranged to write

$$\sum_{j=1}^N b_j \left[\int_0^1 e^x \psi_j(x)\,dx + \sum_{i=1}^N a_i \int_0^1 \left(\frac{d\psi_j}{dx}\right)\left(\frac{d\psi_i}{dx}\right) dx\right] = 0.$$

Since the formulation is applicable to a single node and any arbitrary values of b_j, it follows then that the integration kernel shown within square brackets must be zero, which yields

$$\sum_{i=1}^{N} a_i \int_0^1 \left(\frac{d\psi_j}{dx}\right)\left(\frac{d\psi_i}{dx}\right) dx = -\int_0^1 e^x \psi_j(x)\,dx.$$

This equation may be rewritten in the matrix form shown in Eq. (1.24), namely,

$$\sum_{i=1}^{N} K_{j,i}\, a_i = F_j,$$

where

$$K_{j,i} = \int_0^1 \left(\frac{d\psi_j}{dx}\right)\left(\frac{d\psi_i}{dx}\right) dx,$$

and

$$F_j = -\int_0^1 e^x \psi_j(x)\,dx.$$

The solution to the matrix form will yield all undermined coefficients a_i, which may then be substituted back into the basis function expansion to yield $\phi(x)$. The next task is to choose the basis functions so that the stiffness matrix and load vector can be determined.

In this example, we choose linear basis functions of the top-hat type, mathematically written as

$$\psi_i(x) = \begin{cases} \dfrac{x - x_{i-1}}{\Delta x} & \text{if } x_{i-1} \le x \le x_i \\ \dfrac{x_{i+1} - x}{\Delta x} & \text{if } x_i \le x \le x_{i+1} \\ 0 & \text{otherwise} \end{cases}$$

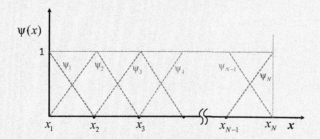

The shape of these basis functions is illustrated in the previous sketch. Based on the choice of the basis function, since $\phi(0) = 0$ and $\psi_1(0) = 1$, it follows that $a_1 = 0$. Similarly, since $\phi(1) = 1$ and $\psi_N(1) = 1$, it follows that $a_N = 1$.

Substituting the basis function shown previously into the expression for the stiffness matrix, we obtain

$$K_{j,i} = \int_0^1 \left(\frac{d\psi_j}{dx}\right)\left(\frac{d\psi_i}{dx}\right)dx = \begin{cases} 2/\Delta x & \text{if } i = j \\ -1/\Delta x & \text{if } i \neq j \end{cases},$$

where $\Delta x = 1/(N-1)$ and $x_j = (j-1)\Delta x$, such that $x_1 = 0$ and $x_N = 1$. Likewise, substituting the basis function into the expression for the load vector, followed by integration by parts, yields

$$F_j = -\int_0^1 e^x \psi_j(x)\, dx = \frac{1}{\Delta x}\left(2e^{x_j} - e^{x_{j-1}} - e^{x_{j+1}}\right).$$

The previous expressions are valid for $2 \leq j \leq N-1$. For the top-hat basis function used here, the stiffness matrix is tridiagonal. The solution to these equations will yield a_2 though a_{N-1}, which can then be used to compute $\phi(x)$. The solution is tabulated as follows for the case when just 5 nodes ($N = 5$) are used.

x	$\phi_{analytical}$	$\phi_{numerical}$	$\phi_{analytical} - \phi_{numerical}$
0.00	0.0000000	0.0000000	0.00×10^{-16}
0.25	0.1044550	0.1044550	0.27×10^{-16}
0.50	0.2895804	0.2895804	0.55×10^{-16}
0.75	0.5782886	0.5782886	1.11×10^{-16}
1.00	1.0000000	1.0000000	0.00×10^{-16}

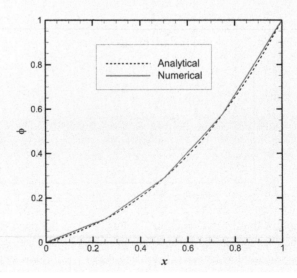

The results show that the solution is accurate up to the 15th decimal place, and the errors are comparable to round-off errors. One of the remarkable features of the finite element method is that the solution accuracy can be readily controlled by an appropriate choice of basis functions. If the solution can be represented accurately by a polynomial of degree n, then using a basis function that is of degree n-1 will yield exact answers. In this particular case, the solution contains an exponential function, which, of course, can only be exactly represented by a polynomial of infinite degree. However, since the solution domain is [0,1], the exponential can be represented fairly accurately by a quadratic function, for which the linear basis functions used here are expected to yield exact answers. The smaller the value of x, the more accurate the quadratic representation. This is reflected in the error distribution, which shows that although the error is negligibly small, it is increasing with x. It is important to note that the exact solution is replicated only at the nodes that were used to obtain the solution. At other locations (values of x) the solution may have much larger deviation from the exact solution as shown in the previous figure. In this case 5 nodes were used, and it is clear from the previous figure, that while the exact solution is replicated at the nodes, far away from the nodes, the solution follows the shape of the linear basis functions used in this case.

One of the fascinating features of the finite element method is its flexibility in the choice of the basis function – both type and degree. The accuracy of the solution can be enhanced either by decreasing the size of the elements – so-called *h-adaptivity* – or by increasing the degree (sometimes referred to as the order) of the basis function – so-called *p-adaptivity*. Over decades of research, practitioners of the finite element method have realized that the two approaches can be hybridized, and it is not necessary to use a basis function of the same degree on all of the elements. This has led to a powerful paradigm known as *hp-adaptivity*. Such flexibility is inconceivable in either the finite difference or the finite volume method, wherein increasing the order of the method requires extension of the radius of influence of a node or cell or the so-called stencil. As we shall see in subsequent chapters, extension of the stencil brings additional challenges to the numerical procedure, and is not always desirable.

Historically, the finite element method was introduced for problems in structural mechanics, an area in which geometric complexity is the driver. Consequently, the finite element formulation is particularly suited to solution of PDEs in domains of complex shape. Courtesy of this attractive feature of the finite element method, researchers have endeavored to use it for problems in other physical disciplines, such as fluid dynamics and heat transfer. Unfortunately, the conservation property, discussed in the preceding section, of the finite element method comes under scrutiny for such problems, wherein conservation is the topmost priority. The standard continuous Galerkin finite element method neither guarantees global conservation nor local conservation. There is no mechanism in the continuous Galerkin formulation to enforce either local or global conservation. In realization of this important shortcoming of the standard continuous Galerkin formulation, significant effort has gone into making the finite element method amenable for use in fluid dynamics and heat and mass transfer calculations. Two methods emanating from these efforts have, arguably, outshone other methods. These include the so-called control volume finite element method (CVFEM) and the discontinuous Galerkin finite element method (DGFEM).

The CVFEM, first proposed by Winslow [23] in 1966, is a hybrid between the standard Galerkin finite element method and the finite volume method (Section 1.3.2 and Chapters 6 and 7). In order to enforce local conservation, the governing PDE is first integrated over a control volume that is constructed by joining the midpoints of the edges or faces of the elements in an unstructured finite element mesh. This results in an equation similar to Eq. (1.15). Rather than use Taylor series expansions (as in the finite volume method), the gradients and integrals in the resulting equation are then approximated using the finite element procedure. Notable contributors to this method have been Baliga [24,25] and Patankar [24,26]. More recently, the method has been used for the solution of the drift-diffusion equation on a 2D unstructured mesh [27], and the Poisson–Nernst–Planck equation on a 3D unstructured mesh [28]. An excellent review of the CVFEM method, including useful tidbits for code writing, may be found in the research monograph authored by Voller [29].

While the CVFEM has witnessed a fair amount of success in the solution of PDEs representing conservation laws, when it comes to the application of the finite element method to such PDEs, the DGFEM has, arguably, outshone all other finite element variants. It has been used extensively for the solution of all three canonical types of PDEs. Application areas have included fluid flow both in low Mach number (incompressible) regimes [30], compressible supersonic flows [31], reacting flows [32], and computational electromagnetics [33], among many others. Like the CVFEM, the DGFEM also hybridizes the finite volume and the finite element method in order to enforce local conservation. However, the basic mathematical formulation is somewhat different. Unlike the CVFEM in which continuous piecewise functions are used to approximate the solution (known as the conformal finite element method), the DGFEM uses discontinuous functions (nonconformal finite element method) to construct approximations to the solution as well as the test function. This allows treatment of discontinuities across control volume faces. Its downside includes an increase in the number of degrees of freedom (the number of unknown coefficients one has to solve for) and the resulting stiffness matrix may often be ill-conditioned (see Chapter 4 for a discussion of matrix conditions), occasionally leading to poor convergence and large solution times. Nonetheless, the DGFEM continues to be the frontrunner among finite element variants for solving advection–diffusion equations that are at the heart of PDEs representing conservation laws.

Other popular variants of the finite element method include the generalized finite element method (GFEM), the extended finite element method (XFEM), and the least square finite element method (LSFEM). The GFEM [34–36] uses basis functions that do not have to be polynomials necessarily. They can be quite arbitrary and may be selected based on the known symptoms of the solution. Furthermore, different functions may be used in different regions of the computational domain. The XFEM

[34–36] extends the power of the GFEM to include the ability to capture discontinues and singularities in the solution, such as cracks in a solid and its propagation. The LS-FEM [37,38] – a relatively new variant of the finite element method – is conceptually somewhat different from traditional finite element methods. It poses an optimization problem in which the optimal solution to the governing equation is found subject to the constraints posed by the boundary conditions. In the past two decades, its popularity has grown tremendously in realization of its ability to simulate problems in fluid dynamics and electromagnetics with relative ease compared with other finite element methods. It is believed that the method can also be used to solve the Navier–Stokes equations across all regimes ranging all the way from incompressible to hypersonic flows with minimal or no adjustments to the underlying formulation and algorithms [37]. For details, the reader is referred to texts dedicated to this method [37].

1.3.4 OTHER DETERMINISTIC METHODS

Outside of the three volume discretization based methods discussed in the preceding three subsections, a myriad of other methods for solving PDEs exist and continue to be used. It is probably fair to state, however, that these "other" methods are either still in their infancy – relatively speaking – or have been used only for a special class of problems for which they were designed. In terms of usage, there is no doubt that the three methods discussed in subsections 1.3.1 through 1.3.3 far outweigh any other method for solving PDEs. While construction of an exhaustive list of other methods for solving PDEs is quite daunting and, perhaps, of little use in the opinion of this author, at least a few of these other methods deserve a mention. These include the so-called meshless or meshfree method, the boundary element method, and the spectral method.

The name *meshless* or *meshfree* method is quite broad, and is applicable to any method that does not rely on a mesh. These may also include statistical or spectral methods. In the specific case of solution of PDEs, however, the name is reserved for methods where a *cloud* of points is used to develop an algebraic approximation to the PDE. These points have domains of influence, but do not have rigid interconnections, as in a traditional mesh. One of the notable features of meshless methods is that the points (or nodes) can be allowed to move freely, which, in the context of a traditional mesh, may end up twisting the mesh to a state where it cannot be used, and may require remeshing – a step that brings a new set of difficulties to the table. In light of this feature, meshless methods have grown in popularity for the solution of problems that involve moving boundaries, such as fluid–structure interaction [39,40], multiphase flow [41,42], moving body problems, and any problem where nodes may be created or destroyed due to material addition or removal [42]. The meshless method is a close relative of the smooth particle hydrodynamics method [43,44] that was developed for fluid flow. Some researchers contend that, fundamentally, the method is closely related to either variational formulations using radial basis functions or the moving least squares method [45]. The general concept of not using a mesh has been used in a variety of methods that include the element-free Galerkin method [46], the meshless local Petrov–Galerkin method [47], the diffuse element method [48], and the immersed boundary method [49], among others. A good review of meshless methods, as it applies to the solution of advection–diffusion type PDEs, has been provided by Pepper [50].

The boundary element method (BEM) is a derivative of the finite element method. Its attractive feature is that it does not require a volumetric mesh, but only a surface mesh. This is particularly convenient for 2D geometries, wherein the only geometric input required is a set of coordinates of points. BEM is particularly suited to problems for which the interior solution is not of practical interest, but the quantities at the boundary are, for example, a heat transfer problem in which one is primarily interested in the heat flux and temperature at the boundaries. The method is significantly more efficient than the conventional finite element method. Unfortunately, the method has several serious limitations. BEM can only be used for PDEs with constant coefficients or in which the variation of the coefficients obey a certain format. This is because it is only applicable to problems for which Green's functions can be computed easily. Nonlinear PDEs, PDEs with strong inhomogeneities, and 3D problems are generally outside the realm of pure BEM. The application of BEM to such scenarios usually requires a volumetric mesh and combination of BEM with a volume discretization method, which undermines its advantage of being an efficient method. On account of these limitations BEM has found limited use except for some specific problems in mechanics [51,52], heat conduction [53,54], and electrostatics [51].

Spectral methods constitute a class of methods in which the Laplace eigenfunctions are used as basis functions rather than polynomials. These include sine and cosine functions in Cartesian coordinates, Bessel functions in cylindrical coordinates, and Legendre polynomials in spherical coordinates. As opposed to the convectional finite element procedure in which basis functions are used locally, in spectral methods, the basis functions are used globally by constructing a Fourier series out of the basis functions and using that as an approximation to the solution. The method then solves for the undetermined coefficients of the Fourier series. As a result, spectral methods produce extremely accurate results for situations where the solution is smooth. In cases where the solutions have discontinuities, such as shocks, on account of the well-known Gibbs phenomenon [8], spectral methods yield poor solutions. Also, they are difficult to use when the geometry is complex. In the finite element arena, efforts at using very high–degree basis functions have resulted in the so-called spectral element method [55,56]. In this method, the basis functions are still used locally, and therefore, such an approach mitigates the problems associated with global approximations. A similar philosophy has also resulted in the spectral difference method [57,58] – the finite difference counterpart. As discussed earlier, neither the finite difference method nor the traditional Galerkin finite element method guarantees local conservation. In an effort to attain spectral-like accuracy while concurrently enforcing local conservation, the so-called spectral finite volume method has also been developed and demonstrated [59,60] in recent years for solution of the advection–diffusion equation on an unstructured mesh. The spectral method is attractive for its extreme accuracy. Consequently, it continues to be developed and explored for problems in which high accuracy is sought with the use of a relatively coarse mesh.

1.4 **OVERVIEW OF MESH TYPES**

Since the focus of this text is the finite difference and finite volume methods, and both of those methods require a mesh for discretization of the governing PDE, a discussion of the mesh – both its type and attributes – is a prerequisite for understanding the discretization procedures presented in later chapters. The word, *grid*, usually refers to the interconnected nodes and lines (see Fig. 1.4), while the word, *mesh*, usually refers to the spaces enclosed by the lines. Throughout this text, this subtle distinction will be ignored since it only creates confusion in the mind of the reader. Henceforth, the words mesh and grid will be used interchangeably.

The history of discretization-based procedures for solving boundary value problems can be traced back to the works of Courant around 1928. In those days, when computers were almost nonexistent, solving an elliptic PDE on a square mesh with a few handful of node points was already considered ground breaking. The need for a complex procedure or algorithm to generate a mesh was never felt. As a matter of fact, this trend continued into the early 1960s, during which period, most reported computations were performed on rectangular domains with a perfectly orthogonal (Cartesian) mesh, and mostly for 2D geometries, as shown in Fig. 1.6(a). The seminal paper by Patankar and Spalding, published in 1972, which introduced the world to the famous SIMPLE algorithm [61], reports solution of the Navier–Stokes equation performed in a square domain with a 16×16 Cartesian mesh. With rapidly advancing computer technology, by the late 1970s, the feasibility of performing slightly larger scale computations was realized.

Along with the desire to perform larger computations came the desire to perform computations in irregular geometries. In the earliest computations of this type, angled

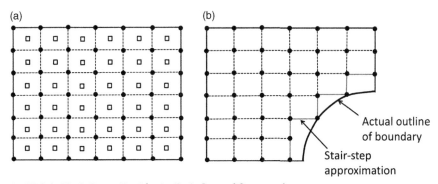

(a)　(b)

Actual outline of boundary

Stair-step approximation

FIGURE 1.6 Mesh Types Used in the Early Days of Computation

(a) A structured Cartesian mesh used for finite difference or finite volume computations. The nodes used for finite difference computations are denoted by solid circles, while the cell centers used for finite volume computations are denoted by hollow squares. The cells used in the finite volume method are the regions enclosed by the dotted lines. (b) Stair-step approximation of a curved boundary, as used in finite difference computations in the early days of computation.

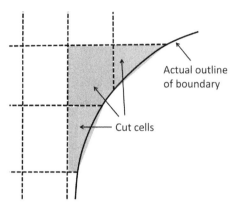

FIGURE 1.7 Cut Cells Resulting from Cutting the Original Rectangular Cells Along Straight Edges that Approximately Fit the Outline of the Boundary

or curved boundaries were addressed using a *stair-step* grid [5,26], as shown in Fig. 1.6(b). With the growing emphasis on satisfying conservation laws, and the resulting proliferation of the finite volume method, the stair-step approach was soon replaced by the *cut cell* approach, which produced polygonal cells (control volumes) at curved boundaries, as depicted in Fig. 1.7. This was considered both a reasonable as well as convenient approximation – reasonable because with a fine mesh one can always approximate a curve with piecewise linear segments (it is a practice followed even today!), and convenient because the core finite volume procedure required little modification to accommodate one slanted edge. Unfortunately, it so happens that the resulting cut cells may be of arbitrary polygonal shape, and the shape is not known *a priori*. The example depicted in Fig. 1.7 shows that three different types of cut cells, namely triangular, quadrilateral, and pentagonal, could be produced by cutting three different cells. If the mesh is refined or coarsened, the cutting pattern will change completely. As a result, using the cut cell approach became quite tedious, especially in 3D. One has to realize that all these improvements were being attempted at a time when the PDE solver itself was being modified to accommodate these cut cells and no mesh generators were available. Based on the very limited success of these preliminary attempts to solve PDEs on irregular domains, it became clear to the computational community that focused effort needs to be directed toward mesh generation itself. Over time, mesh generation and PDE solution have parted ways and become disciplines of their own. While there are groups who still perform both tasks, they generally do so either because their PDE solver requires a special type of mesh, that off-the-shelf mesh generators do not produce (e.g., Voronoi cells [62]), or because they make use of solution adaptive meshes.

The history of mesh generation can be traced back to the early 1980s when computers became powerful enough for numerical analysts to endeavor to solve PDEs in complex domains. One of the figures who led early efforts was Thompson, who put together a multiinstitution initiative known as the National Grid Project [63]. Much of the early impetus in mesh generation came from the aerospace engineering community whose primary interest was in the solution of the Navier–Stokes equation for the simulation of external flow over airfoils. As a result, rapid progress was made in the 1980s and 1990s in the area of structured body-fitted mesh generation. Parallel development of the finite element method for structural mechanics applications prompted development of grid generation algorithms capable of generating unstructured grids (primarily triangular or tetrahedral). With the rapid growth of the finite element method and its proliferation into disciplines outside of structural mechanics, such as fluid dynamics and electromagnetics, the focus shifted more toward unstructured mesh generation in the latter half of the 1990s and early 2000s. Grid generation became a discipline of its own – perhaps, bigger than the discipline of solution of PDEs itself. Today, computational geometry and grid generation are considered specialty areas in applied mathematics and computer science, with the primary focus being unstructured automatic grid generation. Although structured grids continue to be used, its popularity and usage is gradually fading because of its limited applicability.

In the preceding discussion, two types of mesh were mentioned: structured and unstructured. The general notion shared by many is that a structured mesh is comprised of a set of ordered curves and surfaces that result in either quadrilaterals (in 2D) or hexahedrons (in 3D). On the other hand, an unstructured mesh is disordered and is comprised of either triangular (in 2D) or tetrahedral (in 3D) elements. While the majority of this notion is true, some parts are not. In the discussion to follow we aim to clarify a few important attributes of these two major types of grids.

A *structured mesh* is one that is always ordered, generally using three counters or indices (i, j, k). If the grid lines are aligned with the Cartesian coordinate directions, one may think of the indices (i, j, k) as being counters that count the number of grid points in the x, y, and z directions, respectively. If a generalized curvilinear coordinate system is used, then a similar interpretation also holds true: (i, j, k) would then be counters in the ξ_1, ξ_2, and ξ_3 directions, where ξ_i denotes curvilinear coordinates (see Section 2.8 for details). The two scenarios are shown in Fig. 1.8 for a 2D geometry. One of the important attributes of a structured grid is that the connectivity

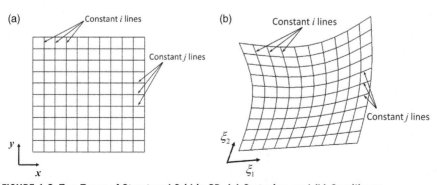

FIGURE 1.8 Two Types of Structured Grid in 2D: (a) Cartesian, and (b) Curvilinear

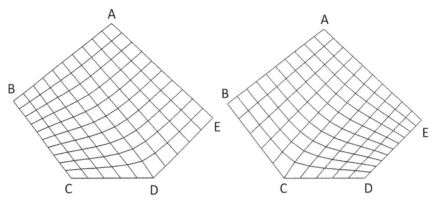

FIGURE 1.9 Two Ways of Creating a Structured Mesh in a Pentagonal Domain

is implicit. A node with index (5,32,6) will have (5,33,6) and (5,31,6) as its two neighbors in the y (for Cartesian) or ξ_2 (for curvilinear) direction. One does not have to store such neighboring node information explicitly – it is obvious from the indexing scheme due to the (i, j, k) ordering of a structured mesh.

While the implicit or automatic connectivity due to the (i, j, k) ordering of a structured grid is very convenient, it also brings to the table several downsides. The most critical downside is that in a structured mesh, the number of nodes on opposite edges (in 2D) and faces (in 3D) must match. While "opposite" edges are relatively straightforward to determine in a region bounded by four edges, it is not straightforward for irregular or curved regions. Let us, for example, consider creating a structured mesh in a pentagonal region. In this particular case, there are at least two ways to match the opposite edges, as illustrated in Fig. 1.9. In the left figure, the edge AB has been matched with a composite edge CDE, while in the right figure, edge AE has been matched with the composite edge BCD. In other words, human intervention is needed in deciding which choice is to be used, and the process cannot be automated (in an algorithmic sense) since there is no unique solution. In complex 3D geometries, such choices are much more numerous, and serious human interaction is necessary to generate a complex structured mesh. While the resulting mesh quality is generally satisfactory in the end, for decades, because of the aforementioned reason, structured mesh generation has been considered a major bottleneck [63] for the numerical solution of PDEs using structured meshes.

Two other difficulties arise as offshoots of the need to match opposite edges (or faces) in a structured mesh. First, curves that would otherwise be considered a single geometric entity, need to be split in order to generate a structured mesh. For example, to generate a structured mesh in a circle, the circle needs to be split into at least four segments. Second, a structured mesh is almost impossible to create for triangular shapes or shapes with sharp corners (small solid angles in 3D). Examples of these two scenarios are illustrated in Fig. 1.10. In the first case, the circle had to be split into four segments to realize opposite edges. The mesh created is a valid

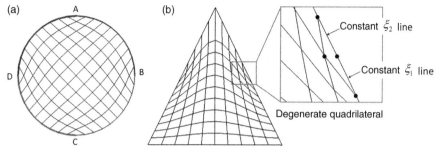

FIGURE 1.10 Some Geometries for Which a Structured Mesh is Difficult to Create

(a) A circle showing the four split segments used to generate a structured mesh, and (b) a triangle, with a zoomed-in illustration of the mesh at one of its sides.

mesh, although of poor quality near the four splitting points because of the extreme obtuseness of the angles in the quadrilaterals generated in that area. In the case of the triangle, the two sides had to be split, and the top halves of the two split sides were considered to be a composite edge that matched the edge at the base. In this case, the mesh is an invalid mesh. As shown in the zoomed image in Fig. 1.10(b), the ξ_1 and ξ_2 lines become coaligned (this means they are locally no longer independent variables) at the splitting point, and a degenerate quadrilateral is produced. This implies that the transformation from Cartesian to curvilinear coordinates will fail in this particular case, as will become clear upon perusal of Section 2.8.

Local or selective grid refinement is another important need in modern-day computing that structured grids cannot cater to very effectively. Once again, since matching of opposite edges is mandated in this paradigm, the refinement is forced to be global, resulting in undesirably large node or cell counts. This is depicted in the two examples shown in Fig. 1.11. In the first example, Fig. 1.11(a), flow around a square obstacle is considered. The mesh has been clustered near the surface of the square in an effort to resolve the thin boundary layers on the surfaces of the square, and vortex structures behind it. However, as depicted in Fig. 1.11(a), if a structured mesh is used, the refined regions propagate all the way to the edges of the computational domain. This results in small grid spacing in regions that do not need a fine mesh, as labeled by the elliptical tags. In the second example depicted in Fig. 1.11(b), two reservoirs are connected by a narrow channel. In order to resolve the flow through this narrow channel, 11 grid points (which is still quite coarse!) are used across the channel. This refinement, however, propagates into the reservoir if a structured mesh is used. In fact, the grid spacing changes dramatically in the reservoir in the vertical direction. Ideally, the semicircular regions shown in the figure should be refined, and the grid should coarsen outward gradually from the edge of the semicircle. However, such local or clustered refinement is beyond the capability of a structured mesh.

The second example, shown in Fig 1.11(b), brings to light another important aspect of modeling: multiscale modeling. While the governing PDEs we use today to model various physical processes have remained unchanged from a century ago, the

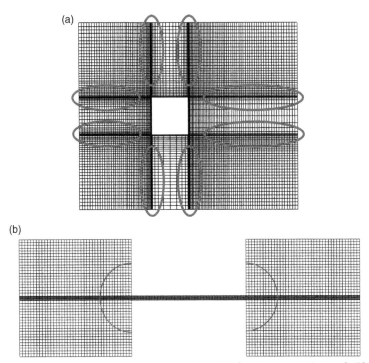

FIGURE 1.11 Two Examples Illustrating the Downside of Using Structured Meshes in Situations Where Local Refinement is Desirable

(a) Square obstacle in a flow path, and (b) two reservoirs connected by a narrow channel.

length and timescales over which they are applied, or, at least, expected to be successfully applied today, span several orders of magnitude. For example, if a ceramic dinner plate is to be heated in a furnace, the plate may be a few millimeters thick, while the furnace may be a few feet in size. How do we mesh and model both scales within the same framework? The refinement and coarsening algorithms of modern mesh generators are put to severe test by such multiscale problems. In the unstructured mesh generation arena, significant progress has been made in the past decade to address this important need, as will be demonstrated through some examples shortly. However, as pointed out earlier, due to the inherent constraint of opposite edge/face matching for structured grids, progress has been at a standstill.

Thus far, the focus of the discussion has been primarily on the disadvantages of using a structured mesh. Of course, structured meshes have some distinct advantages as well, because of which they have had widespread use, especially in computational fluid dynamics. However, their discussion would be more meaningful once the reader has had a chance to review the features of an unstructured mesh. Any mesh can be thought of as an unstructured mesh. A structured mesh may be treated as an unstructured mesh, and is, therefore, a subset of the broader category of an unstructured mesh. The main difference between a structured and an unstructured mesh is that while a structured mesh is (i, j, k) ordered, an unstructured mesh has no specific ordering. This attribute immediately removes some of the aforementioned shortcomings associated with matching opposite faces. In principle, the volume in question can be filled with a combination of polygons or polyhedrons of arbitrary shape and size. Since no specific order exists in an unstructured mesh, the connectivity, i.e., how nodes, elements, or cells are connected to each other, is not obvious. Such information has to be explicitly generated and provided as output by the mesh generator so that the PDE solver has access to this information. This may be cited as the most critical difficulty of using an unstructured mesh over a structured mesh when it comes to developing unstructured PDE solvers.

Aside from the convenience of having automatic connectivity, the (i, j, k) ordering of a structured mesh offers one other important advantage over an unstructured mesh. The (i, j, k) ordering results in coefficient matrices that are banded. As a result, easy-to-use but powerful iterative linear algebraic equation solvers that rely on the banded structure of the matrix can be used. These include line-by-line or the alternating direction implicit (ADI) method, and the Stone's strongly implicit method. The reader is referred to Chapter 4 for a detailed discussion of these solvers. While many powerful and advanced linear algebraic equation solvers are available today, the aforementioned two solvers continue to be used for structured grid computations for their ease of use and low-memory properties.

While coordinate transformation and curvilinear coordinates enabled use of structured meshes for bodies of arbitrary shape – so-called *body-fitted* meshes – many complex computational domains require use of multiple blocks with each block having its own (i, j, k) ordering. Such structured meshes are referred to as *block-structured* meshes. As a matter of fact, the meshes shown in Fig. 1.11 are both block structured. An example of a body-fitted, block-structured mesh is shown in Fig. 1.12. The mesh is body fitted because the grid lines are aligned to the airfoil (the body under consideration) outline. It is block structured because the mesh is comprised of multiple blocks, as shown in Fig. 1.12. Although connectivity is automatic in structured grids, in block-structured grids, additional information needs to be stored

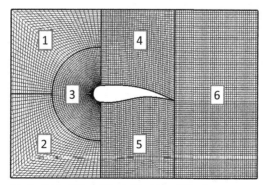

FIGURE 1.12 Example of a Body-Fitted, Block-Structured Mesh

at the interface between blocks. Since each block has its own (i, j, k) ordering, one must know and store the connectivity between blocks, as well as the pairing of the (i, j, k) lines on the two sides of the interface. This increases bookkeeping and memory requirements of the PDE solver, and also adversely affects convergence since each block still has to be treated in isolation. Thus, some of the aforementioned advantages of using a structured grid are negated.

As discussed earlier, the choice of edge or face pairs in a structured mesh can lead to different grids. For example, the grid shown in block 3 of the mesh shown in Fig. 1.12 is known a *C-grid*, evidently because the grid looks like the letter C. If a mirror image of block 3 were created about the vertical line separating block 3 from blocks 4 and 5, the resulting grid would look like the letter O, and is known as an *O-grid*. C- and O-grids are suitable for domains that have the approximate shape of a circle (in 2D) or sphere (in 3D). However, they can only be generated if the center of the circular shape is not part of the computational domain. Extending the grid to the center of the O would create a singularity (faces with zero area). For such cases, an *H-grid* is preferred. Figure 1.13 shows examples of an O-grid and a hybrid O-H-grid (H in center and O in the periphery). An H-grid is the most common type of structured grid, and includes any grid that is not a C- or O-grid.

An *unstructured* grid or mesh is one in which there is no specific ordering of the nodes or cells. Any structured mesh may be treated as an unstructured mesh – one in which the ordering is of a certain specific type, and the resulting cells are either quadrilaterals (in 2D) or hexahedron (in 3D). Historically, unstructured meshes have their roots in the finite element arena where using triangular (in 2D) or tetrahedral (in 3D) elements is standard practice. Unstructured mesh generation is now a discipline of its own, and considerable progress has been made in this area. The finite volume method was adapted to unstructured mesh topology in the mid-1990s, thereby further expanding its spectrum of use to include computational fluid dynamics and related applications. In the modern era, most commercial computational fluid dynamics codes use an unstructured mesh formulation. Although structured-looking meshes

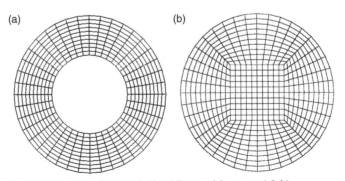

FIGURE 1.13 Examples of Two Commonly Used Types of Structured Grid

(a) O-grid of a hollow disc, and (b) hybrid O-H-grid of a solid disc.

continue to be used, the PDE solution algorithms treat such meshes as unstructured meshes.

An unstructured mesh can accommodate all element or cell types. Common element types include triangles and quadrilaterals in 2D, and tetrahedrons, hexahedrons, prisms, and pyramids in 3D. In addition, arbitrary polygons (in 2D) or polyhedrons (in 3D) may also be generated under special circumstances. The fact that the unstructured formulation (see Chapter 7 for details of unstructured finite volume) can accommodate arbitrary polygons or polyhedrons, makes them particularly attractive for local adaptive mesh refinement or coarsening, as will be discussed shortly. The element types used in an unstructured mesh are shown schematically in Fig. 1.14. In general, for a given computational domain, the element types may be mixed – referred to as a *hybrid mesh*. Almost all commercially available mesh generations allow the user to select regions of the computational domain, along with the specific type of element desired in that region. Many modern-day mesh generators also allow selection of the dominant cell type. For example, one may choose to generate a quad-dominant mesh, in which case most elements will be quadrilaterals, and a

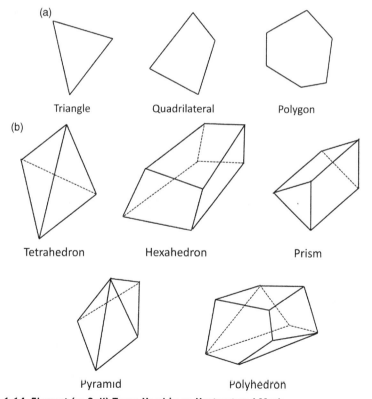

FIGURE 1.14 Element (or Cell) Types Used in an Unstructured Mesh

(a) 2D element types, and (b) 3D element types.

small number of other element types may be used, if needed, to produce a mesh of the best quality.

Two of the disadvantages of a structured grid, cited earlier (see Fig. 1.11 and associated discussion), are its inability to produce local refinement and its inability to effectively address geometries that have multiscale features. These difficulties can be mitigated by unstructured grids to a large extent. Figure 1.15 shows unstructured meshes for the same two cases depicted in Fig. 1.11. It is evident from Fig. 1.15 that with an unstructured mesh, tremendous local refinement is possible, enabling resolution of small-scale features within the context of a larger length scale.

With increasing use of unstructured meshes in computational fluid dynamics, another type of mesh that has become quite popular is the so-called *Cartesian mesh*. A Cartesian mesh has many of the features of a structured Cartesian mesh, such as having control volumes whose faces are aligned to the Cartesian planes, but it is fundamentally unstructured because the cells can be arbitrarily bisected or agglomerated

(a)

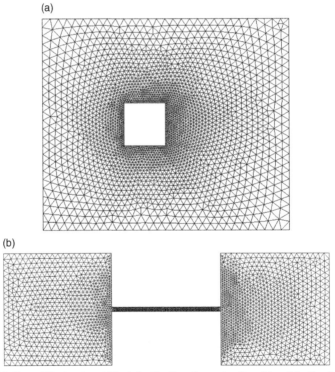

(b)

FIGURE 1.15 Use of Unstructured Mesh for the Two Examples Considered in Fig. 1.11

(a) Square obstacle in a flow path, showing local mesh refinement around the obstacle, and (b) two reservoirs connected by a narrow channel which is filled using very small elements that gradually coarsen in other regions of the computational domain.

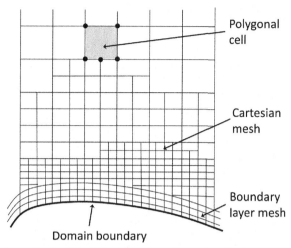

Polygonal cell

Cartesian mesh

Boundary layer mesh

Domain boundary

FIGURE 1.16 Schematic of a Hybrid Boundary Layer/Cartesian Mesh

to form finer or coarser cells, thereby facilitating local refinement or coarsening. The local bisection or agglomeration of the cells results in polygonal (in 2D) or polyhedral (in 3D) cells, as shown in Fig. 1.16. The polygonal cells are created by bisecting one or more of the edges of the mother edge. In the specific example shown in Fig. 1.16, the polygonal cell has five edges and five vertices. Cartesian meshes are generated using an octree-based algorithm [64]. As opposed to the finite element method in which using triangular or tetrahedral meshes is the standard practice, finite volume formulations prefer Cartesian-type meshes. As we shall see in Chapter 7, one of the determining factors for fast convergence of finite volume–based solutions of PDEs is the so-called mesh skewness. The skewness is essentially a measure of how closely the surface normals, at cell faces, are aligned with the vector joining the cell centers of two adjacent cells. If the alignment is perfect, as in the case of a Cartesian mesh or a mesh comprised of equilateral triangles (in 2D) or equal-sized tetrahedrons (in 3D), the convergence is the best. The impetus of Cartesian meshes in recent times has, in fact, been prompted by this attribute of the finite volume formulation.

For viscous fluid-flow computations, it is imperative that the boundary layer next to walls or solid bodies be resolved adequately with a mesh, and that the mesh be smooth. By "smooth," we mean a mesh whose element or cell size does not vary significantly along the surface. If the element or cell size varies along the surface, the solution of the PDE along the surface will, consequently, not be smooth. This is particularly true for postprocessed quantities that require derivatives (gradients) of the dependent (solved) variable. For example, in a fluid-flow computation, the computed shear stress at the wall, which requires the normal gradient of the velocity, may exhibit spurious oscillations due to a nonsmooth mesh next to the surface. In order to avoid such numerical artifacts, structured body-fitted meshes are preferred around solid bodies – often referred to as a *boundary layer mesh*. In realization of the efficacy of such meshes close to the solid surface, it is now common practice to hybridize a boundary layer mesh with Cartesian, or other unstructured topology to create a mesh that gradually coarsens out from the body to fill up the entire computational domain, as depicted in Fig. 1.16. By this strategy, the benefits of both structured and unstructured meshes are reaped.

Over the past several decades, many other types of meshes have been used for deterministic solution of PDEs. The application under consideration usually drives the choice of mesh type to be used. For example, for moving body problems, use of overset grids or Chimera technology [65] is quite common. Moving body or sliding surface simulations make use of so-called *nonconformal meshes* in which cell faces at an interface between two regions of the mesh may not exactly match, and information is exchanged between the two regions using advanced flux interpolation (transfer) mechanisms. In contrast, the meshes discussed thus far all fall under the category of a conformal mesh. The topic of mesh types and mesh generation is extremely broad. In this section, only a brief overview was presented. The advanced reader is referred to texts on mesh generation [66,67].

1.5 VERIFICATION AND VALIDATION

As discussed earlier, in the vast majority of situations, PDEs represent mathematical models of certain physical phenomena. Therefore, their solutions are worthwhile and of practical value only when they resemble reality, i.e., the observed physical phenomena. In order to lend credibility to the solution of the mathematical model – PDEs, in our particular case – it is important to identify the errors that may be incurred in various stages of the modeling/solution process, and quantify and analyze them in a systematic manner. Only then can one have a certain level of confidence in the predicted results in the event that the solution to the mathematical model does not exactly match measured data or observed behavior of the system being modeled. In order to identify and quantify the errors associated with various stages of the modeling process, we begin with an overview of these stages. A flowchart depicting the various stages of the modeling process is shown in Fig. 1.17.

The analysis of any problem begins with a *physical description* of the real-life problem. This description usually constitutes some text and, perhaps, a few accompanying illustrations that describes the problem. For example, a car manufacturer may be interested in exploring the difference between two spoiler designs. In this case, the physical description will constitute typical range of air flow speeds over the spoiler and other physical details, along with illustrations detailing the car body and spoiler geometries. In addition, the description will include what the goal of the analysis is and what is sought. Is it the drag? Is it the structural deformation? Is it the wind noise?

The second step in the analysis is to create a *physical model* from the available physical description. The creation of the physical model entails making important decisions on what physical phenomena need to be included. These decisions are

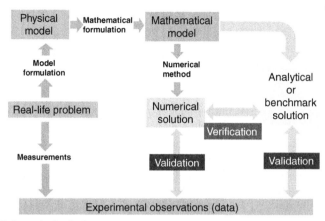

FIGURE 1.17 Steps in Analysis

driven not only by the goals of the study, but also the feasibility of including or excluding a certain physical phenomenon based on the resources and time available to complete the task. For the specific example at hand, important questions that need to be addressed may include (a) is the unsteadiness of the flow important, or is it sufficient to find steady-state solutions? Perhaps, for noise predictions, considering unsteadiness is important, while for drag predictions, it is not. (b) Is it sufficient to model the problem as isothermal? (c) Is it important to account for the unsteady nature of the turbulence? (d) For noise and drag predictions, it is adequate to treat the spoiler as a rigid body? It should be noted that the answers to some of these questions may not be known *a priori*, and one may have to explore the various possibilities. This is where a survey of past work (literature review), in addition to the experience of the analyst, can assist in narrowing down the possibilities. In most disciplines, prior knowledge and training in the field aids in answering many of these questions with a sufficient level of confidence. In emerging fields, this step may be cited as one of the most challenging steps in the overall analysis process.

Once the physical model has been created, the next step is to develop the *mathematical model*. A summary of the various types of mathematical models have already been provided in Section 1.1. Here, we will assume that the relevant mathematical model constitutes PDEs. Consequently, the two most critical questions are: (a) what PDEs should we solve, and (b) what are the relevant initial and boundary conditions? As in the case of the physical model's development, the answers to these questions are often deeply rooted in the state-of-the-knowledge in the field in question and requires formal training and experience. For example, a fluid dynamics expert might easily be able to pinpoint the deficiencies of a particular non-Newtonian model for modeling blood flow. In mature disciplines, the PDEs that need to be solved are generally well known. What is not often clear is what boundary or initial conditions best suit the physical problem at hand. For example, in the actual physical setup, a body may have been immersed in a bucket of ice. What is the appropriate thermal boundary condition? Is it appropriate to assume that the entire surface of the body is at the freezing point of water, or should one pose an external convective heat transfer boundary condition? More often than not, these decisions and associated uncertainties lead to discrepancies between the model's predictions and observed behavior.

Only a limited few PDEs have closed-form analytical solutions. Irregular geometry, space or time dependent coefficients, nonhomogenous boundary conditions, and nonlinearities in the governing equation or boundary conditions, are some of the primary reasons why numerical solutions are warranted. Once a numerical procedure is introduced into the solution process, additional errors creep into the solution. These errors are associated with how the governing PDE is solved numerically, and may be borne out of the discretization process, as in the mesh-based deterministic methods discussed in Section 1.3, or due to other kinds of errors discussed in subsequent chapters in this text. Without going into details at this juncture, it should suffice to state that the numerical solution may have some discrepancies with the closed-form exact analytical solution to the PDE, if available, and the degree of discrepancy is generally dependent on the numerical method being used. *Verification* is the process of comparing the numerical solution of the governing mathematical model (PDE, along with initial and boundary conditions) to a "well-established" solution of the same equations. By well-established we mean either a proven closed-form analytical solution or a solution obtained by a proven numerical technique, agreed upon by the community at large, so that it may be deemed a *benchmark* solution. For example, benchmark solution to the Navier–Stokes equation for laminar fluid flow is available in the literature for the so-called driven-cavity problem [68,69] and the backward-facing step [70,71] problem. These benchmark solutions usually employ very fine mesh and/or high-order discretization schemes to eliminate errors due to discretization, so that the resulting solution has analytical-like accuracy. Verification is an extremely important step in numerical analysis. It ensures that the solution to a given equation, obtained *via* numerical means, is actually the correct solution, pending quantifiable, well-understood errors that are inherent in the numerical solution process. It lends credibility to the numerical method and its implementation, i.e., the computer program. It rules out the possibility that an unidentifiable or inadvertent error was made in obtaining the solution to the governing equation. Verification of a code (implementation of a certain numerical method), however, does not guarantee that the results will match experimental observations.

Benchmark solutions are not available for all problems, and so, verification is not always possible. This difficulty may be somewhat mitigated by performing one or more *sanity checks* on the numerical solution (or code). A sanity check is a check performed on a limiting case of the actual problem at hand. It can be conducted by altering one or more boundary conditions, initial conditions, or other inputs to the problem such that the altered problem has an obvious solution and can be used for comparison with the actual numerical solution. For example, let us consider a situation where we seek to verify the numerical solution to the PDE, $\nabla \bullet (\Gamma \nabla \phi) = 0$, with $\Gamma = \Gamma(x, y, z)$. The analytical solution to the PDE is not known for arbitrary Γ. Therefore, the numerical solution cannot be verified directly. However, by setting

Γ as a constant, one can compare the numerical solution with the known analytical solution of the resulting Laplace equation. This constitutes a sanity check. Another way to perform a sanity check of the same code would be to set the same Dirichlet (prescribed ϕ) boundary condition on all boundaries. Irrespective of the value of Γ – constant or not – the altered problem has a solution equal to whatever value of ϕ the boundaries are set to, and the code must produce that limiting solution. It is worth pointing out that a successful sanity check is a necessary, but not sufficient, condition for correctness of the numerical solution or code. However, it reduces the likelihood that there is an error in the code.

The process of quantitative comparison between the solution of the mathematical model and experimental observations is known as *validation*. Validation is necessary whether or not the solution was obtained analytically or by numerical means, whether the solution is that of a PDE, an ordinary differential equation (ODE), or a simple algebraic equation. Anytime a model has been created following the route sketched in Fig. 1.17, there is an obligation to compare the predictions of the model with experimental data to ascertain that the results are meaningful, and of practical value. The process of verification rules out errors in the numerical solution to the governing equations. However, the governing equations themselves, or the boundary conditions, may have been flawed to begin with. In other words, errors may have been introduced in developing the mathematical model from the physical model. Similarly, errors may also have crept in during the process of creating the physical model. Certain important physics may have been ignored. Additional errors may have been injected into the solution due to improper inputs, such as material properties.

The differences between verification and validation, and the various sources of error in the model prediction are best elucidated by an example. Let us consider a scenario in which an engineer is interested in developing a model that predicts the steady-state temperature distribution in a long steel rod of square cross section. Let us also assume that experimental measurements have been made and the temperature is known at a handful of locations inside the slab. Validation entails comparing these data to the numerically computed results. Upon comparison, the engineer finds that the discrepancy between the two solutions ranges between 10% and 25%. In order to trace back to the root of the discrepancy, the engineer must now reexamine every step of the modeling process. First, he considers the correctness of his numerical procedure and code. He verifies his code against a known benchmark solution, and rules out any possibilities of error in his code. In other words, his verification study succeeds, but his validation study failed. Tracing back to the sources of the failure, he realizes that there could be two sources of error: either in the physical model or in the mathematical model. The physical model assumes that the only mode of heat transfer inside the rod is conduction. Since the measured temperature has no such assumptions, this assumption must be closely examined. Since the rod is made of solid steel, and steel is opaque to radiation, the assumption of heat transfer in the rod by pure conduction is justified. Therefore, the engineer rules out errors borne out of the physical model. Next, he examines the mathematical model. His mathematical model is comprised of the 2D steady-state heat conduction equation (a Poisson equation) with constant thermal conductivity (Γ) and Dirichlet boundary conditions (prescribed temperature) on all four boundaries. Upon reassessing the model, the engineer realizes that errors could have been introduced into the solution due to the following assumptions: (a) the thermal conductivity of steel is constant; in reality, it is temperature dependent; (b) edge effects in the axial direction were neglected (2D assumption); and (c) the temperature at the four edges (boundaries) of the rod may not have been constant during the experimental measurements. Any one of these assumptions, or a combination thereof, could have caused the discrepancy between the measured data and the model's predictions. In summary, even for seemingly simple problems, validation can be tricky! It involves meticulous research in establishing a good physical and mathematical model, as well as determining all the necessary inputs to the model. For a numerical analyst whose starting point is usually the mathematical model, verification is, perhaps, the more important and relevant step. It pinpoints errors in the numerical solution rather than elsewhere in the whole modeling framework.

With the impetus in modeling and simulations, verification and validation have begun to be considered as critical steps in analysis. Many consortiums have been created worldwide in individual disciplines to assimilate benchmark data – both numerical and experimental. It is worth noting that experimental data also need to be scrutinized and approved by the community at large prior to their acceptance as benchmark data. For example, the Sandia National Laboratory now hosts a big database – referred to as Sandia Flames [72] – that contains experimental data for a large number of laboratory-scale flames that may be used to validate combustion codes. Likewise, the National Program for Applications-oriented Research in CFD (NPARC) alliance was created to guide and monitor verification and validation of computational fluid dynamics (CFD) codes. It provides an online archive [73] of benchmark data for verification and validation of CFD codes. The American Institute of Aeronautics and Astronautics (AIAA) has created a guide on verification and validation [74]. A number of benchmark test cases for the verification of CFD codes are also available in a recent AIAA conference paper [75]. Nowadays, technical conferences routinely host sessions and panel discussions on verification and validation, as well. Most journals also mandate verification studies for manuscripts reporting new numerical methods, prior to publication. The American Society of Mechanical Engineering has, in fact, recently launched a new journal entitled the *Journal of Verification, Validation and Uncertainty Quantification*, which disseminates high-quality data for verification and validation of computational models for a wide range of disciplines related to fluid flow and heat transfer.

To summarize, this chapter provided a brief overview of some of the most important aspects of analysis, and in particular, numerical solution of PDEs. A brief history of computing was followed by a discussion of why and how computations could be beneficial. The procedure to classify PDEs was presented, along with examples of some commonly encountered PDEs in scientific and engineering analysis. This was followed by an overview of the various methods that are currently used to solve PDEs, with an emphasis on deterministic methods. Since the grid or mesh is an integral part of the solution process, an overview of the various commonly used mesh types and the rationale for their use was also presented. The chapter concluded with a discussion of the steps involved in solving a problem of practical interest, and the steps that may be undertaken to ascertain the correctness of the computed solution, namely sanity checks, verification, and validation. The present chapter is meant to provide only a broad flavor of the topic of modeling and simulations. The subsequent chapters delve deep into individual topics. The chapter to follow introduces the reader to the finite difference method.

REFERENCES

[1] Moore GE. Cramming more components onto integrated circuits. Electronics 1965;38(8):114–7.

[2] Whitaker S. Fundamental principles of heat transfer. Krieger Publishing Company; Malabar, Florida; 1983.

[3] Bird RB, Stewart WE, Lightfoot EN. Transport phenomena. 2nd ed. New York: Wiley; 2001.

[4] White FM. Fluid Mechanics. 7th ed. New York: McGraw Hill; 2010.

[5] Ferziger JH, Perić M. Computational methods for fluid dynamics. 3rd ed. Berlin Springer; 2002.

[6] Pletcher RH, Tannehill JC, Anderson D. Computational fluid mechanics and heat transfer. 3rd ed. Boca Raton, Florida CRC Press; 2011.

[7] Zucrow MJ, Hoffman JD. Gas dynamics, vol. 1. New York: John Wiley and Sons; 1975.

[8] Kreyzig E. Advanced engineering mathematics. 10th ed. New York: John Wiley and Sons; 2010.

[9] Fishman GS. Monte Carlo: concepts, algorithms, and applications. New York: Springer; 1995.

[10] Modest MF. Radiative heat transfer. 3rd ed. New York Academic Press; 2013.

[11] Mazumder S, Kersch A. A fast Monte-Carlo scheme for thermal radiation in semiconductor processing applications. Numer Heat Transfer Part B 2000;37(2):185–99.

[12] Mazumder S, Majumdar A. Monte Carlo study of phonon transport in solid thin films including dispersion and polarization. J Heat Transfer 2001;123:749–59.

[13] Lacroix D, Joulain K, Lemonnier D. Monte Carlo transient phonon transport in Silicon and Germanium at nanoscale. Phys Rev B 2005;72. 064305(1–11).

[14] Mittal A, Mazumder S. Monte Carlo study of phonon heat conduction in silicon thin films including contributions of optical phonons. J Heat Transfer 2010;132. Article number 052402.

[15] Fischetti MV, Laux SE. Monte Carlo study of electron transport in silicon inversion layers. Phys Rev B 1993;48(4):2244–74.

[16] Jacoboni C, Reggiani L. The Monte Carlo method for the solution of charge transport in semiconductors with applications to covalent materials. Rev Mod Phys 1983;55(3): 642–705.

[17] Bird GA. Molecular gas dynamics and the direct simulation of gas flows. Oxford: Clarendon; 1994.

[18] Pope S. PDF methods for turbulent reactive flows. Prog Energy Combust Sci 1985;11: 119–92.

[19] Mazumder S, Modest MF. A probability density function approach to modeling turbulence-radiation interactions in nonluminous flames. Int J Heat Mass Transf 1999;42(6):971–91.

[20] Neimarlija N, Demirdzic I, Muzaferija S. Finite volume method for calculation of electrostatic fields in electrostatic precipitators. J Electrostat 2009;67(1):37–47.

[21] Johnson T, Jakobsson S, Wettervik B, Andersson B, Mark A, Edelvik F. A finite volume method for negative three species negative corona discharge simulations with application to externally charged powder bells. J Electrostat 2015;74:27–36.

[22] Zienkiewicz OC, Taylor RL, Zhu JZ. The finite element method: its basis and fundamentals. 7th ed. Waltham, MA Butterworth-Heinemann; 2013.

[23] Winslow AM. Numerical solution of the quasilinear Poisson equation in a nonuniform triangular mesh. J Comput Phys 1966;1:149–72.

[24] Baliga BR, Patankar SV. A new finite element formulation for convection-diffusion problems. Numer Heat Transfer 1980;3:393–409.

[25] Baliga BR, Atabaki N. Control-volume-based finite-difference and finite-element methods. In: Minkowycz WJ, Sparrow EM, Murthy JY, editors. Handbook of numerical heat transfer. Hoboken: Wiley; 2006.

[26] Patankar SV. Numerical heat transfer and fluid flow. Washington: Hemisphere; 1980.

[27] Bochev P, Peterson K, Gao X. A new control volume finite element method for the stable and accurate solution of the drift-diffusion equations on general unstructured grids. Comput Methods Appl Mech Eng 2013;254:126–45.

[28] Pittino F, Selmi L. Use and comparative assessment of the CVFEM method for Poisson–Boltzmann and Poisson–Nernst–Planck three dimensional simulations of impedimetric nano-biosensors operated in the DC and AC small signal regimes. Comput Methods Appl Mech Eng 2014;278:902–23.

[29] Voller VR. Basic control volume finite element methods for fluids and solids. Singapore: World Scientific Publishing Co.; 2009. (IISc Research Monographs Series).

[30] Schotzau D, Schwab C, Toselli A. Mixed hp-DGFEM for incompressible flows. SIAM J Numer Anal 2003;40(6):2171–94.

[31] Feistauer M, Felcman J, Straskraba I. Mathematical and computational methods for compressible flow. New York: Oxford University Press; 2003.

[32] Zhu J, Zhang YT, Newman SA, Alber M. Application of discontinuous Galerkin methods for reaction-diffusion systems in developmental biology. J Sci Comput 2009;40:391–418.

[33] Houston P, Perugia I, Schotzau D. *hp*-DGFEM for Maxwell's equations in Numerical mathematics and advanced applications. F. Brezzi et al. Eds.; Springer Verlag Italia, Milano, 2003. pp. 785–794.

[34] Fries TP, Belytschko T. The extended/generalized finite element method: an overview of the method and its applications. Int J Numer Method Eng 2010;84(3):253–304.

[35] Khoei AR. Extended finite element method: theory and applications. John Wiley and Sons: Chichester, UK; 2015.

[36] Efendiev Y, Hou TY. Multiscale finite element methods: theory and applications. Springer; 2009; in "Surveys and Tutorials in the Applied Mathematical Sciences, Vol. 4". see http://www.springer.com/us/book/9780387094953

[37] Jiang B. The least-squares finite element method: theory and applications in computational fluid dynamics and electromagnetics. Springer: Berlin; 1998.

[38] Reddy JN, Gartling DK. The finite element method in heat transfer and fluid dynamics. 3rd ed. CRC Press: Boca Raton, FL; 2010.

[39] Zhang LT, Wagner GJ, Liu WK. Modelling and simulation of fluid structure interaction by meshfree and FEM. Commun Numer Methods Eng 2003;19(8):615–21.

[40] Tiwary S, Antonov S, Hietel D, Kuhnert J, Olawsky F, Wegener R. A meshfree method for simulations of interactions between fluids and flexible structures. Meshfree Methods Partial Differential Equations III 2007;57:249–64. (Lecture notes in computational science and engineering).

[41] Fu L, Jin YC. Simulating gas-liquid multiphase flow using meshless Lagrangian method. Int J Numer Meth Fluids 2014;76(11):938–59.

[42] Li S, Liu WK. Meshfree and particle methods and their applications. Appl Mech Rev 2002;55(1):1–34.

[43] Benz W. Smooth particle hydrodynamics: a review. Numerical modeling of non-linear stellar pulsation: problems and prospects. Boston: Kluwer Academic; 1990.

[44] Liu GR, Liu MB. Smoothed particle hydrodynamics: a meshfree particle method. Singapore: World Scientific Publishing Co; 2003.

[45] Fasshauer, G.F. Meshfree approximation methods with Matlab in Interdisciplinary Mathematical Sciences, Vol. 6, Hackensack, NJ. World Scientific Publishing Co., New Jersey.

[46] Belytschko T, Lu YY, Gu L. Element-free Galerkin methods. Int J Numer Meth Eng 1994;37:229–56.

[47] Atluri SN, Zhu T. A new meshless local Petrov–Galerkin (MLPG) approach in computational mechanics. Comput Mech 1998;22:117–27.

[48] Nayroles B, Touzot G, Villon P. Generalizing the finite element method: diffuse approximation and diffuse elements. Comput Mech 1992;10(5):307–18.

[49] Mittal R, Iaccarino G. Immersed boundary methods. Ann Rev Fluid Mech 2005;37: 239–61.

[50] Pepper DW. Meshless methods. In: Minkowycz WJ, Sparrow EM, Murthy JY, editors. Handbook of numerical heat transfer. Hoboken: Wiley; 2006.

[51] Banerjee PK, Wilson B. Developments in boundary element methods: industrial applications. CRC Press: Boca Raton, FL; 2005.

[52] Beskos DE. Boundary element methods in dynamic analysis. Appl Mech Rev 1987;40(1):1–23.

[53] Wrobel LC, Brebbia CA. A formulation of the boundary element method to axisymmetric transient heat conduction. Int J Heat Mass Transfer 1981;24(5):843–60.

[54] Sutradhar A, Paulino GH. The simple boundary element method for transient heat conduction in functionally graded materials. Comput Method Appl Mech Eng 2004;193:4511–39.

[55] Bernardi C, Maday Y. Spectral element methods. In: Ciarlet PG, Lions JL, editors. Handbook of numerical analysis. Amsterdam: North-Holland; 1996.

[56] Lee U. Spectral element methods in structural dynamics. John Wiley and Sons: New York; 2009.

[57] Liu Y, Vinokur M, Wang ZJ. Spectral difference method for unstructured grids: I. Basic formulation. J Comput Phys 2006;216(2):780–801.

[58] Wang ZJ, Liu Y, May G, Jameson A. Spectral difference method for unstructured grids: II. Extension to the Euler equations. J Sci Comput 2007;32(1):45–71.

[59] Wang ZJ. Spectral (finite) volume method for conservation laws on unstructured grids: basic formulation. J Comput Phys 2002;178(1):210–51.

[60] Wang ZJ, Liu Y. Spectral (finite) volume method for conservation laws on unstructured grids: II. Extension to two-dimensional scalar equation. J Comput Phys 2002;179(2): 665–97.

[61] Patankar SV, Spalding DB. A calculation procedure for heat, mass and momentum transfer in three-dimensional parabolic flows. Int J Heat Mass Transfer 1972;15(10): 1787–806.

[62] Du Q, Gunzburger M. Grid generation and optimization based on centroidal Voronoi tessellations. Appl Math Comput 2002;133:591–607.

[63] Thompson JF. The national grid project. Comput Syst Eng 1992;3:393–9.

[64] Schneiders R. Octree-based hexahedral mesh generation. Int J Comput Geom Appl 2000;10(4):383–98.

[65] Chan, W.M. Advances in software tools for pre-processing and post-processing of over-set grid computations. In: Proceedings of the 9th International Conference on Numerical Grid Generation in Computational Field Simulations, San Jose, California; 2005.

[66] Thompson JF, Warsi ZUA, Mastin CW. Numerical grid generation: foundations and applications. North-Holland: Amsterdam; 1985.

[67] Liseikin VD. Grid generation methods. 2nd ed. Springer: Berlin; 2010.

[68] Ghia U, Ghia KN, Shin CT. High-re solutions for incompressible flow using the Navier-Stokes equations and a multigrid method. J Comput Phys 1982;48:387–411.

[69] Botella O, Peyret R. Benchmark spectral results on the lid-driven cavity flow. Comput Fluids 1998;27(4):421–33.

[70] Perez Guerrero JS, Cotta RM. Benchmark integral transform results for flow over a backward-facing step. Comput Fluids 1996;25:527.

[71] Leone JM Jr. Open boundary condition symposium benchmark solution: stratified flow over a backward-facing step. Int J Numer Meth Eng 2005;11(7):969–84.

[72] Barlow R. Proceedings of the international workshop on measurement and computation of turbulent non-premixed flames; 2014. Available from: http://www.sandia.gov/TNF/abstract.html.

[73] Slater J.W. NPARC alliance verification and validation archive; 2015. Available from: http://www.grc.nasa.gov/WWW/wind/valid/archive.html.

[74] AIAA. Guide for the verification and validation of computational fluid dynamics simulations. Paper number AIAA G-077-1998; 1998.

[75] Ghia U, Bayyuk S, Habchi S, Roy C, Shih T, Conlisk AT, Hirsch C, Powers JM, The AIAA code verification project – test cases for CFD code verification. Proceedings of the 48th AIAA aerospace sciences meeting and exhibit, Orlando, Florida, AIAA Paper Number 2010-0125; 2010.

EXERCISES

1.1 Classify the following PDEs into the three canonical types or a combination thereof. In each case, ϕ is the dependent variable.

a. $\dfrac{\partial \phi}{\partial t} + \mathbf{U} \bullet \nabla \phi = 0$, where $\mathbf{U}$ is a vector (such as velocity).

b. $\dfrac{\partial \phi}{\partial t} + \mathbf{U} \bullet \nabla \phi = \nabla \bullet (\Gamma \nabla \phi)$, where Γ is a positive real number.

c. $\dfrac{\partial}{\partial x}\left[\Gamma_1 \dfrac{\partial \phi}{\partial x} - \Gamma_2 \dfrac{\partial \phi}{\partial y} \right] + \dfrac{\partial}{\partial y}\left[\Gamma_1 \dfrac{\partial \phi}{\partial y} - \Gamma_2 \dfrac{\partial \phi}{\partial x} \right] = 0$, where Γ_1 and Γ_2 are real numbers, either positive or negative.

d. $\dfrac{\partial \phi}{\partial t} - \dfrac{1}{r}\dfrac{\partial}{\partial r}\left(r\dfrac{\partial \phi}{\partial r}\right) = 0$

e. $\dfrac{\partial^2 \phi}{\partial t^2} + \dfrac{\partial \phi}{\partial t} - \nabla \cdot (\Gamma \nabla \phi) = \phi$, where Γ is a positive real number.

1.2 Consider the following second-order linear ODE and boundary conditions:

$$\dfrac{d^2\phi}{dx^2} = \cos x, \qquad \phi(0) = 0, \phi(1) = 1.$$

Following Example 1.1, discretize the equation using the Galerkin finite element procedure and linear basis functions. Use equal nodal spacing with five nodes (Case 1) and nine nodes (Case 2). Solve the resulting discrete equations. Also, derive the analytical solution. Plot the solution ($\phi(x)$ vs. x) and also the error between the analytical and numerical solution for both cases.

For both cases also compute the fluxes at the left and right boundaries as follows:

Left boundary: flux, $J_L = \left.\dfrac{d\phi}{dx}\right|_{x=0}$

Right boundary: flux, $J_R = \left.\dfrac{d\phi}{dx}\right|_{x=1}$

To do so, differentiate the basis function expansions analytically and then use the coefficients you already computed. Next, compute the net source in the domain by integration:

$S = \displaystyle\int_0^1 \cos x\, dx$. Calculate the same quantities from the analytical solution that you derived, and then tabulate your results for both cases in the following manner:

Fluxes for N = …nodes		
	Analytical	Galerkin FEM
J_L		
J_R		
S		
$I = J_R - J_L - S$		

Comment on your results. Can you determine if your solution satisfies local conservation?

The Finite Difference Method

Although it is difficult to credit a particular individual with the discovery of what is known today as the finite difference method (FDM), it is widely believed that the seminal work of Courant, Friedrichs, and Lewy [1] in 1928 is the first published example of using five-point difference approximations to derivatives for solving the elliptic Laplace equation. Using the same method, the same researchers also proceeded to solve the hyperbolic wave equation in the same paper [1] and established the so-called CFL criterion for the stability of hyperbolic partial differential equations (PDEs). For additional details on the history of the FDM and how it influenced the development of the finite element method, the reader is referred to the excellent article by Thomée [2]. Even though the FDM has had several descendants since 1928, most of whom have, arguably, outshined the FDM in many regards, the FDM remains the method of choice for solving PDEs in many application areas in science and engineering. Its success can be attributed to its simplicity as well as its adaptability to PDEs of any form. Disciplines where it still finds prolific usage are computational heat transfer and computational electromagnetics – the so-called finite difference time domain (FDTD) method.

As discussed in Section 1.3.1, one of the important features of the FDM, in contrast with other popular methods for solving PDEs, is that the PDE is solved in its original form, i.e., without modifying the governing equation into an alternative form. In other words, the original differential form of the governing equation is satisfied at various locations within the computational domain. The solution, hence obtained, is called the strong form solution. In contrast, other popular methods, such as the finite volume method and the finite element method, provide the so-called weak form solution, in which a modified form, usually an integral form of the governing equation, is satisfied within the computational domain. Which type of solution is desirable depends of the underlying physical problem from which the governing equation was conceived. It is worth clarifying here that the words "strong" and "weak" should not be interpreted by the reader as "good" and "bad," respectively. Further discussion of the *pros* and *cons* of these two types of solution is postponed until Chapter 6.

This chapter commences with a discussion of the basic procedure for deriving difference approximations and applying such approximations to solving PDEs. The fundamental concepts developed here will also be used later in Chapters 6 and 7, when the finite volume method is presented. The current chapter begins with simple one-dimensional differential equations (i.e., ODEs) and eventually extends the concepts to multidimensional problems (i.e., PDEs), including irregular geometry. The treatment of three canonical boundary condition types, i.e., Dirichlet, Neumann, and Robin, is also discussed in the context of the FDM.

2.1 DIFFERENCE APPROXIMATIONS AND TRUNCATION ERRORS

In order to understand the application of the finite difference method to the solution of differential equations, let us consider a simple second-order linear ordinary differential equation (ODE) of the following form:

$$\frac{d^2\phi}{dx^2} = S_\phi, \tag{2.1}$$

where ϕ is the dependent variable or unknown that needs to be determined *via* solution of the differential equation. The independent variable, on the other hand, is denoted by x. The quantity, S_ϕ, on the right-hand side of Eq. (2.1), could be, in general, a function of both the independent and dependent variables, i.e., $S_\phi = S_\phi(x,\phi)$. This term is often referred to as the *source term* or *source*. If the functional dependence of S_ϕ is nonlinear, then Eq. (2.1) would be nonlinear. Otherwise, Eq. (2.1) would be a linear ODE. For the discussion at hand, let us assume that S_ϕ is such that Eq. (2.1) is linear. Since Eq. (2.1) is a second-order ODE, it will also require two boundary conditions in x. Let us assume that the boundary conditions are given by

$$\begin{aligned}\phi(x_L) &= \phi_L, \\ \phi(x_R) &= \phi_R,\end{aligned} \tag{2.2}$$

where x_L and x_R denote values of x at the left and right ends of the domain of interest, respectively. For the discussion at hand, the type of boundary condition being applied is not relevant. Rather, the fact that these conditions are posed on the two ends is the only point to be noted.

One critical point to note is that Eq. (2.1) is valid in the open interval (x_L, x_R), not in the closed interval $[x_L, x_R]$. At the end points of the domain, only the boundary conditions are valid. Let us now consider the steps that are undertaken to obtain a closed-form analytical solution to Eq. (2.1) subject to the boundary conditions given by Eq. (2.2). For simplicity, we will assume that $S_\phi = 0$. Integrating Eq. (2.1) twice, we obtain $\phi(x) = C_1 x + C_2$, where C_1 and C_2 are the undetermined constants of integration. To determine these two constants, we generally substitute the two boundary conditions [Eq. (2.2)] into $\phi(x) = C_1 x + C_2$ and solve for C_1 and C_2. This procedure has the underlying assumption that the second derivative of ϕ is continuous in the closed interval $[x_L, x_R]$, and therefore, the validity of the governing equation is implied to extend to the end points of the domain. In principle, this may not always be the case, and strictly speaking, the boundary conditions [Eq. (2.1)] should be interpreted as follows:

$$\begin{aligned}\lim_{x \to x_L} \phi &= \phi_L \\ \lim_{x \to x_R} \phi &= \phi_R\end{aligned}. \tag{2.3}$$

If the limiting value for ϕ, as obtained from solution of the governing equation, or its derivative (for Neumann boundary conditions) at the boundary [cf. Eq. (2.3)] is singular or undefined, the problem is ill-posed, and a unique solution cannot be found. We shall see later that for cylindrical or spherical coordinate systems, the distinction between open and closed intervals comes into play, and limit-based interpretation of the boundary condition needs to be utilized. For the vast majority of physical problems, though, the dependent variable and its higher derivatives are continuous right up to and including the boundary, and therefore, taking the limiting value is equivalent to substituting the governing equation into the boundary condition.

As mentioned earlier, the fundamental premise behind the finite difference method is that the governing differential equation (ordinary or partial) is to be satisfied at various points within the computational domain. These various points are prescribed and are referred to as *nodes*. The nodes may be placed at equal or unequal distances from each other, and the spacing (or distance) between them is referred to as *grid spacing* or *grid size*. As a starting point, we will use an arrangement in which the nodes are spaced at equal distances from each other, as shown in Fig. 2.1. It is clear from Fig. 2.1 that N nodes result in $N-1$ intervals. Thus, for equally spaced nodes, we get

$$\Delta x = \frac{L}{N-1}, \tag{2.4a}$$

$$x_i = x_\text{L} + (i-1)\Delta x, \tag{2.4b}$$

where Δx is the aforementioned grid spacing, and x_i is the spatial location of the i-th node. As shown in Fig. 2.1, the nodes are numbered $i=1,2,...,N$, where the two nodes $i=1$ and $i=N$ are so-called boundary nodes.

A digital computer can neither understand differential operators nor can it perform operations on them. It can only perform floating point arithmetic operations, and therefore, can be instructed using a programming language to solve a system of linear algebraic equations since solution of algebraic equations only involves floating point operations. Based on this premise, the first objective of any deterministic method for solving differential equations is to convert the differential operators to their algebraic equivalents. It is in the manner (mathematical formulation) by which this conversion is attained that distinguishes one deterministic method from another. In general, to derive an algebraic equivalent of a differential operator, one has to make use of the value of the dependent variable at two or more locations. For example, we know that the derivative

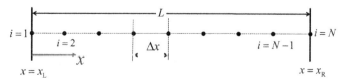

FIGURE 2.1 Schematic Representation Showing Arrangement of Nodes (Solid Circles) in a 1D Computational Domain

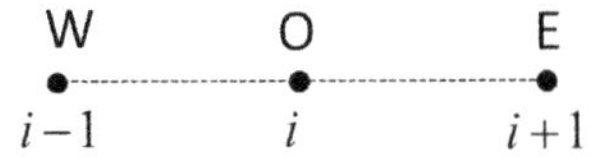

W   O   E

FIGURE 2.2 **1D Stencil Showing Three Nodes With Equal Nodal Spacing**

dy/dx may be approximated as $[(y_2 - y_1)/(x_2 - x_1)]$. In this case, two locations or nodes have been used. In general, for each derivative, several nodes may be used. The process of expressing the differential equation at each node in terms of its algebraic equivalent (discrete approximation) is referred to as *discretization*. It is termed so because the domain has been split into a set of nodes or discrete points at which the solution to the differential equation is now being sought rather than at infinite many points.

Based on the preceding discussion, for the problem at hand, the objective of the finite difference method is to express the second derivative on the left-hand side of Eq. (2.1) in terms of the values of ϕ at nodes $i = 1, 2, ..., N$. Let us consider just three nodes, as shown in Fig. 2.2. Such a diagram is known as a stencil diagram or *stencil*, in short. In Fig. 2.2, the central node is denoted by i, while the nodes to its left and right are denoted by $i-1$ and $i+1$, respectively. It is also customary to denote the nodes by uppercase letters: the central node is denoted by O, while the nodes to its left (west) and right (east) are denoted by W and E, respectively. Let us now perform two Taylor series expansions as follows:

$$\phi_{i+1} = \phi_i + (x_{i+1} - x_i)\frac{d\phi}{dx}\Big|_i + \frac{(x_{i+1} - x_i)^2}{2!}\frac{d^2\phi}{dx^2}\Big|_i + \frac{(x_{i+1} - x_i)^3}{3!}\frac{d^3\phi}{dx^3}\Big|_i + \frac{(x_{i+1} - x_i)^4}{4!}\frac{d^4\phi}{dx^4}\Big|_i + ..., \quad (2.5a)$$

$$\phi_{i-1} = \phi_i + (x_{i-1} - x_i)\frac{d\phi}{dx}\Big|_i + \frac{(x_{i-1} - x_i)^2}{2!}\frac{d^2\phi}{dx^2}\Big|_i + \frac{(x_{i-1} - x_i)^3}{3!}\frac{d^3\phi}{dx^3}\Big|_i + \frac{(x_{i-1} - x_i)^4}{4!}\frac{d^4\phi}{dx^4}\Big|_i + ... \quad (2.5b)$$

Noting that $x_{i+1} - x_i = \Delta x$ and $x_{i-1} - x_i = -\Delta x$, Eq. (2.5) can be alternatively written as

$$\phi_{i+1} = \phi_i + (\Delta x)\frac{d\phi}{dx}\Big|_i + \frac{(\Delta x)^2}{2!}\frac{d^2\phi}{dx^2}\Big|_i + \frac{(\Delta x)^3}{3!}\frac{d^3\phi}{dx^3}\Big|_i + \frac{(\Delta x)^4}{4!}\frac{d^4\phi}{dx^4}\Big|_i + ..., \quad (2.6a)$$

$$\phi_{i-1} = \phi_i - (\Delta x)\frac{d\phi}{dx}\Big|_i + \frac{(\Delta x)^2}{2!}\frac{d^2\phi}{dx^2}\Big|_i - \frac{(\Delta x)^3}{3!}\frac{d^3\phi}{dx^3}\Big|_i + \frac{(\Delta x)^4}{4!}\frac{d^4\phi}{dx^4}\Big|_i + ... \quad (2.6b)$$

Since our objective is to derive an expression for the second derivative in terms of the nodal values of ϕ, the two equations need to be manipulated such that the first derivative is eliminated and the second derivative is not eliminated. Retaining the first derivative does not serve the purpose of expressing the second derivative solely in terms of nodal values of ϕ. For example, if we were to use just one of the two Taylor series expansions, we could also express the second derivative in terms of nodal values, but that expression, in addition, would also contain the first derivative. Therefore, this strategy is not acceptable. Of course, even if the first derivative is eliminated, the higher derivatives may still remain in the resulting expression. This issue will be addressed shortly. For now, in order to eliminate the first derivative, let us add Eqs (2.6a) and (2.6b). This yields

$$\phi_{i+1} + \phi_{i-1} = 2\phi_i + (\Delta x)^2 \left.\frac{d^2\phi}{dx^2}\right|_i + \frac{(\Delta x)^4}{12} \left.\frac{d^4\phi}{dx^4}\right|_i + \dots \tag{2.7}$$

Rearranging, we get

$$\left.\frac{d^2\phi}{dx^2}\right|_i = \frac{\phi_{i+1} + \phi_{i-1} - 2\phi_i}{(\Delta x)^2} - \frac{(\Delta x)^2}{12} \left.\frac{d^4\phi}{dx^4}\right|_i + \dots \tag{2.8}$$

As expected, the expression for the second derivative has not only nodal values of ϕ but also higher-order derivatives. These higher-order derivatives are, obviously, unknown, and therefore, retaining them in the approximation does not serve our objective. Due to a lack of a better alternative, we neglect all higher-order derivatives containing terms, resulting in

$$\left.\frac{d^2\phi}{dx^2}\right|_i \approx \frac{\phi_{i+1} + \phi_{i-1} - 2\phi_i}{(\Delta x)^2}, \tag{2.9}$$

with the understanding that in doing so, we introduce an error that may be written as

$$\varepsilon_i = -\frac{(\Delta x)^2}{12} \left.\frac{d^4\phi}{dx^4}\right|_i + \dots \tag{2.10}$$

The error incurred in approximating a derivative by an algebraic expression is referred to as the *discretization* or *truncation error*. It is called the discretization error because a finite number of discrete points or nodes are used to approximate a continuous function. The name, truncation error, stems from the fact that the infinite Taylor series was truncated after a few terms to derive the expression for the derivative that is being approximated.

The introduction of the truncation error sacrifices the accuracy of the numerical solution to the differential equation. Therefore, it is important to understand the attributes of this error both qualitatively and quantitatively. We first discuss some of the qualitative attributes of the truncation error:

1. In this particular case, $\varepsilon_i \propto (\Delta x)^2$. This implies that as the nodal spacing, Δx, is decreased, i.e., as more and more nodes are used, the error will become vanishingly small. This is the reason why mesh or grid refinement is looked upon as a favorable step toward a more accurate numerical solution to any differential equation. In general, if the truncation error scales as $\varepsilon_i \propto (\Delta x)^n$, it will vanish with mesh refinement as long as $n > 0$. Such a scheme is known as a *consistent scheme*, i.e., a discretization scheme in which the local truncation error tends to zero as the grid spacing tends to zero. Consistency is a necessary property of any discretization scheme. The larger the value of n, the better the approximation, since larger values of n will cause the errors to decrease more rapidly. For example, in this particular case, as the grid spacing is halved, the error will decrease by a

factor of 4. Thus far, our discussion has focused only on the first or leading term of the error expression. Clearly, the error expression contains other higher-order terms. How do those error terms behave? The higher-order terms have higher powers of Δx premultiplying them, implying that they will vanish more rapidly than the leading order term, and therefore, are not a cause of concern as long as n > 1. For example, the next term in the error expression shown in Eq. (2.10) will have $(\Delta x)^4$ premultiplying it, which implies that halving the grid spacing will reduce this term by a factor of 16. Thus, in future discussions in this text, we will concern ourselves only with the leading order term.

2. The truncation error is different at different nodes or spatial locations. Even for a uniform grid, the error depends on the magnitude of the derivative (in this particular case, fourth derivative) that it is proportional to. However, the scaling behavior of the error remains unchanged from node to node if a uniform grid is used.

3. For the same nodal arrangement and the same differential equation form, the truncation error may be different. This is because of the fact that the derivative that it scales as could be different, depending on what the source term S_ϕ is. For example, if $S_\phi = x$, the truncation error would be zero because the fourth derivative would be zero. This is easily established by differentiating Eq. (2.1) twice to obtain the fourth derivative. On the other hand, if $S_\phi = e^x$, the truncation error would not only be nonzero but would also vary from node to node.

4. Since we are employing Taylor series expansions to approximate derivatives, the underlying assumption is that ϕ is infinitely differentiable. Therefore, no matter what order derivative is contained in the leading order error term, the truncation error will be bounded. Therefore, as the mesh is refined, the truncation error will approach zero.

Another type of error that is unavoidable in any numerical computation is the so-called *round-off error*. Round-off errors are a manifestation of the fact that on a digital computer, real numbers are stored only approximately, i.e., up to a certain number of decimal places, often referred to as the *precision* of the number. The precision depends on how many bytes are used for storage of the real number in question. For example, so-called single precision requires 4 bytes for memory and has accuracy up to approximately 8 decimal places. On the other hand, double precision requires 8 bytes of memory and enables storage of a real number accurately up to roughly its 16th decimal place. Most modern-day deterministic computations are conducted in double precision so that extremely small or large numbers can be treated simultaneously in the same computation without having to worry about loss of accuracy.

Truncation errors are unavoidable in any discretization scheme unless the higher derivatives are zero, which rarely occurs in problems of practical interest. Therefore, these errors must be kept as small as possible. As discussed earlier, one way to achieve this goal is to refine the mesh to a point where the truncation error becomes negligible. By negligible, we mean smaller than the round-off error. Such a solution is known as a *grid-independent solution*. In practice, this formal criterion cannot always be strictly met. Most often, a much more relaxed criterion is employed. For example, one might deem the solution grid independent when the solution changes less than 0.1% when the grid spacing is halved. The criterion to be used in practice to judge grid independence depends largely on the problem at hand. If one were solving a PDE to determine temperature, which typically has a value in the tens and hundreds, it would not be worthwhile to keep refining the mesh until the temperature stops changing in the 16th decimal place or even in the 8th decimal place. On the other hand, the same criterion may be appropriate for monitoring grid independence of the solution for concentration of a pollutant, which, for example, may already be of the order 10^{-6} or lower to begin with. Many numerical analysts suggest converting the governing equation and boundary conditions to the nondimensional form to mitigate such issues.

As discussed earlier, in a valid difference approximation to a derivative, the truncation error must scale as some positive power of the grid spacing. The power (or exponent) to which the grid spacing is raised in the leading order error term is also known as the *order* of the truncation error. In the particular case of Eq. (2.10), the error is second order. Another way of expressing the same concept is to state that the differencing scheme that was used to approximate the derivative is second-order accurate in space. We will see later that the order of the scheme is often not self-evident if a nonuniform grid is used. In practice, the lowest order is first order.

Another observation worth making at this point is that the difference approximation given by Eq. (2.9) was derived using two Taylor series expansions – one in the forward direction $(+\Delta x)$ and the other in the backward direction $(-\Delta x)$. The node i (or O), for which we derived an approximation to the second derivative, was located centrally. Hence, such differencing schemes are named *central difference schemes*. Alternatively, expressions for the second derivative could have been derived using two Taylor series expansions in the forward direction $(+\Delta x, +2\Delta x)$, resulting in the so-called *forward difference scheme*, or two expansions in the backward direction $(-\Delta x, -2\Delta x)$, resulting in the so-called *backward difference scheme*. Which scheme is appropriate or desirable depends on the truncation error incurred in each case and also the physical problem underlying the governing equations. Further discussion of these issues is deferred until the end of the section to follow.

With a better understanding of the discretization error incurred in finite difference approximation of derivatives, we now return to the task of constructing algebraic equations for the nodes. Since the governing equation [Eq. (2.1)] must be satisfied everywhere in the open interval (x_L, x_R), it follows then that it must be satisfied for all nodes other than the two boundary nodes, i.e., $i = 2, 3,..., N - 1$. These nodes are called *interior nodes*. Substituting Eq. (2.9) into Eq. (2.1) for all interior nodes, we obtain

$$\frac{\phi_{i+1} + \phi_{i-1} - 2\phi_i}{(\Delta x)^2} = S_i \qquad \forall i = 2, 3,...N - 1, \qquad (2.11)$$

where S_i is the source term evaluated at node i, i.e., $S_i = S_\phi(x_i, \phi_i)$. Equation (2.11) represents a set of $N - 2$ equations with N unknowns. If the value of i is set to 2 in Eq. (2.11), the resulting equation will contain ϕ_1. Similarly, if the value of i is set to $N-1$ in Eq. (2.11), the resulting equation will contain ϕ_N. In other words, all N unknowns are present in Eq. (2.11). In order to attain closure to the set of equations, i.e., have the same number of equations as unknowns, two additional equations are necessary. These two equations are the boundary conditions. Boundary conditions of three canonical types will be discussed in Section 2.3 after a more general treatment of the finite difference procedure.

2.2 GENERAL PROCEDURE FOR DERIVING DIFFERENCE APPROXIMATIONS

In the preceding section, derivation of a central difference approximation to the second derivative was presented. The procedure was relatively straightforward to grasp because the nodal spacing used was uniform, and only three points in the stencil were used. In general, however, several important questions need to be addressed before one can concoct a foolproof recipe for deriving difference approximations. These questions include the following:

1. Is there a relationship between the order of the derivative to be derived and the number of Taylor series expansions that may be used?
2. What should be the pivot point in the Taylor series expansions? For example, in Eq. (2.6), why did we expand ϕ_{i+1} and ϕ_{i-1} about ϕ_i and not the other way around?
3. Given the derivative order and the number of Taylor series expansions to be used, can we predict *a priori* what the order of the resulting error will be?

There are a minimum number of Taylor series expansions one must use to derive an expression for the derivative of a certain order. For example, with just two nodes of the stencil and one Taylor series expansion, it is impossible to derive an expression for the second derivative. A minimum of two Taylor series expansions and three nodes must be used. In general, to derive an expression for the m-th derivative, at least m Taylor series expansions and $m+1$ nodes must be used. The order of the error, n, using the minimum number of Taylor series expansions would be greater than or equal to 1. For example, the central difference scheme employed two Taylor series expansions, i.e., the bare minimum, but resulted in a second-order error. This is because in this particular case, with equal grid spacing the third derivative containing terms fortuitously cancelled out. If more than the bare minimum number of Taylor series expansions is used, it is possible to increase the order of the error by manipulating the expansions to cancel higher derivative containing terms. The pivot point, i.e., the point about which the Taylor series expansion is performed, is dictated by the objective of the derivation. For example, in the derivation presented in Section 2.1, the pivot point used was the node i because the second derivative was sought at that node.

Some of the rules of thumb just discussed are brought to light by the example that follows. It highlights a general procedure for deriving a difference approximation.

EXAMPLE 2.1

Consider the stencil shown in the following figure. Derive a finite difference approximation for $\dfrac{d^3\phi}{dx^3}\bigg|_O$ using the nodes O, E, W, and WW. Also, derive an error expression. The node spacing is uniform and equal to Δx.

Following the rules of thumb discussed earlier, in order to derive the third derivative, we need a minimum of three Taylor series expansions. Also, since the third derivative is sought at node O, all Taylor series expansions must be performed about node O. Hence,

$$\phi_E = \phi_O + (\Delta x)\frac{d\phi}{dx}\bigg|_O + \frac{(\Delta x)^2}{2!}\frac{d^2\phi}{dx^2}\bigg|_O + \frac{(\Delta x)^3}{3!}\frac{d^3\phi}{dx^3}\bigg|_O + \frac{(\Delta x)^4}{4!}\frac{d^4\phi}{dx^4}\bigg|_O + \dots$$

$$\phi_W = \phi_O - (\Delta x)\frac{d\phi}{dx}\bigg|_O + \frac{(\Delta x)^2}{2!}\frac{d^2\phi}{dx^2}\bigg|_O - \frac{(\Delta x)^3}{3!}\frac{d^3\phi}{dx^3}\bigg|_O + \frac{(\Delta x)^4}{4!}\frac{d^4\phi}{dx^4}\bigg|_O + \dots$$

$$\phi_{WW} = \phi_O - (2\Delta x)\frac{d\phi}{dx}\bigg|_O + \frac{(2\Delta x)^2}{2!}\frac{d^2\phi}{dx^2}\bigg|_O - \frac{(2\Delta x)^3}{3!}\frac{d^3\phi}{dx^3}\bigg|_O + \frac{(2\Delta x)^4}{4!}\frac{d^4\phi}{dx^4}\bigg|_O + \dots$$

For the derivation of the second derivative, it was quite obvious that the first two equations needed to be added. In this case, such manipulation is not obvious. Instead, we apply the following general procedure. First, we multiple each equation by an unknown constant, for example, A, B, and C. This yields

$$A \times \left[\phi_E = \phi_O + (\Delta x)\frac{d\phi}{dx}\bigg|_O + \frac{(\Delta x)^2}{2!}\frac{d^2\phi}{dx^2}\bigg|_O + \frac{(\Delta x)^3}{3!}\frac{d^3\phi}{dx^3}\bigg|_O + \frac{(\Delta x)^4}{4!}\frac{d^4\phi}{dx^4}\bigg|_O + \dots \right]$$

$$B \times \left[\phi_W = \phi_O - (\Delta x)\frac{d\phi}{dx}\bigg|_O + \frac{(\Delta x)^2}{2!}\frac{d^2\phi}{dx^2}\bigg|_O - \frac{(\Delta x)^3}{3!}\frac{d^3\phi}{dx^3}\bigg|_O + \frac{(\Delta x)^4}{4!}\frac{d^4\phi}{dx^4}\bigg|_O + \dots \right]$$

$$C \times \left[\phi_{WW} = \phi_O - (2\Delta x)\frac{d\phi}{dx}\bigg|_O + \frac{(2\Delta x)^2}{2!}\frac{d^2\phi}{dx^2}\bigg|_O - \frac{(2\Delta x)^3}{3!}\frac{d^3\phi}{dx^3}\bigg|_O + \frac{(2\Delta x)^4}{4!}\frac{d^4\phi}{dx^4}\bigg|_O + \dots \right]$$

Next, we construct equations based on objectives, which are as follows: (a) the coefficients of the first derivative must add up to zero, i.e., the first derivatives cancel out; (b) the coefficients of the second derivative must add up to zero; and (c) the coefficients of the third derivative must add up to unity. Satisfaction of the three objectives results in

$$A\Delta x - B\Delta x - 2C\Delta x = 0 \qquad \Rightarrow A - B - 2C = 0$$

$$A\frac{(\Delta x)^2}{2} + B\frac{(\Delta x)^2}{2} + C\frac{(2\Delta x)^2}{2} = 0 \Rightarrow A + B + 4C = 0$$

$$A\frac{(\Delta x)^3}{6} - B\frac{(\Delta x)^3}{6} - C\frac{(2\Delta x)^3}{6} = 1 \Rightarrow A - B - 8C = \frac{6}{(\Delta x)^3}$$

Solving the three simultaneous equations, we obtain $A = 1/(\Delta x)^3$, $B = 3/(\Delta x)^3$, and $C = -1/(\Delta x)^3$. Further, adding the three equations yields

$$A\phi_E + B\phi_W + C\phi_{WW} - (A + B + C)\phi_O = \left.\frac{d^3\phi}{dx^3}\right|_O + \left(A\frac{(\Delta x)^4}{4!} + B\frac{(\Delta x)^4}{4!} + C\frac{(2\Delta x)^4}{4!} \right)\left.\frac{d^4\phi}{dx^4}\right|_O + \dots$$

Rearranging to express the third derivative in terms of nodal values of ϕ, we obtain

$$\left.\frac{d^3\phi}{dx^3}\right|_O = A\phi_E + B\phi_W + C\phi_{WW} - (A + B + C)\phi_O - (A + B + 16C)\frac{(\Delta x)^4}{24}\left.\frac{d^4\phi}{dx^4}\right|_O + \dots,$$

which, upon substituting $A = 1/(\Delta x)^3$, $B = 3/(\Delta x)^3$, and $C = -1/(\Delta x)^3$, yields

$$\left.\frac{d^3\phi}{dx^3}\right|_O = \frac{1}{(\Delta x)^3}\phi_E + \frac{3}{(\Delta x)^3}\phi_W - \frac{1}{(\Delta x)^3}\phi_{WW} - \frac{3}{(\Delta x)^3}\phi_O + \frac{1}{2}\Delta x\left.\frac{d^4\phi}{dx^4}\right|_O + \dots$$

Thus,

$$\left.\frac{d^3\phi}{dx^3}\right|_O \approx \frac{1}{(\Delta x)^3}\phi_E + \frac{3}{(\Delta x)^3}\phi_W - \frac{1}{(\Delta x)^3}\phi_{WW} - \frac{3}{(\Delta x)^3}\phi_O$$

$$\varepsilon = \frac{1}{2}\Delta x\left.\frac{d^4\phi}{dx^4}\right|_O + \dots$$

As discussed earlier, the order of the error will always be greater than or equal to 1. In this case, with the bare minimum number of Taylor series expansions being used, the order is exactly 1. The order could be improved by using additional Taylor series expansions. For example, if four expansions were used instead of three, we would have four unknown constants and four objectives to meet. In addition to the three already used, the fourth objective would be to cancel out the fourth derivative. In such a case, the resulting error would be second order.

In the process of deriving the central difference formula in Section 2.1, equal nodal spacing was used. This resulted in a second-order error. Having established a general procedure for deriving difference approximations, we now apply the procedure to derive a central difference formula for the case when the nodal spacing is nonuniform, as shown in Fig. 2.3.

Performing the two necessary Taylor series expansions of the $i+1$ and $i-1$ nodes about node i, we obtain

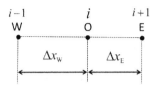

FIGURE 2.3 1D Stencil Showing Three Nodes With Unequal Nodal Spacing

$$\phi_{i+1} = \phi_i + (\Delta x_E)\frac{d\phi}{dx}\bigg|_i + \frac{(\Delta x_E)^2}{2!}\frac{d^2\phi}{dx^2}\bigg|_i + \frac{(\Delta x_E)^3}{3!}\frac{d^3\phi}{dx^3}\bigg|_i + \frac{(\Delta x_E)^4}{4!}\frac{d^4\phi}{dx^4}\bigg|_i + ...,\quad (2.12a)$$

$$\phi_{i-1} = \phi_i - (\Delta x_W)\frac{d\phi}{dx}\bigg|_i + \frac{(\Delta x_W)^2}{2!}\frac{d^2\phi}{dx^2}\bigg|_i - \frac{(\Delta x_W)^3}{3!}\frac{d^3\phi}{dx^3}\bigg|_i + \frac{(\Delta x_W)^4}{4!}\frac{d^4\phi}{dx^4}\bigg|_i + ...\quad (2.12b)$$

It is evident that in this case simply adding the two expansions will not eliminate the first derivative. Following the same procedure, as described in Example 2.1, if we multiply Eq. (2.12a) by A and Eq. (2.12b) by B, set the objective that the coefficients premultiplying the first derivative must cancel, and the coefficients premultiplying the second derivative must sum to unity, we get $A = 2/[\Delta x_E(\Delta x_E + \Delta x_W)]$ and $B = 2/[\Delta x_W(\Delta x_E + \Delta x_W)]$. The resulting second derivative at node i is then derived as

$$\frac{d^2\phi}{dx^2}\bigg|_i \approx \frac{2\Delta x_W\phi_{i+1} + 2\Delta x_E\phi_{i-1} - 2(\Delta x_W + \Delta x_E)\phi_i}{\Delta x_W\Delta x_E(\Delta x_W + \Delta x_E)},\quad (2.13)$$

with the corresponding truncation error given by

$$\varepsilon_i = -\frac{1}{3}(\Delta x_E - \Delta x_W)\frac{d^3\phi}{dx^3}\bigg|_i + ...\quad (2.14)$$

It is easy to show that in the special case when the nodal spacing is equal, i.e., $\Delta x_E = \Delta x_W$, Eq. (2.13) reduces to Eq. (2.9). Also, the truncation error, as shown by Eq. (2.14), would vanish under such a scenario, implying that the leading error term would no longer be the third derivative containing term, but rather, the fourth derivative containing term, as shown by Eq. (2.10). As far as the order of the truncation error shown by Eq. (2.14) is concerned, at first glance, it might appear to be of first order since the term premultiplying the third derivative contains $\Delta x_E - \Delta x_W$. However, having the truncation error scale as Δx_E or Δx_W is quite different than having it scale as the difference between these two quantities. This is because $\Delta x_E - \Delta x_W$ is generally a small fraction of the larger of the two quantities, Δx_E or Δx_W. Most mesh generation algorithms use a so-called *stretching factor* – defined as the ratio between adjacent nodal spacings – to stretch the mesh. It is customary to use a stretching factor close to unity in order to keep the resulting coefficient matrix relatively well conditioned, as we shall see in Chapter 3. Be that as it may, it is clear that in this case, the order of the truncation error is bounded between 1 and 2. The order is 2 if the grid is uniform, and it is close to 1 if the grid is severely stretched. For more practical scenarios, where moderate stretching factors are used, the order is closer to 2 than 1. In conclusion, the central difference scheme on a nonuniform mesh is between first- and second-order accurate, but the exact order is mesh dependent.

2.3 APPLICATION OF BOUNDARY CONDITIONS

As discussed in Section 2.1, in order to attain closure to the system of discrete algebraic equations, two additional equations (for the one-dimensional (1D) problem discussed in Section 2.1) are needed. These two additional equations come from the two boundary conditions at the two ends of the 1D domain. Although spatial boundary conditions encountered in science and engineering disciplines can take a myriad of forms, the vast majority of them can arguably be classified under three canonical types: Dirichlet, Neumann, and Robin (alternatively known as Robbins) boundary conditions. In Chapter 5, we will also discuss various types of initial conditions, i.e., conditions posed temporally. From a strictly mathematical perspective, these temporal conditions may also be classified as boundary conditions. Among these boundary conditions in time (or initial conditions), the Cauchy condition is probably one that deserves special coverage. In this text, however, we reserve the terminology boundary condition for conditions applied to spatial operators and initial condition for conditions applied to temporal operators.

2.3.1 DIRICHLET BOUNDARY CONDITION

The Dirichlet boundary condition, credited to the German mathematician Dirichlet,* is also known as the boundary condition of the first kind. In this case, the value of the dependent variable, ϕ, is prescribed on the boundary, as shown mathematically in Eq. (2.2). In the finite difference method, since nodes are located on the boundary, the Dirichlet boundary condition is straightforward to apply. For the nodal system depicted in Fig. 2.1, the boundary conditions given by Eq. (2.2) may be written as

$$\begin{aligned} \phi_1 &= \phi_L \\ \phi_N &= \phi_R \end{aligned}. \tag{2.15}$$

Equation (2.15), in conjunction with Eq. (2.11), represents a system of N linear algebraic equations with N unknowns. Strictly speaking, in the case of Dirichlet boundary conditions, two of the unknowns are actually known directly [Eq. (2.15)] since these equations are not coupled to any other equation. However, in order to solve for $N-2$ unknowns only, the boundary conditions [Eq. (2.15)] need to

---

*__Johann Peter Gustav Lejeune Dirichlet__ (1805–1855) was a German mathematician who made major contributions to number theory, convergence of the Fourier series, and to boundary value problems (Laplace equation and its solution). He is often credited with the formal definition of the term "function."

be substituted into Eq. (2.11) for $i = 2$ and $i = N-1$ and the resulting equations rearranged prior to solution. These, and other related issues, will be discussed in Section 2.4 and in more detail in Chapter 3.

For scientific or engineering applications, it is often desirable to determine the first derivative of the dependent variable at the boundaries. This is because the derivative is related to the so-called *flux* at the boundary. By flux, we mean the amount transferred or transported per unit area per unit time (see Section 1.3.2 and Chapters 6 and 7 for additional details). These concepts are best illustrated by a physical example. Let us consider a heat conduction problem in a 1D slab of constant thermal conductivity, κ, and length L. The two ends of the slab, $x = 0$ and $x = L$, are maintained at fixed temperatures T_{L} and T_{R}, respectively, i.e., Dirichlet boundary conditions are applied to the ends of the slab. Let us also assume that the heat generation rate per unit volume within the slab, perhaps due to an electrical heater installed in the slab, is $\dot{q}$. For simplicity, we assume that the heat generation rate is uniform. The objective is to determine the temperature distribution, $T(x)$, within the slab as well as the heat transfer rate per unit area (heat flux) at the left end ($x = 0$) of the slab. The governing equation for the physical problem just described is $d^2T/dx^2 = -\dot{q}/\kappa$, where T is the temperature [3]. The heat flux at the left end is given by $q(0) = -\kappa \, dT/dx\big|_{x=0}$ [3]. Comparing with Eq. (2.1) clearly shows that the governing equation is essentially in the canonical form given by Eq. (2.1), and the boundary conditions are in the canonical form shown in Eq. (2.2). Calculation of the heat flux, however, requires determination of the derivative of the dependent variable at the boundary. This is a postprocessing step that needs to be conducted once the problem has been solved numerically and all nodal values been determined.

In order to determine the first derivative (the terms flux and first derivative will be used interchangeably throughout this text even though, strictly speaking, the flux is the first derivative times a constant), we will use the same Taylor series-based procedure used to derive the difference approximation to the second derivative. Let us consider the nodal arrangement at the left boundary, as shown in Fig. 2.4.

Recalling that our objective is to derive an expression for the first derivative at $i = 1$, we use $i = 1$ as our pivot point, and write

$$\phi_2 = \phi_1 + (\Delta x)\frac{d\phi}{dx}\bigg|_1 + \frac{(\Delta x)^2}{2!}\frac{d^2\phi}{dx^2}\bigg|_1 + \frac{(\Delta x)^3}{3!}\frac{d^3\phi}{dx^3}\bigg|_1 + \frac{(\Delta x)^4}{4!}\frac{d^4\phi}{dx^4}\bigg|_1 + \dots \qquad (2.16)$$

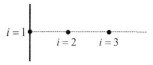

FIGURE 2.4 1D Stencil Diagram Showing One Boundary Node and Two Other Interior Nodes Close to the Boundary

Based on the previously stated thumb rule, it is possible to derive an expression for the first derivative using a single Taylor series expansion. If we choose to do so, we obtain

$$\left.\frac{d\phi}{dx}\right|_1 = \frac{\phi_2 - \phi_1}{\Delta x} - \frac{\Delta x}{2}\left.\frac{d^2\phi}{dx^2}\right|_1 + \dots = \frac{\phi_2 - \phi_L}{\Delta x} - \frac{\Delta x}{2}\left.\frac{d^2\phi}{dx^2}\right|_1 + \dots \quad (2.17)$$

In the second part of Eq. (2.17), the boundary condition at the left end [first equation of Eq. (2.15)] has been substituted. While the expression shown in Eq. (2.17) is a perfectly valid choice for computing the flux at the boundary, it is worth considering that it is only first-order accurate. Since our overall solution procedure is second-order accurate, second-order accuracy for flux calculations is also warranted. Second-order accuracy may be attained by using an additional Taylor series expansion, as follows:

$$\phi_3 = \phi_1 + (2\Delta x)\left.\frac{d\phi}{dx}\right|_1 + \frac{(2\Delta x)^2}{2!}\left.\frac{d^2\phi}{dx^2}\right|_1 + \frac{(2\Delta x)^3}{3!}\left.\frac{d^3\phi}{dx^3}\right|_1 + \frac{(2\Delta x)^4}{4!}\left.\frac{d^4\phi}{dx^4}\right|_1 + \dots \quad (2.18)$$

Since we now have two Taylor series expansions [Eqs (2.16) and (2.18)] at our disposal, the objective of canceling the second derivative can be met. To do so, we multiply Eq. (2.16) by 4 and then subtract Eq. (2.18), yielding

$$4\phi_2 - \phi_3 = 3\phi_1 + (2\Delta x)\left.\frac{d\phi}{dx}\right|_1 - \frac{2}{3}(\Delta x)^3\left.\frac{d^3\phi}{dx^3}\right|_1 + \dots \quad (2.19)$$

Rearranging, we obtain

$$\left.\frac{d\phi}{dx}\right|_1 = \frac{4\phi_2 - \phi_3 - 3\phi_1}{2\Delta x} + \frac{1}{3}(\Delta x)^2\left.\frac{d^3\phi}{dx^3}\right|_1 + \dots = \frac{4\phi_2 - \phi_3 - 3\phi_L}{2\Delta x} + \frac{1}{3}(\Delta x)^2\left.\frac{d^3\phi}{dx^3}\right|_1 + \dots \quad (2.20)$$

As expected, the error in this flux expression is second-order. For the sake of consistency, if the second-order central difference scheme is used to solve the governing PDE, then Eq. (2.20) is the flux expression that should be used to compute the flux at the left boundary.

2.3.2 NEUMANN BOUNDARY CONDITION

The Neumann boundary condition, credited to the German mathematician Neumann,** is also known as the boundary condition of the second kind. In this type of boundary condition, the value of the gradient of the dependent variable normal to the boundary, $\partial\phi/\partial n$, is prescribed on the boundary. The reader is referred to Chapter 7 for the general vectorial representation of this type of boundary condition. In the context of the 1D

Karl Gottfried Neumann (1832–1925) was a German mathematician. He made pioneering contributions to the theory of integral equations and infinite geometric series formulations for repeated mathematical operations, often referred to as the Neumann series.

problem at hand, the Neumann boundary condition at the left boundary, for example, may be written as

$$\left.\frac{d\phi}{dx}\right|_{x=x_{L}} = \left.\frac{d\phi}{dx}\right|_{i=1} = J_{L}, \tag{2.21}$$

where J_{L} is the prescribed value of the derivative. As discussed earlier, the first derivative is a measure of the flux at the boundary. Therefore, in the physics and engineering literature, the Neumann boundary condition is often referred to as a flux boundary condition. For the heat transfer example, discussed in Section 2.3.1, a Neumann boundary condition is tantamount to a prescribed heat flux boundary condition.

In the context of the finite difference method, the boundary condition serves the purpose of providing an equation for the boundary node so that closure can be attained for the system of equations. Thus, one approach to treatment of the Neumann boundary condition is to derive a discrete equivalent to Eq. (2.21) and use that as the nodal equation for $i = 1$. To do so, we simply take the flux expression derived in Eq. (2.20) and set it equal to the prescribed flux with the understanding that in this case, ϕ_{1} is an unknown. This yields

$$\frac{4\phi_{2} - \phi_{3} - 3\phi_{1}}{2\Delta x} = J_{L}. \tag{2.22}$$

The treatment of the flux in this manner yields a nodal equation that is second-order accurate.

As discussed in Section 2.1, the governing equation must be enforced everywhere in the open interval (x_{L}, x_{R}). In practice, since ϕ and its higher derivatives are continuous in problems arising out of physics or engineering, the governing equation could, equivalently, be enforced in the closed interval $[x_{L}, x_{R}]$. In the approach discussed in the preceding paragraph, the governing equation is only satisfied at the interior nodes. This implies that the governing equation is satisfied in a domain that is actually significantly smaller than the open interval (x_{L}, x_{R}). In fact, how much smaller this domain is depends on how close to the boundaries the nodes $i = 2$ and $i = N - 1$ are placed. In any case, this approximation is likely to introduce additional errors (in addition to truncation errors) in the solution, as will be demonstrated in a numerical example to follow. An alternative approach to treatment of the Neumann boundary condition is to ensure that both the boundary condition and the governing equation are satisfied at the boundary. To implement this alternative approach, as before, we first consider the two Taylor series expansions given by Eqs (2.16) and (2.18). Next, the boundary condition [Eq. (2.21)] is substituted into both equations to yield

$$\phi_{2} = \phi_{1} + (\Delta x)J_{L} + \frac{(\Delta x)^{2}}{2!}\left.\frac{d^{2}\phi}{dx^{2}}\right|_{1} + \frac{(\Delta x)^{3}}{3!}\left.\frac{d^{3}\phi}{dx^{3}}\right|_{1} + \frac{(\Delta x)^{4}}{4!}\left.\frac{d^{4}\phi}{dx^{4}}\right|_{1} + ..., \tag{2.23}$$

and

$$\phi_{3} = \phi_{1} + (2\Delta x)J_{L} + \frac{(2\Delta x)^{2}}{2!}\left.\frac{d^{2}\phi}{dx^{2}}\right|_{1} + \frac{(2\Delta x)^{3}}{3!}\left.\frac{d^{3}\phi}{dx^{3}}\right|_{1} + \frac{(2\Delta x)^{4}}{4!}\left.\frac{d^{4}\phi}{dx^{4}}\right|_{1} + ... \tag{2.24}$$

Now, instead of trying to derive an expression for the first derivative, we derive an expression for the second derivative, with the foresight that this will ultimately be substituted into the governing equation. Thus, in this case, our objective is to cancel the next higher derivative, namely the third derivative. To do so, we multiply Eq. (2.23) by 8 and subtract Eq. (2.24), yielding

$$8\phi_2 - \phi_3 = 7\phi_1 + (6\Delta x)J_L + 2(\Delta x)^2 \left.\frac{d^2\phi}{dx^2}\right|_1 - \frac{(\Delta x)^4}{3} \left.\frac{d^4\phi}{dx^4}\right|_1 + \dots \tag{2.25}$$

Rearranging, we obtain

$$\left.\frac{d^2\phi}{dx^2}\right|_1 = \frac{8\phi_2 - \phi_3 - 7\phi_1 - (6\Delta x)J_L}{2(\Delta x)^2} + \frac{(\Delta x)^2}{6} \left.\frac{d^4\phi}{dx^4}\right|_1 + \dots \tag{2.26}$$

As in the case of the interior nodes, the approximation for the second derivative at the left boundary node is also second order. Note that in deriving this expression for the second derivative, the boundary condition has already been satisfied. Next, to satisfy the governing equation at the boundary, for reasons discussed earlier, we substitute the difference approximation in Eq. (2.26) into Eq. (2.1) to yield

$$\frac{8\phi_2 - \phi_3 - 7\phi_1 - (6\Delta x)J_L}{2(\Delta x)^2} = S_1. \tag{2.27}$$

Equation (2.27) is an alternative nodal equation [alternative to Eq. (2.22)] for $i = 1$. To illustrate the nuances of the two approaches just described to apply Neumann boundary conditions, namely one that satisfies just the boundary condition [Eq. (2.22)] versus one that satisfies the boundary condition as well as the governing equation at the boundary [Eq. (2.27)], a numerical example is considered next.

EXAMPLE 2.2

In this example, we consider solution of the differential equation given by Eq. (2.1) in the interval [0,1] subject to a Dirichlet boundary condition $\phi = 1$ at the right end and the Neumann boundary $d\phi/dx = 1$ at the left end. The source term we will consider is $S_\phi = e^x$. The analytical solution to this problem can be obtained by integrating the differential equation twice and then using the two boundary conditions to yield $\phi(x) = e^x + 1 - e$.

In the first approach (named Approach 1), in which only the boundary condition is satisfied, the discrete equations to be solved are as follows:

$$\frac{4\phi_2 - \phi_3 - 3\phi_1}{2\Delta x} = 1 \text{ (Left node)}$$

$$\frac{\phi_{i+1} + \phi_{i-1} - 2\phi_i}{(\Delta x)^2} = \exp(x_i) \text{ (Interior nodes)}$$

$$\phi_N = 1 \text{ (Right node)}$$

In the second approach (named Approach 2), in which both the boundary condition and the governing equation are satisfied at the left end, the following equation is used at the leftmost node:

$$\frac{8\phi_2 - \phi_3 - 7\phi_1}{2(\Delta x)^2} = 1 + \frac{3}{\Delta x}\text{(Left node)}$$

The interior and right node equations remain unchanged. The table below shows the results obtained using the two approaches with 5, 9, and 17 nodes, i.e., with $\Delta x = 1/4$, $\Delta x = 1/8$, and $\Delta x = 1/16$. The percentage errors have been computed using

$$\%E = 100 \times \frac{\phi_{\text{analytical}} - \phi_{\text{numerical}}}{\phi_{\text{right boundary}}}$$

x	$\phi_{\text{analytical}}$	$\phi_{\text{numerical}}$ (Approach 1)	$\%E$ (Approach 1)	$\phi_{\text{numerical}}$ (Approach 2)	$\%E$ (Approach 2)
			$N = 5$		
0.00	−0.71828	−0.73966	2.13788	−0.71599	−0.22900
0.25	−0.43426	−0.44954	1.52784	−0.43178	−0.24732
0.50	−0.06956	−0.07916	0.95969	−0.06732	−0.22375
0.75	0.39872	0.39427	0.44532	0.40018	−0.14640
1.00	1.00000	1.00000	0.00000	1.00000	0.00000
			$N = 9$		
0.00	−0.71828	−0.72307	0.47844	−0.71752	−0.07634
0.13	−0.58513	−0.58921	0.40802	−0.58436	−0.07742
0.25	−0.43426	−0.43766	0.33990	−0.43349	−0.07619
0.38	−0.26329	−0.26603	0.27439	−0.26257	−0.07235
0.50	−0.06956	−0.07168	0.21185	−0.06891	−0.06555
0.63	0.14996	0.14844	0.15266	0.15052	−0.05539
0.75	0.39872	0.39775	0.09727	0.39913	−0.04143
0.88	0.68059	0.68013	0.04619	0.68083	−0.02316
1.00	1.00000	1.00000	0.00000	1.00000	0.00000
			$N = 17$		
0.00	−0.71828	−0.71941	0.11307	−0.71807	−0.02129
0.06	−0.65379	−0.65483	0.10461	−0.65357	−0.02136
0.13	−0.58513	−0.58610	0.09628	−0.58492	−0.02129
0.19	−0.51205	−0.51293	0.08810	−0.51184	−0.02107
0.25	−0.43426	−0.43506	0.08007	−0.43405	−0.02071
0.31	−0.35144	−0.35217	0.07220	−0.35124	−0.02017
0.38	−0.26329	−0.26394	0.06451	−0.26310	−0.01947
0.44	−0.16945	−0.17002	0.05700	−0.16927	−0.01858
0.50	−0.06956	−0.07006	0.04969	−0.06939	−0.01749
0.56	0.03677	0.03635	0.04259	0.03694	−0.01620
0.63	0.14996	0.14961	0.03571	0.15011	−0.01468
0.69	0.27046	0.27017	0.02907	0.27059	−0.01292
0.75	0.39872	0.39849	0.02268	0.39883	−0.01091
0.81	0.53525	0.53509	0.01656	0.53534	−0.00863
0.88	0.68059	0.68049	0.01073	0.68065	−0.00607
0.94	0.83531	0.83526	0.00520	0.83534	−0.00320
1.00	1.00000	1.00000	0.00000	1.00000	0.00000

The following observations may be made from the results presented in the table above:

1. The error in Approach 1 is larger than the error in Approach 2 at all nodes no matter what the grid spacing.
2. The error in Approach 1 is always largest at the left boundary, whereas the error in Approach 2 is largest at an interior node. Essentially, the large error at the boundary in Approach 1 has contaminated the rest of the solution.
3. In both approaches, error roughly scales as the square of the grid spacing. Consider, for example, the node at $x = 0.5$. For Approach 1, the error goes from 0.95969 to 0.21185 to 0.04969. The ratios of the errors are 4.53 and 4.26. The reason the ratio of the errors is not exactly 4 is because the error also contains the higher-order terms, not just the leading order term. When the mesh has been refined far enough that the higher-order terms are truly negligible, the ratio will be exactly 4, as indicated by its value approaching 4. For Approach 2, the error goes from 0.22375 to 0.06555 to 0.01749, implying ratios of 3.41 and 3.75. Here again, the ratio is approaching a value of 4 with mesh refinement.

In conclusion, an effort should be made to satisfy both the boundary condition and the governing equation at the boundary node when applying boundary conditions that are not of the Dirichlet type. This strategy extends application of the governing equation right up to the edge of the domain of solution, and is, in fact, the same strategy followed to obtain a closed-form analytical solution to the problem. On a final note, there are often situations in which the fourth derivative of the solution may be zero, but the third derivative is not zero. For example, if the source term were a linear function of x, the third derivative of ϕ would be a constant, and the fourth derivative would be zero. In such a scenario, Approach 2 would have no error since the truncation error at all its nodes is proportional to the fourth derivative [see Eqs (2.8) and (2.26)]. On the other hand, Approach 1 would have errors because of the truncation error at the boundary node being proportional to the third derivative [see Eq. (2.20)]. If the source term were constant, both solutions would be truncation error free, as if to suggest that they are equivalent, when, in fact, it is only a fortuitous coincidence.

2.3.3 ROBIN BOUNDARY CONDITION

In this type of boundary condition, a linear combination of the value of the dependent variable and its normal gradient is specified at the boundary. This type of boundary condition is credited to the French mathematician Gustave Robin[†]. It is also known

---

[†]**Victor Gustave Robin** (1855–1897) was a French applied mathematician who is known for his contributions to single- and double-layered boundary value problems in electrostatics and his contributions to the field of thermodynamics. Although a mathematician by training, his lifelong endeavor was to use mathematics as a means to an end, that is, to use mathematics to solve important problems in science.

as the boundary condition of the third kind and sometimes referred to as the Robbins boundary condition. A general form of this boundary condition using vector notations is presented in Chapter 7. For the 1D problem at hand, when applied to the left boundary, the Robin boundary condition may be written as

$$\alpha \phi(x_L) + \beta \frac{d\phi}{dx}\bigg|_{x_L} = \gamma, \tag{2.28}$$

where α, β, and γ are prescribed constants. To apply the boundary condition, it is first rewritten as follows:

$$\frac{d\phi}{dx}\bigg|_1 = \frac{\gamma}{\beta} - \frac{\alpha}{\beta}\phi_1 \tag{2.29}$$

Substituting Eq. (2.29) into Eqs (2.16) and (2.18), we obtain

$$\phi_2 = \phi_1 + (\Delta x)\left(\frac{\gamma}{\beta} - \frac{\alpha}{\beta}\phi_1\right) + \frac{(\Delta x)^2}{2!}\frac{d^2\phi}{dx^2}\bigg|_1 + \frac{(\Delta x)^3}{3!}\frac{d^3\phi}{dx^3}\bigg|_1 + \frac{(\Delta x)^4}{4!}\frac{d^4\phi}{dx^4}\bigg|_1 + \dots, \tag{2.30}$$

and

$$\phi_3 = \phi_1 + (2\Delta x)\left(\frac{\gamma}{\beta} - \frac{\alpha}{\beta}\phi_1\right) + \frac{(2\Delta x)^2}{2!}\frac{d^2\phi}{dx^2}\bigg|_1 + \frac{(2\Delta x)^3}{3!}\frac{d^3\phi}{dx^3}\bigg|_1 + \frac{(2\Delta x)^4}{4!}\frac{d^4\phi}{dx^4}\bigg|_1 + \dots \tag{2.31}$$

As before, the next objective is to cancel the third derivative containing terms, so that the resulting expression for the second derivative is second-order accurate. To do so, Eq. (2.30) is multiplied by 8, and Eq. (2.31) is subtracted from the result. This yields

$$8\phi_2 - \phi_3 = 7\phi_1 + 6\Delta x\left(\frac{\gamma}{\beta} - \frac{\alpha}{\beta}\phi_1\right) + 2(\Delta x)^2\frac{d^2\phi}{dx^2}\bigg|_1 - \frac{(\Delta x)^4}{3}\frac{d^4\phi}{dx^4}\bigg|_1 + \dots \tag{2.32}$$

Rearranging, we obtain

$$\frac{d^2\phi}{dx^2}\bigg|_1 = \frac{8\phi_2 - \phi_3 - \left(7 - 6\Delta x\frac{\alpha}{\beta}\right)\phi_1 - 6\Delta x\left(\frac{\gamma}{\beta}\right)}{2(\Delta x)^2} + \frac{(\Delta x)^2}{6}\frac{d^4\phi}{dx^4}\bigg|_1 + \dots \tag{2.33}$$

It is evident from Eq. (2.33) that, as before, the error in this difference approximation is second order. As a final step, Eq. (2.33) is substituted into the governing equation [Eq. (2.1)] to obtain the final nodal equation for the leftmost node:

$$\frac{8\phi_2 - \phi_3 - \left(7 - 6\Delta x\frac{\alpha}{\beta}\right)\phi_1 - 6\Delta x\left(\frac{\gamma}{\beta}\right)}{2(\Delta x)^2} = S_1. \tag{2.34}$$

The preceding description completes the discussion of the treatment of boundary conditions. The treatment of Dirichlet boundary conditions in the context of the finite difference method is straightforward, while the treatment of Neumann and Robin boundary conditions requires derivation of a nodal equation for the boundary. Depending on the order of the scheme sought, either one or two (or more) Taylor series expansions may be employed to derive the nodal equation at the boundary. As demonstrated in Example 2.2, it is desirable to satisfy both the boundary condition and the governing equation at the boundary node in order to obtain a consistent and accurate overall solution. Now that the pertinent nodal equations are available at our fingertips, in the next section, we discuss how to assemble these equations in matrix form.

2.4 ASSEMBLY OF NODAL EQUATIONS IN MATRIX FORM

The ODE considered in the discussion in the preceding three sections is a linear ODE. Accordingly, the resulting discrete algebraic equations are also linear. Chapter 3 will present strategies for addressing nonlinear source terms, while Chapter 8 will present strategies for addressing nonlinearities in the rest of the differential equation. Chapter 3 will also present general algorithms for the manipulation and solution of the system of linear or nonlinear equations. In this section, a preliminary discussion on how to assemble the system of linear algebraic equations, arising out of discretization of the governing ODE and boundary conditions, is included to lay the groundwork for the detailed coverage in Chapter 3.

Prior to solution of a system of linear algebraic equations, the equations must be assembled in the general matrix form:

$$[A][\phi] = [Q], \tag{2.35}$$

where $[A]$ is the so-called *coefficient matrix*. If there are N unknowns (or equations) then the coefficient matrix is of size $N \times N$. $[\phi]$ is a column matrix (or vector) and is comprised of the unknown nodal values of ϕ as its elements, i.e., $[\phi] = \begin{bmatrix} \phi_1 & \phi_2 & ... & \phi_N \end{bmatrix}^T$, where the superscript "T" denotes the transpose of a matrix. Similarly, $[Q]$ is the right-hand side vector. When expanded, Eq. (2.35) may be written as

$$A_{1,1}\phi_1 + A_{1,2}\phi_2 + ... + A_{1,i-1}\phi_{i-1} + A_{1,i}\phi_i + A_{1,i+1}\phi_{i+1} + ... + A_{1,N-1}\phi_{N-1} + A_{1,N}\phi_N = Q_1$$
$$A_{2,1}\phi_1 + A_{2,2}\phi_2 + ... + A_{2,i-1}\phi_{i-1} + A_{2,i}\phi_i + A_{2,i+1}\phi_{i+1} + ... + A_{2,N-1}\phi_{N-1} + A_{2,N}\phi_N = Q_2$$
$$...$$
$$A_{i,1}\phi_1 + A_{i,2}\phi_2 + ... + A_{i,i-1}\phi_{i-1} + A_{i,i}\phi_i + A_{i,i+1}\phi_{i+1} + ... + A_{i,N-1}\phi_{N-1} + A_{i,N}\phi_N = Q_i$$
$$...$$
$$A_{N,1}\phi_1 + A_{N,2}\phi_2 + ... + A_{N,i-1}\phi_{i-1} + A_{N,i}\phi_i + A_{N,i+1}\phi_{i+1} + ... + A_{N,N-1}\phi_{N-1} + A_{N,N}\phi_N = Q_N$$
$$\tag{2.36}$$

Once the nodal equations have been derived, the final step in solving a differential equation using any deterministic method is to write a computer program that fills up the matrices $[A]$ and $[Q]$. To identify the values of the individual elements of the matrices $[A]$ and $[Q]$, we now compare Eq. (2.36) with the difference equations derived in the preceding sections. As an example, let us consider Eq. (2.27) as the nodal equation for the leftmost node, Eq. (2.11) for the nodal equation for the interior nodes, and the second equation of Eq. (2.15) as the nodal equation for the rightmost node. Rearranging these equations yields

$$i = 1 : -\frac{7}{2(\Delta x)^2}\phi_1 + \frac{8}{2(\Delta x)^2}\phi_2 - \frac{1}{2(\Delta x)^2}\phi_3 = S_1 + \frac{(6\Delta x)J_L}{2(\Delta x)^2}, \tag{2.37a}$$

$$i = 2, 3, ... N - 1 : \frac{-2}{(\Delta x)^2}\phi_i + \frac{1}{(\Delta x)^2}\phi_{i+1} + \frac{1}{(\Delta x)^2}\phi_{i-1} = S_i, \tag{2.37b}$$

$$i = N : \phi_N = \phi_R. \tag{2.37c}$$

Comparing Eq. (2.37) with Eq. (2.36) yields

$$A_{1,1} = -\frac{7}{2(\Delta x)^2}; A_{1,2} = \frac{8}{2(\Delta x)^2}; A_{1,3} = -\frac{1}{2(\Delta x)^2}; Q_1 = S_1 + \frac{(6\Delta x)J_L}{2(\Delta x)^2}$$

$$A_{i,i} = -\frac{2}{(\Delta x)^2}; A_{i,i+1} = \frac{1}{(\Delta x)^2}; A_{i,i-1} = \frac{1}{(\Delta x)^2}; Q_i = S_i \quad \forall i = 2, 3, ..., N - 1 \tag{2.38}$$

$$A_{N,N} = 1; Q_N = \phi_R$$

All other elements of the coefficient matrix [A] are zero. Eq. (2.37), when written in matrix form, assumes the following form:

$$\begin{bmatrix} \frac{-7}{2(\Delta x)^2} & \frac{8}{2(\Delta x)^2} & \frac{-1}{2(\Delta x)^2} & 0 & \cdots & & & 0 \\ \frac{1}{(\Delta x)^2} & \frac{-2}{(\Delta x)^2} & \frac{1}{(\Delta x)^2} & 0 & \cdots & & & 0 \\ \cdots & 0 & \frac{1}{(\Delta x)^2} & \frac{-2}{(\Delta x)^2} & \frac{1}{(\Delta x)^2} & 0 & \cdots & 0 \\ & 0 & & \frac{1}{(\Delta x)^2} & \frac{-2}{(\Delta x)^2} & \frac{1}{(\Delta x)^2} & 0 & \\ & \cdots & 0 & & \frac{1}{(\Delta x)^2} & \frac{-2}{(\Delta x)^2} & \frac{1}{(\Delta x)^2} & 0 & \cdots \\ 0 & & & \cdots & 0 & \frac{1}{(\Delta x)^2} & \frac{-2}{(\Delta x)^2} & \frac{1}{(\Delta x)^2} \\ 0 & & & & \cdots & & 0 & 1 \end{bmatrix} \begin{bmatrix} \phi_1 \\ \phi_2 \\ \vdots \\ \phi_{i-1} \\ \phi_i \\ \phi_{i+1} \\ \vdots \\ \phi_{N-1} \\ \phi_N \end{bmatrix} = \begin{bmatrix} S_1 + \frac{3}{\Delta x}J_L \\ S_2 \\ \vdots \\ S_{i-1} \\ S_i \\ S_{i+1} \\ \vdots \\ S_{N-1} \\ \phi_R \end{bmatrix}. \tag{2.39}$$

As evident from Eq. (2.39), the coefficient matrix [A] has only a few nonzero elements. These types of matrices are known as sparse matrices. In general, every node is influenced by only a few neighboring nodes. The exact number depends on what order scheme is used, the dimensionality of the problem (1D, 2D, or 3D), and whether the mesh is structured or unstructured. In all of these cases, however, the number of nonzero elements in a given row, which is representative of how many other nodes influence a given node, is usually very small compared to the total number of nodes in the computational domain, thereby making the coefficient matrix sparse. As we shall see later in Chapter 3, the sparseness of the coefficient matrix has serious implications both in terms of memory usage as well as computational efficiency. The Code Snippet 2.1 illustrates how the coefficient matrix can be filled in an efficient manner for the particular case shown by Eq. (2.39).

CODE SNIPPET 2.1: DEMONSTRATION OF HOW TO FILL COEFFICIENT MATRIX [A] AND RIGHT-HAND SIDE VECTOR [Q] BASED ON EQ. (2.39).

```
A(:,:) = 0 ! Set entire [A] to zero. Later overwrite non-zero elements
Q(:) = 0   ! Set [Q] to zero.
dx = L/(N-1) ! Grid spacing
dx2 = dx*dx ! Calculating dx^2 here eliminates extra multiplications
! Left Boundary Condition
A(1,1) = -7/(2*dx2)
A(1,2) = 8/(2*dx2)
A(1,3) = -1/(2*dx2)
Q(1) = S(1) + 3*JL/dx
! Interior Nodes
For i = 2 : N-1   ! Loop over all interior nodes
   A(i,i) = -2/dx2
   A(i,i-1) = 1/dx2
   A(i,i+1) = 1/dx2
   Q(i) = S(i)
End
! Right Boundary Condition
A(N,N) = 1
Q(N) = phi_R
```

In the Code Snippet 2.1, the matrix is filled row by row in a sequential manner. The first index of the array, A, represents the row number, while the second index represents the column number. Each row actually represents a nodal equation, and thus, filling the matrices row-wise is the most systematic way of achieving the task. Any other way of filling the matrices is likely to introduce errors and is not recommended.

One final point to note is that the final nodal equation, which corresponds to the Dirichlet boundary condition at the right end, is actually decoupled from the rest of the linear system. Conceptually, the value of ϕ_N is already known (equal to ϕ_R) and should not have to be solved for. However, if that approach is chosen, the total number of unknowns becomes $N - 1$, and the nodal equation at $N - 1$ needs to be modified as follows prior to setting up the matrix:

$$\frac{\phi_{N-2} - 2\phi_{N-1}}{(\Delta x)^2} = S_{N-1} - \frac{\phi_R}{(\Delta x)^2}. \tag{2.40}$$

In contrast, in the strategy shown earlier [Eq. (2.39)], the nodal equation for $i = N$ is treated as part of the solution procedure and it is set up in a way that recovers the boundary condition at the right end. This way, the aforementioned complication of rearranging the nodal equation for $i = N - 1$ is completely avoided. Another advantage of solving for N unknowns rather than $N - 1$ unknowns is that the solution vector (output vector of the linear algebraic equation solver) does not have to be allocated temporarily and then the data copied to the $[\phi]$ vector. The solution vector itself can be directly used as the $[\phi]$ vector.

2.5 MULTIDIMENSIONAL PROBLEMS

Thus far, the focus of our discussion has been ordinary differential equations, as encountered in 1D systems. In this section, we build upon the discussion by applying the concepts that were developed earlier to multidimensional systems, i.e., to PDEs. For simplicity, we will restrict our derivations and discussions in this section to Cartesian (orthogonal) structured grids. In Section 2.8, the concepts developed in this section will be further extended to nonorthogonal structured grids.

As a natural extension to the governing equation considered in Section 2.1 for 1D problems, here, we consider solution of the Poisson equation in a two-dimensional (2D) framework, for which the Poisson equation is written as

$$\nabla^2 \phi = \frac{\partial^2 \phi}{\partial x^2} + \frac{\partial^2 \phi}{\partial y^2} = S_\phi, \tag{2.41}$$

where the symbols carry the same meanings as in previous sections. Here, of course, ϕ is a function of two independent variables, i.e., $\phi = \phi(x, y)$. Likewise, in general, the source term could be a function of three variables, i.e., $S_\phi = S_\phi(x, y, \phi)$. If the source term is zero, the governing equation reduces to the Laplace equation. Finite difference approximations to the derivatives will now be derived on the orthogonal grid, depicted in Fig. 2.5. The nodes are now denoted by the double index, (i, j), where i represents the nodal index in the x direction and j represents the nodal index in the y direction. Thus, the index i starts from a value of 1 at the $x = 0$ line and goes up to a value of N at the $x = L$ line. On the other hand, the index j starts from a value of 1 at the $y = 0$ line and goes up to a value of M at the $y = H$ line. The nodes in the stencil are also denoted by O, E, W, N, and S, where the northern (N) and southern (S) nodes have now been introduced, in addition to the three that existed for the 1D stencil (see Fig. 2.2).

For simplicity, we will assume that the nodes are uniformly spaced both in the x and y directions, such that

$$\Delta x = \frac{L}{N-1}, \tag{2.42a}$$

$$\Delta y = \frac{H}{M-1}. \tag{2.42b}$$

In order to derive difference approximations to the second derivative, once again, we employ Taylor series expansions in space. However, in this case, as stated earlier,

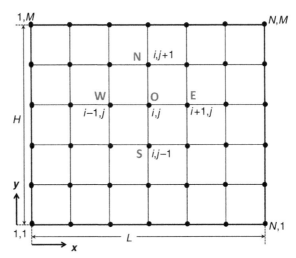

FIGURE 2.5 A Schematic Representation of a 2D Cartesian (Orthogonal) Grid Showing a Five-Point Stencil

the dependent variable is a 2D function, i.e., $\phi = \phi(x, y)$. Therefore, the Taylor series expansions must be 2D. Expanding $\phi_{i+1,j}$ about $\phi_{i,j}$, we get

$$\phi_{i+1,j} = \phi_{i,j} + (x_{i+1,j} - x_{i,j})\frac{\partial\phi}{\partial x}\bigg|_{i,j} + (y_{i+1,j} - y_{i,j})\frac{\partial\phi}{\partial y}\bigg|_{i,j} +$$

$$+ \frac{(x_{i+1,j} - x_{i,j})^2}{2!}\frac{\partial^2\phi}{\partial x^2}\bigg|_{i,j} + \frac{(y_{i+1,j} - y_{i,j})^2}{2!}\frac{\partial^2\phi}{\partial y^2}\bigg|_{i,j} + \frac{(x_{i+1,j} - x_{i,j})(y_{i+1,j} - y_{i,j})}{2!}\frac{\partial^2\phi}{\partial x\partial y}\bigg|_{i,j}$$

$$+ \frac{(x_{i+1,j} - x_{i,j})^3}{3!}\frac{\partial^3\phi}{\partial x^3}\bigg|_{i,j} + \frac{(y_{i+1,j} - y_{i,j})^3}{3!}\frac{\partial^3\phi}{\partial y^3}\bigg|_{i,j} + \frac{(x_{i+1,j} - x_{i,j})^2(y_{i+1,j} - y_{i,j})}{3!}\frac{\partial^3\phi}{\partial x^2\partial y}\bigg|_{i,j} +$$

$$+ \frac{(x_{i+1,j} - x_{i,j})(y_{i+1,j} - y_{i,j})^2}{3!}\frac{\partial^3\phi}{\partial x\partial y^2}\bigg|_{i,j} + \dots$$

$$\text{(2.43)}$$

One of the unique features of an orthogonal Cartesian grid is that $y_{i+1,j} = y_{i,j}$. Also, noting that $x_{i+1,j} - x_{i,j} = \Delta x$, Eq. (2.43) reduces to

$$\phi_{i+1,j} = \phi_{i,j} + \Delta x\frac{\partial\phi}{\partial x}\bigg|_{i,j} + \frac{(\Delta x)^2}{2!}\frac{\partial^2\phi}{\partial x^2}\bigg|_{i,j} + \frac{(\Delta x)^3}{3!}\frac{\partial^3\phi}{\partial x^3}\bigg|_{i,j} + \frac{(\Delta x)^4}{4!}\frac{\partial^4\phi}{\partial x^4}\bigg|_{i,j} + \dots \quad \text{(2.44)}$$

Essentially, Eq. (2.44) is the same as Eq. (2.6a), with the exception that the total derivatives have been replaced by partial derivatives. It is worth emphasizing that the reduction from Eq. (2.43) to Eq. (2.44) was possible only because the grid is orthogonal. If the grid were not orthogonal, the cross-derivative containing terms in Eq. (2.43) would remain, making the derivation of the finite difference approximation considerably more complicated, as we will see in Section 2.8. Following the same procedure as in Section 2.1 for deriving the central difference scheme, we also expand $\phi_{i-1,j}$ about $\phi_{i,j}$, yielding

$$\phi_{i-1,j} = \phi_{i,j} - \Delta x \frac{\partial \phi}{\partial x}\bigg|_{i,j} + \frac{(\Delta x)^2}{2!} \frac{\partial^2 \phi}{\partial x^2}\bigg|_{i,j} - \frac{(\Delta x)^3}{3!} \frac{\partial^3 \phi}{\partial x^3}\bigg|_{i,j} + \frac{(\Delta x)^4}{4!} \frac{\partial^4 \phi}{\partial x^4}\bigg|_{i,j} - .... \quad (2.45)$$

Adding Eqs (2.44) and (2.45) yields

$$\frac{\partial^2 \phi}{\partial x^2}\bigg|_{i,j} = \frac{\phi_{i+1,j} + \phi_{i-1,j} - 2\phi_{i,j}}{(\Delta x)^2} - \frac{(\Delta x)^2}{12} \frac{\partial^4 \phi}{\partial x^4}\bigg|_{i,j} + ..., \quad (2.46)$$

which is identical to Eq. (2.8). Expanding $\phi_{i,j+1}$ and $\phi_{i,j-1}$ about $\phi_{i,j}$ and following the same procedure yields

$$\frac{\partial^2 \phi}{\partial y^2}\bigg|_{i,j} = \frac{\phi_{i,j+1} + \phi_{i,j-1} - 2\phi_{i,j}}{(\Delta y)^2} - \frac{(\Delta y)^2}{12} \frac{\partial^4 \phi}{\partial y^4}\bigg|_{i,j} + ... \quad (2.47)$$

It is clear from Eqs (2.46) and (2.47) that the errors associated with finite difference approximations of the partial derivatives are second order in both x and y directions. Substitution of Eqs (2.46) and (2.47) into Eq. (2.41) after neglecting the higher-order derivatives yields

$$\frac{\phi_{i+1,j} + \phi_{i-1,j} - 2\phi_{i,j}}{(\Delta x)^2} + \frac{\phi_{i,j+1} + \phi_{i,j-1} - 2\phi_{i,j}}{(\Delta y)^2} \approx S_{i,j}. \quad (2.48)$$

As in the 1D case, Eq. (2.48) is valid only for the interior nodes, i.e., for $i = 2, 3, ..., N-1$ and $j = 2, 3, ..., M-1$. For the boundary nodes, one must apply boundary conditions. The procedure to apply boundary conditions is the same as that followed in the 1D case. To demonstrate application of the boundary conditions, let us consider a case where we apply the following boundary conditions on the four boundaries shown in Fig. 2.5:

$$\text{Left} : \phi(0, y) = \phi_L, \quad (2.49a)$$

$$\text{Right} : \frac{\partial \phi}{\partial x}\bigg|_{L,y} = J_R, \quad (2.49b)$$

$$\text{Bottom} : \alpha\phi(x, 0) + \beta \frac{\partial \phi}{\partial y}\bigg|_{x,0} = \gamma, \quad (2.49c)$$

$$\text{Top} : \phi(x, H) = \phi_T. \quad (2.49d)$$

First of all, it is customary to let the Dirichlet boundary condition override all other boundary conditions. So, the left boundary condition [Eq. (2.49a)] is to be applied to the bottom left corner node (1,1), i.e., it overrides the Robin boundary condition applied to the bottom boundary. Similarly, the top boundary condition is to be applied to the top right corner node (N,M). We have a mathematical discontinuity at the corner nodes where the two Dirichlet boundary conditions meet. Therefore, either of the two boundary values, or even an average value, may be used. This is justified because the value chosen for the corner node does not impact the rest of the

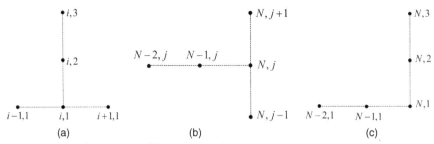

FIGURE 2.6 Stencils for Three Different Scenarios

(a) Bottom boundary node, (b) right boundary node, and (c) bottom right corner node.

solution since the two nodes adjacent to it also have Dirichlet conditions. The value of ϕ at such a corner node has no physical meaning and is only used for postprocessing the results, for example, for making contour plots of ϕ.

In order to apply the Robin boundary condition [Eq. (2.49c)] at the bottom wall, we employ the exact same procedure described in Section 2.3.3, with the exception that the Taylor series expansions now have to be performed in the y direction instead of in the x direction (see Fig. 2.6a). This yields [cf. Eq. (2.34)]

$$\left.\frac{\partial^2 \phi}{\partial y^2}\right|_{i,1} = \frac{8\phi_{i,2} - \phi_{i,3} - \left(7 - 6\Delta y\frac{\alpha}{\beta}\right)\phi_{i,1} - 6\Delta y\left(\frac{\gamma}{\beta}\right)}{2(\Delta y)^2}. \tag{2.50}$$

As shown in Fig. 2.6(a), any node on the bottom boundary, with the exception of the two corners, has nodes both to its right as well as to its left. Therefore, the same procedure that was used earlier to derive the second derivative in x can still be applied to these boundary nodes. Therefore, we obtain

$$\left.\frac{\partial^2 \phi}{\partial x^2}\right|_{i,1} = \frac{\phi_{i+1,1} + \phi_{i-1,1} - 2\phi_{i,1}}{(\Delta x)^2}. \tag{2.51}$$

In order to satisfy the governing equation at the boundary, we now substitute Eqs (2.50) and (2.51) into Eq. (2.41) to yield

$$\frac{\phi_{i+1,1} + \phi_{i-1,1} - 2\phi_{i,1}}{(\Delta x)^2} + \frac{8\phi_{i,2} - \phi_{i,3} - \left(7 - 6\Delta y\frac{\alpha}{\beta}\right)\phi_{i,1} - 6\Delta y\left(\frac{\gamma}{\beta}\right)}{2(\Delta y)^2} = S_{i,1}, \tag{2.52}$$

where the validity of Eq. (2.52) is for $i = 2, 3, ..., N - 1$. The finite difference approximations used for both second derivatives are second order accurate in space. Thus, the order of accuracy of the nodal equation at the bottom boundary [Eq. (2.52)] is consistent with the nodal equation for the interior nodes.

The derivation of the nodal equation for the nodes on the right boundary (excluding corner nodes) requires two backward Taylor series expansions [following Fig. 2.6(b)] to approximate the second derivative in x, as follows:

$$\phi_{N-1,j} = \phi_{N,j} - \Delta x \frac{\partial \phi}{\partial x}\bigg|_{N,j} + \frac{(\Delta x)^2}{2!} \frac{\partial^2 \phi}{\partial x^2}\bigg|_{N,j} - \frac{(\Delta x)^3}{3!} \frac{\partial^3 \phi}{\partial x^3}\bigg|_{N,j} + \frac{(\Delta x)^4}{4!} \frac{\partial^4 \phi}{\partial x^4}\bigg|_{N,j} + ..., \quad (2.53a)$$

$$\phi_{N-2,j} = \phi_{N,j} - 2\Delta x \frac{\partial \phi}{\partial x}\bigg|_{N,j} + \frac{(2\Delta x)^2}{2!} \frac{\partial^2 \phi}{\partial x^2}\bigg|_{N,j} - \frac{(2\Delta x)^3}{3!} \frac{\partial^3 \phi}{\partial x^3}\bigg|_{N,j} + \frac{(2\Delta x)^4}{4!} \frac{\partial^4 \phi}{\partial x^4}\bigg|_{N,j} + ... \quad (2.53b)$$

Next, we substitute Eq. (2.49b) into Eq. (2.53) for the first derivative and use the same procedure as described in Section 2.3.2 to cancel the third derivative. This yields a second-order accurate approximation for the second derivative, written as

$$\frac{\partial^2 \phi}{\partial x^2}\bigg|_{N,j} = \frac{8\phi_{N-1,j} - \phi_{N-2,j} - 7\phi_{N,j} + (6\Delta x)J_R}{2(\Delta x)^2} + \frac{(\Delta x)^2}{6} \frac{\partial^4 \phi}{\partial x^4}\bigg|_{N,j} + ... \quad (2.54)$$

Substituting Eq. (2.54) and the standard expression for the interior nodes for the y derivative [Eq. (2.47)], we obtain

$$\frac{8\phi_{N-1,j} - \phi_{N-2,j} - 7\phi_{N,j} + (6\Delta x)J_R}{2(\Delta x)^2} + \frac{\phi_{N,j+1} + \phi_{N,j-1} - 2\phi_{N,j}}{(\Delta y)^2} = S_{N,j}. \quad (2.55)$$

Equation (2.55) is valid for all right boundary nodes excluding the two corner nodes, i.e., $j = 2, 3, ..., M - 1$. The derivation of the nodal equation for the right bottom corner [Fig. 2.6(c)] essentially entails a combination of two steps – one that was used for the bottom boundary and another that was used for the right boundary. The result is a combination of the two expressions derived for the two boundaries and may be written as

$$\frac{8\phi_{N-1,1} - \phi_{N-2,1} - 7\phi_{N,1} + (6\Delta x)J_R}{2(\Delta x)^2} + \frac{8\phi_{N,2} - \phi_{N,3} - \left(7 - 6\Delta y \dfrac{\alpha}{\beta}\right)\phi_{N,1} - 6\Delta y \left(\dfrac{\gamma}{\beta}\right)}{2(\Delta y)^2} = S_{N,1}. \quad (2.56)$$

We set out with the objective of solving the Poisson equation [Eq. (2.41)], subject to the boundary conditions given by Eq. (2.49), on the regular Cartesian 2D grid depicted in Fig. 2.5. At this juncture, the finite difference equations for all relevant node groups have been derived and are summarized as follows:

- Eq. (2.48) for interior nodes ($i = 2, 3, ..., N - 1$ and $j = 2, 3, ..., M - 1$);
- Nodal equivalent of Eq. (2.49a), i.e., $\phi_{1,j} = \phi_L$ on the left boundary ($j = 1, 2, 3, ..., M$);
- Eq. (2.55) for the right boundary ($j = 2, 3, ..., M - 1$);
- Eq. (2.52) for the bottom boundary ($i = 2, 3, ..., N - 1$);
- Nodal equivalent of Eq. (2.49d), i.e., $\phi_{i,M} = \phi_T$ on the top boundary ($i = 2, 3, ..., N$); and
- Eq. (2.56) for the right bottom corner node.

In this case, prior to assembling the nodal equations in matrix form, each node needs to be assigned a unique number. This is because the solution to the system shown by Eq. (2.35) is a vector ($[\phi]$), i.e., it is a 1D column matrix, not a 2D matrix (or array).

Essentially, this implies combining the i and j indices into a single index, which, in fact, is the aforementioned unique number assigned to the node. One way to number the nodes is to start from the left bottom corner and number the nodes along the bottommost row 1 through N, then turn around and number the second row of nodes $N + 1$ through $2N$, and so on. This is illustrated in Fig. 2.7. The unique number, k, assigned to each node by this procedure can actually be derived by using the node's i and j indices, as follows:

$$k = (j-1)N + i. \tag{2.57}$$

It is to be noted that Eq. (2.57) represents just one way of converting the two indices to a single index and is, by no means, the only way. Based on this numbering pattern, the five nodes in the stencil now have the indices k (or O), $k - 1$ (or W), $k + 1$ (or E), $k - N$ (or S), and $k + N$ (or N), as also shown in Fig. 2.7.

Once the nodes have been numbered uniquely, all nodal equations must first be rearranged in the following general form:

$$A_{k,k}\phi_k + A_{k,k-1}\phi_{k-1} + A_{k,k+1}\phi_{k+1} + \ldots + A_{k,k-N}\phi_{k-N} + \ldots + A_{k,k+N}\phi_{k+N} = Q_k. \tag{2.58}$$

Of course, for the boundary nodes, one or more of the terms in Eq. (2.58) may be missing. Writing the nodal equations in the form shown in Eq. (2.58) lets us easily determine the elements of the coefficient matrix $[A]$ and the right-hand side vector $[Q]$.

The advantage of using an automatic formula, such as Eq. (2.57), for generation of the unique node number is that it allows one to still formulate all nodal equations using the i and j indices, as done previously, while making assembly of the system of linear equations in matrix form relatively easy. To illustrate this point, let us consider a code snippet for filling up the coefficient matrix $[A]$ and the right-hand side vector $[Q]$ with just the equations for the interior nodes, i.e., Eq. (2.48):

CODE SNIPPET 2.2: DEMONSTRATION OF HOW TO FILL COEFFICIENT MATRIX [A] AND RIGHT-HAND SIDE VECTOR [Q] FOR JUST THE INTERIOR NODES IN A 2D FDM CODE FOR SOLVING THE POISSON EQUATION.

```
A(:,:) = 0 ! Set entire [A] to zero. Later overwrite non-zero elements
Q(:) = 0  ! Set [Q] to zero.
dx = L/(N-1) ! Grid spacing in x
dy = H/(M-1) ! Grid spacing in y
dx2 = dx*dx; dy2 = dy*dy
! Interior Nodes
For i = 2 : N-1  ! Loop over all interior nodes in i direction
  For j = 2 : M-1  ! Loop over all interior nodes in j direction
    k = (j-1)*N+i ! Use Equation (2.57) to construct global index, k
    A(k,k) = -(2/dx2+2/dy2)  ! Diagonal
    A(k,k-1) = 1/dx2  ! Link to Western node
    A(k,k+1) = 1/dx2  ! Link to Eastern node
    A(k,k-N) = 1/dy2  ! Link to Southern node
    A(k,k+N) = 1/dy2  ! Link to Northern node
    Q(k) = S(i,j)
  End
End
```

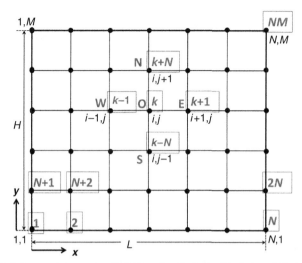

FIGURE 2.7 2D Orthogonal Cartesian Grid Showing Both Double-Indexed Node Numbers and Renumbered Single-Indexed (Enclosed in Boxes) Nodes

Execution of Code Snippet 2.2 fills up all the matrix coefficients of both [A] and [Q] that are in the rows that correspond to the interior nodes. For example, if $N = 7$ and $M = 6$, as shown in Fig. 2.7, the matrix [A] will be of size 42×42, and [Q] will be of size 42×1. As soon as Code Snippet 2.2 is executed, rows 9–13, 16–20, 23–27, and 30–34 of both [A] and [Q] will be populated with nonzeroes in appropriate locations. Based on the code snippet, for a given row, k, the elements corresponding to the following column numbers will receive nonzero values: k, $k-1$, $k+1$, $k-N$, and $k+N$. For example, if $k = 25$, then in the 25th row there will be nonzero values in columns 25, 24, 26, 18, and 32. The rest of the elements will stay unchanged at zero. The code snippet automates the populating process without having to manually determine what columns should receive nonzero values.

Proceeding with the bulleted list presented above for the various equation types, the next task is to populate the matrices with the left boundary condition. This may be accomplished with Code Snippet 2.3:

CODE SNIPPET 2.3: DEMONSTRATION OF HOW TO FILL COEFFICIENT MATRIX [A] AND RIGHT-HAND SIDE VECTOR [Q] FOR THE LEFT BOUNDARY NODES (DIRICHLET CONDITION) IN A 2D FDM CODE FOR SOLVING THE POISSON EQUATION. THIS CODE IS TO IMMEDIATELY FOLLOW CODE SNIPPET 2.2.

```
! Left Boundary Nodes
i = 1   ! Left Boundary index for i
For j = 1 : M   ! Loop over all left boundary nodes in j direction
   k = (j-1)*N+i   ! Use Equation (2.57) to construct global index, k
   A(k,k) = 1   ! Diagonal
   Q(k) = phi_R
End
```

When Code Snippet 2.3 is executed, all rows corresponding to the left boundary will be populated. It is worth noting that the index k will not assume continuous values within the loop shown in Code Snippet 2.3. Rather, it will automatically skip every N rows and assume values of 1, 8, 15, 22, 29, and 36. In other words, one does not have to manually seek out these rows of the matrices and populate them. They will be automatically accessed and populated because of the formula used to compute k from i and j. The next task is to employ Eq. (2.55) and populate the rows corresponding to the right boundary nodes. This may be accomplished using Code Snippet 2.4.

CODE SNIPPET 2.4: DEMONSTRATION OF HOW TO FILL COEFFICIENT MATRIX [A] AND RIGHT-HAND SIDE VECTOR [Q] FOR THE RIGHT BOUNDARY NODES (NEUMANN CONDITION) IN A 2D FDM CODE FOR SOLVING THE POISSON EQUATION. THIS CODE IS TO IMMEDIATELY FOLLOW CODE SNIPPET 2.3.

```
! Right boundary Nodes
i = N   ! Right Boundary index for i
For j = 2 : M-1   ! Loop over all Right boundary nodes in j direction
    k = (j-1)*N+i   ! Use Equation (2.57) to construct global index, k
    A(k,k) = -(7/(2*dx2)+2/dy2)   ! Diagonal
    A(k,k-1) = 4/dx2   ! Link to Western node
    A(k,k-2) = -1/(2*dx2)   ! Link to Western Western node
    A(k,k-N) = 1/dy2   ! Link to Southern node
    A(k,k+N) = 1/dy2   ! Link to Northern node
    Q(k) = S(i,j)-3*JL/dx   ! Known flux term transposed to right hand side
End
```

The remaining rows, corresponding to the bottom and top boundaries as well as the right bottom corner node, may be filled up in a similar manner; this is left as an exercise for the reader. Once the matrices have been fully populated, the system of linear equations can be solved using a variety of methods. These methods will be presented in Chapter 3. In closure, a few comments pertaining to the matrix [A] are warranted. First, in an attempt to solve the linear system, if a so-called "NaN" (Not a Number) is generated, it indicates that a division by zero has been encountered. This typically happens when one or more of the diagonal elements of the matrix are zero, implying that the matrix is singular, i.e., its determinant is zero, and the system of equations cannot be solved. At this point, it would be judicious to carefully investigate why one of the diagonal elements, $A(k, k)$, has not been populated with a nonzero value. Secondly, as discussed earlier, the solution vector will be a 1D column matrix of size $N \times M$. This structure is not amenable to postprocessing, such as contour plotting. Furthermore, it is difficult to identify what node each value in the matrix corresponds to. To facilitate contour plotting and to interpret the solution easily, it is best to map the data back to a (i, j)-indexed 2D array. This may be accomplished using the following Code Snippet:

CODE SNIPPET 2.5: DEMONSTRATION OF HOW TO TRANSFORM THE 1D SOLUTION VECTOR (PHI1D) TO A 2D ARRAY (PHI2D).

```
For i = 1 : N  ! Loop over all nodes in i direction
  For j = 1 : M  ! Loop over all nodes in j direction
    k = (j-1)*N+i  ! Use Equation (2.57) to construct global index, k
    phi2D(i,j) = phi1D(k)
  End
End
```

The preceding discussion lays out the necessary steps needed to derive finite difference equations in two dimensions for both interior and boundary nodes and assembly of the resulting algebraic equations in matrix form. The formulations presented thus far are directly extendable to three-dimensional (3D) geometry. In 3D, the stencil extends in the third direction and includes two additional nodes, F (for forward) and B (for backward), in addition to W, E, S, and N, as illustrated in Fig. 2.8. In this case, the global index or unique node number, k, of any node can be constructed from the three individual indices (i, j, l) using the following relation:

$$k = [(j-1)N + i] + (l-1)MN, \qquad (2.59)$$

where i, j, and l are individual nodal indices in the x, y, and z directions, respectively, as shown in Fig. 2.8. As far as deriving finite difference approximations are concerned, the procedure to be followed is exactly the same as for 2D: Taylor series

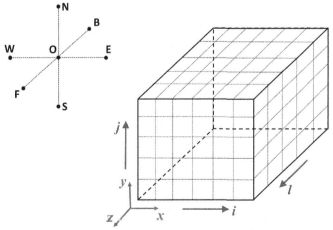

FIGURE 2.8 Schematic Representation of a 3D Cartesian (Orthogonal) Grid Showing a Seven-Point Stencil

expansions must be performed in individual directions to derive expressions for each of the three partial second derivatives and subsequently substituted into the governing equation to obtain the discrete nodal equation.

2.6 HIGHER-ORDER APPROXIMATIONS

Thus far, our discussion of the finite difference method has been restricted to second-order accurate schemes. In many applications, such as computation of turbulent flow [4] or acoustics calculations [5], second-order accuracy is often considered inadequate. This is because with a second-order scheme, the number of nodes required to make the computed solution either grid independent or able to capture the physics at all relevant length scales is often too large, thereby making the computations too resource and time intensive. In such a scenario, a higher-order scheme is warranted so that the same accuracy can be attained with a coarser grid.

Following our basic procedure, higher-order difference approximations can clearly be derived by using additional Taylor series expansions and extending the stencil beyond the three points (for 1D) that we have been using thus far. Such a derivation procedure, although conceptually straightforward, can quickly become tedious. Instead, we use a procedure here that recursively makes use of expressions that we have already derived in preceding sections. The procedure used here is adapted from the general procedure used to derive so-called compact difference schemes. For additional details on compact difference schemes beyond what is discussed here, the reader is referred to the popular article by Lele [6] and the references cited therein.

Our objective is to derive a difference approximation for the second derivative that shall have accuracy better than second order on a uniform grid. To begin the derivation, we first write Eq. (2.8) in an alternative form, as follows

$$\phi_j^{ii} = \frac{\phi_{j+1} - 2\phi_j + \phi_{j-1}}{\delta^2} - \frac{\delta^2}{12}\phi_j^{iv} - \frac{\delta^4}{360}\phi_j^{vi} + ..., \tag{2.60}$$

where the subscripts $j, j+1$, and $j-1$ have been used to denote nodal locations, and the superscript ii has been used to denote the order of the derivative. The last two terms in Eq. (2.60) represent the truncation error. Thus far, we have been concerning ourselves only with the leading order (fourth derivative containing) truncation error term. Here, since our objective is to eliminate that term, it is important to also consider the next term in the truncation error expression, as has been done in Eq. (2.60). The grid spacing, Δx, has been denoted by δ for simplicity of notation. By mathematical induction, from Eq. (2.60) it follows that

$$\phi_j^{iv} = \frac{\phi_{j+1}^{ii} - 2\phi_j^{ii} + \phi_{j-1}^{ii}}{\delta^2} - \frac{\delta^2}{12}\phi_j^{vi} - \frac{\delta^4}{360}\phi_j^{viii} + ... \tag{2.61}$$

Substituting Eq. (2.61) into Eq. (2.60) for the fourth derivative, and simplifying, we obtain

$$\phi_j^{ii} = \frac{\phi_{j+1} - 2\phi_j + \phi_{j-1}}{\delta^2} - \frac{1}{12}(\phi_{j+1}^{ii} - 2\phi_j^{ii} + \phi_{j-1}^{ii}) + \frac{\delta^4}{240}\phi_j^{vi} + \frac{\delta^6}{4320}\phi_j^{viii} + ... \tag{2.62}$$

Next, we apply mathematical induction again to express the second derivatives at the nodes $j+1$ and $j-1$ to yield

$$\phi_{j+1}^{ii} = \frac{\phi_{j+2} - 2\phi_{j+1} + \phi_j}{\delta^2} - \frac{\delta^2}{12}\phi_{j+1}^{iv} - \frac{\delta^4}{360}\phi_{j+1}^{vi} + ..., \qquad (2.63a)$$

$$\phi_{j-1}^{ii} = \frac{\phi_j - 2\phi_{j-1} + \phi_{j-2}}{\delta^2} - \frac{\delta^2}{12}\phi_{j-1}^{iv} - \frac{\delta^4}{360}\phi_{j-1}^{vi} + ... \qquad (2.63b)$$

Combining Eqs (2.60) and (2.63), followed by simplification, results in

$$\phi_{j+1}^{ii} - 2\phi_j^{ii} + \phi_{j-1}^{ii} = \frac{\phi_{j+2} - 4\phi_{j+1} + 6\phi_j - 4\phi_{j-1} + \phi_{j-2}}{\delta^2}$$
$$- \frac{\delta^2}{12}(\phi_{j+1}^{iv} - 2\phi_j^{iv} + \phi_{j-1}^{iv}) - \frac{\delta^4}{360}(\phi_{j+1}^{vi} - 2\phi_j^{vi} + \phi_{j-1}^{vi}) + ... \qquad (2.64)$$

Again, applying mathematical induction to Eq. (2.60), we obtain

$$\phi_j^{vi} = \frac{\phi_{j+1}^{iv} - 2\phi_j^{iv} + \phi_{j-1}^{iv}}{\delta^2} - \frac{\delta^2}{12}\phi_j^{viii} - \frac{\delta^4}{360}\phi_j^{x} + ..., \qquad (2.65)$$

which, upon rearrangement, yields

$$\phi_{j+1}^{iv} - 2\phi_j^{iv} + \phi_{j-1}^{iv} = \delta^2\phi_j^{vi} + \frac{\delta^4}{12}\phi_j^{viii} + \frac{\delta^6}{360}\phi_j^{x} + ... \qquad (2.66)$$

By applying mathematical induction to Eq. (2.66), one can also write

$$\phi_{j+1}^{vi} - 2\phi_j^{vi} + \phi_{j-1}^{vi} = \delta^2\phi_j^{viii} + \frac{\delta^4}{12}\phi_j^{x} + \frac{\delta^6}{360}\phi_j^{xii} + ... \qquad (2.67)$$

Substitution of Eqs (2.66) and (2.67) into Eq. (2.64) yields

$$\phi_{j+1}^{ii} - 2\phi_j^{ii} + \phi_{j-1}^{ii} = \frac{\phi_{j+2} - 4\phi_{j+1} + 6\phi_j - 4\phi_{j-1} + \phi_{j-2}}{\delta^2} - \frac{\delta^2}{12}(\delta^2\phi_j^{vi} + \frac{\delta^4}{12}\phi_j^{viii} + \frac{\delta^6}{360}\phi_j^{x} + ...)$$
$$- \frac{\delta^4}{360}(\delta^2\phi_j^{viii} + \frac{\delta^4}{12}\phi_j^{x} + \frac{\delta^6}{360}\phi_j^{xii} + ...) + ... \qquad (2.68)$$

Finally, substituting Eq. (2.68) into the right-hand side of Eq. (2.62), followed by simplification, yields

$$\phi_j^{ii} = \frac{\phi_{j+1} - 2\phi_j + \phi_{j-1}}{\delta^2} - \frac{1}{12}\left(\frac{\phi_{j+2} - 4\phi_{j+1} + 6\phi_j - 4\phi_{j-1} + \phi_{j-2}}{\delta^2}\right) + \frac{\delta^4}{90}\phi_j^{vi} + \frac{\delta^6}{960}\phi_j^{viii} + ..., \qquad (2.69)$$

where only the first two leading order error terms have been retained. It is evident from Eq. (2.69) that the difference approximation for the second derivative is now fourth-order accurate. The scheme derived here is also a central difference scheme because an equal number of points to the left and right of the node in question have been utilized, except that it is a fourth-order accurate central difference scheme. However, this has come at the expense of extending the stencil to five nodes – two in each direction from the central node.

The fact that the stencil has now extended beyond three nodes has two major repercussions. First, applying boundary conditions becomes significantly more complex. Second, the sparseness of the coefficient matrix decreases. We will see in Chapter 3 that with the inclusion of additional nodes in the stencil, the structure of the coefficient matrix becomes such that certain solution algorithms that could be used with the second-order central difference scheme can no longer be used directly.

To appreciate the increased complexity in applying boundary conditions, let us consider a simple scenario where a Dirichlet boundary condition is applied to the leftmost node [first equation of Eq. (2.15)]. If the second-order central difference scheme is used for the second node ($j = 2$), no special treatment is necessary, i.e., the nodal equation derived for the interior node [Eq. (2.11)] is valid for all interior nodes right up to the one that is adjacent to the boundary. This is because the formula contains ϕ_{j-1}, which in this particular case is ϕ_1. If, however, the fourth-order central difference scheme [Eq. (2.69)] is used, the formula contains ϕ_{j-1} as well as ϕ_{j-2}. For $j = 2$, ϕ_{j-1} is the boundary value, but ϕ_{j-2} is the value at a nonexistent node. This implies that Eq. (2.69) cannot be used for all interior nodes. It can only be used for interior nodes that are at least two nodes away from the boundary. Attempting to derive a fourth-order accurate scheme for $j = 2$ using one backward ($j = 1$) and three forward expansions ($j = 3, 4$, and 5) will actually result in a scheme that is only third-order accurate (this is left as an exercise for the reader), not to mention that the derivation is quite tedious. On account of these difficulties in applying boundary conditions, higher-order finite difference or compact difference schemes tend to be used in applications where the boundary conditions are either symmetry or periodic boundary conditions. If, for example, the symmetry boundary condition, which is essentially the same as a Neumann boundary condition with a zero derivative, is used, then $\phi_0 = \phi_2$, i.e., the value at the nonexistent 0th node that is located at a distance Δx to the left of the left boundary is the same as the value at the node located at a distance Δx to the right of the left boundary. Using this symmetry condition in Eq. (2.69), the nodal equation for $j = 2$ becomes

$$\frac{\phi_3 - 2\phi_2 + \phi_L}{\delta^2} - \frac{1}{12}\left(\frac{\phi_4 - 4\phi_3 + 6\phi_2 - 4\phi_L + \phi_2}{\delta^2}\right) = S_2. \qquad (2.70)$$

In other words, no additional difference equation has to be derived for interior nodes adjacent to the boundary.

In the event that the boundary condition is not of the symmetry or periodic type, it is customary to revert back to a lower-order scheme for nodes adjacent to the boundary. It goes without saying that such an approach will contaminate the accuracy of the nodes where the fourth-order scheme is being used. As to how much the solution will be contaminated is problem dependent (depending on the nonlinearity of the solution). An example is considered next to shed further light on this issue.

EXAMPLE 2.3

In this example, we consider solution of the differential equation given by Eq. (2.1) in the interval [0,1] subject to Dirichlet boundary conditions $\phi = 0$ at the left end and $\phi = 1$ at the right end. The source term we will consider is $S_\phi = e^x$. The analytical solution to this problem can be obtained by integrating the differential equation twice and then using the two boundary conditions to yield

$$\phi(x) = e^x + (2-e)x - 1$$

The problem is solved using the second-order and fourth-order central difference (CD) schemes. In the case of the fourth-order scheme, the second-order scheme is used for the two nodes adjacent to the boundaries in keeping with the preceding discussion. The table below shows the results obtained using the two schemes with five and nine nodes, i.e., with $\Delta x = 1/4$ and $\Delta x = 1/8$. The percentage errors have been computed using

$$\%E = 100 \times \frac{\phi_{\text{analytical}} - \phi_{\text{numerical}}}{\phi_{\text{right boundary}}}$$

		$\phi_{\text{numerical}}$	$\%E$	$\phi_{\text{numerical}}$	$\%E$
x	$\phi_{\text{analytical}}$	(2nd Order CD)	(2nd Order CD)	(4th Order CD)	(4th Order CD)
			$N = 5$		
0.00	0.00000	0.00000	0.00000	0.00000	0.00000
0.25	0.10446	0.10521	−0.07557	0.10498	−0.05245
0.50	0.28958	0.29067	−0.10925	0.29021	−0.06301
0.75	0.57829	0.57918	−0.08915	0.57895	−0.06603
1.00	1.00000	1.00000	0.00000	1.00000	0.00000
			$N = 9$		
0.00	0.00000	0.00000	0.00000	0.00000	0.00000
0.13	0.04336	0.04347	−0.01062	0.04339	−0.00284
0.25	0.10446	0.10464	−0.01894	0.10449	−0.00338
0.38	0.18564	0.18588	−0.02464	0.18567	−0.00375
0.50	0.28958	0.28985	−0.02738	0.28962	−0.00412
0.63	0.41932	0.41959	−0.02676	0.41937	−0.00450
0.75	0.57829	0.57851	−0.02234	0.57834	−0.00485
0.88	0.77038	0.77052	−0.01361	0.77043	−0.00487
1.00	1.00000	1.00000	0.00000	1.00000	0.00000

The following observations may be made from the results presented in the table above:

1. When five grid points are used, only one of those grid points (at $x = 0.5$) is the beneficiary of the fourth-order scheme. Even so, the errors are smaller at all nodes in that case than the case when a second-order scheme is used throughout.

2. When nine grid points are used, the average error with the fourth-order scheme is significantly smaller than with the second-order scheme even for the nodes adjacent to the boundaries for which the second-order scheme has been used.

3. In the second-order scheme, the error roughly scales by a factor of 4 when the mesh size is halved, while it scales roughly by a factor of slightly less than 16 in the fourth-order scheme. Consider, for example, the node at $x = 0.5$. For the second-order scheme, the error goes from 0.10925 to 0.02738 – a ratio of exactly 4. On the other hand, for the fourth-order scheme, the error goes from 0.06301 to 0.00412 – a ratio of 15.3. The reason the ratio is not exactly 16 is perhaps due to the fact that the solution is somewhat contaminated by the second-order scheme used for two out of the seven interior nodes.

Based on the quantitative results presented in Example 2.3, it is fair to conclude that even though using the second-order scheme for the nodes adjacent to the boundaries contaminates the solution, the benefits of using a fourth-order scheme for the remaining nodes outweighs the shortcomings. As a general principle, using a higher-order scheme is, therefore, desirable. The application of the higher-order scheme to multidimensional problems is straightforward. Essentially, second derivatives in y and z, which will assume similar form as Eq. (2.69), are to be added to the second derivative in x and then substituted into the governing equation to obtain the final nodal equation.

2.7 DIFFERENCE APPROXIMATIONS IN THE CYLINDRICAL COORDINATE SYSTEM

Many scientific and engineering computations involve cylindrical shapes. In fact, outside of a rectangular prism (brick) shape, the cylindrical shape is probably the second most commonly encountered regular shape in engineering applications. Furthermore, in many applications, angular symmetry is valid, making the computations so-called 2D axisymmetric. In this section, we will develop finite difference approximations to the Poisson equation in cylindrical coordinates using the same principles discussed earlier but with an emphasis on issues that are unique to the cylindrical coordinate system. The cylindrical coordinate system has three independent variables: $r, \theta,$ and z, as shown in Fig. 2.9. The three coordinates are generally referred to as the radial, azimuthal (or simply, angular), and axial coordinates, respectively.

In the cylindrical coordinate system, the Poisson equation, which has been our governing equation thus far, is written as

$$\frac{1}{r}\frac{\partial}{\partial r}\left(r\frac{\partial \phi}{\partial r}\right) + \frac{1}{r^2}\frac{\partial^2 \phi}{\partial \theta^2} + \frac{\partial^2 \phi}{\partial z^2} = S_\phi. \tag{2.71}$$

As in the case of Cartesian coordinates, one can formulate 1D, 2D, and 3D problems in the cylindrical coordinate system. Formulating a 1D problem in the

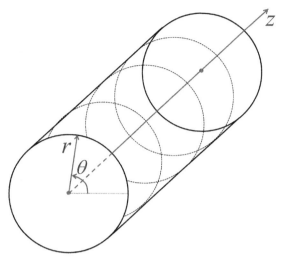

FIGURE 2.9 The Cylindrical Coordinate System

z direction is no different than formulating a 1D problem in the Cartesian coordinate system, i.e., Eq. (2.71) would reduce to Eq. (2.1). Such a scenario has been discussed at length already and requires no further discussion. A 1D problem in the θ direction is rarely encountered for practical applications. The most common scenario for which a 1D problem may be formulated is one where the solution in the radial direction is sought. For example, trying to determine the temperature distribution inside a long nuclear fuel rod would require solution of the 1D heat conduction equation in radial coordinates. Therefore, we start our discussion with an ordinary differential equation in radial coordinates, written as

$$\frac{1}{r}\frac{d}{dr}\left(r\frac{d\phi}{dr}\right) = \frac{d^2\phi}{dr^2} + \frac{1}{r}\frac{d\phi}{dr} = S_\phi. \tag{2.72}$$

The second part of Eq. (2.72) was obtained by expanding the derivative of the product of two terms. The nodal system in this particular case is shown in Fig. 2.10.

In order to derive finite difference approximations for the first and second derivatives, we perform two Taylor series expansions as before – one forward and one backward – as follows:

$$\phi_{i+1} = \phi_i + (\Delta r)\left.\frac{d\phi}{dr}\right|_i + \frac{(\Delta r)^2}{2!}\left.\frac{d^2\phi}{dr^2}\right|_i + \frac{(\Delta r)^3}{3!}\left.\frac{d^3\phi}{dr^3}\right|_i + \frac{(\Delta r)^4}{4!}\left.\frac{d^4\phi}{dr^4}\right|_i + \cdots, \tag{2.73a}$$

$$\phi_{i-1} = \phi_i - (\Delta r)\left.\frac{d\phi}{dr}\right|_i + \frac{(\Delta r)^2}{2!}\left.\frac{d^2\phi}{dr^2}\right|_i - \frac{(\Delta r)^3}{3!}\left.\frac{d^3\phi}{dr^3}\right|_i + \frac{(\Delta r)^4}{4!}\left.\frac{d^4\phi}{dr^4}\right|_i + \cdots, \tag{2.73b}$$

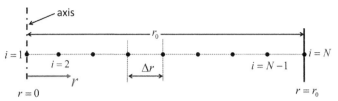

FIGURE 2.10 Arrangement of Nodes in a 1D Computational Domain in the Radial Direction of a Cylindrical Geometry

where $\Delta r = r_0 / (N-1)$. In order to derive an approximation for the second derivative, as before, Eqs (2.73a) and (2.73b) are added to yield

$$\left.\frac{d^2\phi}{dr^2}\right|_i = \frac{\phi_{i+1} + \phi_{i-1} - 2\phi_i}{(\Delta r)^2} - \frac{(\Delta r)^2}{12}\left.\frac{d^4\phi}{dr^4}\right|_i + \dots \tag{2.74}$$

The first derivative, on the other hand, is derived by subtracting Eq. (2.73b) from Eq. (2.73a) to yield

$$\left.\frac{d\phi}{dr}\right|_i = \frac{\phi_{i+1} - \phi_{i-1}}{2\Delta r} - \frac{(\Delta r)^2}{6}\left.\frac{d^3\phi}{dr^3}\right|_i + \dots \tag{2.75}$$

It is evident from Eq. (2.75) that, as in the case of the second derivative, the difference approximation for the first derivative is also second-order accurate. Substituting Eq. (2.74) and Eq. (2.75) into Eq. (2.72) yields the necessary finite difference equation for the interior nodes:

$$\frac{\phi_{i+1} + \phi_{i-1} - 2\phi_i}{(\Delta r)^2} + \frac{1}{r_i}\frac{\phi_{i+1} - \phi_{i-1}}{2\Delta r} = S_i \tag{2.76}$$

As far as boundary conditions are concerned, any of the three boundary conditions discussed in Section 2.3 may be applied at $r = r_0$. The procedure for the treatment of the boundary conditions is almost identical to the Cartesian coordinate system. The extra first derivative term is easy to accommodate because it is directly specified in the case of the Neumann boundary condition and can be expressed directly in terms of the nodal value in the case of the Robin boundary condition. The second boundary condition must be applied at the axis, i.e., at $r = 0$. Since $r = 0$ represents the axis of symmetry, the symmetry boundary condition is appropriate at that location and is written as

$$\left.\frac{d\phi}{dr}\right|_{r=0} = 0. \tag{2.77}$$

As discussed in Section 2.3.2, this boundary condition may be applied by itself, or it may be substituted into the governing equation, and the governing equation also satisfied in a limiting sense at the boundary. It was shown in Section 2.3.2 that the latter approach yields significantly better accuracy, especially if the grid is coarse. Therefore, we aim to follow the latter strategy here. An attempt to satisfy the governing equation at the boundary reveals a difficulty: the first derivative containing term, $(1/r)(d\phi/dr)$, becomes undefined ($=0/0$) at $r = 0$. As discussed in Section 2.3.2, strictly speaking, the governing equation is not valid at the boundary and must be satisfied at the boundary only in a limiting sense [cf. Eq. (2.3)]. Thus, instead of substituting $r = 0$ in $(1/r)(d\phi/dr)$, we take the limit of this term as r approaches zero. To do so, we apply L'Hospital's rule, as follows:

$$\lim_{r \to 0} \frac{1}{r}\frac{d\phi}{dr} = \lim_{r \to 0} \frac{\frac{d}{dr}\left(\frac{d\phi}{dr}\right)}{\frac{d}{dr}(r)} = \left.\frac{d^2\phi}{dr^2}\right|_{r=0}. \tag{2.78}$$

Substitution of Eq. (2.78) into Eq. (2.72) yields the following governing equation at the axis of the cylinder, i.e., at $r = 0$:

$$2\frac{d^2\phi}{dr^2} = S_\phi. \tag{2.79}$$

In order to derive a nodal equation from Eq. (2.79) for the leftmost node ($i = 1$), the same procedure described in Section 2.3.2 may be used.

A scenario often encountered in engineering analysis is the so-called 2D axisymmetric scenario – one where there is no azimuthal (or angular) variation of the dependent variable and only radial and axial variations. Flow through a pipe is the most common example of angular symmetry because no-slip (zero velocity) conditions prevail around the circumference of the pipe. To extend the 1D radial calculations to 2D axisymmetric calculations, one only needs to include the second derivative with respect to z in addition. Since this term is identical to any of the derivatives in the Cartesian coordinate system, it can be treated in exactly the same manner, as discussed in the preceding sections.

2.8 COORDINATE TRANSFORMATION TO CURVILINEAR COORDINATES

Until this point, the derivations of finite difference approximations have been limited to a regular Cartesian mesh. As shown in Section 2.5, if the nodes are not placed along the Cartesian grid lines, Taylor series expansions result in nonzero cross-derivative containing terms. Hence, derivation of finite difference approximations becomes quite complicated and tedious. To the best of this author's knowledge, body-fitted curvilinear meshes have their early roots in computational fluid dynamics, particularly for aerospace applications involving external flow. This happened at a time when researchers realized that the only accurate way to compute velocity distributions and, subsequently, drag and lift on an airfoil, is to construct a mesh that aligned with the profile of the airfoil. Such ideas were driven by conformal maps (angle preserving coordinate transformations) introduced by the Russian aerodynamicist Zhukovsky (transliterated into Joukowsky in the western scientific literature in the English language) for the analysis of airfoils, in which he transformed a circle into the fish-like shape of an airfoil. Since the early 1980s, the use of body-fitted structured meshes has seen tremendous growth and still continues to find prolific usage in numerical solution of PDEs, although it is slowly giving way to unstructured meshes, especially in the context of the finite volume method.

The basic idea behind a coordinate transformation is to transform the governing differential equation written in Cartesian coordinates to a new coordinate system. This new coordinate system may have axes that are either straight lines or curves. For the general case of curved axes, the new coordinate is known as a *curvilinear coordinate system*. When these curved axes align exactly with the outer contour of the computational domain (or body), the resulting mesh is called a *body-fitted mesh* or *body-fitted grid*. The main difference between the Cartesian coordinate system and a curvilinear coordinate system is that in the Cartesian coordinate system the unit vectors (or basis vectors) are global, while in a curvilinear system, the basis vectors are local (locally tangential to the curve) and are known as covariant basis vectors. Figure 2.11 shows a Cartesian coordinate system (x_1, x_2, x_3) and a general curvilinear coordinate system (ξ_1, ξ_2, ξ_3). In the case of the Cartesian coordinate system, the unit vectors $\hat{x}_1$, $\hat{x}_2$, and $\hat{x}_3$ are global, while in the case of the curvilinear coordinate system, the unit vectors $\hat{\xi}_1$, $\hat{\xi}_2$, and $\hat{\xi}_3$ are local.

In general, the transformation from (x_1, x_2, x_3) to (ξ_1, ξ_2, ξ_3) is the so-called forward transformation and may be written as

$$
\begin{aligned}
x_1 &= x_1(\xi_1, \xi_2, \xi_3) \\
x_2 &= x_2(\xi_1, \xi_2, \xi_3). \\
x_3 &= x_3(\xi_1, \xi_2, \xi_3)
\end{aligned}
\tag{2.80}
$$

In most cases, the forward transformation can be written in explicit functional form. The opposite transformation, the so-called backward transformation, cannot usually be written in explicit form. It usually appears as an implicit relationship. As an

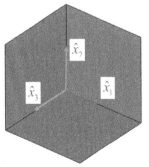

FIGURE 2.11 The Cartesian Coordinate System (Left) and a General Curvilinear Coordinate System (Right)

example, if we transform from Cartesian to spherical coordinates, we can write the following relationships: $x = r \sin\theta \cos\psi$, $y = r \sin\theta \sin\psi$, and $z = r \cos\theta$, i.e., the forward transformation relationships are explicit. However, the backward transformation cannot be written in such explicit form. This is an important issue, as we shall see shortly.

Since PDEs contain derivatives, our first task is to express each partial derivative in the Cartesian coordinate system in terms of partial derivatives in the curvilinear coordinate system. Using the chain rule of partial derivatives, we may write

$$
\begin{aligned}
\frac{\partial \phi}{\partial x_1} &= \frac{\partial \phi}{\partial \xi_1} \frac{\partial \xi_1}{\partial x_1} + \frac{\partial \phi}{\partial \xi_2} \frac{\partial \xi_2}{\partial x_1} + \frac{\partial \phi}{\partial \xi_3} \frac{\partial \xi_3}{\partial x_1} \\
\frac{\partial \phi}{\partial x_2} &= \frac{\partial \phi}{\partial \xi_1} \frac{\partial \xi_1}{\partial x_2} + \frac{\partial \phi}{\partial \xi_2} \frac{\partial \xi_2}{\partial x_2} + \frac{\partial \phi}{\partial \xi_3} \frac{\partial \xi_3}{\partial x_2} , \\
\frac{\partial \phi}{\partial x_3} &= \frac{\partial \phi}{\partial \xi_1} \frac{\partial \xi_1}{\partial x_3} + \frac{\partial \phi}{\partial \xi_2} \frac{\partial \xi_2}{\partial x_3} + \frac{\partial \phi}{\partial \xi_3} \frac{\partial \xi_3}{\partial x_3}
\end{aligned}
\tag{2.81}
$$

which may be written in matrix form as

$$
\begin{bmatrix} \dfrac{\partial \phi}{\partial x_1} \\[2mm] \dfrac{\partial \phi}{\partial x_2} \\[2mm] \dfrac{\partial \phi}{\partial x_3} \end{bmatrix} = \begin{bmatrix} \dfrac{\partial \xi_1}{\partial x_1} & \dfrac{\partial \xi_2}{\partial x_1} & \dfrac{\partial \xi_3}{\partial x_1} \\[2mm] \dfrac{\partial \xi_1}{\partial x_2} & \dfrac{\partial \xi_2}{\partial x_2} & \dfrac{\partial \xi_3}{\partial x_2} \\[2mm] \dfrac{\partial \xi_1}{\partial x_3} & \dfrac{\partial \xi_2}{\partial x_3} & \dfrac{\partial \xi_3}{\partial x_3} \end{bmatrix} \begin{bmatrix} \dfrac{\partial \phi}{\partial \xi_1} \\[2mm] \dfrac{\partial \phi}{\partial \xi_2} \\[2mm] \dfrac{\partial \phi}{\partial \xi_3} \end{bmatrix}.
\tag{2.82}
$$

Equation (2.82) expresses the components of the gradient of the scalar ϕ in the Cartesian coordinate system in terms of the components of the gradient in the curvilinear coordinate system. However, the use of Eq. (2.82) requires partial derivatives of the backward transformation. Since the backward transformation is not known in explicit form, it is not always possible to determine the partial derivatives $\partial \xi_1 / \partial x_1$, $\partial \xi_2 / \partial x_1$, and so on. The 3×3 matrix containing the partial derivatives of the backward transformation is known as the *Jacobian* of the backward transformation. To determine this matrix, one has to make use of the forward transformation relationships that are available in explicit form. This is described next.

In general, for a 3D system, the forward transformation may be written as $x_1 = x_1(\xi_1, \xi_2, \xi_3)$, $x_2 = x_2(\xi_1, \xi_2, \xi_3)$, and $x_3 = x_3(\xi_1, \xi_2, \xi_3)$. Therefore, we may write

$$
dx_1 = \frac{\partial x_1}{\partial \xi_1} d\xi_1 + \frac{\partial x_1}{\partial \xi_2} d\xi_2 + \frac{\partial x_1}{\partial \xi_3} d\xi_3 = \frac{\partial x_1}{\partial \xi_i} d\xi_i,
\tag{2.83a}
$$

$$
dx_2 = \frac{\partial x_2}{\partial \xi_1} d\xi_1 + \frac{\partial x_2}{\partial \xi_2} d\xi_2 + \frac{\partial x_2}{\partial \xi_3} d\xi_3 = \frac{\partial x_2}{\partial \xi_i} d\xi_i,
\tag{2.83b}
$$

$$
dx_3 = \frac{\partial x_3}{\partial \xi_1} d\xi_1 + \frac{\partial x_3}{\partial \xi_2} d\xi_2 + \frac{\partial x_3}{\partial \xi_3} d\xi_3 = \frac{\partial x_3}{\partial \xi_i} d\xi_i.
\tag{2.83c}
$$

In the last part of Eq. (2.83), Cartesian tensor notations have been used for the sake of compactness. Tensor notations are discussed in detail in Appendix C. Similarly, the backward transformation may be written as $\xi_1 = \xi_1(x_1, x_2, x_3)$, $\xi_2 = \xi_2(x_1, x_2, x_3)$, and $\xi_3 = \xi_3(x_1, x_2, x_3)$, although, as stated earlier, this transformation is not expressible in explicit form. Nonetheless, using this general description, we may write

$$d\xi_1 = \frac{\partial \xi_1}{\partial x_1} dx_1 + \frac{\partial \xi_1}{\partial x_2} dx_2 + \frac{\partial \xi_1}{\partial x_3} dx_3 = \frac{\partial \xi_1}{\partial x_j} dx_j, \tag{2.84a}$$

$$d\xi_2 = \frac{\partial \xi_2}{\partial x_1} dx_1 + \frac{\partial \xi_2}{\partial x_2} dx_2 + \frac{\partial \xi_2}{\partial x_3} dx_3 = \frac{\partial \xi_2}{\partial x_j} dx_j, \tag{2.84b}$$

$$d\xi_3 = \frac{\partial \xi_3}{\partial x_1} dx_1 + \frac{\partial \xi_3}{\partial x_2} dx_2 + \frac{\partial \xi_3}{\partial x_3} dx_3 = \frac{\partial \xi_3}{\partial x_j} dx_j. \tag{2.84c}$$

Substituting Eq. (2.84) into Eq. (2.83), we obtain

$$dx_1 = \frac{\partial x_1}{\partial \xi_1}\left(\frac{\partial \xi_1}{\partial x_j} dx_j\right) + \frac{\partial x_1}{\partial \xi_2}\left(\frac{\partial \xi_2}{\partial x_j} dx_j\right) + \frac{\partial x_1}{\partial \xi_3}\left(\frac{\partial \xi_3}{\partial x_j} dx_j\right) = \frac{\partial x_1}{\partial \xi_i}\left(\frac{\partial \xi_i}{\partial x_j} dx_j\right), \tag{2.85a}$$

$$dx_2 = \frac{\partial x_2}{\partial \xi_1}\left(\frac{\partial \xi_1}{\partial x_j} dx_j\right) + \frac{\partial x_2}{\partial \xi_2}\left(\frac{\partial \xi_2}{\partial x_j} dx_j\right) + \frac{\partial x_2}{\partial \xi_3}\left(\frac{\partial \xi_3}{\partial x_j} dx_j\right) = \frac{\partial x_2}{\partial \xi_i}\left(\frac{\partial \xi_i}{\partial x_j} dx_j\right), \tag{2.85b}$$

$$dx_3 = \frac{\partial x_3}{\partial \xi_1}\left(\frac{\partial \xi_1}{\partial x_j} dx_j\right) + \frac{\partial x_3}{\partial \xi_2}\left(\frac{\partial \xi_2}{\partial x_j} dx_j\right) + \frac{\partial x_3}{\partial \xi_3}\left(\frac{\partial \xi_3}{\partial x_j} dx_j\right) = \frac{\partial x_3}{\partial \xi_i}\left(\frac{\partial \xi_i}{\partial x_j} dx_j\right). \tag{2.85c}$$

Grouping terms containing dx_1, dx_2, and dx_3 separately on the right-hand side of Eq. (2.85) yields

$$dx_1 = \left(\frac{\partial x_1}{\partial \xi_i}\frac{\partial \xi_i}{\partial x_1}\right) dx_1 + \left(\frac{\partial x_1}{\partial \xi_i}\frac{\partial \xi_i}{\partial x_2}\right) dx_2 + \left(\frac{\partial x_1}{\partial \xi_i}\frac{\partial \xi_i}{\partial x_3}\right) dx_3, \tag{2.86a}$$

$$dx_2 = \left(\frac{\partial x_2}{\partial \xi_i}\frac{\partial \xi_i}{\partial x_1}\right) dx_1 + \left(\frac{\partial x_2}{\partial \xi_i}\frac{\partial \xi_i}{\partial x_2}\right) dx_2 + \left(\frac{\partial x_2}{\partial \xi_i}\frac{\partial \xi_i}{\partial x_3}\right) dx_3, \tag{2.86b}$$

$$dx_3 = \left(\frac{\partial x_3}{\partial \xi_i}\frac{\partial \xi_i}{\partial x_1}\right) dx_1 + \left(\frac{\partial x_3}{\partial \xi_i}\frac{\partial \xi_i}{\partial x_2}\right) dx_2 + \left(\frac{\partial x_3}{\partial \xi_i}\frac{\partial \xi_i}{\partial x_3}\right) dx_3. \tag{2.86c}$$

Since x_1, x_2, and x_3 are independent variables the following relationships are true:

$$\frac{\partial x_1}{\partial \xi_i}\frac{\partial \xi_i}{\partial x_1} = 1; \frac{\partial x_1}{\partial \xi_i}\frac{\partial \xi_i}{\partial x_2} = 0; \frac{\partial x_1}{\partial \xi_i}\frac{\partial \xi_i}{\partial x_3} = 0, \tag{2.87a}$$

$$\frac{\partial x_2}{\partial \xi_i}\frac{\partial \xi_i}{\partial x_1} = 0; \frac{\partial x_2}{\partial \xi_i}\frac{\partial \xi_i}{\partial x_2} = 1; \frac{\partial x_2}{\partial \xi_i}\frac{\partial \xi_i}{\partial x_3} = 0,$$ (2.87b)

$$\frac{\partial x_3}{\partial \xi_i}\frac{\partial \xi_i}{\partial x_1} = 0; \frac{\partial x_3}{\partial \xi_i}\frac{\partial \xi_i}{\partial x_2} = 0; \frac{\partial x_3}{\partial \xi_i}\frac{\partial \xi_i}{\partial x_3} = 1.$$ (2.87c)

Equations (2.87) may be written in matrix form as

$$\begin{bmatrix} \dfrac{\partial x_1}{\partial \xi_1} & \dfrac{\partial x_1}{\partial \xi_2} & \dfrac{\partial x_1}{\partial \xi_3} \\ \dfrac{\partial x_2}{\partial \xi_1} & \dfrac{\partial x_2}{\partial \xi_2} & \dfrac{\partial x_2}{\partial \xi_3} \\ \dfrac{\partial x_3}{\partial \xi_1} & \dfrac{\partial x_3}{\partial \xi_2} & \dfrac{\partial x_3}{\partial \xi_3} \end{bmatrix} \begin{bmatrix} \dfrac{\partial \xi_1}{\partial x_1} & \dfrac{\partial \xi_1}{\partial x_2} & \dfrac{\partial \xi_1}{\partial x_3} \\ \dfrac{\partial \xi_2}{\partial x_1} & \dfrac{\partial \xi_2}{\partial x_2} & \dfrac{\partial \xi_2}{\partial x_3} \\ \dfrac{\partial \xi_3}{\partial x_1} & \dfrac{\partial \xi_3}{\partial x_2} & \dfrac{\partial \xi_3}{\partial x_3} \end{bmatrix} = [I],$$ (2.88)

where $[I]$ is the identity matrix. The first matrix on the left-hand side of Eq. (2.88) is the *Jacobian of the forward transformation*, while the second matrix is the *Jacobian of the backward transformation*, which is the matrix we had set out to determine. From Eq. (2.88) it is evident that the Jacobian of the backward transformation is the inverse of the Jacobian of the forward transformation. Denoting the Jacobian of the forward transformation by $[J]$, Eq. (2.82) may be written as

$$\begin{bmatrix} \dfrac{\partial \phi}{\partial x_1} \\ \dfrac{\partial \phi}{\partial x_2} \\ \dfrac{\partial \phi}{\partial x_3} \end{bmatrix} = \left[[J]^{-1} \right]^T \begin{bmatrix} \dfrac{\partial \phi}{\partial \xi_1} \\ \dfrac{\partial \phi}{\partial \xi_2} \\ \dfrac{\partial \phi}{\partial \xi_3} \end{bmatrix}.$$ (2.89)

Since $[J]$ is at most a 3×3 matrix, its inverse can be determined easily using Cramer's rule, and subsequently, Eq. (2.89) may be written as

$$\frac{\partial \phi}{\partial x_i} = \frac{\beta_{ij}}{J}\frac{\partial \phi}{\partial \xi_j},$$ (2.90)

where β_{ij} are the cofactors of the matrix $[J]$, and J its determinant. An important point to note from Eq. (2.90) is that the right-hand side term, by virtue of the tensor summation, has derivatives in all three curvilinear coordinates. In other words, the derivative in the x_1 direction alone depends on the derivatives in all three coordinates in the curvilinear (transformed) coordinate system. It can be shown after some tedious algebra [see derivation of Eq. (C.12) in Appendix C] that the cofactors can be inserted inside the derivatives in Eq. (2.90) to yield

$$\frac{\partial \phi}{\partial x_i} = \frac{1}{J} \frac{\partial (\beta_{ij} \phi)}{\partial \xi_j}.$$

(2.91)

We will now use Eqs (2.90) and (2.91) to transform the Poisson equation to the curvilinear coordinate system. To do so, we begin by writing the Poisson equation in Cartesian tensor notations:

$$\nabla^2 \phi = \frac{\partial^2 \phi}{\partial x_1^2} + \frac{\partial^2 \phi}{\partial x_2^2} + \frac{\partial^2 \phi}{\partial x_3^2} = \frac{\partial}{\partial x_i} \left(\frac{\partial \phi}{\partial x_i} \right) = S_\phi.$$

(2.92)

Substituting Eq. (2.90) for the inner (first) derivative and Eq. (2.91) for the outer (second) derivative into Eq. (2.92), we obtain

$$\frac{1}{J} \frac{\partial}{\partial \xi_k} \left(\frac{\beta_{ik} \beta_{ij}}{J} \frac{\partial \phi}{\partial \xi_j} \right) = S_\phi,$$

(2.93)

where the quantity $\beta_{ik} \beta_{ij}$ represents a matrix–matrix multiplication (or tensor product) and is often written using a new symbol as

$$B_{jk} = \beta_{ik} \beta_{ij}.$$

(2.94)

Equation (2.94), when substituted into Eq. (2.93) and expanded out, may be written as

$$\frac{\partial}{\partial \xi_1} \left(\frac{B_{11}}{J} \frac{\partial \phi}{\partial \xi_1} + \frac{B_{21}}{J} \frac{\partial \phi}{\partial \xi_2} + \frac{B_{31}}{J} \frac{\partial \phi}{\partial \xi_3} \right) + \frac{\partial}{\partial \xi_2} \left(\frac{B_{12}}{J} \frac{\partial \phi}{\partial \xi_1} + \frac{B_{22}}{J} \frac{\partial \phi}{\partial \xi_2} + \frac{B_{32}}{J} \frac{\partial \phi}{\partial \xi_3} \right)$$
$$+ \frac{\partial}{\partial \xi_3} \left(\frac{B_{13}}{J} \frac{\partial \phi}{\partial \xi_1} + \frac{B_{23}}{J} \frac{\partial \phi}{\partial \xi_2} + \frac{B_{33}}{J} \frac{\partial \phi}{\partial \xi_3} \right) = JS_\phi.$$

(2.95)

It is clear from Eq. (2.95) that the governing equation in transformed coordinates will have cross-derivatives when the off-diagonal components of the [B] matrix are nonzero. This will happen when the ξ_1, ξ_2, and ξ_3 lines are not perpendicular to each other. As we will see shortly in an example, finding finite difference approximations to cross-derivatives is somewhat difficult. For this reason, many mesh generation algorithms make use of conformal maps or angle preserving transformations in which if one starts with an orthogonal mesh on a regular rectangle or brick and then transforms it to a nonregular shape, the curves still intersect perpendicular to each other, although the grid lines become curves. For further details on these issues, the reader is referred to texts on mesh generation and more advanced texts on computations using curvilinear grids [7,8].

In order to bring to light some of the important aspects of the transformed equation as well as to elucidate the details of finding the solution to a PDE on a nonregular geometry using the transformed equations, we next consider an example problem.

EXAMPLE 2.4

In this example, our objective is to find the solution of the Laplace equation on a rhombus shown in the following figure. The sides of the rhombus have lengths of one unit. For simplicity, the boundary conditions chosen are of the Dirichlet type, as shown in the figure. Equal grid spacing is used, and a typical stencil in transformed coordinates is shown for reference.

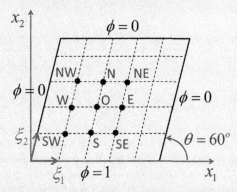

The forward transformation from (x_1, x_2) to (ξ_1, ξ_2) is given by

$$\begin{bmatrix} x_1 \\ x_2 \end{bmatrix} = \begin{bmatrix} \xi_1 + \xi_2 \cos\theta \\ \xi_2 \sin\theta \end{bmatrix}.$$

In this case, the Jacobian of the forward transformation is invariant with space and is written as

$$[J] = \begin{bmatrix} \dfrac{\partial x_1}{\partial \xi_1} & \dfrac{\partial x_1}{\partial \xi_2} \\ \dfrac{\partial x_2}{\partial \xi_1} & \dfrac{\partial x_2}{\partial \xi_2} \end{bmatrix} = \begin{bmatrix} 1 & \cos\theta \\ 0 & \sin\theta \end{bmatrix}$$

and the determinant of the Jacobian is $J = \det([J]) = \sin\theta$. Therefore, the co-factors of the Jacobian matrix are $\beta_{11} = \sin\theta, \beta_{12} = 0, \beta_{21} = -\cos\theta, \beta_{22} = 1$. Using Eq. (2.94), for 2D, we can write

$$B_{jk} = \beta_{ik}\beta_{ij} = \beta_{1k}\beta_{1j} + \beta_{2k}\beta_{2j}.$$

Therefore, the four components of the [B] matrix may be written as

$$B_{11} = \beta_{11}\beta_{11} + \beta_{21}\beta_{21} = \sin^2\theta + \cos^2\theta = 1$$
$$B_{21} = \beta_{11}\beta_{12} + \beta_{21}\beta_{22} = -\cos\theta$$
$$B_{12} = \beta_{12}\beta_{11} + \beta_{22}\beta_{21} = \cos\theta$$
$$B_{22} = \beta_{12}\beta_{12} + \beta_{22}\beta_{22} = 1$$

(Continued)

The governing equation in transformed coordinate, from Eq. (2.95), may be written as

$$\frac{\partial}{\partial \xi_1}\left(\frac{B_{11}}{J}\frac{\partial \phi}{\partial \xi_1} + \frac{B_{21}}{J}\frac{\partial \phi}{\partial \xi_2} \right) + \frac{\partial}{\partial \xi_2}\left(\frac{B_{12}}{J}\frac{\partial \phi}{\partial \xi_1} + \frac{B_{22}}{J}\frac{\partial \phi}{\partial \xi_2} \right) = 0.$$

Substituting the relationships derived above, we obtain

$$\frac{\partial}{\partial \xi_1}\left(\frac{1}{\sin \theta}\frac{\partial \phi}{\partial \xi_1} - \frac{\cos \theta}{\sin \theta}\frac{\partial \phi}{\partial \xi_2} \right) + \frac{\partial}{\partial \xi_2}\left(\frac{-\cos \theta}{\sin \theta}\frac{\partial \phi}{\partial \xi_1} + \frac{1}{\sin \theta}\frac{\partial \phi}{\partial \xi_2} \right) = 0, \text{ or}$$

$$\frac{\partial^2 \phi}{\partial \xi_1^2} - 2\cos \theta \frac{\partial^2 \phi}{\partial \xi_1 \partial \xi_2} + \frac{\partial^2 \phi}{\partial \xi_2^2} = 0.$$

As expected, because the grid lines in transformed coordinates are no longer perpendicular to each other, the transformation results in cross-derivatives. The derivatives along the coordinate directions may be approximated using the procedures described earlier for orthogonal grids. The cross-derivative may be approximated as follows:

$$\frac{\partial^2 \phi}{\partial \xi_1 \partial \xi_2} = \frac{\partial}{\partial \xi_1}\left(\frac{\partial \phi}{\partial \xi_2} \right) \approx \frac{\left(\left.\frac{\partial \phi}{\partial \xi_2}\right|_E - \left.\frac{\partial \phi}{\partial \xi_2}\right|_W \right)}{2\Delta \xi_1} \approx \frac{\left(\frac{\phi_{NE} - \phi_{SE}}{2\Delta \xi_2} - \frac{\phi_{NW} - \phi_{SW}}{2\Delta \xi_2} \right)}{2\Delta \xi_1}$$

$$= \frac{\left(\frac{\phi_{k+N+1} - \phi_{k-N+1}}{2\Delta \xi_2} - \frac{\phi_{k+N-1} - \phi_{k-N-1}}{2\Delta \xi_2} \right)}{2\Delta \xi_1}.$$

Based on these finite difference approximations, the following coefficients may be obtained for interior nodes:

$$A(k,k) = -\frac{2}{(\Delta \xi_1)^2} - \frac{2}{(\Delta \xi_2)^2}, A(k,k+1) = \frac{1}{(\Delta \xi_1)^2}, A(k,k-1) = \frac{1}{(\Delta \xi_1)^2},$$

$$A(k,k+N) = \frac{1}{(\Delta \xi_2)^2}, A(k,k-N) = \frac{1}{(\Delta \xi_2)^2}, A(k,k-N-1) = \frac{-2\cos \theta}{4(\Delta \xi_1)(\Delta \xi_2)},$$

$$A(k,k-N+1) = \frac{2\cos \theta}{4(\Delta \xi_1)(\Delta \xi_2)}, A(k,k+N+1) = \frac{-2\cos \theta}{4(\Delta \xi_1)(\Delta \xi_2)},$$

$$A(k,k+N-1) = \frac{2\cos \theta}{4(\Delta \xi_1)(\Delta \xi_2)}$$

The following figure shows a contour of the solution obtained on a 41 × 41 grid by solving the resulting finite difference equations subject to the Dirichlet boundary conditions shown in the previous figure. Although the error in the solution cannot be assessed in this case because a closed-form analytical solution to the problem is not available, it is clear that the solution is physically meaningful and demonstrates the power of coordinate transformation.

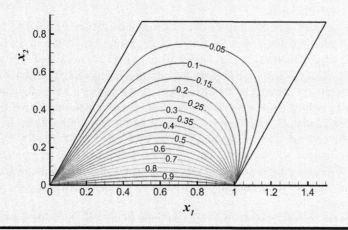

Example 2.4 demonstrates how to solve a PDE using transformed coordinates. One of the limitations of Example 2.4 was that the grid lines were not curved. As a result, the partial derivatives, also known as the *metrics* of the transformation, did not change from node to node. In the general case, when the metrics of transformation change from node to node, special care must be exercised in approximating derivatives. For example,

$$
\frac{\partial}{\partial \xi_1}\left(\frac{B_{11}}{J}\frac{\partial \phi}{\partial \xi_1}\right) \approx \frac{\left.\frac{B_{11}}{J}\frac{\partial \phi}{\partial \xi_1}\right|_{i+\frac{1}{2}} - \left.\frac{B_{11}}{J}\frac{\partial \phi}{\partial \xi_1}\right|_{i-\frac{1}{2}}}{\Delta \xi_1} = \frac{\left.\frac{B_{11}}{J}\right|_{i+\frac{1}{2}}\frac{\phi_E - \phi_O}{\Delta \xi_1} - \left.\frac{B_{11}}{J}\right|_{i-\frac{1}{2}}\frac{\phi_O - \phi_W}{\Delta \xi_1}}{\Delta \xi_1}
$$

$$
\approx \frac{\frac{1}{2}\left(\left.\frac{B_{11}}{J}\right|_i + \left.\frac{B_{11}}{J}\right|_{i+1}\right)\frac{\phi_E - \phi_O}{\Delta \xi_1} - \frac{1}{2}\left(\left.\frac{B_{11}}{J}\right|_i + \left.\frac{B_{11}}{J}\right|_{i-1}\right)\frac{\phi_O - \phi_W}{\Delta \xi_1}}{\Delta \xi_1}
$$

$$(2.96)$$

It is evident from Eq. (2.96) that the [B] matrix and the [J] matrix need to be computed locally at each node. This requires significant storage if computations on the fly are to be avoided. Also, as will become evident from the discussions in Chapter 3, the cross-derivative containing terms slow down convergence significantly when iterative solvers are used to solve the final set of algebraic equations, because they are usually treated explicitly. One final point to note is that the forward transformation is not always known. This is because this information is generally internal to the mesh generation software, and the developer of the PDE solver is often not privy to such information. In such a case, the metrics must be computed at each node using the finite difference procedure. For example, $(\partial x_1 / \partial \xi_1)|_O \approx (x_1|_E - x_1|_W)/2\Delta\xi_1$. In conclusion, although coordinate transformation enables us to solve PDEs in complex nonregular geometry, there is a penalty to be paid both in terms of memory usage and computational efficiency.

In this chapter, a detailed discussion of the finite difference method for solving differential equations was presented. The general procedure to derive finite difference approximations to derivatives of any order was outlined. The primary focus of this chapter was on boundary value problems and the treatment of spatial derivatives. Errors introduced by difference approximations were elucidated, and strategies to reduce them were discussed. This chapter also introduced the readers to three canonical boundary condition types that are encountered frequently in scientific computations of boundary value problems, i.e., Dirichlet, Neumann, and Robin boundary conditions. The mathematical procedure for application of the finite difference method to solving multidimensional problems in Cartesian, cylindrical, and arbitrary curvilinear coordinate systems was also presented and demonstrated.

One of the important issues not touched upon in this chapter is the issue of how the final linear algebraic equations, resulting from discretization of the governing PDE, are to be solved. In the examples shown, it was assumed that the direct matrix inversion procedure would be used. In the chapter to follow, the difficulties associated with direct solution of such linear systems are discussed, followed by an in-depth presentation of a large number of iterative solvers.

REFERENCES

[1] Courant R, Friedrichs KO, Lewy H. Über die partiellen Differenzengleichungen der Mathematischen Physik. Math. Ann. 1928;100:32–74. English translation, with commentaries by Lax PB, Widlund OB, Parter, SV, in IBM J Res Develop 11 (1967).

[2] Thomée V. From finite differences to finite elements: a short history of numerical analysis of partial differential equations. J Comp Appl Math 2001;128:1–54.

[3] Incropera FP, Dewitt DP, Bergman TL, Lavine AS. Introduction to heat transfer. 6th ed. New York: Wiley; 2011.

[4] Morinishi Y, Lund TS, Vasilyev OV, Moin P. Fully conservative higher order finite difference schemes for incompressible flow. J Comp Phys 1998;143:90–124.

[5] Visbal MR, Gaitonde DV. Very high-order spatially implicit schemes for computational acoustics on curvilinear meshes. J Comp Acous 2001;9(4):1259.

[6] Lele SK. Compact finite difference schemes with spectral-like resolution. J Comp Phys 1992;103:16–42.

[7] Liseikin VD. Grid generation methods. 2nd ed. Springer: Berlin; 2010.

[8] Thompson JF, Warsi ZUA, Mastin CW. Numerical grid generation: foundations and applications, North-Holland: Amsterdam; 1985.

EXERCISES

2.1 Derive a second order accurate expression for the second derivative using only forward Taylor series expansions. Also, derive an expression for the resulting truncation error. Comment on what implication the final result would have on the structure of the coefficient matrix if you were solving a Poisson equation in 2D.

2.2 In Section 2.6, a remark was made that for the node $i = 2$, using Taylor series expansions with nodes $i = 1, 3, 4$, and 5 would result in an expression for the second derivative that is only third order accurate. Show that this is indeed the case.

2.3 Consider the top left corner node shown in the following figure. For this node, derive a nodal equation that is first order accurate. The governing equation to be solved is the Laplace equation in 2D Cartesian coordinates on a regular orthogonal mesh.

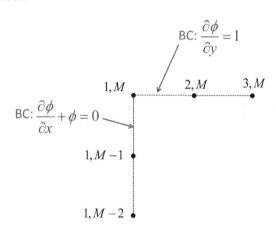

2.4 Consider the following second order ordinary differential equation and boundary conditions:

$$\frac{d^2\phi}{dx^2} = \exp(-x),$$
$$\phi(0) = 0, \phi(1) = 1$$

a. Solve the equation analytically to determine $\phi(x)$. Plot $\phi(x)$ versus x.

b. Discretize the equation using the finite-difference method. Use a second-order accurate central difference scheme. Use equal node spacing. Solve the resulting discrete equations using a standard matrix solver. Use

$N = 6, 11, 21$, and 41, where N is the number of nodes. For each mesh (value of N), plot the percentage error distribution (E vs. x) between the analytical and numerical solution, defined as

$$E(x) = \frac{\phi_{\text{analytical}}(x) - \phi_{\text{numerical}}(x)}{\phi_{\text{analytical}}(x)} \times 100.$$ Exclude boundary nodes in this plot.

c. Comment on your solutions and what you have learned from this exercise.

2.5 Consider the following ordinary differential equation and boundary conditions:

$$\frac{d^2\phi}{dx^2} = 2x - 1,$$
$$\phi(0) = 0, \phi(1) = 1$$

a. Solve the equation analytically to determine the solution for ϕ. Also determine the flux at the left and right ends, where the flux is given by

$$J_\phi = -\frac{d\phi}{dx}$$

b. Discretize the equation using the three-point central difference scheme. Use unequal grid spacing. The grid is to be generated using a power law, as explained below. If the length in the x direction is L (= 1, in this case), then $\sum_{i=1}^{N-1} \Delta x_i = L$, where N is the number of nodes, and Δx_i is the distance between the i-th node and the $(i + 1)$-th node. If we now assume that $\Delta x_{i+1} = s\Delta x_i$, where s is a stretching factor greater than 1, then we obtain the relation

$$\frac{s - s^N}{1 - s} = \frac{L}{\Delta x_0}$$

Thus, if s and N are specified, one can obtain all the Δx_i. For this exercise, use $s = 1.02$. Solve the resulting discrete equations for various values of N, starting from 11. Increase N to 21, 41, and 81. Plot the error between the numerical and analytical solution on a single graph for all four mesh sizes. For each value of N, also calculate and compare the flux at the right and left ends with each other and with those obtained analytically. Comment on your results.

2.6 Consider the following second order PDE and boundary conditions:

$$\frac{\partial^2\phi}{\partial x^2} + \frac{\partial^2\phi}{\partial y^2} = S_\phi = 50000 \cdot \exp[-50\{(1-x)^2 + y^2\}] \cdot [100\{(1-x)^2 + y^2\} - 2]$$

$$\phi(1, y) = 100(1 - y) + 500\exp(-50y^2)$$
$$\phi(0, y) = 500\exp(-50\{1 + y^2\})$$
$$\phi(x, 0) = 100x + 500\exp(-50(1 - x)^2)$$
$$\phi(x, 1) = 500\exp(-50\{(1 - x)^2 + 1\})$$

Note that this is a linear system of equations. The above system has an analytical solution, given by

$$\phi(x, y) = 500\exp(-50\{(1 - x)^2 + y^2\}) + 100x(1 - y)$$

a. Develop the finite-difference equations for the above PDE using a uniform mesh in both x and y directions, and write a computer program to solve the resulting algebraic equations using a standard off-the-shelf matrix solver. Use 21 grids points in each direction.

b. Make contour plots of the numerical solution and the absolute error between the analytical and numerical solutions.

c. Repeat Parts (a) and (b) with 41 grid points in each direction.

d. Comment on your results.

2.7 Repeat Example 2.4 for a parallelogram of length 10 units. The height is to remain the same as in Example 2.4. Use 101 nodes in the ξ_1 direction, and 11 nodes in the ξ_2 direction. Consider now the analytical solution to the corresponding 1D problem with the length of the parallelogram very large compared to its height. Compare the 2D numerical solution along the center ($\xi_1 = 5$) of the parallelogram to the analytical solution. Comment on your findings. Under what limiting condition will the two solutions match?

Solution to a System of Linear Algebraic Equations

3

In the preceding chapter, the finite difference method was introduced and subsequently used to develop linear algebraic equations for a discrete set of nodes. These linear algebraic equations were then assembled into matrix form and solved to obtain the final solution to the governing differential equation subject to a set of boundary conditions. One important issue that had been deferred to Chapter 3 is the method to be used to solve the final set of linear algebraic equations. The present chapter addresses the issue of how this set of linear algebraic equations is to be solved.

To appreciate why a chapter needs to be dedicated to the solution of the set of linear algebraic equations arising out of discretization of a differential equation, two key attributes of a typical matrix system, derived in Chapter 2, need to be brought to light: (a) the coefficient matrix arising out of discretization of a PDE is sparse and, often, banded, and (b) the coefficient matrix is large. As we shall see in the remainder of this chapter, these two attributes play a critical role in deciding the type of algorithm to be used in solving the linear system of equations arising out of discretization of a PDE. The chapter commences with a discussion of so-called *direct solvers*. This discussion establishes the rationale for using so-called *iterative solvers* to which the majority of the rest of the chapter is dedicated.

3.1 DIRECT SOLVERS

Direct solution to a set of linear algebraic equations is obtained by the method of substitution in which equations are successively substituted into other equations to reduce the number of unknowns until, finally, one unknown remains. If performed on a computer, the solution obtained by this method has no errors other than round-off errors. Hence, the solution obtained by a direct solver is often referred to as the *exact numerical solution*. It is termed "numerical" because the governing PDE still has to be discretized and solved numerically. It is termed "exact" because the algebraic equations resulting from discretization of the PDE are solved exactly. Depending on the structure of the coefficient matrix – whether it is full or sparse or banded – the number of substitutions needed may vary. In this subsection, we discuss two algorithms based on the general philosophy of substitution. The first of these methods, known as *Gaussian elimination* (or *Gauss–Jordan elimination*), is the most general method to solve a system of linear equations directly without any assumption with regard to the nature of the coefficient matrix, i.e., the method allows for the fact that the coefficient matrix may be full. Later in this section, we discuss a class of direct solvers in which the coefficient matrix is banded.

3.1.1 GAUSSIAN ELIMINATION

The process of solving a set of linear algebraic equations by Gaussian elimination involves two main steps: *forward elimination* and *backward substitution*. To understand each of these steps, let us consider a system of K linear algebraic equations of the general form shown in Eq. (2.35), which, in expanded form, may be written as

$$
\begin{aligned}
A_{1,1}\phi_1 &+ A_{1,2}\phi_2 &+ ... &+ A_{1,K}\phi_K &= Q_1 \\
A_{2,1}\phi_1 &+ A_{2,2}\phi_2 &+ ... &+ A_{2,K}\phi_K &= Q_2 \\
&\vdots & \vdots \\
A_{i,1}\phi_1 &+ A_{i,2}\phi_2 &+ ... &+ A_{i,K}\phi_K &= Q_i \\
&\vdots & \vdots & & \vdots \\
A_{K,1}\phi_1 &+ A_{K,2}\phi_2 &+ ... &+ A_{K,K}\phi_K &= Q_K
\end{aligned}
\tag{3.1}
$$

where $A_{i,j}$ are the elements of the coefficient matrix $[A]$, Q_i are the elements of the right-hand side vector $[Q]$, and ϕ_i are the unknowns. In the forward elimination step, we start from the first (topmost) equation, and express ϕ_i in terms of all the other ϕ's. This yields the following equation:

$$
\phi_1 = \frac{Q_1}{A_{1,1}} - \frac{A_{1,2}}{A_{1,1}}\phi_2 - \frac{A_{1,3}}{A_{1,1}}\phi_3 - ... - \frac{A_{1,K}}{A_{1,1}}\phi_K.
\tag{3.2}
$$

Next, we substitute Eq. (3.2) into each of the equations in Eq. (3.1) except the first (topmost) one. This, after some simplification, yields

$$
\begin{aligned}
A_{1,1}\phi_1 &+ & A_{1,2}\phi_2 &+ ... + A_{1,K}\phi_K & &= Q_1 \\
&\left(A_{2,2} - A_{1,2}\frac{A_{2,1}}{A_{1,1}}\right)\phi_2 &+ ... &+ \left(A_{2,K} - A_{1,K}\frac{A_{2,1}}{A_{1,1}}\right)\phi_K & &= Q_2 - \frac{A_{2,1}}{A_{1,1}}Q_1 \\
&\vdots & & \vdots & & \vdots \\
&\left(A_{i,2} - A_{1,2}\frac{A_{i,1}}{A_{1,1}}\right)\phi_2 &+ ... &+ \left(A_{i,K} - A_{1,K}\frac{A_{i,1}}{A_{1,1}}\right)\phi_K & &= Q_i - \frac{A_{i,1}}{A_{1,1}}Q_1 \\
&\vdots & & \vdots & & \vdots \\
&\left(A_{K,2} - A_{1,2}\frac{A_{K,1}}{A_{1,1}}\right)\phi_2 &+ ... &+ \left(A_{K,K} - A_{1,K}\frac{A_{K,1}}{A_{1,1}}\right)\phi_K & &= Q_K - \frac{A_{K,1}}{A_{1,1}}Q_1
\end{aligned}
\tag{3.3}
$$

The first equation, which was used to obtain Eq. (3.2), remains unchanged. It is known as the *pivot equation* in the steps shown above. Examination of Eq. (3.3) reveals that a certain pattern has emerged in the manner the original coefficients have changed. Essentially, the following transformation has occurred:

$$A_{i,j}\big|_{new} = A_{i,j}\big|_{old} - \left(\frac{A_{i,p}\big|_{old}}{A_{p,p}\big|_{old}}\right) A_{p,j}\big|_{old} \qquad \forall j = 1,2,...,K, \text{and } \forall i = p+1,...,K$$

$$Q_i\big|_{new} = Q_i\big|_{old} - \left(\frac{A_{i,p}\big|_{old}}{A_{p,p}\big|_{old}}\right) Q_p\big|_{old} \qquad \forall i = p+1,...,K,$$

(3.4)

where the subscripts "old" and "new" denote values of the elements before and after the substitution of Eq. (3.2) into Eq. (3.1). The process of forward elimination can be continued by sequential substitution. The subscript p denotes the pivot equation. So, when the substitution is done for the very first time, the denominator will have $A_{1,1}$, as evident in Eq. (3.3). In the next step, an expression for ϕ_2 needs to be derived from the second equation of Eq. (3.3) and substituted into the equations below it. When repeated $K-1$ times (it is not required for the last equation), the forward elimination step will have been completed, and the resulting matrix will assume the following form:

$$
\begin{aligned}
A_{1,1}\phi_1 + A_{1,2}\phi_2 + \quad &\cdots \quad +A_{1,K}\phi_K = Q_1 \\
A_{2,2}^*\phi_2 + \quad &\cdots \quad +A_{2,K}^*\phi_K = Q_2^* \\
&\vdots \\
A_{i,i}^*\phi_i + \cdots \quad &+A_{i,K}^*\phi_K = Q_i^* \\
&\ddots \\
A_{K,K}^*\phi_K &= Q_K^*
\end{aligned}
$$

(3.5)

where the superscript "*" denotes the fact that these are not the original coefficients, but modified coefficients – modified according to Eq. (3.4). In the second equation, the coefficients will have been modified only once, while in the last equation, they will have been modified $K-1$ times successively. The modified coefficient matrix has an upper triangular shape, and hence, is referred to as an *upper triangular matrix*. Its diagonal lower half only has zeroes, which are not shown in Eq. (3.5). A code snippet designed to execute the forward elimination algorithm is shown below.

CODE SNIPPET 3.1: FORWARD ELIMINATION ALGORITHM

```
For k = 1 : K-1  ! Number of times substitution is done
  For i = k+1 : K  ! Modify coefficients of all equations starting from pivot equation, k
    const = A(i,k)/A(k,k) ! This factor is independent of column number [see Eq. (3.4)]
    For j = k+1 : K ! Modify coefficients of all columns in a given row
      A(i,j) = A(i,j) + const*A(k,j)  ! Eq. (3.4)
    End
    A(i,k) = const
    Q(i) = Q(i) – const*Q(k)  ! Eq. (3.4)
  End
End
```

It is instructive at this juncture to count the number of long operations that would be needed to execute the forward elimination algorithm shown in Code Snippet 3.1. By long operations, we mean multiplication and division operations. These operations require substantially more computing time than addition or subtraction operations, and therefore, are of interest when estimating the computational efficiency of any algorithm. It is also assumed that a multiplication and a division require the same amount of time, although, in reality, a division is slightly more time consuming than a multiplication. The innermost (j) loop in Code Snippet 3.1 runs from $K+1$ to K, and there is one multiplication in it. This implies a total of $K-(K+1)+1=K-k$ multiplications within that loop. These many multiplications are repeated $K-(K+1)+1=K-k$ times within the second (i) loop, resulting in $(K-k)^2$ multiplications. In addition, there is a division operation that is performed within the i loop, bringing the total number of long operations within the second (i) loop to $(K-k)^2+(K-k)$. Finally, in the outermost loop, k spans 1 through $K-1$. Thus, the total number of long operations, n_{long}, becomes

$$n_{\text{long}} = \sum_{k=1}^{K-1}(K-k)^2 + (K-k) = \left[(K-1)^2 + (K-2)^2 + ...1\right] + \left[K-1+K-2+...+1\right].$$
(3.6)

Evaluating the two arithmetic series on the right-hand side of Eq. (3.6), we obtain

$$n_{\text{long}} = \left[\frac{1}{3}K^3 - \frac{1}{2}K^2 + \frac{1}{6}K + 1\right] + \left[\frac{1}{2}K^2 - \frac{1}{2}K\right] = \frac{1}{3}K^3 - \frac{1}{3}K + 1.$$
(3.7)

Since $K^3 \gg K$ for systems of equations of practical interest, it is fair to conclude that the number of long operations in the forward elimination step scales as K^3, i.e., $n_{\text{long}} \sim K^3$. The implications of this finding will be discussed after we have completed a discussion of the next phase of the Gaussian elimination process, namely the backward substitution phase.

In the backward substitution phase, the unknowns, ϕ_i, are obtained using Eq. (3.5) starting from the last (bottommost) equation. Thus,

$$\phi_K = \frac{Q_k^*}{A_{K,K}^*}.$$
(3.8)

Next, the ϕ_K, obtained using Eq. (3.8), is substituted into the second last equation of Eq. (3.5), and rearranged to obtain

$$\phi_{K-1} = \frac{Q_{K-1}^* - A_{K-1,K}^* \phi_K}{A_{K-1,K-1}^*}.$$
(3.9)

The process continues until all K unknowns have been determined. The backward substitution algorithm may be generalized by the following equation:

$$\phi_i = \frac{Q_i^* - \sum\limits_{j=i+1}^{K} A_{i,j}^* \phi_j}{A_{i,i}^*} \qquad \forall i = K-1, K-2, ..., 1. \tag{3.10}$$

A code snippet designed to execute the entire backward substitution algorithm, Eqs (3.8) and (3.10), is shown below.

CODE SNIPPET 3.2: BACKWARD SUBSTITUTION ALGORITHM

```
phi(K) = Q(K)/A(K,K)  ! Solve last equation [Eq. (3.8)]
For i = K-1 : 1 (with backward step)  ! Number of times substitution is done
    numerator = Q(i)   ! Initialize computation of numerator of Eq. (3.10)
    For j = i+1 : K  ! loop to compute summation
        numerator = numerator – A(i,j)*phi(j)
    End
    phi(i) = numerator/A(i,i)
End
```

In this case, the number of long operations may be expressed as

$$n_{\text{long}} = \sum_{i=1}^{K-1}\left[K-(i+1)+1\right] = \sum_{i=1}^{K-1} K - i = K(K-1) - \frac{1}{2}K(K-1) = \frac{1}{2}K^2 - \frac{1}{2}K. \tag{3.11}$$

For problems of practical interest, especially for 2D and 3D computational domains, the number of equations is likely to far exceed several hundreds. Consequently, we may assume that $K^3 \gg K^2 \gg K$. Thus, the number of long operations will be dominated by the forward elimination phase of the overall algorithm. The long operation count, and consequently, the computational time, is expected to roughly scale in a cubic fashion with the total number or unknowns (or equations). In summary, the number of long operations in the two phases of the Gaussian elimination algorithm scale as follows:

Forward elimination: $n_{\text{long}} \sim K^3$
Backward substitution: $n_{\text{long}} \sim K^2$

The fact that the number of long operations scales as K^3 implies that Gaussian elimination is prohibitive for solution of large systems. For example, if the number of equations is increased by a factor of 10, the computational time will increase by a factor of 1000. To illustrate this nonlinear (cubic) scaling of the Gaussian elimination algorithm's computational efficiency, we consider an example next.

EXAMPLE 3.1

In this example we consider solution of the Poisson equation, Eq. (2.41), in a square of unit length. The source term is assumed to be

$$S_\phi = 1000\left[2\sinh(x-\tfrac{1}{2})+4(x-\tfrac{1}{2})\cosh(x-\tfrac{1}{2})+(x-\tfrac{1}{2})^2\sinh(x-\tfrac{1}{2})\right]+$$
$$1000\left[2\sinh(y-\tfrac{1}{2})+4(y-\tfrac{1}{2})\cosh(y-\tfrac{1}{2})+(y-\tfrac{1}{2})^2\sinh(y-\tfrac{1}{2})\right].$$

The boundary conditions are as follows:

$$\phi(0,y)=1000\left[\tfrac{1}{4}\sinh(-\tfrac{1}{2})+(y-\tfrac{1}{2})^2\sinh(y-\tfrac{1}{2})\right],$$
$$\phi(1,y)=1000\left[\tfrac{1}{4}\sinh(\tfrac{1}{2})+(y-\tfrac{1}{2})^2\sinh(y-\tfrac{1}{2})\right],$$
$$\phi(x,0)=1000\left[(x-\tfrac{1}{2})^2\sinh(x-\tfrac{1}{2})+\tfrac{1}{4}\sinh(-\tfrac{1}{2})\right],$$
$$\phi(x,1)=1000\left[(x-\tfrac{1}{2})^2\sinh(x-\tfrac{1}{2})+\tfrac{1}{4}\sinh(\tfrac{1}{2})\right].$$

The analytical solution to the problem is given by $\phi(x,y)=1000$ $\left[(x-\tfrac{1}{2})^2\sinh(x-\tfrac{1}{2})+(y-\tfrac{1}{2})^2\sinh(y-\tfrac{1}{2})\right]$.

Equal mesh spacing is used in both directions. In this case, since we have Dirichlet boundary conditions on all boundaries, finite difference equations are needed only for the interior nodes, and these are given by Eq. (2.48). The resulting linear system is solved using Gaussian elimination on a 20×20 mesh. Contour plots of the numerical solution, $\phi(x, y)$, as well as the error between the analytical and numerical solutions, are shown in the figure below.

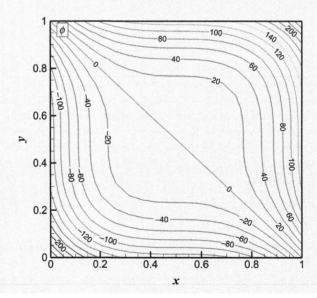

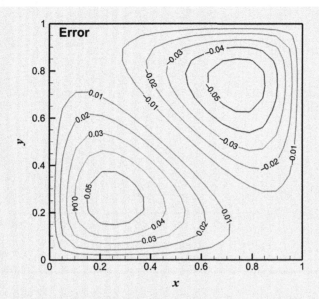

The results show that despite the strong gradients in the solution, the errors are quite small even for the relatively coarse mesh used in this case.

In order to assess the computational efficiency of the Gaussian elimination algorithm, the same computations were repeated for the following mesh sizes: 20, 30, 40, 50, 60, and 70 nodes in each direction, resulting in $K = 400, 900, 1600, 2500, 3600,$ and 4900, respectively. The plot below shows the CPU times required to complete the calculation for each case. The calculations were performed on an Intel core i7 2.2 GHz processor with 8GB of RAM.

The figure shows that, as expected, the CPU time scales nonlinearly with the number of equations. Whereas the solution on the 20×20 mesh requires less than 1 s to compute, the solution on the 70×70 mesh requires about 15 min. The exact scaling behavior is best understood by fitting a polynomial to the data, as shown by the dotted curve. The equation to this curve is given by CPU time $= 10^{-8} K^3 - 10^{-5} K^2 + 0.0068K - 1.5918$. This clearly indicates that the CPU time scales with all three powers of K, as derived theoretically in Eqs (3.7) and (3.11), with the cubic term dominating the nonlinear behavior.

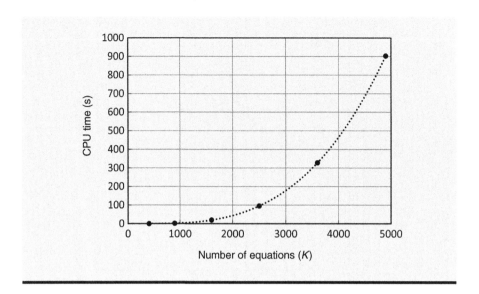

In summary, although Gaussian elimination is routinely used for the solution of a linear system of equations, it is prohibitively inefficient for the numerical solution of PDEs even on a relatively medium-sized mesh. The Gaussian elimination algorithm discussed above is also known as *naïve Gaussian elimination*. It is referred to as naïve because it disregards problems associated with floating point precision or inexact arithmetic. It would produce acceptable results on the computer only if the elements in a given row of the coefficient matrix are numbers that are within a reasonable range in terms of how many orders of magnitude they span. Typically, for double-precision calculations, several orders of magnitude variation is acceptable. If, on the other hand, one element is 20 orders of magnitude larger than another, arithmetic precision issues may come into play, and the algorithm could produce erroneous results. Such coefficient matrices are often referred to as *stiff*, and some reordering and manipulation of the system of equations is necessary prior to applying Gaussian elimination, resulting in so-called Gaussian elimination with *scaling* and *pivoting*. These issues, although important for numerical computations, are not discussed here any further because, as mentioned earlier, Gaussian elimination is not the method of choice for the task at hand.

Another important issue that pertains to the solution of the linear system arising out of discretization of a PDE is memory usage. Recalling that the coefficient matrix is of size $K \times K$, the memory necessary to store this matrix can quickly grow. For example, if a 3D computation with just 40 nodes in each direction were to be conducted, then $K = 40^3 = 64{,}000$. Thus, use of Gaussian elimination would require storage of a coefficient matrix of size $K^2 \approx 4 \times 10^9$, i.e., 4 billion real numbers. If double precision is used for each number, the memory required to store them would be 32 billion bytes, or 32 GB of RAM – a requirement that is clearly beyond the scope of a modern-day computer unless parallel computing is employed. Of course, computations of problems of practical interest would require far more than 64,000 nodes, implying that from a memory requirement standpoint also, Gaussian elimination is prohibitive for the solution of the relatively large linear systems arising out of discretization of PDEs.

3.1.2 BANDED LINEAR SYSTEM SOLVERS

As discussed in the preceding section, Gaussian elimination has two major shortcomings. Both shortcomings stem from the fact that the entire coefficient matrix, including all zeroes, is stored and used. As we have seen in Chapter 2, coefficient matrices, arising out of discretization of any PDE, are sparse. This is true irrespective of the type of mesh used, as will be corroborated further in Chapter 7. Unfortunately, a computer cannot distinguish between a zero and a nonzero automatically. It treats a zero as any other real number, and multiplications by zero are as time consuming as multiplications by nonzeroes. Therefore, a better approach would be to store only the nonzeroes and the locations in the matrix where they are situated. At the very least, this will reduce the memory usage significantly. In the example discussed in the last paragraph of the preceding section, for a 64,000-node mesh, the real number storage will reduce dramatically from $64,000^2$ to at most $64,000 \times 7$, assuming a hexahedral mesh is being used along with a second-order central difference scheme. Not storing the zeroes will also improve computational efficiency dramatically. This is because multiplications by zeroes will not be performed in the first place. For example, instead of performing a row times column multiplication involving $64,000^2$ long operations, such as in the computation of $\sum_{j=1}^{K} A_{i,j}\phi_j$, only $7 \times 64,000$ long operations will be performed since each row will have at most seven nonzero elements. To summarize, when it comes to solution of the linear system arising out of discretization of PDEs, using the sparseness of the coefficient matrix is warranted, as it can result in significant reduction in both memory usage and computational time.

A special class of sparse matrices arises out of discretization of a PDE if a structured mesh is used: banded matrices. This has already been alluded to in Chapter 2, although detailed discussion of this matter had been deferred to the present chapter. To elucidate the matrix structures represented by the equations derived in Chapter 2, let us consider the solution of Eq. (2.1) subject to the boundary conditions shown by Eq. (2.2). The resulting coefficient matrix, following Eqs (2.11) and (2.15), may be written as

$$
\begin{bmatrix}
1 & 0 & \cdots & & & & & & 0 \\
\frac{1}{(\Delta x)^2} & \frac{-2}{(\Delta x)^2} & \frac{1}{(\Delta x)^2} & 0 & \cdots & & & & 0 \\
& \ddots & \ddots & \ddots & & & & & \\
\cdots & 0 & \frac{1}{(\Delta x)^2} & \frac{-2}{(\Delta x)^2} & \frac{1}{(\Delta x)^2} & 0 & \cdots & & 0 \\
& & 0 & \frac{1}{(\Delta x)^2} & \frac{-2}{(\Delta x)^2} & \frac{1}{(\Delta x)^2} & 0 & & \\
& \cdots & & 0 & \frac{1}{(\Delta x)^2} & \frac{-2}{(\Delta x)^2} & \frac{1}{(\Delta x)^2} & 0 & \cdots \\
& & & & \ddots & \ddots & \ddots & & \\
0 & & & \cdots & & 0 & \frac{1}{(\Delta x)^2} & \frac{-2}{(\Delta x)^2} & \frac{1}{(\Delta x)^2} \\
0 & & & & & \cdots & & 0 & 1
\end{bmatrix}
\begin{bmatrix}
\phi_1 \\ \phi_2 \\ \vdots \\ \phi_{i-1} \\ \phi_i \\ \phi_{i+1} \\ \vdots \\ \phi_{N-1} \\ \phi_N
\end{bmatrix}
=
\begin{bmatrix}
\phi_L \\ S_2 \\ \vdots \\ S_{i-1} \\ S_i \\ S_{i+1} \\ \vdots \\ S_{N-1} \\ \phi_R
\end{bmatrix}.
$$

$$(3.12)$$

Clearly, the coefficient matrix is banded with three bands (or diagonals) clustered in the middle. This type of matrix is known as a *tridiagonal matrix*. The three diagonals are usually referred to as the *central diagonal*, the *superdiagonal* (the one on the top right), and the *subdiagonal* (the one on the bottom left). All other elements are zero, and need not be stored.

An algorithm for solution of a tridiagonal system can make use of the banded structure in order to simplify the elaborate Gaussian elimination process. First, instead of storing the full $[A]$ matrix, only the three diagonals are stored in the following form:

$$
\begin{bmatrix}
d_1 & c_1 & 0 & \cdots & & & & 0 \\
a_1 & d_2 & c_2 & 0 & \cdots & & & 0 \\
 & \ddots & \ddots & \ddots & & & & \\
\cdots & 0 & a_{i-2} & d_{i-1} & c_{i-1} & 0 & \cdots & 0 \\
 & \cdots & 0 & a_{i-1} & d_i & c_i & 0 & \cdots \\
 & & \cdots & 0 & a_i & d_{i+1} & c_{i+1} & 0 & \cdots \\
 & & & & \ddots & \ddots & \ddots & \\
0 & & & \cdots & 0 & a_{N-2} & d_{N-1} & c_{N-1} \\
0 & & & & \cdots & & a_{N-1} & d_N
\end{bmatrix}
\begin{bmatrix}
\phi_1 \\ \phi_2 \\ \vdots \\ \phi_{i-1} \\ \phi_i \\ \phi_{i+1} \\ \vdots \\ \phi_{N-1} \\ \phi_N
\end{bmatrix}
=
\begin{bmatrix}
Q_1 \\ Q_2 \\ \vdots \\ Q_{i-1} \\ Q_i \\ Q_{i+1} \\ \vdots \\ Q_{N-1} \\ Q_N
\end{bmatrix},
\quad (3.13)
$$

where the central diagonal, superdiagonal, and subdiagonals are denoted by d, c, and a, respectively. The diagonal is of size N, while the superdiagonals and subdiagonals are of size $N-1$. The solution to the general set of equations shown in Eq. (3.13) follows the same procedure as the Gaussian elimination algorithm, albeit a significantly simplified form. First, the topmost equation is rearranged to express ϕ_1 in terms of ϕ_2. The result is then substituted into all the other equations, and the process is repeated until the last equation is reached. This yields the following general relationships:

$$
\begin{aligned}
d_i\big|_{\text{new}} &= d_i\big|_{\text{old}} - \left(\frac{a_{i-1}}{d_{i-1}}\right) c_{i-1} \quad \forall i = 2,3,...,N \\[2mm]
Q_i\big|_{\text{new}} &= Q_i\big|_{\text{old}} - \left(\frac{a_{i-1}}{d_{i-1}}\right) Q_{i-1}\big|_{\text{old}} \quad \forall i = 2,3,...,N
\end{aligned}
\quad (3.14)
$$

Equation (3.14) essentially represents the forward elimination process. At the end of this process, the new coefficient matrix will assume an upper triangular shape with only two diagonals. The next step is to solve this system using backward substitution starting with the last equation. This results in the following general relationships

$$
\begin{aligned}
\phi_N &= \frac{Q_N^*}{d_N^*} \\[2mm]
\phi_i &= \frac{Q_i^* - c_i^* \phi_{i+1}}{d_i^*} \quad \forall i = N-1, N-2,...,1
\end{aligned}
\quad (3.15)
$$

where, as before, the superscript "*" indicates values of coefficients after repeated modifications using Eq. (3.14). The combined process of forward elimination and backward substitution is captured in the code snippet shown as follows.

CODE SNIPPET 3.3: TRIDIAGONAL MATRIX SOLVER

```
For i = 2 : N  ! Forward Elimination
    const = a(i-1)/D(i-1)
    d(i) = d(i) – const*c(i-1)  ! Eq. (3.14), diagonal
    Q(i) = Q(i) – const*Q(i-1)  ! Eq. (3.14), right hand side
End
phi(N) = Q(N)/d(N)  ! Solve last equation [Eq. (3.15)]
For i = N-1 : 1 (with backward step)  ! Backward substitution
    phi(i) = (Q(i)-c(i)*phi(i+1))/d(i)  ! Eq. (3.15)
End
```

The simplified Gaussian elimination algorithm for a tridiagonal matrix system is also commonly known as the *tridiagonal matrix algorithm* (TDMA) or the *Thomas algorithm*, named after Llewellyn Thomas.*

It is fairly straightforward to determine the number of long operations that would be required to execute the aforementioned algorithm. Code Snippet 3.3 clearly shows that in the forward elimination phase, there are $3(N-1)$ long operations, while in the backward substitution phase, there are $2(N-1)$ long operations, N being the number of equations solved in this case. Thus, the total number of long operations scales linearly as the number of unknowns (or equations), i.e., $n_{\text{long}} \sim K$ (if N is replaced by K for generality). Obviously, such linear scaling is ideal for large-scale scientific computations. In summary, by making use of the banded structure of the coefficient matrix, it is possible to reduce both memory usage and computational cost dramatically. Of course, tridiagonal matrices only arise in the case of 1D problems, and numerical computation of 1D problems is rarely of significant practical interest. However, we shall see in Section 3.2.4, that this 1D framework and the tridiagonal matrix solver, just described, can be conveniently used to develop low-memory and efficient solvers for multidimensional problems.

In Section 2.6, it was shown that even for a 1D problem, use of higher-order schemes extends the stencil beyond the three points used in a second-order central difference scheme. For example, use of the fourth-order central difference scheme results in the use of five nodes, [see Eq. (2.69)]. In such a scenario, the resulting matrix has five diagonals – two on each side of the central diagonal. The five diagonals (or bands), however, are all clustered in the middle. This is in contrast with the matrices arising out of 2D PDEs discretized on a structured mesh, in which case, we also get

---

*__Llewellyn Hilleth Thomas__ (1903–1992) was a British physicist and applied mathematician known famously for development of the Thomas algorithm and his contributions to quantum physics. He laid the foundation for what later became known as the density functional theory. He migrated to the United States in 1929, and held appointments at the Ohio State University, Columbia University, and North Carolina State University. He was elected into the National Academy of Science in 1958.

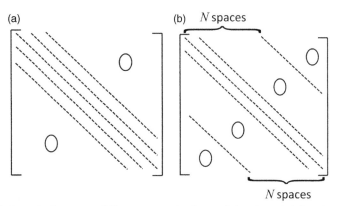

FIGURE 3.1 Generic Structure of Five-Banded Matrices Arising Out of Finite Difference Discretization of PDEs on Structured Grids

(a) 1D problem with fourth-order central difference, and (b) 2D problem with second-order central difference.

five diagonals [see Code Snippet 2.2], but two of the diagonals, corresponding to the nodes $k - N$ and $k+N$ reside N spaces away from the central diagonal. The difference between the two scenarios is depicted pictorially in Fig. 3.1, and is an important one. When the diagonals are clustered in the middle, the forward elimination phase of the Gaussian elimination algorithm retains only the five diagonals, and no nonzero elements are generated beyond the two upper diagonals. On the other hand, when the diagonals are not clustered, the entire upper triangle gets populated with nonzeroes once the forward elimination phase has been completed. This implies that in the latter case, simply allocating memory for the five diagonals is not sufficient. Due to this reason, the five-banded matrices arising out of discretization of a 2D PDE is generally not referred to as a pentadiagonal matrix. The term *pentadiagonal* is only reserved for matrices in which all five diagonals are clustered in the middle, as shown in Fig. 3.1(a).

Pentadiagonal matrix systems may be solved using the exact same procedure as tridiagonal systems. Let us consider the following general pentadiagonal system of equations:

$$
\begin{bmatrix}
d_1 & c_1 & f_1 & \cdots & & & & 0 \\
a_1 & d_2 & c_2 & f_2 & \cdots & & & 0 \\
e_1 & \ddots & \ddots & \ddots & \ddots & & & \\
\cdots & \ddots & & & & \cdots & & \\
& \cdots & e_{i-2} & a_{i-1} & d_i & c_i & f_i & \cdots \\
& & \cdots & & & \ddots & & \cdots \\
& & & & \ddots & \ddots & \ddots & f_{N-2} \\
0 & & & \cdots & & a_{N-2} & d_{N-1} & c_{N-1} \\
0 & & & & & e_{N-2} & a_{N-1} & d_N
\end{bmatrix}
\begin{bmatrix}
\phi_1 \\
\phi_2 \\
\vdots \\
\\
\phi_i \\
\\
\vdots \\
\phi_{N-1} \\
\phi_N
\end{bmatrix}
=
\begin{bmatrix}
Q_1 \\
Q_2 \\
\vdots \\
\\
Q_i \\
\\
\vdots \\
Q_{N-1} \\
Q_N
\end{bmatrix},
\quad (3.16)
$$

where the two new diagonals are denoted by *e* and *f*. The forward elimination and backward substitution formulae may be derived in a manner similar to what has been described earlier for tridiagonal systems. The following code snippet depicts the overall algorithm for solving a pentadiagonal system:

```
CODE SNIPPET 3.4:  PENTADIAGONAL MATRIX SOLVER

For i = 2 : N-1  ! Forward Elimination
    const1 = a(i-1)/d(i-1)  ! First substitution starts here
    d(i) = d(i) – const1*c(i-1)  ! diagonal
    c(i) = c(i) – const1*f(i-1)   ! super-diagonal
    Q(i) = Q(i) – const1*Q(i-1)  ! right hand side
    const2 = e(i-1)/d(i-1)  ! second substitution starts here
    a(i) = a(i) – const2*c(i-1) ! sub-diagonal
    d(i+1) = d(i+1) – const2*f(i-1) ! diagonal
    Q(i+1) = Q(i+1) – const2*b(i-1) ! right hand side
End
const3 = a(N-1)/d(N-1) ! Backward substitution starts here
d(N) = d(N) – const3*c(N-1)
phi(N) = (Q(N)-const3*Q(N-1))/d(N)
phi(N-1) = (Q(N-1)-c(N-1)*Q(N))/d(N-1)
For i = N-2 : 1 (with backward step)  ! Backward substitution
    phi(i) = (Q(i)-c(i)*phi(i+1)-f(i)*phi(i+2))/d(i)
End
```

It is straightforward to show from Code Snippet 3.4 that the number of long operations needed in the solution of the pentadiagonal system also scales linearly with the number of equations being solved. The pentadiagonal solver, just presented, is handy for developing solution algorithms to solve multidimensional PDEs when higher order schemes are used, as will be demonstrated in later sections of this chapter.

To summarize, in Section 3.1, solution algorithms for finding the exact numerical solution to a linear system of equations has been presented. Gaussian elimination, which is the most routine method for solving a linear system of equations was found to be wasteful and prohibitively time consuming for the solution of a large system of equations, as frequently encountered in the solution of multidimensional PDEs. The matrices arising out of discretization of PDEs are usually very sparse, and the sparseness must be capitalized upon to develop low-memory efficient solvers. Since structured meshes often result in banded coefficient matrices, two direct solvers specifically designed to obtain direct solution to banded systems, namely tridiagonal and pentadiagonal systems, were also presented. These banded system solvers were found to be significantly more efficient than the full-blown Gaussian elimination algorithm.

3.2 ITERATIVE SOLVERS

Direct solution to a linear system of equations produces the exact numerical solution to the PDE in question. Barring truncation and round-off errors, this solution has no other error when compared with the closed-form analytical solution to the same PDE. Therein lies its strength. Unfortunately, as shown in the preceding section, direct solution is prohibitive both from a memory usage and a computational efficiency standpoint. This leaves us with only one other alternative: solving the equations iteratively.

The philosophy of an iterative solution to a linear system is best explained by a simple example. Let us consider the following 3×3 system of equations:

$$5x + 2y + 2z = 9, \tag{3.17a}$$

$$2x - 6y + 3z = -1, \tag{3.17b}$$

$$x + 2y + 7z = 10. \tag{3.17c}$$

The exact solution to the system is [1, 1, 1]. The Gaussian elimination method, by which this exact solution is obtained, is also known as an *implicit* method. An implicit method of solution is one where all unknowns are treated simultaneously during the solution procedure.

The solution to the above system may also be obtained using the following iterative procedure:

1. Start with a guess for the solution.
2. Solve Eq. (3.17a) for x and update its value using old (guessed) values of y and z.
3. Solve Eq. (3.17b) for y and update its value using old (guessed) values of x and z.
4. Solve Eq. (3.17c) for z and update its value using old (guessed) values of x and y.
5. Replace old guess of solution by new solution.
6. Repeat steps 2–5 (i.e., iterate) until the solution has stopped changing beyond a certain number of significant digits, to be prescribed *a priori*.

In each of the steps 2 through 4, two of the three unknowns are guessed, or treated explicitly, i.e., they are treated as known quantities and transposed to the right-hand side of the equation, while the unknowns are retained on the left-hand side. The fact that some terms in the equation receive *explicit* treatment implies that at each step, the solution obtained is only approximate, and therefore, iterations are necessary to correct the approximation. It will be shown later that how quickly the correct solution is approached, i.e., how many iterations are needed to arrive at the correct solution, depends on how many terms are treated explicitly, or the so-called *degree of explicitness*. Table 3.1 shows how the values of [x, y, z] change when the above iteration procedure is executed. The initial guess used in this example calculation was [x, y, z] = [0, 0, 0]. It is clear that it takes 12 iterations before the solution is accurate up to 3 decimal places, and 24 iterations before it is accurate up to 6 decimal places (not shown).

Table 3.1 Results of the Iterative Solution of the System of Equations Shown in Eq. (3.17)

Iteration	x	y	z
1	1.800000	0.166667	1.428571
2	1.161905	1.480952	1.123810
3	0.758095	1.115873	0.839456
4	1.017868	0.839093	1.001451
5	1.063782	1.006682	1.043421
6	0.979959	1.042971	0.988979
7	0.987220	0.987809	0.990586
8	1.008642	0.991033	1.005309
9	1.001463	1.005535	1.001327
10	0.997255	1.001152	0.998209
11	1.000256	0.998190	1.000063
12	1.000699	1.000117	1.000481

The number of iterations required, of course, depends on the initial guess. The farther the initial guess is from the correct solution, the more iterations it will take. One important distinction between the solutions obtained by a direct solver (Gaussian elimination) versus an iterative solver is that in the latter case, the accuracy of the solution obtained is dependent upon when the iterations are stopped. In this particular example, if the iterations were terminated after 12 iterations, the solution would only be an approximate solution – one that is approximate in the 4th decimal place. The process of the solution approaching the exact numerical solution with successive iterations is termed *convergence*, and the error between the exact numerical solution and the partially converged solution is termed *convergence error*. In any iterative solver, convergence error will be always be present. The only way to eliminate convergence error is to continue the iteration until so-called machine accuracy has been reached. This means that the convergence error is comparable with the round-off error, and further iterations will simply make the solution oscillate. For example, in this particular case, if the solution is taken to machine accuracy, it will oscillate around the value of 1 in the 16th decimal place, since the present calculation used double precision to store all real numbers. In practice, taking the solution to machine accuracy is unnecessary, and the iterations are generally terminated before that. The criteria that may be used for termination of iterations will be discussed shortly. One final point to note is that convergence is not always monotonic. Successive iterations may not always lead to an answer that is closer to the correct answer. Depending on the type of equation being solved and the iteration scheme, which could be different from the one described earlier, the answers may move away from the correct solution before approaching it again. In some cases, with successive iterations, the answer may continually deviate from the correct answer. Such a scenario is known as *divergence*.

The preceding discussion alludes to the fact that convergence is not always guaranteed. Two factors contribute to convergence or divergence: (a) the type of coefficient matrix, and (b) the iterative scheme. To illustrate the fact that convergence is not always guaranteed, let us consider the solution to the same system as Eq. (3.17), except that we interchange the x and z coefficients in Eq. (3.17c), such that $7x + 2y + z = 10$ rather than $x + 2y + 7z = 10$. If the same iterative scheme, as described before, is now used to find a solution, the following is observed: the x values go from 1.8 in the first iteration to -64.32 in the 10th iteration, the y values go from 0.16 in the first iteration to 51.25 in the 10th iteration, and the z values go from 10 in the first iteration to 212.31 in the 10th iteration. In other words, the iteration diverges. Simply interchanging two coefficients in one of the three equations caused the iteration to diverge. Since the iteration scheme was unchanged, we can safely conclude that the small change in the coefficient matrix led to a change in a certain property of the coefficient matrix, which in turn led to divergence. In Chapter 4, we will study the exact mathematical property of the matrix that is responsible for this behavior. Regardless, without delving into any new theory, there is one observation we can make immediately with regard to the two matrices considered in the present example. In the original matrix, Eq. (3.17), we note that the following is true:

$$\left|A_{i,i}\right| \geq \sum_{\substack{j=1 \\ j \neq i}}^{K} \left|A_{i,j}\right| \qquad \forall i = 1, 2, ..., K, \tag{3.18a}$$

$$\left|A_{i,i}\right| > \sum_{\substack{j=1 \\ j \neq i}}^{K} \left|A_{i,j}\right| \qquad \text{for at least one } i. \tag{3.18b}$$

In other words, the absolute value of the diagonal is larger than or equal to the sum of the absolute value of the off-diagonals for all three equations and there is at least one equation where the absolute value of the diagonal is greater than the sum of the off-diagonals. In the modified system, on the other hand, the third equation does not satisfy Eq. (3.18a). The property of the coefficient matrix stated in Eq. (3.18) is known as *diagonal dominance*, and matrices which obey the criteria shown in Eq. (3.18) are known as diagonally dominant. Eq. (3.18) is also known as the *Scarborough criterion* [1]. The Scarborough criterion is not a necessary condition for obtaining convergence. Even if it is violated, convergence may still be attainable using an iterative solver. Conversely, if it is satisfied, convergence is guaranteed. As to why this is the case will become clear in Chapter 4.

3.2.1 THE RESIDUAL AND CORRECTION FORM OF EQUATIONS

In Chapter 2, we derived finite difference equations for the 2D Poisson equation subject to various types of boundary conditions. Irrespective of the type of boundary condition used, or the PDE that served as the starting point for the development of the finite difference equations, the finite difference equation, for a node k, can be written in general form as:

$$a_k\phi_k + \sum_{\substack{j=1 \\ j \neq k}}^{N_{nb,k}} a_j\phi_j = Q_k \qquad \forall k = 1, 2, ..., K, \tag{3.19}$$

where, as before, K denotes the total number of nodes. $N_{nb,k}$ denotes the total number of neighboring nodes to node k. The quantity a_j denotes the factor that premultiplies ϕ_j in the finite difference equation. As we have seen before, these premultipliers are, in fact, the same as the elements of the $[A]$ matrix if the entire set of equations were written in matrix form. Henceforth, we will refer to these premultipliers as *link coefficients* because they represent the link or interconnection between nodes. For the five-band system represented by Eq. (2.48), for example, Eq. (3.19) may be written as

$$\begin{aligned} a_k\phi_k + a_{k+1}\phi_{k+1} + a_{k-1}\phi_{k-1} + a_{k+N}\phi_{k+N} + a_{k-N}\phi_{k-N} &= Q_k \\ = a_O\phi_O + a_E\phi_E + a_W\phi_W + a_N\phi_N + a_S\phi_S &= Q_O \end{aligned}, \tag{3.20}$$

where two alternative forms of the same equation are shown – one using the index k, and the other using the directional naming convention. It is easy to show, by comparing Eq. (3.20) with Eq. (2.48), that the following relationships hold:

$$\begin{aligned} a_k = a_O &= -\left(\frac{2}{(\Delta x)^2} + \frac{2}{(\Delta y)^2} \right), \\ a_{k+1} = a_E &= \frac{1}{(\Delta x)^2}, \\ a_{k-1} = a_W &= \frac{1}{(\Delta x)^2}, \\ a_{k+N} = a_N &= \frac{1}{(\Delta y)^2}, \\ a_{k-N} = a_S &= \frac{1}{(\Delta y)^2}, \\ Q_k = Q_O &= S_k = S_O. \end{aligned} \tag{3.21}$$

It is worth noting that Eq. (3.20) has only five terms, and is often referred to as the five-band form of the finite difference equation, for reasons discussed in the preceding section. This five-band form will be discussed extensively in later sections when iterative sparse matrix solvers, specifically designed for banded matrices, are introduced. For now, we revert back to the general form of the discrete algebraic equation shown in Eq. (3.19).

As discussed earlier, in an iterative solver, iterations are performed to update the values of ϕ at all nodes. Let the value of ϕ at the n-th iteration be denoted by $\phi^{(n)}$. This solution will not necessarily satisfy the set of equations given by Eq. (3.19). In other words, an imbalance will result between the left- and right-hand sides of Eq. (3.19). This imbalance is known as the *residual*, and may be written as

$$R_k^{(n)} = Q_k - a_k \phi_k^{(n)} - \sum_{\substack{j=1 \\ j \neq k}}^{N_{nb,k}} a_j \phi_j^{(n)} \quad \forall k = 1, 2, ..., K, \tag{3.22}$$

where $R^{(n)}$ is the so-called *residual vector* at the n-th iteration. It is, in fact, a column matrix or a hyperdimensional vector. The residual is a measure of nonconvergence. If the entire residual vector is zero, it implies that each of the equations that we set out to solve have been satisfied exactly. Even if one of the elements of the residual vector is nonzero, the system of equations has not been satisfied.

Let us now consider a case, where starting from the n-th iteration, we are trying to find the solution at the $(n+1)$-th iteration. Let us also assume that Eq. (3.19) is satisfied exactly at the $(n+1)$-th iteration, or, at least, that is the goal. Therefore, for the $(n+1)$-th iteration, we may write

$$a_k \phi_k^{(n+1)} + \sum_{\substack{j=1 \\ j \neq k}}^{N_{nb,k}} a_j \phi_j^{(n+1)} = Q_k. \tag{3.23}$$

Henceforth, we will refer to the $(n+1)$-th and n-th iterations as the current and previous iterations, respectively. Let the change (or correction) in the value of ϕ from the previous to the current iteration be denoted by ϕ'. Therefore, it follows that

$$\phi^{(n+1)} = \phi^{(n)} + \phi'. \tag{3.24}$$

Substitution of Eq. (3.24) into Eq. (3.23) yields

$$a_k \left(\phi_k^{(n)} + \phi_k' \right) + \sum_{\substack{j=1 \\ j \neq k}}^{N_{nb,k}} a_j \left(\phi_j^{(n)} + \phi_j' \right) = Q_k. \tag{3.25}$$

Rearranging, we get

$$a_k \phi_k' + \sum_{\substack{j=1 \\ j \neq k}}^{N_{nb,k}} a_j \phi_j' = Q_k - a_k \phi_k^{(n)} - \sum_{\substack{j=1 \\ j \neq k}}^{N_{nb,k}} a_j \phi_j^{(n)}. \tag{3.26}$$

Substituting Eq. (3.22) into (3.26), we obtain

$$a_k \phi_k' + \sum_{\substack{j=1 \\ j \neq k}}^{N_{nb,k}} a_j \phi_j' = R_k^{(n)}. \tag{3.27}$$

Equation (3.27) is known as the *correction form* of the algebraic equation. It bears a strong resemblance, at least structurally, to the original algebraic equation, Eq. (3.19). Only, the ϕ in the left-hand side has been replaced by ϕ', and the right-hand side vector has been replaced by the residual vector. There is, however, an important attribute of Eq. (3.27) that Eq. (3.19) does not possess. In Eq. (3.19), at convergence, the left-hand side becomes equal to the right-hand side. In Eq. (3.27), in addition to the left and right-hand sides becoming equal, at convergence, each term is zero, since ϕ' is equal to zero at all nodes at convergence, as is the residual vector. In Chapter 4, we will discuss strategies to artificially improve diagonal dominance by making use of this special attribute of the discrete equations in correction form.

When the residual vector is zero, the implication is that the algebraic equations have been satisfied exactly. Therefore, the residual can be used to monitor convergence. Unfortunately, examining the residual at every node to see if it has gone below a certain prescribed small tolerance is computationally very expensive since it would involve logical ("IF") checks within the innermost loops. These logical checks, when used inside the innermost loops, slow down the execution of any code considerably due to so-called branch mispredictions. Noting that the logical checks have to be performed for each node and at each iteration, makes checking of the residual of individual nodes prohibitive. A better approach is to check the accumulated residual over all nodes. However, simply summing the residuals of all nodes (equations) will not serve the purpose. This is because the residual at some nodes may be positive, while being negative at other nodes. Consequently, summation may lead to a zero net (accumulated) residual, and a false indication that the equations have been satisfied, when, in fact, they have not been satisfied, as evident from the fact that the individual residuals are not zero.

There are two approaches that are commonly used to compute accumulated residuals, namely the L^1Norm and the L^2Norm. They are defined as follows:

$$L^1 Norm : R1 = \sum_{k=1}^{K} |R_k|, \tag{3.28a}$$

$$L^2 Norm : R2 = \sqrt{\sum_{k=1}^{K} (R_k)^2}. \tag{3.28b}$$

The L^2Norm, also known as the *Euclidean norm*, can be easily identified as the square-root of the *inner product* or *scalar product* of the residual vector, and is often written as $R2 = \sqrt{[R]^T[R]}$. Of the two methods to monitor convergence, it is the more commonly used, and will also be adopted for this text. It is worth noting that the accumulated residual, as defined by Eq. (3.28b), will not be zero even if only one of the elements of the residual vector is nonzero. For example, if the residual vector is [0,0,0,1] for a 4-node calculation, then the value of the L^2Norm is 1. Using the aforementioned definition of the accumulated residual, we now define *convergence* as the state when the following *criterion* has been satisfied:

$$R2 < \varepsilon_{tol}, \tag{3.29}$$

where ε_{tol} is the prescribed tolerance below which the L^2Norm must reach. As to what numerical value should be prescribed for the tolerance depends on the problem at hand, and also on how much accuracy is desired. For example, if ϕ represents temperature in a heat transfer calculation, then typical values of ϕ may range between

100–1000 Kelvin. If the solution domain is 1 m long, then, with 101 grid points, the grid spacing, Δx, will be 0.01 m. Thus, the order of magnitude of each term in the algebraic equation will be $\sim T/(\Delta x)^2 \sim 10^7$. In this case, specifying a tolerance, ε_{tol}, equal to 10^{-3} would imply reduction in the residual by approximately 10 orders of magnitude, which would produce temperatures that are accurate approximately to the 7th decimal place. On the other hand, if the computations were performed using nondimensional equations in which nondimensional temperature and length both varied between 0 and 1, then, an individual term in the equation would be of order of magnitude $\sim 10^4$. In such a case, specifying a tolerance, ε_{tol}, equal to 10^{-3} would imply reduction in the residual by approximately 7 orders of magnitude, and nondimensional temperatures accurate in the 7th decimal place or dimensional temperatures accurate in the 4th decimal place. Of course, the latter case would require fewer iterations since the residuals are forced to decrease by 7 orders of magnitude, as opposed to 10. However, it would also yield poorer accuracy. A final point to note is that the magnitude of the residual is also dependent on grid spacing and the number of nodes.

In order to remove problem dependency on the choice of the prescribed tolerance, it is preferable to monitor the normalized residual rather than the raw residual. The normalized residual is computed as follows:

$$R2^* = \frac{R2}{R2_{max}}, \tag{3.30}$$

where $R2_{max}$ is the maximum value of $R2$ up to the current iteration. If the prescribed tolerance is now applied to $R2^*$ rather than $R2$, the convergence criterion becomes problem independent. For example, in the scenario discussed in the preceding paragraph, if $\varepsilon_{tol} = 10^{-6}$, in both cases, the residual would be reduced by 6 orders of magnitude, and the corresponding dimensional temperatures would be accurate in the 3rd decimal place. On account of its generality, the scaled or normalized residual method is used routinely for monitoring convergence in general-purpose codes for solving linear systems.

There are other approaches to monitoring convergence that do not make use of the residual. After deriving Eq. (3.27), we noted that the change (or correction), ϕ', also goes to zero at convergence. Therefore, one could compute the L^2Norm of the ϕ' vector rather than the residual vector and monitor convergence based on it. However, this is not a desirable approach to monitoring convergence. It is easy to encounter a situation, where, due to a programming error, ϕ' may have been computed as zero at some node (often referred to as stalling of convergence), whereas, in reality, the nodal equation has not been satisfied. This will trigger a false message that convergence has been reached. However, the solution, thus produced, will be incorrect because the algebraic equations have not been satisfied in the first place. It is important to realize that the only goal of the iterative solver is to satisfy the algebraic equations that we have set out to solve, and therefore, any method for monitoring convergence that does not enforce that goal is not foolproof.

As mentioned earlier, Gaussian elimination has the highest degree of implicitness because none of the unknowns are assumed, and the solution is obtained simultaneously without any iteration or after 1 iteration if an iteration is thought of as one full step through the algorithm. This is the best-case scenario in terms of the number of iterations required to solve a system of equations. Unfortunately, even though the solution is obtained only in 1 iteration, the work per iteration is extremely high, as discussed earlier. When developing an iterative solver, the objective is always to obtain the correct solution with as few iterations as possible and with as little work per iteration as possible. Clearly, Gaussian elimination provides the clue that the more implicit the iteration procedure, the fewer the number of iterations will be. With this philosophy in mind, several iterative schemes have been developed over the past century. In the next sections, we discuss several such popular iterative solvers and examine their performance relative to each other.

3.2.2 JACOBI METHOD

The iterative scheme described earlier for the 3×3 system, Eq. (3.17), is in fact known as the Jacobi method. It is named after the German mathematician Carl Jacobi.** In this section, we will discuss the Jacobi method in the context of solution of the equations that arise out of finite difference discretization of a multidimensional PDE. As discussed in the preceding section, the discrete equations arising out of discretization of a multidimensional PDE may be written in the general form shown by Eq. (3.19). In the Jacobi method, this equation is first rewritten such that all terms except the diagonal term are transposed to the right-hand side of the equation. Next, the entire equation is divided by the diagonal coefficient to yield the so-called Jacobi update formula:

$$\phi_k^{(n+1)} = \frac{Q_k - \sum_{\substack{j=1 \\ j\neq k}}^{N_{nb,k}} a_j \phi_j^{(n)}}{a_k}. \tag{3.31}$$

One of the salient features of the Jacobi update process is that all terms on the right-hand side of the formula are computed using values of ϕ at the previous iteration, as indicated by the use of $\phi^{(n)}$ on the right-hand side of Eq. (3.31). In other words, all off-diagonal terms, i.e., the contributions of all neighboring nodes, are treated explicitly, as shown schematically in Fig. 3.2 for the solution of a 2D problem on a structured mesh. Hence, the Jacobi method has the least possible degree of implicitness, and consequently, its convergence is expected to be slow.

The Jacobi method is a so-called *point-wise iteration* method because the solution is updated sequentially node by node or point by point. Since the right-hand side of the update formula uses only previous iteration values, the pattern used to sweep through the nodes in the computational domain is not relevant. All sweeping patterns will yield the same convergence behavior, and may be cited as one of its strengths. Since the iteration is point-wise, the method can be applied to any mesh topology,

---

Carl Gustav Jacob Jacobi (1804–1851) was a German mathematician. He is best known for his development of the theory of elliptic functions, and their applications to solving inverse problems involving periodicity. He also made significant contributions to the theory of differential equations and number theory.

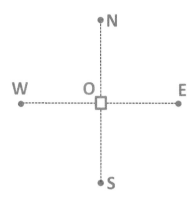

FIGURE 3.2 Pictorial Representation of Explicitness Versus Implicitness in the Jacobi Method

The nodes denoted by solid circles are treated explicitly while the node denoted by a hollow square is treated implicitly.

both structured and unstructured. There is absolutely no reliance on the structure of the matrix. This may be cited as another one of its strengths. A final point to note is that the current and previous iteration values of ϕ must be stored separately and not mixed. The algorithm to use the Jacobi method for solution of the set of linear algebraic equations arising out discretization of a 2D PDE, for example, Eq. (2.48), is presented below.

ALGORITHM: JACOBI METHOD

Step 1: Guess values of ϕ at all nodes, i.e., $\phi_{i,j}$ $\forall i = 1,...,N$ and $\forall j = 1,...,M$. We denote these values as $\phi^{(0)}$. If any of the boundaries have Dirichlet boundary conditions, the guessed values for the boundary nodes corresponding to that boundary must be equal to the prescribed boundary values.

Step 2: Apply the Jacobi update formula, Eq. (3.31). For the interior nodes, this yields

$$\phi_{i,j}^{(n+1)} = \frac{S_{i,j} - a_E\phi_{i+1,j}^{(n)} - a_W\phi_{i-1,j}^{(n)} - a_N\phi_{i,j+1}^{(n)} - a_S\phi_{i,j-1}^{(n)}}{a_O},$$ where the link coefficients

are given by Eq. (3.21). For Dirichlet boundaries, replace $\phi_{i,j}^{(n+1)}$ by $\phi_{i,j}^{(n)}$. For other types of boundary conditions, appropriate values of the link coefficients must be derived from the nodal equation at that boundary, and an update formula must be used.

Step 3: Compute the residual vector using $\phi^{(n+1)}$, and then compute $R2^{(n+1)}$.

Step 4: Monitor convergence, i.e., check if $R2^{(n+1)} < \varepsilon_{\text{tol}}$? If YES, then go to Step 7. If NO, then go to Step 5.

Step 5: Replace old guess by new value: $\phi^{(n)} = \phi^{(n+1)}$.

Step 6: Go to Step 2.

Step 7: Stop iteration and postprocess the results.

In the above algorithm, every iteration requires 10 long operations per node – 5 in the update formula, and 5 in the computation of the residual. For K nodes, this amounts to $10K$ long operations. In addition, there are approximately K long operations in the computation of $R2$, and one square-root operation. Assuming that the square-root operation counts as 2 long operations, the total number of long operations is approximately $11K + 2$ per iteration. Let us now consider a 2D problem, such as in Example 3.1, computed on a 80×80 mesh ($K = 6400$). For this particular case, the total number of long operations per iteration would be $11 \times 6400 + 2 = 70,402$. The total number of long operations, if Gaussian elimination were to be used, would be approximately $K^3/3 = 8.73 \times 10^{10}$. Thus, as long as the Jacobi method requires less than $8.73 \times 10^{10}/70,402 = 1.24 \times 10^6$ iterations, it would be computationally superior to Gaussian elimination. In the section to follow, an example will be considered, and it will be shown that, typically, only a few thousand iterations are necessary rather than a few million, thereby making the Jacobi method far superior to Gaussian elimination from a computational efficiency standpoint even though it is the least implicit of all methods.

3.2.3 GAUSS–SEIDEL METHOD

The Gauss–Seidel method is also a point-wise iteration method and bears a strong resemblance to the Jacobi method, but with one notable exception. In the Gauss–Seidel method, instead of always using previous iteration values for all terms of the right-hand side of Eq. (3.31), whenever an updated value becomes available, it is immediately used. Thus, for the 3×3 example system considered earlier [Eq. (3.17)] when x is determined using Eq. (3.17a), both y and z assume previous iteration values. However, when y is determined using Eq. (3.17b), only z assumes a previous iteration value. For x, the most recent value, which happens to be the current iteration value (since it has already been updated), is used. In the context of solution of a 2D PDE on a structured mesh, if the node by node update pattern (or sweeping pattern) is from left to right and bottom to top, as shown in Fig. 3.3(a), then, by the time it is node O's turn to get updated, nodes W and S have already been updated, and these updated values must be used. Essentially, this implies that only two out of the four terms on the right-hand side of the update formula are treated explicitly, as shown in Fig. 3.3(b). In general, the update formula for the Gauss–Seidel method may be written as

$$\phi_k^{(n+1)} = \frac{Q_k - \sum_{\substack{j=1 \\ j \neq k}}^{N_{nbu,k}} a_j \phi_j^{(n+1)} - \sum_{\substack{j=1 \\ j \neq k}}^{N_{nb,k}-N_{nbu,k}} a_j \phi_j^{(n)}}{a_k}, \tag{3.32}$$

where $N_{nbu,k}$ denotes the number of neighboring nodes to node k that have already been updated, and $N_{nb,k} - N_{nbu,k}$ is the number of neighboring nodes to node k that have not been updated and are treated explicitly. It is clear from the preceding discussion and Fig. 3.3(b) that the Gauss–Seidel scheme has a higher degree of

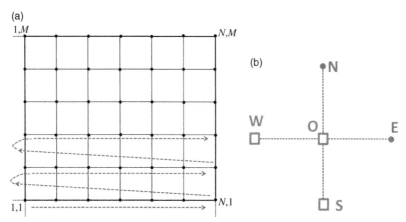

FIGURE 3.3 Pictorial Representation of the Gauss-Seidel Scheme

(a) left-to-right and bottom-to-top sweeping pattern in the Gauss–Seidel method, and (b) explicitness versus implicitness with the sweeping pattern shown in (a). The nodes denoted by solid circles are treated explicitly while nodes denoted by hollow squares are treated implicitly.

implicitness than the Jacobi method, and is, therefore, expected to yield faster convergence. However, the added implicitness would manifest itself only if the sweeping pattern is strictly adhered to, whatever that might be. Otherwise, the convergence behavior may revert back to that of the Jacobi method. The algorithm to use the Gauss–Seidel method for solution of the set of linear algebraic equations arising out discretization of the 2D Poisson equation [Eq. (2.48)] is presented below.

ALGORITHM: GAUSS–SEIDEL METHOD

Step 1: Guess values of ϕ at all nodes, i.e., $\phi_{i,j}$ $\forall i = 1,...,N$ and $\forall j = 1,...,M$. We denote these values as $\phi^{(0)}$. If any of the boundaries have Dirichlet boundary conditions, the guessed values for the boundary nodes corresponding to that boundary must be equal to the prescribed boundary values.

Step 2: Set $\phi^{(n+1)} = \phi^{(n)}$ and apply the Gauss–Seidel update formula, Eq. (3.32). For the interior nodes, this yields $\phi_{i,j}^{(n+1)} = \dfrac{S_{i,j} - a_E\phi_{i+1,j}^{(n)} - a_W\phi_{i-1,j}^{(n+1)} - a_N\phi_{i,j+1}^{(n)} - a_S\phi_{i,j-1}^{(n+1)}}{a_O}$,

where the link coefficients are given by Eq. (3.21). For boundary conditions other than the Dirichlet type, appropriate values of the link coefficients must be derived from the nodal equation at that boundary, and an update formula must be used.

Step 3: Compute the residual vector using $\phi^{(n+1)}$, and then compute $R2^{(n+1)}$.

Step 4: Monitor convergence, i.e., check if $R2^{(n+1)} < \varepsilon_{tol}$? If YES, then go to Step 5. If NO, then go to Step 2.

Step 5: Stop iteration and postprocess the results.

As opposed to the Jacobi method, in the Gauss–Seidel method, it is not necessary to store values of ϕ at both previous and current iterations. The same array may store a mixture of old and new values. As a matter of fact, in the update formula, it is not necessary to distinguish between old and new values. Within the same array, old values will be automatically replaced by new values as soon as they become available, and subsequently used in the update formula. Code Snippet 3.5 for the Gauss–Seidel scheme highlights some of these issues.

CODE SNIPPET 3.5: GAUSS–SEIDEL ALGORITHM

```
phi(:,:) = 0  ! Initial Guess
phi(1,:) = phi_left  ! Boundary Conditions. Dirichlet BCs used here as example.
phi(N,:) = phi_right
phi(:,1) = phi_bottom
phi(:,M) = phi_top
For n = 1 : Maximum_iterations  ! Start Iteration loop
  For i = 2 : N-1
    For j = 2 : M-1
      phi(i,j) = (S(i,j)-aE*phi(i+1,j)-aW*phi(i-1,j)-aN*phi(i,j+1)-aS*phi(i,j-1))/aO  ! Gauss-Seidel Update
    End
  End
  R2 = 0  ! Initialize Residual
  For i = 2 : N-1
    For j = 2 : M-1
      R(i,j) = S(i,j)-aE*phi(i+1,j) -aW*phi(i-1,j)-aN*phi(i,j+1) -aS*phi(i,j-1)-aO*phi(i,j)  ! Residual at each node
      R2 = R2 + R(i,j)*R(i,j)
    End
  End
  R2 = sqrt(R2)
  IF (R2 < tolerance) BREAK  ! Exit from loop if convergence is reached
End  ! Next iteration
```

The fact that the same array can be used in the Gauss–Seidel method to store both previous and current iteration values is an additional advantage of the Gauss–Seidel method over the Jacobi method. As in the case of the Jacobi method, the Gauss–Seidel method, being a point-wise iterative method, can be used for both structured and unstructured meshes. The number of long operations in the Gauss–Seidel method is identical to that of the Jacobi method. To highlight the differences, especially in convergence behavior, between the Jacobi and the Gauss–Seidel method, a numerical example is considered next.

EXAMPLE 3.2

In this example we consider solution of the Poisson equation, Eq. (2.41), in a square of unit length. The source term is assumed to be

$$S_\phi = 2\sinh[10(x-\tfrac{1}{2})] + 40(x-\tfrac{1}{2})\cosh[10(x-\tfrac{1}{2})] + 100(x-\tfrac{1}{2})^2\sinh[10(x-\tfrac{1}{2})] +$$
$$2\sinh[10(y-\tfrac{1}{2})] + 40(y-\tfrac{1}{2})\cosh[10(y-\tfrac{1}{2})] + 100(y-\tfrac{1}{2})^2\sinh[10(y-\tfrac{1}{2})] + \;.$$
$$4(x^2+y^2)\exp(2xy)$$

The boundary conditions are as follows:

$$\begin{aligned}
\phi(0,y) &= \tfrac{1}{4}\sinh(-5) + (y-\tfrac{1}{2})^2\sinh[10(y-\tfrac{1}{2})] + 1,\\
\phi(1,y) &= \tfrac{1}{4}\sinh(5) + (y-\tfrac{1}{2})^2\sinh[10(y-\tfrac{1}{2})] + \exp(2y)\\
\phi(x,0) &= \tfrac{1}{4}\sinh(-5) + (x-\tfrac{1}{2})^2\sinh[10(x-\tfrac{1}{2})] + 1\\
\phi(x,1) &= \tfrac{1}{4}\sinh(5) + (x-\tfrac{1}{2})^2\sinh[10(x-\tfrac{1}{2})] + \exp(2x).
\end{aligned}$$

The analytical solution to system is given by

$$\phi(x,y) = (x-\tfrac{1}{2})^2\sinh[10(x-\tfrac{1}{2})] + (y-\tfrac{1}{2})^2\sinh[10(y-\tfrac{1}{2})] + \exp(2xy).$$

Equal mesh spacing is used in both directions. In this case, since we have Dirichlet boundary conditions on all boundaries, finite difference equations are needed only for the interior nodes, and these are given by Eq. (2.48). The resulting linear system is solved using both the Jacobi and Gauss–Seidel methods for various mesh sizes: 41×41, 81×81, and 161×161. An initial guess equal to 0 was used for all interior nodes. For convergence, the tolerance was set to 10^{-6}. The figure below shows the numerical solution obtained using the Jacobi method on the 81×81 mesh, as well as the error between the analytical and the numerical solution for two different mesh sizes.

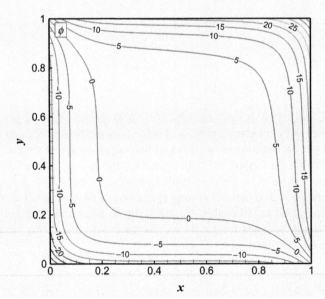

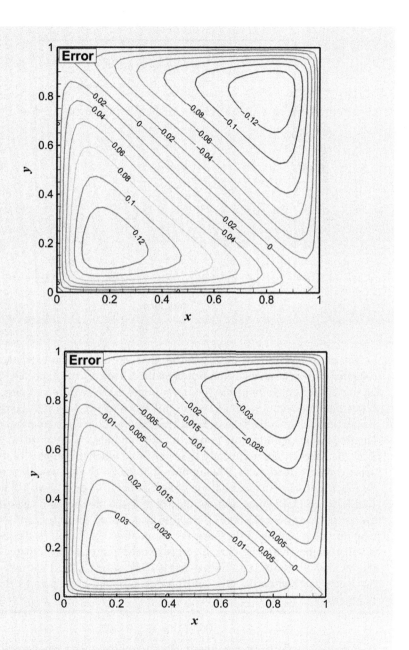

The error distributions [Top: 41×41, Bottom: 81×81] show that the error at each node decreases by a factor of 4 when the grid spacing is halved (maximum error goes from 0.1388 to 0.0348), once again highlighting the fact that the second-order central difference scheme was used. The Gauss–Seidel method yielded identical results. The plot below shows the convergence behavior of the two methods on the aforementioned three different mesh sizes (41×41 is not labeled).

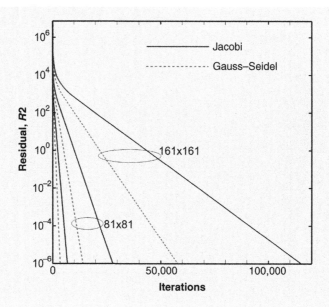

The convergence plot clearly shows that the Gauss–Seidel method is roughly twice as efficient as the Jacobi method. This is to be expected based on our earlier contention that the more implicit the iterative scheme, the faster the convergence. Since two out of the four off-diagonal terms are treated implicitly in the Gauss–Seidel method, its convergence is superior. One important point to note is that in both methods, the number of iterations increases by approximately a factor of four when the number of nodes approximately quadruples. This implies that the CPU time would increase by a factor of 16 with quadrupling of the number of nodes. Ideally, it is desirable to have CPU time scaling linearly with the number of nodes. However, this is not the case here. It roughly scales as K^2, which is still better than the cubic scaling of the Gaussian elimination algorithm. In Chapter 4, it will become clear why the convergence deteriorates with an increase in the number of nodes (i.e., a finer mesh). The actual CPU time taken by the Jacobi method on the 161×161 mesh was about 25 s on an Intel core i7 processor.

In summary, two popular point-wise iteration schemes have been presented and demonstrated. The Gauss–Seidel method was found to be twice as effective as the Jacobi method. Both schemes have the advantage that they are simple to implement and are applicable to any mesh topology. Although the convergence is slow, the cost per iteration of both methods is also very low, making them attractive choices. However, their major shortcoming is that both schemes scale poorly, and the number of iterations go up by a factor of four when the number of nodes is increased by a factor of four.

3.2.4 LINE-BY-LINE METHOD

When a structured mesh is used, the computational domain could be thought of as a series of constant i lines and a series of constant j lines. For any given line, finding the solution at the nodes belonging to that line represents a 1D calculation provided the other nodes that do not belong to the line are treated explicitly. To illustrate this concept, let us consider the finite difference equation given by Eq. (2.48), which may be rewritten in the following manner:

$$-\left(\frac{2}{(\Delta x)^2} + \frac{2}{(\Delta y)^2}\right)\phi_{i,j}^{(n+1)} + \frac{1}{(\Delta x)^2}\phi_{i+1,j}^{(n+1)} + \frac{1}{(\Delta x)^2}\phi_{i-1,j}^{(n+1)} = S_{i,j} - \frac{1}{(\Delta y)^2}\phi_{i,j+1}^{(n)} - \frac{1}{(\Delta y)^2}\phi_{i,j-1}^{(n+1)},$$

(3.33)

where the equation has been rearranged such that all terms contributed by nodes belonging to a constant j line (or row) have been retained on the left-hand side of the equation, and all terms contributed by nodes belonging to other j lines ($j+1$ and $j-1$) have been transposed to the right-hand side of the equation. Further, all terms on the left-hand side of Eq. (3.33) are treated implicitly (simultaneous unknowns), as indicated by the iteration index $n+1$, while all terms on the right-hand side are treated explicitly using previous iteration values. Equation (3.33) essentially represents a tridiagonal system of equations. When solved, the solution to all nodes belonging to a constant j line is obtained simultaneously. Once the solution to the nodes in a certain row (constant j line) has been obtained, the equations for the next row are solved. The process may continue until the solution for all rows have been obtained. Thus, in this approach, the computational domain is being swept *row-wise*. Similarly, it is also possible to sweep the computational domain *column-wise*, in which case, the tridiagonal system of equations would be as follows:

$$-\left(\frac{2}{(\Delta x)^2} + \frac{2}{(\Delta y)^2}\right)\phi_{i,j}^{(n+1)} + \frac{1}{(\Delta y)^2}\phi_{i,j+1}^{(n+1)} + \frac{1}{(\Delta y)^2}\phi_{i,j-1}^{(n+1)} = S_{i,j} - \frac{1}{(\Delta x)^2}\phi_{i+1,j}^{(n)} - \frac{1}{(\Delta x)^2}\phi_{i-1,j}^{(n+1)}.$$

(3.34)

Irrespective of row-wise or column-wise sweeps, since some terms are always treated explicitly, iterations are still necessary to obtain the correct solution. In reality, only one of the two nodal contributions on the right-hand side of Eq. (3.33) or (3.34) is treated explicitly. Let us consider a row-wise sweeping pattern starting from the bottommost row and moving up (increasing values of j). When it is time to obtain the solution of the j-th row, the solution to the $(j-1)$-th row (southern row) has already been updated, and therefore, the nodes in the $(j-1)$-th row have the solution at the current iteration. The same concept applies to column-wise sweeps. This is the reason the $(j-1)$ and the $(i-1)$-th nodes in Eqs (3.33) and (3.34) have been assigned the iteration index $(n+1)$, rather than (n). The implicit versus explicit treatment of the various neighboring nodes for the line-by-line sweeping pattern is depicted in Fig. 3.4.

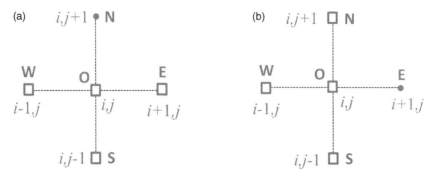

FIGURE 3.4 Pictorial Representation of Explicitness Versus Implicitness in Line-By-Line Methods
(a) Row-wise sweep, and (b) column-wise sweep. The node denoted by a solid circle is treated explicitly while the nodes denoted by hollow squares are treated implicitly.

The overall solution algorithm for row-wise iterative solution of the five-band linear system, arising out of discretization of a 2D PDE, is shown below.

ALGORITHM: LINE-BY-LINE METHOD WITH ROW-WISE SWEEP FOR A 2D PROBLEM

Step 1: Guess values of ϕ at all nodes, i.e., $\phi_{i,j}$ $\forall i = 1,...,N$ and $\forall j = 1,...,M$. We denote these values as $\phi^{(0)}$. If any of the boundaries have Dirichlet boundary conditions, the guessed values for the boundary nodes corresponding to that boundary must be equal to the prescribed boundary values. Set $\phi^{(n+1)} = \phi^{(n)}$.

Step 2: Set up a tridiagonal matrix system to solve Eq. (3.33) for the j-th row. This entails filling up matrix elements for d, c, a, and Q [see Eq. (3.13)]. For boundary conditions other than Dirichlet type, appropriate values of the matrix elements must be derived from the nodal equation at that boundary.

Step 3: Solve the tridiagonal system of equations using the algorithm described in Section 3.2.2 and Code Snippet 3.3, and update the value of $\phi^{(n+1)}$ of the j-th row.

Step 4: Repeat Steps 2 and 3 for all rows (values of j).

Step 5: Compute the residual vector using $\phi^{(n+1)}$, and then compute $R2^{(n+1)}$.

Step 6: Monitor convergence, i.e., check if $R2^{(n+1)} < \varepsilon_{tol}$? If YES, then go to Step 7. If NO, then go to Step 2.

Step 7: Stop iteration and postprocess the results.

The exact same algorithm may be used for an iterative solver based on column-wise sweeps. In that case, Steps 2 and 3 would entail setting up and solving a tridiagonal matrix solver for each column. When a line-by-line iterative solver is developed, it is desirable to write the tridiagonal matrix solver as a function (or subroutine or procedure) so that it can be used repeatedly within an iteration loop. This results in a more compact and error-free code.

As to which type of sweep (row versus column) is more effective, depends on a number of factors. These include the global aspect ratio of the geometry (whether it is long or short or fairly isotropic), the aspect ratio of the grid, and the boundary conditions in the problem. As we shall see later in Chapter 4, the aforementioned parameters, in conjunction, affect, in a rather convoluted way, the so-called condition number of the iteration matrix, which in turn, is responsible for altering the convergence. In order to investigate the impact of the two types of sweeping patterns on convergence, a numerical example is considered next.

EXAMPLE 3.3

The same problem, as considered in Example 3.2, is considered again in this example. The resulting linear system is first solved using both row and column sweeps on a 81×81 mesh. The residuals are shown below. As evident from the figure below, the convergence is identical in the two cases – so much so that the two residual curves are indistinguishable. This is because the problem is perfectly identical in the x and y directions from the perspective of all of the aforementioned parameters. If, however, the solution is computed on a 41×81 mesh, since the global aspect ratio is still unity, the grid aspect ratio becomes equal to 2. In this case, the two sweeping patterns result in vastly different convergence behavior as shown in the second figure below. Essentially, column-wise sweeps are far more effective in this case than row-wise sweeps. This implies that implicitness in the direction of fine mesh is preferable over implicitness in the direction of coarse mesh in order to attain fast convergence. This is because the information propagates instantly from boundary to boundary in the implicit direction while it takes many iterations to travel in the explicit direction. Thus, with 41 nodes in the explicit direction, convergence is faster than with 81 nodes. The number of nodes in the implicit direction essentially does not impact the convergence.

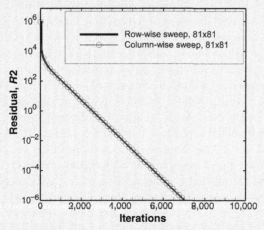

Convergence using row-wise and column-wise sweeps on a 81×81 mesh (grid aspect ratio = 1).

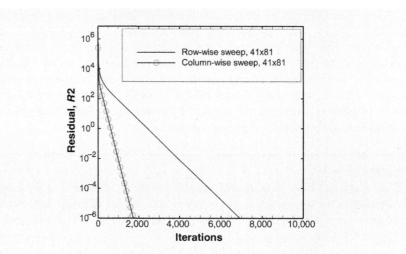

Convergence using row-wise and column-wise sweeps on a 41×81 mesh (grid aspect ratio = 2).

The findings of the preceding example can actually be used to design a line-by-line solver that is optimal in performance by choosing the appropriate sweeping direction. Unfortunately, while designing a general-purpose line-by-line solver, the developer has no prior knowledge of what geometry the solver might be applied to, what mesh might be used, or what boundary conditions might be applied. Even in a code designed to solve a single problem (invariant geometry), parametric studies, such as variation in the boundary values or mesh, may lead to situations where the solver may produce vastly different convergence. To avoid drastic changes in convergence behavior with direction of sweep, as in Example 3.3, it is customary to alter the direction of sweep at each iteration. Hence, a row-wise sweep of the whole domain is followed by a column-wise sweep and vice versa, where each row-wise or column-wise sweep of the entire computational domain counts as one iteration. This method is known as the *alternating direction implicit* (ADI) method. This method is not be confused with the ADI method for time-advancing schemes (see Section 5.2.4) in which the domain is swept in alternating directions at successive time steps [2,3], although the general philosophy is similar. The ADI method is an extremely popular method for iterative solution of the linear system arising out of discretization on a structured mesh, in which alternating (or cyclic) i, j, and k sweeps can be used easily. The method is popular because it is easy to implement, while at the same time, yields reasonably rapid convergence at a low computational cost. The following example demonstrates the performance of the ADI method for a boundary value problem.

EXAMPLE 3.4

The same problem, as considered in Example 3.2, is considered again in this example. The same grid sizes, namely 41×41, 81×81 and 161×161, are used again so that a one-on-one comparison can be made with the point-wise iterative solvers investigated in Example 3.2. The figure below shows the convergence behavior of the ADI method for the three mesh sizes. When compared with the convergence plots shown in Example 3.2, it is clear that the ADI method converges approximately twice as fast as the Gauss–Seidel method. This is to be expected since the degree of implicitness of the ADI method is better than that of the Gauss–Seidel method. However, the scaling behavior is roughly the same as the point-wise iterative solvers: when the number of nodes is quadrupled, the number of iterations also roughly quadrupled. In order to assess the performance of the ADI method for problems with asymmetry (as discussed in Example 3.3), computations were also performed on a 41×81 mesh, as in Example 3.3, and the convergence is shown in the second figure. The performance, as expected, is bounded by the performance of the row-wise and column-wise sweep methods. Essentially, of the 2780 iterations required for convergence, half (the column-wise sweeps) are not as effective, rendering the method less effective than the pure row-wise sweep method. However, in cases where the nature of the problem in not known *a priori*, using the ADI method removes excessive uncertainty in the iteration count, and may, therefore, be deemed a more robust procedure, overall.

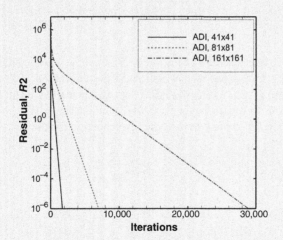

Convergence of the ADI method for three different mesh sizes of unity aspect ratio.

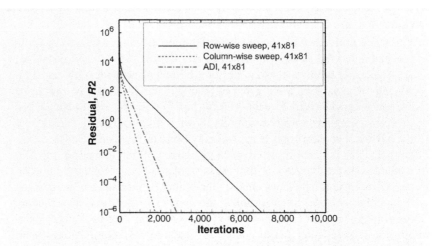

Convergence of the ADI method compared with row and column sweeps for a problem with non-unity grid aspect ratio.

The line-by-line methods, just described, can, just as easily, be applied to scenarios where a higher-order scheme is used on a structured mesh. To illustrate this point, let us extend Eq. (2.69) to a 2D mesh. This yields

$$
\begin{aligned}
&\frac{\phi_{i+1,j} - 2\phi_{i,j} + \phi_{i-1,j}}{(\Delta x)^2} - \frac{1}{12}\left(\frac{\phi_{i+2,j} - 4\phi_{i+1,j} + 6\phi_{i,j} - 4\phi_{i-1,j} + \phi_{i-2,j}}{(\Delta x)^2}\right) \\
&+ \frac{\phi_{i,j+1} - 2\phi_{i,j} + \phi_{i,j-1}}{(\Delta y)^2} - \frac{1}{12}\left(\frac{\phi_{i,j+2} - 4\phi_{i,j+1} + 6\phi_{i,j} - 4\phi_{i,j-1} + \phi_{i,j-2}}{(\Delta y)^2}\right) = S_{i,j}
\end{aligned}
\tag{3.35}
$$

In order to use line-by-line sweeps, Eq. (3.35) may be rearranged in two possible ways: (a) set up a tridiagonal matrix and treat the contribution of any node beyond the three central diagonals explicitly, and (b) set up a pentadiagonal matrix, and treat the contribution of any node beyond the five central diagonals explicitly. The two strategies yield the following two row-wise sweep iterative schemes:

$$
\begin{aligned}
&-\left(\frac{5}{2(\Delta x)^2} + \frac{5}{2(\Delta y)^2}\right)\phi_{i,j} + \frac{4}{3(\Delta x)^2}\phi_{i+1,j} + \frac{4}{3(\Delta x)^2}\phi_{i-1,j} = S_{i,j} \\
&+ \frac{1}{12}\left(\frac{\phi_{i,j+2} - 4\phi_{i,j+1} - 4\phi_{i,j-1} + \phi_{i,j-2}}{(\Delta y)^2}\right) - \frac{\phi_{i,j+1} + \phi_{i,j-1}}{(\Delta y)^2} + \frac{1}{12}\left(\frac{\phi_{i+2,j} + \phi_{i-2,j}}{(\Delta x)^2}\right),
\end{aligned}
\tag{3.36a}
$$

$$
\begin{aligned}
&-\left(\frac{5}{2(\Delta x)^2} + \frac{5}{2(\Delta y)^2}\right)\phi_{i,j} + \frac{4}{3(\Delta x)^2}\phi_{i+1,j} + \frac{4}{3(\Delta x)^2}\phi_{i-1,j} - \frac{1}{12(\Delta x)^2}\phi_{i+2,j} - \frac{1}{12(\Delta x)^2}\phi_{i-2,j} \\
&= S_{i,j} + \frac{1}{12}\left(\frac{\phi_{i,j+2} - 4\phi_{i,j+1} - 4\phi_{i,j-1} + \phi_{i,j-2}}{(\Delta y)^2}\right) - \frac{\phi_{i,j+1} + \phi_{i,j-1}}{(\Delta y)^2},
\end{aligned}
\tag{3.36b}
$$

where Eq. (3.36a) is for tridiagonal row-wise sweeps, while Eq. (3.36b) is for pentadiagonal row-wise sweeps. In the former case, a tridiagonal matrix solver must be used, while in the latter case, a pentadiagonal solver, as described in Section 3.1.2 (Code Snippet 3.4), must be used. The latter approach may be preferable if the off-diagonal coefficients are all of the same sign (the reason for this will become clear in Chapter 4). In such a scenario, the pentadiagonal iterative scheme is expected to yield faster convergence, albeit at a slightly higher cost per iteration. The column-wise sweep equations may be written in the same manner.

Thus far, three iterative solvers with increasing degrees of implicitness have been discussed. The convergence was found to improve with an increase in the degree of implicitness. Out of the three solvers, the line-by-line method is applicable only to structured mesh computations. The three solvers, discussed thus far, are often referred to as *classical iterative solvers*, in which the explicit versus implicit terms are clearly identifiable. Next, we discuss a solver that is also designed specifically for banded matrix systems and falls under the category of so-called *fully implicit solvers*.

3.2.5 STONE'S STRONGLY IMPLICIT METHOD

This strongly implicit method for banded systems was first proposed by Herbert Stone [4] in 1968. Since then, it has grown tremendously in popularity, and is now considered one of the most efficient solvers for banded systems arising out of discretization of a multidimensional PDE. In this section, we will discuss its formulation in the context of the five-band system arising out of discretization of a 2D PDE.

Let us consider the system of equations shown in Eq. (2.48), rewritten using the global nodal index k:

$$\left(\frac{2}{(\Delta x)^2} + \frac{2}{(\Delta y)^2}\right)\phi_k - \frac{1}{(\Delta x)^2}\phi_{k-1} - \frac{1}{(\Delta x)^2}\phi_{k+1} - \frac{1}{(\Delta y)^2}\phi_{k-N} - \frac{1}{(\Delta y)^2}\phi_{k+N} = -S_k, \quad (3.37)$$

where the equation has been written such that the diagonal is positive, as is customarily done. Following Stone, we further write this equation in the following five-band form:

$$E_k\phi_k + F_k\phi_{k+1} + D_k\phi_{k-1} + B_k\phi_{k-N} + H_k\phi_{k+N} = Q_k. \quad (3.38)$$

Written in matrix form, Eq. (3.38) has the following appearance:

$$\begin{bmatrix} E_1 & F_1 & \cdots & H_1 & & & & & \\ D_2 & E_2 & F_2 & \cdots & H_2 & & & & \\ & & & & & & & & \\ & B_k & \cdots & D_k & E_k & F_k & \cdots & H_k & \\ & & & & & & & & \\ & & & B_{K-1} & \cdots & D_{K-1} & E_{K-1} & F_{K-1} \\ & & & & B_K & \cdots & D_K & E_K \end{bmatrix} \begin{bmatrix} \phi_1 \\ \phi_2 \\ \\ \phi_k \\ \\ \phi_{K-1} \\ \phi_K \end{bmatrix} = \begin{bmatrix} Q_1 \\ Q_2 \\ \\ Q_k \\ \\ Q_{K-1} \\ Q_K \end{bmatrix}.$$

$$(3.39)$$

As discussed in Section 2.5, the diagonals E and H are separated by N spaces, as are E and B. Stone first converted the five-band equation to their correction form because he intended to develop an iterative solver from the very outset. Stone denoted the correction vector by $[\delta]$ rather than $[\phi']$, as we did in Section 3.2.1. In this section, we will continue to follow his notations. The correction form of the equations is easily derived starting from the general matrix form of the algebraic equations, namely $[A][\phi] = [Q]$. Following the approach described in Section 3.2.1, using an iterative procedure, this equation may be written as

$$[A][\phi^{(n+1)}] - [A][\phi^{(n)}] = [Q] - [A][\phi^{(n)}], \tag{3.40}$$

which may be simplified to the form

$$[A][\delta^{(n)}] = [R^{(n)}], \tag{3.41}$$

where $[\delta^{(n)}]$ is the correction in the solution from the (n) to the $(n+1)$-th iteration, i.e.,

$$[\delta^{(n)}] = [\phi^{(n+1)}] - [\phi^{(n)}], \tag{3.42}$$

and the residual in the (n)-th iteration is written as

$$[R^{(n)}] = [Q] - [A][\phi^{(n)}]. \tag{3.43}$$

Next, Stone proposed factorizing the matrix $[A]$ into a lower triangular matrix $[L]$, and an upper triangular matrix $[U]$. His motivation for factorization of the coefficient matrix into upper and lower triangular matrices is that he wanted to split the solution procedure into two steps. We will discuss the factorization process shortly. Assuming that the coefficient matrix is factorizable into upper and lower triangular matrices, and that both of these matrices are known, we can write Eq. (3.41) as follows:

$$[L][U][\delta^{(n)}] = [R^{(n)}]. \tag{3.44}$$

Defining a new vector $[Y]$, such that

$$[Y^{(n)}] = [U][\delta^{(n)}], \tag{3.45}$$

Eq. (3.44) may be rewritten as

$$[L][Y^{(n)}] = [R^{(n)}]. \tag{3.46}$$

Equations (3.45) and (3.46) enable the splitting of the solution of Eq. (3.41) into a two-step process. First, one can solve Eq. (3.46) to obtain $[Y^{(n)}]$. Next, Eq. (3.45) can be solved to obtain $[\delta^{(n)}]$.

One of the perceived advantages of solving either Eq. (3.45) or Eq. (3.46) over the solution of Eq. (3.41) is that if Gaussian elimination is used for solution of Eq. (3.41), both forward elimination and backward substitution steps will have to be executed, making the solution procedure computationally inefficient, as discussed in Section 3.1.1. On the other hand, the solution of either Eq. (3.45) or (3.46) can be considerably more efficient because both $[L]$ and $[U]$ matrices are already triangular matrices, and would only require either a forward or backward substitution, but no forward elimination. Recalling that the forward elimination step is the most time-consuming part of the Gaussian elimination algorithm, it becomes clear why solution of either Eq. (3.45) or (3.46) can be considerably more efficient than solution of Eq. (3.41).

Although this LU decomposition (or factorization) based algorithm proposed by Stone appears to be promising from a computational efficiency standpoint, it is based wholly upon the hypothesis that an exact factorization, $[A] = [L][U]$, exists. Based on the observation that the coefficient matrix $[A]$ only has five bands [Eq. (3.39)] Stone proposed a factorization in which the factors of $[A]$ are not only triangular, but also banded, as follows:

$$
[L] = \begin{bmatrix} d_1 & & & & & \\ c_2 & d_2 & & & & \\ & & \ddots & & & \\ b_{N+1} & & \ddots & \ddots & & \\ & & \ddots & & & \\ & & b_K & & c_K & d_K \end{bmatrix}, \quad [U] = \begin{bmatrix} 1 & e_1 & & f_1 & & \\ & 1 & e_2 & & \ddots & \\ & & & & & f_{K-N} \\ & & & \ddots & \ddots & \\ & & & & & e_{K-1} \\ & & & & & 1 \end{bmatrix}.
$$

$$(3.47)$$

The product $[L][U]$ results in a matrix that has seven bands, and has the structure depicted in Fig. 3.5.

The relationship between the seven bands shown in Fig. 3.5 and the coefficients of the $[L]$ and $[U]$ matrices shown in Eq. (3.47) are as follows:

$$
\begin{aligned}
\bar{B}_k &= b_k, \\
\bar{C}_k &= b_k e_{k-N}, \\
\bar{D}_k &= c_k, \\
\bar{E}_k &= b_k f_{k-N} + c_k e_{k-1} + d_k, \\
\bar{F}_k &= d_k e_k, \\
\bar{G}_k &= c_k f_{k-1}, \\
\bar{H}_k &= d_k f_k.
\end{aligned}
$$

$$(3.48)$$

The original matrix $[A]$ has five bands, while the product of $[L]$ and $[U]$ results in seven bands. This implies that Stone's proposition of factorizing $[A]$ exactly into $[L]$ and $[U]$ matrices, with each of these matrices having only three bands, is not realizable. Undaunted by this failure, Stone realized that even if $[A]$ can be factorized approximately into $[L]$ and $[U]$, it still serves the needs of the task at hand. If the factorization $[A] = [L][U]$ were exact, the correction $[\delta^{(n)}]$ would be such that, when added to $[\phi^{(n)}]$, the result would be the exact solution to $[A][\phi] = [Q]$. In other words, the solution would have been obtained by just one update (or iteration). Instead, if the factorization is approximate, the predicted correction $[\delta^{(n)}]$ would be such that more than one iteration would be necessary, and the number of iterations would depend

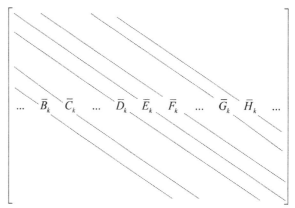

FIGURE 3.5 **Structure of the Seven-Band Matrix Arising Out of the Product [L][U] in Stone's Strongly Implicit Method**

on how approximate (or exact) the factorization is. Therefore, Stone set his goal on approximate factorization of the matrix [A].

Using the matrix structure depicted in Fig. 3.5, upon performing the product $[L][U][\delta]$, we obtain $\bar{E}_k\delta_k + \bar{F}_k\delta_{k+1} + \bar{D}_k\delta_{k-1} + \bar{B}_k\delta_{k-N} + \bar{H}_k\delta_{k+N} + \bar{C}_k\delta_{k-N+1} + \bar{G}_k\delta_{k+N-1}$. In contrast, the product $[A][\delta]$ only has five terms. The last two terms $(\bar{C}_k\delta_{k-N+1} + \bar{G}_k\delta_{k+N-1})$ are missing. Therefore, in order to establish an equivalence between [A] and [L][U], the last two terms shown in the expression above must be somehow eliminated. Stone employed truncated Taylor series expansions to achieve this goal. Referring to the nine-point stencil shown in Fig. 3.6, we may write

$$\delta_{k+N-1} \approx \delta_k - \left.\frac{\partial\delta}{\partial x}\right|_k \Delta x + \left.\frac{\partial\delta}{\partial y}\right|_k \Delta y, \tag{3.49}$$

where higher-order terms have been neglected. Neglecting the higher-order terms is justified because, at convergence, the value of the correction, δ, at all nodes is zero. Hence, at convergence, Eq. (3.49) is an exact relationship rather than an

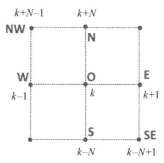

FIGURE 3.6 **Nine-Point Stencil Used in Stone's Method for Approximate Factorization of the Coefficient Matrix**

approximation. Further, the derivatives on the right-hand side of Eq. (3.49) may be approximated as

$$
\begin{aligned}
\left.\frac{\partial \delta}{\partial x}\right|_k &\approx \frac{\delta_k - \delta_{k-1}}{\Delta x}, \\
\left.\frac{\partial \delta}{\partial y}\right|_k &\approx \frac{\delta_{k+N} - \delta_k}{\Delta y}.
\end{aligned}
\tag{3.50}
$$

Substitution of Eq. (3.50) into Eq. (3.49), followed by rearrangement, yields

$$
\delta_{k+N-1} \approx -\delta_k + \delta_{k+N} + \delta_{k-1}.
\tag{3.51}
$$

A similar procedure may also be followed to derive

$$
\delta_{k-N+1} \approx -\delta_k + \delta_{k+1} + \delta_{k-N}
\tag{3.52}
$$

The next step is to substitute Eqs (3.51) and (3.52) into the last two terms of the aforementioned seven-band form. However, realizing that Eqs (3.51) and (3.52) are approximations, Stone proposed using a calibration or fudge factor, α, prior to substitution, such that the seven-band form may be written as

$$
\begin{aligned}
\bar{E}_k \delta_k &+ \bar{F}_k \delta_{k+1} + \bar{D}_k \delta_{k-1} + \bar{B}_k \delta_{k-N} + \bar{H}_k \delta_{k+N} + \bar{C}_k \delta_{k-N+1} + \bar{G}_k \delta_{k+N-1} \\
&= \bar{E}_k \delta_k + \bar{F}_k \delta_{k+1} + \bar{D}_k \delta_{k-1} + \bar{B}_k \delta_{k-N} + \bar{H}_k \delta_{k+N} \\
&\quad + \bar{C}_k [\alpha(-\delta_k + \delta_{k+N} + \delta_{k-1})] + \bar{G}_k [\alpha(-\delta_k + \delta_{k+1} + \delta_{k-N})]
\end{aligned}
\tag{3.53}
$$

Stone argued that this calibration or fudge factor, henceforth referred to as the *Stone's factor*, should have a value close to unity, and could be adjusted arbitrarily to attain faster convergence. The reason for this argument will become clear in due course. Simplifying the right-hand side of Eq. (3.53), and substituting it into the left-hand side of Eq. (3.41) yields the approximately LU-decomposed form of the correction equation, written as

$$
\begin{aligned}
\left[\bar{E}_k - \alpha\bar{C}_k - \alpha\bar{G}_k\right]\delta_k^{(n)} &+ \left[\bar{F}_k + \alpha\bar{G}_k\right]\delta_{k+1}^{(n)} + \left[\bar{D}_k + \alpha\bar{C}_k\right]\delta_{k-1}^{(n)} \\
&+ \left[\bar{B}_k + \alpha\bar{G}_k\right]\delta_{k-N}^{(n)} + \left[\bar{H}_k + \alpha\bar{G}_k\right]\delta_{k+N}^{(n)} = R_k^{(n)}
\end{aligned}
\tag{3.54}
$$

Comparing Eq. (3.54) with the original form of the correction equation, namely,

$$
E_k \delta_k^{(n)} + F_k \delta_{k+1}^{(n)} + D_k \delta_{k-1}^{(n)} + B_k \delta_{k-N}^{(n)} + H_k \delta_{k+N}^{(n)} = R_k^{(n)},
\tag{3.55}
$$

the following relationships may be established:

$$
\begin{aligned}
B_k &= \bar{B}_k + \alpha\bar{C}_k \\
D_k &= \bar{D}_k + \alpha\bar{G}_k \\
E_k &= \bar{E}_k - \alpha(\bar{C}_k + \bar{G}_k). \\
F_k &= \bar{F}_k + \alpha\bar{C}_k \\
H_k &= \bar{H}_k + \alpha\bar{G}_k
\end{aligned}
\tag{3.56}
$$

Further, substitution of Eq. (3.48) into Eq. (3.56), followed by some algebraic manipulations, yield the following relationships:

$$b_k = \frac{B_k}{1 + \alpha e_{k-N}}$$

$$c_k = \frac{D_k}{1 + \alpha f_{k-1}}$$

$$d_k = E_k + \alpha(b_k e_{k-N} + c_k f_{k-1}) - b_k f_{k-N} - c_k e_{k-1} .$$

$$e_k = \frac{F_k - \alpha b_k e_{k-N}}{d_k}$$

$$f_k = \frac{H_k - \alpha c_k f_{k-1}}{d_k}$$

(3.57)

Equation (3.57) provides the relationships necessary to compute the elements of the $[L]$ and $[U]$ matrices [see Eq. (3.47)] from the elements of the original $[A]$ matrix. These relationships are based on the approximate factorization, just described. At first glance, the relationships provided in Eq. (3.57) appear to be implicit. For example, computing b using the first relationship requires e, while computing e using the fourth relationship requires b. This ambiguity can be resolved by the fact that all coefficients with a 0 or negative index must be first assigned a value of zero. Thus, computation of b_1, for example, does not require the value of e, since e_{1-N} is zero. A code snippet designed to compute the coefficients of the $[L]$ and $[U]$ matrices from the elements of the $[A]$ matrix using Eq. (3.57) is shown below.

CODE SNIPPET 3.6: COMPUTATION OF THE ELEMENTS OF THE [L] AND [U] MATRICES FROM THE COEFFICIENTS OF THE [A] MATRIX IN STONE'S STRONGLY IMPLICIT METHOD.

```
d(:) = 0; e(:) = 0; c(:) = 0; b(:) = 0; f(:) = 0 ! Set all elements to zero to begin
d(1) = E(1)  ! Compute row 1 first to begin recursive process
e(1) = F(1)/d(1)
f(1) = H(1)/d(1)
For k = 2 : K  ! Compute all other rows
   IF (k > n) THEN
      b(k) = B(k)/(1 + alpha*e(k-n))
   END IF
   c(k) = D (k)/(1 + alpha*f(k-1))
   IF (k > n) THEN
      d(k) = E(k) + alpha*(b(k)*e(k-n) + c(k)*f(k-1)) - b(k)*f(k-n) - c(k)*e(k-1)
   ELSE
      d(k) = E(k) + alpha*(c(k)*f(k-1)) - c(k)*e(k-1)
   END IF
   f(k) = (H(k) - alpha*c(k)*f(k-1))/d(k)
   IF (k > n) THEN
      e(k) = (F(k) - alpha*b(k)*e(k-n))/d(k)
   ELSE
      e(k) = F(k)/d(k)
   END IF
End
```

The fact that the original coefficient matrix, $[A]$, has only been factorized approximately would appear to suggest that the original equations that we set out to solve have been distorted. However, this is not the case. The approximate factorization is only applied to the correction form of the equations, Eq. (3.44). While it may predict an "incorrect" value of the corrections at each node, since $[\delta]$ tends to zero at convergence, the original equations remain unaffected. In other words, the fact that the factorization is approximate, only affects the path to convergence, not the final solution to the equations.

The overall solution algorithm to use Stone's strongly implicit method to solve the five-band form of the difference equations arising out of finite difference discretization of a 2D PDE is presented beow.

ALGORITHM: STONE'S STRONGLY IMPLICIT METHOD FOR A 2D PROBLEM

Step 1: Guess values of ϕ at all nodes, i.e., $\phi_{i,j} \ \forall i = 1,...,N$ and $\forall j = 1,...,M$. We denote these values as $\phi^{(0)}$. If any of the boundaries have Dirichlet boundary conditions, the guessed values for the boundary nodes corresponding to that boundary must be equal to the prescribed boundary values.

Step 2: Compute link coefficients (A_O, A_E, A_W, A_N, and, A_S) and fill up the elements of vectors E, F, D, B, H, and Q.

Step 3: Compute the elements of the $[L]$ and $[U]$ matrices using Eq. (3.57) [Code Snippet 3.6].

Step 4: Compute residual vector using Eq. (3.43), and then compute $R2^{(n)}$.

Step 5: Solve (forward substitution) Eq. (3.46) to obtain $[Y^{(n)}]$.

Step 6: Solve (backward substitution) Eq. (3.45) to obtain $[\delta^{(n)}]$.

Step 7: Update solution: $[\phi^{(n+1)}] = [\phi^{(n)}] + [\delta^{(n)}]$.

Step 8: Monitor convergence, i.e., check if $R2^{(n)} < \varepsilon_{tol}$? If YES, then go to Step 9. If NO, then go to Step 4 (i.e., proceed to next iteration).

Step 9: Stop iteration and postprocess the results.

The two most time-consuming steps in the above algorithm are the forward and backward substitution steps (Steps 5 and 6 in the above algorithm). Earlier, a remark was made that the two-step solution procedure, using forward and backward substitution steps, is considerably more efficient than the full-blown Gaussian elimination process because of the absence of the time-consuming forward elimination step. According to the discussion in Section 3.1.1, even forward or backward substitutions have long operation counts that scale as $\sim K^2$, which would make Stone's method quite inefficient for large-scale computations. Stone had the foresight to recognize this shortcoming, and hence, proposed that even the $[L]$ and $[U]$ matrices be banded so that the operation count is much lower. Since both the $[L]$ and $[U]$ matrices have only three bands each, the long operation count in the forward substitution step is only $\sim 3K$, while the long operation count in the backward substitution step is only $\sim 2K$. Thus, Stone's method is theoretically linearly scalable as far as the long operation count is concerned, making it one of the most attractive methods for banded linear systems. To demonstrate the efficacy of Stone's strongly implicit method, a numerical example is considered next.

EXAMPLE 3.5

In this example, we consider the solution of the same problem considered in Example 3.2. The same grid sizes, namely 41×41, 81×81, and 161×161, are used again so that a one-on-one comparison can be made with all the solvers considered earlier (Examples 3.2 and 3.4). The figure below shows the convergence behavior of the Stone's method for the three mesh sizes. The value used for Stone's factor is $\alpha = 0.9$.

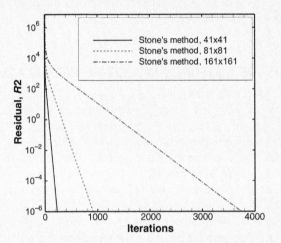

It is clear from the figure above that the convergence of the Stone's strongly implicit method is vastly superior to any of the iterative methods discussed earlier. On the 161×161 mesh, only 3737 iterations were required for convergence. In contrast, the ADI method required 28,908 iterations. As far as scaling is concerned, the number of iterations increased from 913 for the 81×81 mesh, to 3737 for the 161×161 mesh – a factor of approximately 4. Thus, in terms of scaling, the Stone's method is similar to the methods discussed earlier. The convergence of the Stone's method can be further improved by fine-tuning the Stone's factor, α, as shown in the table below. It shows that for this particular case, $\alpha = 0.93$ produces the fastest convergence. In general, a value close to 0.9 is recommended.

Stone's factor, α	41×41	81×81	161×161
0.88	262	1052	4308
0.90	229	913	3737
0.92	193	766	3127
0.93	175	688	2806
0.94	752	diverged	diverged

3.2.6 METHOD OF STEEPEST DESCENT (MSD)

The MSD is built upon a philosophy that is completely different from the methods discussed in the preceding sections. It solves an optimization problem to find the solution to the system of equations, $[A][\phi] = [Q]$. To understand this philosophy, let us consider the following multidimensional scalar function:

$$f(\phi_1, \phi_2, ..., \phi_K) = \frac{1}{2}[\phi]^T[A][\phi] - [Q]^T[\phi] + c, \qquad (3.58)$$

where c is an arbitrary constant. The function is actually a quadratic function, as evident from the first term on the right-hand side of Eq. (3.58). If the matrix $[A]$ is either positive definite or negative definite (see Appendix A for definitions), then the function, given by Eq. (3.58), has a distinct minimum or maximum. In addition, since it is a quadratic function, the function has only one maximum or minimum. This is best demonstrated through a simple example in two variables. Let us consider the following linear system:

$$[A][\phi] = \begin{bmatrix} 3 & 2 \\ 2 & 6 \end{bmatrix} \begin{bmatrix} \phi_1 \\ \phi_2 \end{bmatrix} = [Q] = \begin{bmatrix} 2 \\ -8 \end{bmatrix}. \qquad (3.59)$$

In this case, the matrix $[A]$ is positive definite. For this particular linear system, the function shown in Eq. (3.58) assumes the following form:

$$f(\phi_1, \phi_2) = \frac{3}{2}\phi_1^2 + 2\phi_1\phi_2 + 3\phi_2^2 - 2\phi_1 + 8\phi_2 + c. \qquad (3.60)$$

Figure 3.7(a) shows a contour plot of the function. A value of $c = 10$ has been used to generate Fig. 3.7. If the two diagonal elements of matrix $[A]$ shown in Eq. (3.59) are assigned values of -3 and 6, instead of 3 and 6, the matrix $[A]$ becomes negative definite, and the contour plot of the corresponding function is shown in

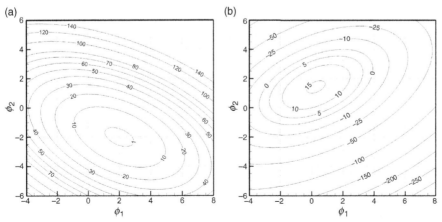

FIGURE 3.7 Contour Plots of the 2D Example of the Function Given by Eq. (3.60) for (a) Positive Definite [A], and (b) Negative Definite [A]

Fig. 3.7(b). It is clear from Fig. 3.7 that when $[A]$ is positive definite, the function has a minimum, while when $[A]$ is negative definite, the function has a maximum.

The next pertinent question is what relation the solution to the linear system has to the maximum or minimum of the function, if any. This question can be answered by differentiating the function with respect to its independent variables and setting the result equal to zero, resulting in a vector equation, as follows:

$$\nabla_\phi f = \begin{bmatrix} \dfrac{\partial f}{\partial \phi_1} \\ \dfrac{\partial f}{\partial \phi_2} \\ \vdots \\ \dfrac{\partial f}{\partial \phi_N} \end{bmatrix} = \frac{1}{2}[A]^T[\phi] + \frac{1}{2}[A][\phi] - [Q] = 0. \tag{3.61}$$

In this equation, $\nabla_\phi f$ denotes the gradient of the function f in hyperdimensional space, and is a hyperdimensional vector. If the matrix $[A]$ is symmetric, then $[A]^T = [A]$, and Eq. (3.61) can be further reduced to

$$\nabla_\phi f = [A][\phi] - [Q] = 0, \tag{3.62}$$

which, in fact, is the linear system whose solution is being sought. Based on the preceding derivation, we may conclude that if the matrix $[A]$ is symmetric and either positive or negative definite, then finding the maximum or minimum of the function shown in Eq. (3.60) is equivalent to solving the linear system, $[A][\phi] = [Q]$.

In the method of steepest descent, the preceding optimization philosophy is utilized to develop an iterative solution procedure. First, we start with a guess for $[\phi]$ and denote this by $[\phi^{(n)}]$. Noting that $[R^{(n)}] = [Q] - [A][\phi^{(n)}]$, it follows from Eq. (3.62) that at any iteration, the gradient vector is equal to the negative of the residual vector, i.e.,

$$\left[\nabla_\phi f\right]^{(n)} = -[R^{(n)}]. \tag{3.63}$$

To find the solution at the next iteration, we follow the function along its gradient vector, such that

$$[\phi^{(n+1)}] = [\phi^{(n)}] - \alpha^{(n)}\left[\nabla_\phi f\right]^{(n)} = [\phi^{(n)}] + \alpha^{(n)}[R^{(n)}], \tag{3.64}$$

where α is an arbitrary scalar constant, whose value changes from iteration to iteration. It represents the distance to be traversed along the gradient vector in hyperdimensional space at each iteration. As to what value is to be used for α will be discussed shortly. The rationale for the iteration scheme shown in Eq. (3.64) is best explained through an example in two variables. We consider, once again, the example shown in Eq. (3.59). The exact solution to this system is $(2, -2)$. To start iterations, we begin with the initial guess $(-2, -2)$. Differentiating Eq. (3.60), the gradient vector is easily calculated as

$$\nabla_\phi f = \frac{\partial f}{\partial \phi_1}\hat{\phi}_1 + \frac{\partial f}{\partial \phi_2}\hat{\phi}_2 = (3\phi_1 + 2\phi_2 - 2)\hat{\phi}_1 + (2\phi_1 + 6\phi_2 + 8)\hat{\phi}_2, \tag{3.65}$$

where $\hat{\phi}_1$ and $\hat{\phi}_2$ are unit vectors along the two independent directions. It is easily confirmed that the gradient vector shown by Eq. (3.65) is, in fact, the negative of the residual vector. Using the initial guess $(-2, -2)$, we obtain

$$\left[\nabla_\phi f\right]^{(n)} = -12\hat{\phi}_1 - 8\hat{\phi}_2. \tag{3.66}$$

Equation (3.64) suggests that to find the next solution, we march along the direction opposite the gradient vector, i.e., along the vector $12\hat{\phi}_1 + 8\hat{\phi}_2$. The direction of this vector – henceforth referred to as the *search direction* – is depicted in Fig. 3.8, along with the initial guess and the final solution.

The direction of the negative of the gradient vector or the direction of the residual vector actually represents the direction of steepest descent (if [A] is positive definite) or ascent (if [A] is negative definite) of the function. The residual vector is locally perpendicular to the isocontour at that point. The most efficient way to reach the next lower isocontour is to follow the negative of the gradient vector.

Examination of Fig. 3.8 might give the impression that following the residual vector is leading us away from the solution. However, at $(-2,2)$, the residual vector direction is the best direction to follow to reach the next lower contour, for example, to go from an isocontour at approximately 22 to one at approximately 20. Clearly, if we continue along this direction indefinitely, we will, in fact, be moving away from the minimum of the function. This observation points to the fact that the search direction needs to be updated periodically so that the minimum of the function is approached "smoothly." The method of steepest descent stipulates that we continue along the search direction, computed at the initial location, until the function starts to increase again. Recalling that the scalar α represents the distance to be traversed, the

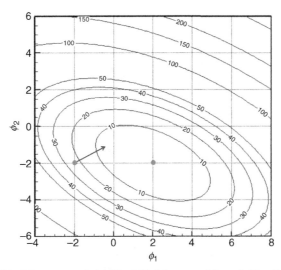

FIGURE 3.8 Graphical Representation of the Initial Guess and Search Direction (shown by the arrow) in the Method of Steepest Descent

The objective is to reach $(-2,2)$ starting from the initial location at $(-2,-2)$.

preceding criterion implies differentiating the function with respect to α and setting it equal to zero:

$$\left.\frac{\partial f}{\partial \alpha}\right|^{(n+1)} = 0. \tag{3.67}$$

Noting that $f = f(\phi_1, \phi_2, ..., \phi_K)$, using the chain rule, Eq. (3.67) may be written as

$$\left[\frac{\partial f}{\partial \phi_1}\frac{\partial \phi_1}{\partial \alpha} + \frac{\partial f}{\partial \phi_2}\frac{\partial \phi_2}{\partial \alpha} + ... + \frac{\partial f}{\partial \phi_K}\frac{\partial \phi_K}{\partial \alpha}\right]^{(n+1)} = 0. \tag{3.68}$$

Using the iteration scheme shown in Eq. (3.64), the derivatives with respect to α may be easily determined, and substituted into Eq. (3.68) to yield

$$\left[\left.\frac{\partial f}{\partial \phi_1}\right|^{(n+1)} R_1^{(n)} + \left.\frac{\partial f}{\partial \phi_2}\right|^{(n+1)} R_2^{(n)} + ... + \left.\frac{\partial f}{\partial \phi_K}\right|^{(n+1)} R_K^{(n)}\right] = 0. \tag{3.69}$$

Substituting Eq. (3.63), Eq. (3.69) may be written as

$$-[R^{(n+1)}]^T[R^{(n)}] = -\left[[Q] - [A][\phi^{(n+1)}]\right]^T[R^{(n)}] = 0, \tag{3.70}$$

where the definition of the residual has been used. Equation (3.70) clearly indicates that the residual vectors at successive iterations are orthogonal to each other. Substituting Eq. (3.64) into Eq. (3.70), we further obtain

$$-[R^{(n+1)}]^T[R^{(n)}] = -\left[[Q] - [A][\phi^{(n)}] - \alpha^{(n)}[A][R^{(n)}]\right]^T[R^{(n)}] = 0. \tag{3.71}$$

Once again, using the definition of the residual, Eq. (3.71) may be written as

$$-\left[[R^{(n)}] - \alpha^{(n)}[A][R^{(n)}]\right]^T[R^{(n)}] = -[R^{(n)}]^T[R^{(n)}] + \alpha^{(n)}\left[[A][R^{(n)}]\right]^T[R^{(n)}] = 0. \tag{3.72}$$

Rearranging, we obtain

$$\alpha^{(n)} = \frac{[R^{(n)}]^T[R^{(n)}]}{\left[[A][R^{(n)}]\right]^T[R^{(n)}]} = \frac{[R^{(n)}]^T[R^{(n)}]}{[R^{(n)}]^T[A][R^{(n)}]}, \tag{3.73}$$

where the matrix identity in Eq. (A.12), along with the fact that $[A]$ is symmetric, has been utilized to derive the last step of Eq. (3.73). Once $\alpha^{(n)}$ has been computed, Eq. (3.64) can be used to find the next (updated) value of $[\phi]$. The process is then continued until the prescribed convergence tolerance has been satisfied. This relatively straightforward algorithm of the method of steepest descent is presented below in a step-wise manner.

ALGORITHM: MSD

Step 1: Guess values of ϕ at all nodes, i.e., $\phi_{i,j}$ $\forall i = 1, ..., N$ and $\forall j = 1, ..., M$. We denote these values as $\phi^{(0)}$. If any of the boundaries have Dirichlet boundary conditions, the guessed values for the boundary nodes corresponding to that boundary must be equal to the prescribed boundary values.

Step 2: Compute the residual vector, $[R^{(n)}]$ using Eq. (3.43), and then compute $R2^{(n)}$.

Step 3: Compute $\alpha^{(n)}$ using Eq. (3.73).
Step 4: Update solution using Eq. (3.64).
Step 5: Monitor convergence, i.e., check if $R2^{(n)} < \varepsilon_{tol}$? If YES, then go to Step 6. If NO, then go to Step 2 (i.e., proceed to next iteration).
Step 6: Stop iteration and postprocess the results.

The method of steepest descent can be applied to any linear system – sparse or full, as long as the matrix $[A]$ is symmetric and positive definite. The most time-consuming step of the algorithm is Step 3, in which Eq. (3.73) is to be used to compute $\alpha^{(n)}$. In reality, Steps 2 and 3 can be combined to make the computations efficient. In addition, when it comes to using the method of steepest descent for solving the sparse matrix systems arising out of discretization of PDEs, the sparseness of the matrix must be utilized to improve efficiency. Code Snippet 3.7 demonstrates how some of the algorithmic steps can be efficiently executed.

CODE SNIPPET 3.7: STEPS 2 THROUGH 4 OF THE ALGORITHM FOR MSD. THE BAND (DIAGONAL) NAMES USED ARE THE SAME AS THE STONE'S METHOD. HERE, DIRICHLET BOUNDARY CONDITIONS ARE USED AT ALL BOUNDARIES.

```
! Step 1
R(:) = 0   ! Initialize residual vector. Residual stays 0 at Dirichlet BCs. Only computed at interior nodes.
For i = 2 : N-1  ! Interior Rows
   For j = 2 : M-1  ! Interior Columns
      k = (j-1)*N+i
      R(k) = Q(k) - E(k)*PHI(k) - F(k)*PHI(k+1) - H(k)*PHI(k+N) - D(k)*PHI(k-1) - B(k)*PHI(k-N)
   End
End
R2sum = 0
For k = 1 : K  ! Calculate square of R2 = [R-transpose]*[R]
   R2sum = R2sum + R(k)*R(k)
End
R2 = sqrt(R2sum) ! L2Norm
! Step 2
c(:) = 0 ! [c] = [A]*[R]
For i = 2 : N-1  ! Interior Rows
   For j = 2 : M-1  ! Interior Columns
      k = (j-1)*N+i
      c(k) = E(k)*R(k) + F(k)*R(k+1) + H(k)*R(k+N) + D(k)*R(k-1) + B(k)*R(k-N)
   End
End
rtc = 0   ! rtc = [R-transpose]*[A]*[R]= [R-transpose]*[c]
For k = 1 : K
   rtc = rtc + R(k)*c(k)
End
alpha = R2sum/rtc
! Step 3 (Solution update)
PHI(:) = PHI(:) + alpha*R(:)
```

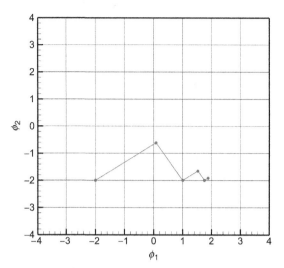

FIGURE 3.9 Scatter Plot of Solutions at Successive Iterations (Solid Circles) and the Path Traversed Between Solutions During Iterative Solution of Eq. (3.59) Using the Method of Steepest Descent

It is clear that Code Snippet 3.7 recognizes and uses the fact that computation of $R2$ and the numerator of the expression for $\alpha^{(n)}$ is synonymous.

Prior to exercising the method of steepest descent for solution of the linear system arising out of discretization of a PDE, it is instructive to exercise it for the solution of the example 2×2 problem given by Eq. (3.59) so that the search process (solution at the next iteration) can be plotted graphically and better understood. Figure 3.9 shows a scatter plot of the solutions obtained in the first five iterations along with the search path that was traversed in moving between solutions at successive iterations. The starting location is $(-2,-2)$. After five iterations, the solution is $(1.88, -1.91)$ with $R2 = 0.33$. Most noticeably, it is seen in Figure 3.9, that after every iteration, the new search direction vector (or residual vector) is orthogonal to the previous search direction vector, as had been shown earlier in Eq. (3.70). Therefore, the path followed is a stair-step path. Since only two directions are traversed, it is likely that, even for this simple 2×2 problem, many iterations may be required before the final solution is reached. This is one of the major shortcomings of the method of steepest descent.

In order to assess the performance of the method of steepest descent in comparison with the iterative methods discussed earlier, an example of the solution to a 2D PDE is considered next.

EXAMPLE 3.6

In this example, we consider the solution of the same problem considered in Examples 3.2–3.5. The same grid sizes, namely 41×41, 81×81, and 161×161, are used again so that a one-on-one comparison can be made with all the solvers considered earlier. The figure below shows the convergence behavior of the MSD for the three mesh sizes under consideration.

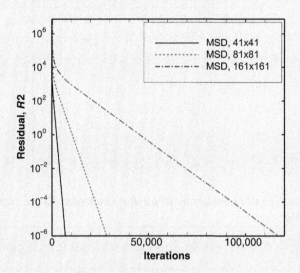

The convergence of the MSD, as expected, is slow. On the 161×161 mesh, 117,044 iterations are necessary to attain convergence. The number of iterations required is even larger than the Jacobi method, with a larger computational cost per iteration compared to the Jacobi method because of the matrix vector multiplication required in Step 3 of the algorithm. As far as scaling is concerned, the method scales similarly to the other methods: approximately four times more iterations are required when the number of nodes is quadrupled.

Although the method of steepest descent is inefficient, it serves as the foundation for other iterative solvers constructed upon the basic philosophy of optimization of a quadratic function. The difference between these various methods is in the algorithm used to conduct the search.

3.2.7 CONJUGATE GRADIENT (CG) AND CONJUGATE GRADIENT SQUARED (CGS) METHODS

In the method of steepest descent, the search directions in successive iterations are orthogonal to each other. This makes the search inefficient because a stair-step pattern has to be followed. The *CG method*, credited to Lanczos [5], combats this shortcoming by proposing to use a search direction vector that is a linear combination

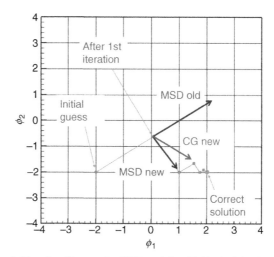

FIGURE 3.10 Search Direction Vectors for MSD and the CG Method for the Example Problem Given by Eq. (3.59)

of the old direction vector and the new residual vector. This is shown graphically in Fig. 3.10 for the example problem given by Eq. (3.59). As discussed earlier, the search direction vectors in the method of steepest descent are the residual vectors. In successive iterations, the residual vectors are orthogonal to each other, as shown in Fig. 3.10. In the CG method, the new direction, being a linear combination of the old search direction vector and the new residual vector, is not necessarily orthogonal to the old direction vector. In the specific example shown in Fig. 3.10, the new direction points almost directly at the correct solution. Even if it does not, it is clear that the solution will be approached more smoothly rather than along a dramatic stair-step pattern.

The exact ratio, β, used to combine the old direction vector with the new residual vector is determined by enforcing the two vectors to be A-orthogonal (see Appendix A for definition). For further mathematical details of this process, the reader is referred to the original article by Lanczos or advanced texts on iterative methods, such as the one by Saad [6]. The algorithm to execute the CG method is given below.

ALGORITHM: CG METHOD

Step 1: Guess values of ϕ at all nodes, i.e., $\phi_{i,j}$ $\forall i = 1,...,N$ and $\forall j = 1,...,M$. We denote these values as $\phi^{(0)}$. If any of the boundaries have Dirichlet boundary conditions, the guessed values for the boundary nodes corresponding to that boundary must be equal to the prescribed boundary values.

Step 2: Compute the residual vector, $[R^{(0)}] = [Q] - [A][\phi^{(0)}]$, and $R2^{(0)}$.

Step 3: Set the initial search direction vector equal to the residual vector: $[D^{(0)}] = R^{(0)}$.

Step 4: Compute $\alpha^{(n+1)}$ using $\alpha^{(n+1)} = \dfrac{[R^{(n)}]^T[R^{(n)}]}{[D^{(n)}]^T[A][D^{(n)}]}$.

Step 5: Update solution using $[\phi^{(n+1)}] = [\phi^{(n)}] + \alpha^{(n+1)}[D^{(n)}]$.

Step 6: Compute the new residual vector, $[R^{(n+1)}] = [Q] - [A][\phi^{(n+1)}]$, and $R2^{(n+1)}$.

Step 7: Compute $\beta^{(n+1)} = \dfrac{[R^{(n+1)}]^T[R^{(n+1)}]}{[R^{(n)}]^T[R^{(n)}]} = \dfrac{(R2^{(n+1)})^2}{(R2^{(n)})^2}$.

Step 8: Update search direction vector: $[D^{(n+1)}] = [R^{(n+1)}] + \beta^{(n+1)}[D^{(n)}]$.
Step 9: Monitor convergence, i.e., check if $R2^{(n+1)} < \varepsilon_{\text{tol}}$? If YES, then go to Step 10. If NO, then go to Step 4 (i.e., proceed to next iteration).
Step 10: Stop iteration and postprocess the results.

Steps 2–3 are known as the preparation phase of the algorithm, while Steps 4–9 are known as the iterative phase. Compared with the method of steepest descent, the conjugate gradient method requires a few additional arithmetic operations in order to compute the mixing parameter, β, and the new search direction. However, the search process is significantly more efficient. For the 2×2 problem discussed earlier, Eq. (3.59), the exact solution is obtained in just two iterations. Table 3.2 shows the outcome of executing the CG algorithm on the 2×2 linear system shown in Eq. (3.59).

It has been shown mathematically by Lanczos [4] that for a linear system comprised of K equations, the conjugate gradient method will result in the exact solution in no more than K iterations. Clearly, the results shown in Table 3.2 confirm this contention. In practice, users of this method have found [6] that due to numerical round-off, the method loses its aforementioned strict convergence property because with increasing iterations, the new direction vectors no longer exactly obey A-orthogonality. This is particularly true for large systems and systems where the matrix $[A]$ is poorly conditioned. On the other hand, rarely does one seek the exact answer. The answer is usually sought within a certain tolerance, and the rate of descent of the residuals in the conjugate gradient method is so large that convergence is reached rapidly even for fairly large linear systems. In order to investigate the convergence of the conjugate gradient method, a numerical example is considered next.

Table 3.2 Results of Successive Iterations Using the Conjugate Gradient Algorithm for the 2×2 Linear System Shown in Eq. (3.59)

Iteration	ϕ_1	ϕ_2	R_1	R_2	L^2Norm, R2
0	−2.000000	−2.000000	12.000000	8.000000	14.422205
1	0.080000	−0.613333	2.986667	−4.480000	5.384290
2	2.000000	−2.000000	0.000000	0.000000	0.000000

EXAMPLE 3.7

The solution to the same problem considered in Examples 3.2–3.6 is attempted here with the conjugate gradient method. Grid sizes of 41×41, 81×81, and 161×161 are used again for one-to-one comparative studies. The figure below shows the convergence behavior of the conjugate gradient method for the three mesh sizes that were considered.

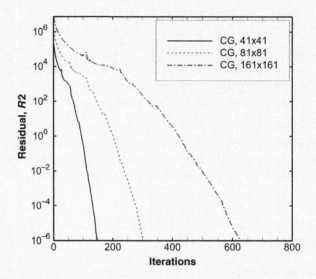

The convergence of the conjugate gradient method is far superior to any of the methods considered in Examples 3.2–3.6. On the 161×161 mesh, only 627 iterations are necessary to attain convergence, which is faster than what the next best method – Stone's method – required for the 81×81 mesh. In fact the convergence is super-linear on a logarithmic scale indicating that as convergence is approached, each iteration becomes more and more effective. In terms of scaling, the CG method is also vastly superior to all other methods considered thus far. When the number of nodes is quadrupled, the iterations needed to attain convergence increases only by a factor of two, as opposed to approximately four, which was noted for all other iterative solvers considered thus far.

One of the disadvantages of the conjugate gradient solver is that the convergence is not monotonic, especially in the initial stages. For many applications, in which multiple coupled PDEs have to be solved, the PDEs are often solved sequentially and are coupled using so-called outer iterations (see Section 8.5). In such cases, only partial convergence of each PDE is sought because the process has to be repeated in any case, rendering full convergence of individual PDEs wasteful. The conjugate gradient solver is not ideal for such scenarios because of its erratic and slow convergence in the initial stages.

Perhaps, the most critical limitation of both the method of steepest descent, as well as the CG method is that the coefficient matrix has to be symmetric. This is a serious limitation because, in the vast majority of scientific and engineering applications, this is not the case. Some common scenarios that will disrupt the symmetry of the coefficient matrix are as follows: (a) nonuniform grid spacing, (b) unstructured mesh, (c) structured curvilinear grids, (d) nonconstant transport coefficients, and (e) inclusion of advection, among many others. On account of this limitation, the CG method finds very little use in practical scientific and engineering computations.

The *CGS method* was developed by Sonneveld [7] in 1989 to combat the serious limitation of the original CG solver, i.e., that of the coefficient matrix having to be symmetric. In fact, Sonneveld referred to his solver as a "Lanczos-type" solver. Between 1952, when the CG method was developed, and 1989, when CGS was proposed, several other methods had been proposed to remove the aforementioned limitation. The most notable among these methods in the so-called bi-conjugate gradient (BCG) method [6], in which two optimization problems were simultaneously solved – one for the original linear system, and another for a companion system that used the transpose of the original coefficient matrix. To understand the motivation behind this, the reader is referred to the derivation of Eq. (3.62) from Eq. (3.61), wherein the symmetry condition was employed. The BCG method will be discussed in the next section. Among other issues, one of its main downsides is that the transpose of the coefficient matrix has to be determined. For a sparse matrix with arbitrary fill, this is not straightforward. The CGS method was the first method to circumvent this issue. It enabled use of asymmetric coefficient matrices without the need to compute its transpose, rendering the method extremely efficient and powerful. Its convergence characteristics are similar to that of the CG method. Like the CG method, in theory, for a linear system comprised of K equations, it is guaranteed to yield the exact solution in no more than K iterations. The algorithm for the CGS method is presented below.

ALGORITHM: CGS METHOD

Step 1: Guess values of ϕ at all nodes, i.e., $\phi_{i,j}$ $\forall i = 1,...,N$ and $\forall j = 1,...,M$. We denote these values as $\phi^{(0)}$. If any of the boundaries have Dirichlet boundary conditions, the guessed values for the boundary nodes corresponding to that boundary must be equal to the prescribed boundary values.

Step 2: Compute the initial residual vector, $[R^{(0)}] = [Q] - [A][\phi^{(0)}]$.

Step 3: Initialize the following vectors.
    Direction vector: $[D^{(0)}] = [R^{(0)}]$.
    Conjugate direction vector: $[D^{(0)}]* = [R^{(0)}]$.

Step 4: Compute $\alpha^{(n+1)}$ using $\alpha^{(n+1)} = \dfrac{[R^{(0)}]^T [R^{(n)}]}{[R^{(0)}]^T [A][D^{(n)}]}$.

Step 5: Compute $[G^{(n+1)}] = [D^{(n)}]^* - \alpha^{(n+1)}[A][D^{(n)}]$.

Step 6: Update solution using $[\phi^{(n+1)}] = [\phi^{(n)}] + \alpha^{(n+1)}\left\{[D^{(n)}]^* + [G^{(n+1)}]\right\}$.

Step 7: Compute the new residual vector, $[R^{(n+1)}] = [Q] - [A][\phi^{(n+1)}]$, and $R2^{(n+1)}$.

Step 8: Compute $\beta^{(n+1)} = \dfrac{[R^{(0)}]^T[R^{(n+1)}]}{[R^{(0)}]^T[R^{(n)}]}$.

Step 9: Update conjugate search direction: $[D^{(n+1)}]^* = [R^{(n+1)}] + \beta^{(n+1)}[G^{(n+1)}]$.

Step 10: Update search direction vector: $[D^{(n+1)}] = [D^{(n+1)}]^* + \beta^{(n+1)} \left\{ [G^{(n+1)}] + \beta^{(n+1)}[D^{(n)}] \right\}$.

Step 11: Monitor convergence, i.e., check if $R2^{(n+1)} < \varepsilon_{tol}$? If YES, then go to Step 12. If NO, then go to Step 4 (*i.e.*, proceed to next iteration).

Step 12: Stop iteration and postprocess the results.

In order to better understand and appreciate the CGS solver, we next apply the above algorithm for the CGS method to a simple 2×2 system that is similar to Eq. (3.59), but in which the coefficient matrix is asymmetric, namely

$$[A][\phi] = \begin{bmatrix} 3 & 2 \\ 1 & 6 \end{bmatrix} \begin{bmatrix} \phi_1 \\ \phi_2 \end{bmatrix} = [Q] = \begin{bmatrix} 2 \\ -10 \end{bmatrix}. \tag{3.74}$$

The exact solution to this system is also $(2, -2)$. As before, we start with an initial guess of $(-2, -2)$. Table 3.3 shows the convergence behavior of the CGS method when applied to the system shown in Eq. (3.74). As in the case of the CG method, the exact solution is obtained in exactly two iterations, indicating that the CGS method is as convergent as the CG method, with the added benefit that the method is valid for asymmetric coefficient matrices as well, thereby opening up its applicability to virtually any problem of practical interest. In order to assess its convergence in comparison with the other solvers discussed earlier, we next consider the same example PDE system as before.

Table 3.3 Results of Successive Iterations Using the CGS Method for the 2×2 Linear System Shown in Eq. (3.74)

Iteration	ϕ_1	ϕ_2	R_1	R_2	L^2 Norm, R2
0	−2.000000	−2.000000	12.000000	4.000000	12.649111
1	1.219955	−2.136054	2.612245	1.596372	3.061409
2	2.000000	−2.000000	0.000000	0.000000	0.000000

EXAMPLE 3.8

The same problem that was solved in Examples 3.2–3.7 is considered here, and solved using the CGS method. Grid sizes of 41×41, 81×81, and 161×161 are used again for one-to-one comparative studies. The figure below shows the convergence of the CGS method for the three mesh sizes that were considered.

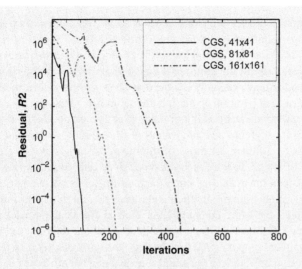

The convergence of the CGS method is slightly better than that of the CG method. On the 161×161 mesh, only 453 iterations are necessary to attain convergence. The fact that its convergence is better than the CG method is somewhat coincidental. In fact, it is extremely erratic during the initial stages of convergence, compared to the CG method. In terms of scaling, the CGS method behaves similarly to the CG method, as well. When the number of nodes is quadrupled, the iterations needed to attain convergence increases only by a factor of two, as opposed to approximately four, which was noted for all other iterative solvers considered thus far. Overall, it is clear that for the specific problem considered in this chapter, the CG and the CGS methods are vastly superior to the other methods, particularly when it pertains to the scaling behavior of the solver. At this juncture, having solved the exact same problem with seven different solvers, it is worthwhile to consider a summary of their relative performances, both in terms of iterations as well as CPU times. The table below provides such a summary.

| | Iterations | | | CPU time (s) | | |
Solver used	41×41	81×81	161×161	41×41	81×81	161×161
Jacobi	6748	27,911	115,262	0.15	2.14	26.42
Gauss–Seidel	3408	14,030	57,786	0.12	1.78	24.1
ADI	1716	7026	28,908	0.12	1.4	21.5
Stone's method	229	913	3737	0.02	0.2	2.9
MSD	6856	28,328	117,044	0.25	3.79	59.84
CG	148	305	627	0.01	0.06	0.29
CGS	109	224	453	0.01	0.07	0.35

The computations were performed on a 2.2 GHz Intel core i7 processor with 8 GB of RAM. As evident from the table above, the CPU times do not always correlate directly with the iteration count. This is because the work per iteration is significantly different for different solvers. Some anomalies are observed in addition. For example, the Jacobi method and the Gauss–Seidel method requires almost the same CPU time although the iteration count of the Jacobi method is approximately double that of the Gauss–Seidel method. We also know that the operation count in both methods is identical. This anomaly is probably due to the fact that in the Gauss–Seidel method, the same memory stack is used for both old and new values of the solution vector, and simultaneous access to different locations of the memory stack may be responsible for the loss of speed. In the Jacobi method, since old values are used on the right-hand side of the update equation, such a situation is not encountered. The ADI method, although more implicit and more efficient in terms of iterations, requires almost the same computational time as the Gauss–Seidel method. Similarly, the CGS method has a lower iteration count than the CG method, but requires more computational time because the work per iteration is slightly higher. Although CPU time is the ultimate measure of the efficiency of a solver, they can be often skewed by inefficient programming, use of poorly designed data structures, and other issues. They are also affected by the machine's architecture, the programming language, and optimization levels used in compiling the program. Hence, one must exercise caution when comparing CPU times.

3.3 OVERVIEW OF OTHER METHODS

Thus far, we have discussed two types of solvers. The Jacobi, Gauss–Seidel, and line-by-line solvers belong to a type known as classical iterative solvers, while the Stone's method, the method of steepest descent, and the CG solver and its variants belong to a type often referred to as fully implicit solvers. The CG and the CGS methods, discussed in the preceding section, also belong to a class of iterative methods known as *Krylov subspace methods*, credited to the Russian mathematician, Alexei Krylov.[†] The Krylov subspace of order m of the matrix $[A]$ is defined as

$$K_m([A],[q]) = \text{span}\left\{[q],[A][q],[A]^2[q],...,[A]^{m-1}[q]\right\}, \tag{3.75}$$

where $[q]$ is an arbitrary vector. In most modern Krylov subspace-based iterative methods, a linear combination of the vectors $[R]$, $[A][R]$, $[A]^2[R]$, and so on are used to find the next updated value of $[\phi]$, as evident from the algorithm of the CG method. This approach is chosen because matrix–vector multiplication is computationally very

[†]**Alexei Nikolaevich Krylov** (1863–1945) was a Russian engineer and applied mathematician. Although known famously for the discovery of Krylov subspaces in mathematics, he was a naval architect by profession. During his lifetime, he published more than 300 books and papers on topics as diverse as ship building, hydrodynamic stability, magnetism, astronomy, and mathematics. In 1898, Krylov received the gold medal from the Royal Institution of Naval Architects, and was the first foreigner to receive this honor. In 1904, he built the first machine capable of integrating ordinary differential equations.

cheap compared with matrix–matrix multiplications, especially if $[A]$ is sparse. The idea behind using Krylov subspaces for finding the solution to linear systems stemmed from the Cayley–Hamilton theorem [6], which states that the inverse of a matrix can be determined in terms of a linear combination of its powers. Since the vectors $[A][R]$, $[A]^2[R]$, and so on, deviate from linear independence due to power iterations, it is necessary to use some type of orthogonalization scheme to modify them, so that the higher order vectors stay independent of the lower order vectors. Many orthogonalization schemes exist. The two most popular ones are the Lanczos scheme, proposed by Lanczos in his development of the CG method [5], and the Arnoldi scheme [6].

In Section 3.2.7, we have already discussed two methods based on the Lanczos scheme. One of the earliest Lanczos scheme–based methods to handle nonsymmetric matrices is the BCG method [6,7]. As mentioned earlier, the BCG method requires determination of the transpose of $[A]$, and this is not always straightforward. Further, the BCG method has been known to be unstable [5,7], and is rarely used today. However, it laid the foundation for the bi-conjugate gradient stabilized (BiCGSTAB) algorithm [8], which is quite popular among state-of-the-art solvers based on the Lanczos scheme. It has been shown to mitigate the erratic convergence behavior exhibited by the CGS method for some problems.

Of the Krylov subspace methods that make use of the Arnoldi scheme for orthogonalization, the most popular is the generalized minimal residual (GMRES) method, developed by Saad and Schultz [9]. GMRES is a generalized version of the minimum residual (MINRES) method [10], which is only applicable to symmetric matrices. The GMRES algorithm has the advantage that the residual can be computed and minimized without determining the next iterate. The final solution can be determined after the residual has been minimized. One of the downsides of GMRES is that it requires storage of a large number of Krylov subspace vectors for efficient convergence, making the method quite memory intensive. In practice, it is common to use between 20–100 Krylov subspaces. Despite some downsides, the GMRES method is one of the most powerful and popular linear algebraic equation solvers available today. An open-source software suite that contains GMRES, known as SPASEKIT, is available from Saad's website [11] for free public use.

3.4 TREATMENT OF NONLINEAR SOURCES

Thus far, the focus of the present chapter has been entirely on linear differential equations. Many problems in science and engineering often result in nonlinear differential equations. A subset of this are problems in which the differential operators in the equation are still linear, but the source term is nonlinear. Examples of such a scenario include heat and mass sources due to chemical reactions in chemically reacting flows [12], sources due to thermal radiation in heat transfer calculations [13], and sources due to subgrid-scale friction forces in modeling flow through a porous media [14], among others. In this section, we will address the general scenario when the source term may not only be a function of the independent variables (linear case) but also an arbitrary function of the dependent variables, i.e., $S_\phi = S_\phi(\mathbf{r}, \phi)$, where $\mathbf{r}$ represents the spatial coordinates. If the source term is a function of the dependent variable, ϕ, then it is an unknown, since ϕ itself is an unknown. Furthermore, if the functional dependence of S_ϕ on ϕ is nonlinear, it renders the differential equation nonlinear. With the exception of a few special cases, the solution to nonlinear equations, require iterations. These iterations are independent of and are not to be confused with the iterations associated with iterative solution of a linear system of equations. To illustrate this point, let us consider solution of Eq. (2.11), but with a nonlinear source, such that

$$\frac{2}{(\Delta x)^2}\phi_i - \frac{1}{(\Delta x)^2}\phi_{i-1} - \frac{1}{(\Delta x)^2}\phi_{i+1} = -\exp(\phi_i) \quad \forall i = 2,3,...N-1, \quad (3.76)$$

where $S_\phi = \exp(\phi)$ has been chosen as the source term for demonstration purposes. Equation (3.76) represents a system of nonlinear equations. The coefficient matrix arising from the left-hand side of Eq. (3.76), as before, is tridiagonal. However, since the system is nonlinear, the linear algebraic solvers discussed earlier in this chapter cannot be used directly. In order to use a linear algebraic equation solver, such as TDMA, to solve Eq. (3.76), approximations have to be made, and iterations would still be necessary. For example, the following iterative algorithm may be used to solve Eq. (3.76):

1. Guess values of ϕ at all nodes. This is denoted by $\phi^{(n)}$.
2. Compute the right-hand side of Eq. (3.76), i.e., compute $-S_i^{(n)} = -\exp(\phi_i^{(n)})$ at all nodes.
3. Use TDMA solver to solve resulting tridiagonal equations, and obtain new values of ϕ at all nodes, i.e., $\phi^{(n+1)}$.
4. Compute residual.
5. Repeat Steps 2–4 until convergence.

Iterations are necessary in this case because the source term is evaluated in Step 2 at previous iteration values, rather than current iteration values, rendering the algorithm partially explicit. While the above algorithm is relatively straightforward to execute, depending on the nonlinearity of the source term, it may or may not converge, or the convergence may be slow. Ideally, the source term should be computed at $\phi^{(n+1)}$ to render Eq. (3.76) fully implicit. Potentially, the implicitness could be improved by computing the source term using a value of ϕ that is closer to the current iteration value, i.e., at $\phi^{(n+1)}$. To do so, we employ a Taylor series expansion as follows:

$$S_i^{(n+1)} = S_i^{(n)} + \frac{dS_i}{d\phi}\bigg|^{(n)} \left(\phi_i^{(n+1)} - \phi_i^{(n)}\right) + ... \quad (3.77)$$

Applying Eq. (3.77) to the specific source term considered here as an example, we obtain

$$\exp(\phi_i^{(n+1)}) = \exp(\phi_i^{(n)}) + \exp(\phi_i^{(n)})\left(\phi_i^{(n+1)} - \phi_i^{(n)}\right) + ... \quad (3.78)$$

Neglecting higher order terms in the Taylor series, and substituting it into Eq. (3.76), yields

$$\frac{2}{(\Delta x)^2}\phi_i^{(n+1)} - \frac{1}{(\Delta x)^2}\phi_{i-1}^{(n+1)} - \frac{1}{(\Delta x)^2}\phi_{i+1}^{(n+1)} = -\exp(\phi_i^{(n+1)})$$
$$\simeq -\left[\exp(\phi_i^{(n)}) + \exp(\phi_i^{(n)})\left(\phi_i^{(n+1)} - \phi_i^{(n)}\right)\right].$$

(3.79)

The right-hand side of Eq. (3.79) is only an approximation for the actual source term. In fact, if the second term on the right-hand side of Eq. (3.79) is neglected, the algorithm, presented earlier, would be recovered. The second term is linear in $\phi_i^{(n+1)}$. Thus, Eq. (3.79) is still a linear equation, and can be solved by a linear algebraic equation solver, such as TDMA. If any higher order term beyond the linear term was to be included, it would render the equation nonlinear, thereby preventing its solution using standard linear algebraic equation solvers. The process of converting a nonlinear source term to its approximate linear form using a Taylor series expansion is referred to as *source term linearization*. Neglecting the higher order terms does not alter the final solution since at convergence, $\phi_i^{(n+1)} = \phi_i^{(n)}$. By that same argument, even the new linear term can be neglected, as was done in the original algorithm without linearization. The inclusion or exclusion of the new linear term simply alters the path to (or rate of) convergence – hopefully, in a favorable manner.

Prior to its solution, Eq. (3.79) must be rearranged with known quantities on one side and unknown quantities on the other side, resulting in

$$\left(\frac{2}{(\Delta x)^2} + \exp(\phi_i^{(n)})\right)\phi_i^{(n+1)} - \frac{1}{(\Delta x)^2}\phi_{i+1}^{(n+1)} - \frac{1}{(\Delta x)^2}\phi_{i-1}^{(n+1)} = \exp(\phi_i^{(n)})(\phi_i^{(n)} - 1). \quad (3.80)$$

Equation (3.80) represents a linear tridiagonal system of equations. Instead of solving Eq. (3.76) in Step 2 of the algorithm described earlier, Eq. (3.80) could be solved.

The pressing question at this juncture is what advantage, if any, is offered by solution of Eq. (3.80) over the solution of the original unlinearized form. This question can be answered by comparing the diagonal of the original matrix, namely $2/(\Delta x)^2$, with the diagonal of the modified matrix, namely $2/(\Delta x)^2 + \exp(\phi_i^{(n)})$, and by using the Scarborough criterion as a guide. Clearly, the latter diagonal has a larger magnitude since both $2/(\Delta x)^2$ and $\exp(\phi_i^{(n)})$ are positive quantities. In other words, linearizing the source term has strengthened the diagonal. As discussed in Section 3.2, a stronger diagonal generally results in a more stable iterative scheme. This issue will be further discussed in Chapter 4. For now, we will abide by this general rule of thumb without rigorous proof.

In the example just discussed, the fact that the diagonal was made stronger by linearizing the source term happens to be specific to that particular example only. This may not always be the case. For example, if the source term was $S_\phi = \exp(-\phi)$, linearizing would result in the diagonal being $2/(\Delta x)^2 - \exp(-\phi_i^{(n)})$. In this case, the diagonal with linearization would be weaker than the one without linearization. Based on our general rule of thumb, this is not desirable, and therefore, the source term should not be linearized at all. In the most general scenario, linearization may make the diagonal stronger for some nodes (nodal equations) and weaker for others. It all depends on how the source term behaves from node to node, and also, from iteration to iteration. In practical scientific computations, there is no foolproof way to judge *a priori* if linearization will strengthen the diagonal.

In order to prevent linearization from making the diagonal weaker, an algorithm that examines the diagonal locally (i.e., node-wise) with and without linearization must be implemented. Such an algorithm is presented below for solution of a general set of algebraic equations with a nonlinear source.

ALGORITHM: MATRIX ASSEMBLY AND SOLUTION WITH NONLINEAR SOURCE TERM

Step 1: Guess values of ϕ at all nodes, i.e., $\phi_{i,j}$ $\forall i = 1,...,N$ and $\forall j = 1,...,M$. We denote these values as $\phi^{(0)}$. If any of the boundaries have Dirichlet boundary conditions, the guessed values for the boundary nodes corresponding to that boundary must be equal to the prescribed boundary values.

Step 2: Write discrete equations in a form such that the diagonal is positive, and compute all link coefficients. The linear system, in its general form, following Eq. (3.19), may be written as:

$$a_k \phi_k^{(n+1)} + \sum_{\substack{j=1 \\ j \neq k}}^{N_{nb,k}} a_j \phi_k^{(n+1)} = Q_k^{(n+1)} = -S_k^{(n+1)} .$$

Step 3: Write the right hand–side vector in linearized form:

$$Q_k^{(n+1)} = Q_k^{(n)} + \left. \frac{dQ_k}{d\phi} \right|^{(n)} \left(\phi_k^{(n+1)} - \phi_k^{(n)} \right) = Q_{C,k} + Q_{P,k} \phi_k^{(n+1)} , \text{ such that}$$

$$Q_{C,k} = Q_k^{(n)} - \left. \frac{dQ_k}{d\phi} \right|^{(n)} \phi_k^{(n)} , \text{ and } Q_{P,k} = \left. \frac{dQ_k}{d\phi} \right|^{(n)} .$$

Step 4: Modify the original diagonal and compute the right hand–side vector as follows:

$$a_k = a_k - MIN(0, Q_{P,k}) , \text{ and}$$

$$Q_k^{(n+1)} = Q_{C,k} + MAX(0, Q_{P,k}) \phi_k^{(n)} .$$

Step 5: Solve the linearized system of algebraic equations using a solver of choice. If an iterative solver is used, it is customary to enforce only partial convergence (few orders of magnitude decrease in residual) since this step has to be repeated anyway.

Step 6: Compute residual ($R2^{(n+1)}$) using the unlinearized form of the algebraic equations, i.e., the equation shown in Step 2.

Step 7: Monitor convergence, i.e., check if $R2^{(n+1)} < \varepsilon_{tol}$? If YES, then go to Step 8. If NO, then go to Step 3 (i.e., proceed to next iteration).

Step 8: Stop iteration and postprocess the results.

Step 4 is a critical step in the above algorithm. It ensures that the component arising out of linearization is added to the diagonal only if it strengthens the diagonal. Otherwise, linearization is essentially bypassed, and $Q_k^{(n+1)}$ is set to $Q_k^{(n)}$. This operation is done locally for each node k. One other point to note is that the residual is computed based on the original nonlinear equation since our final objective is to satisfy this equation. The iterations referred to in the above algorithm are known as *outer iterations*. In contrast, the iterations needed in Step 5 for the linear algebraic equation solver are known as *inner iterations*. The product of these two types of iterations dictates the computational efficiency of the overall solution algorithm. Additional discussion regarding the solution of nonlinear PDEs in general in included in Section 8.1. An example is considered next to demonstrate the linearization strategy and algorithm, just discussed, and to confirm if a stronger diagonal is tantamount to a more stable and convergent iterative scheme.

EXAMPLE 3.9

In this example, we consider solution of the equation $\nabla^2 \phi = 1 + \exp(10\phi)$ in a square of unit length. The boundary conditions are as follows: $\phi(0, y) = \phi(x, 0) = 0$, and $\phi(1, y) = \phi(x, 1) = 1$. The finite difference equations are given by Eq. (2.48). The equation is solved with and without linearization of the source term. The linearized system is solved using the Gauss–Seidel method on a 41×41 mesh. A single sweep (inner iteration) of Gauss–Seidel was performed at each outer iteration. An initial guess equal to 0 was used for all interior nodes. For convergence, the tolerance of the outer iterations was set to 10^{-6}. The figure below shows the residuals with and without linearization of the source term as well as a contour plot of the converged solution.

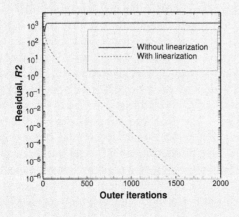

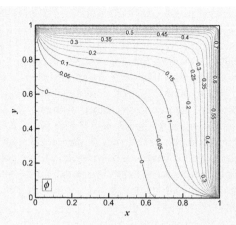

As indicated by the residual plot, when linearization is not performed, convergence is not attained. Although not shown here, convergence was attained both with and without linearization when the source term was taken to be $1 + \exp(5\phi)$. Based on these results, it may be concluded that, in general, linearization is always beneficial. Depending on the degree of nonlinearity of the source term, certain problems may converge even without linearization, but this is not always guaranteed. Although linearization does not guarantee convergence, it is certainly preferable over no linearization.

In this chapter, methods for solving a system of linear algebraic equations were discussed. Direct solution using Gaussian elimination was found to be prohibitive for large matrix sizes, typically encountered in practical computations, both from memory as well as computational workload standpoints. The matrices resulting from discretization of PDEs are typically very sparse and, often, banded. Over the past century, the sparseness of the coefficient matrix has been utilized to develop a wide gamut of iterative solvers that are more effective than direct solution. Several different iterative solvers, with increasing degree of implicitness, were discussed in this chapter. Finally, a class of methods, known as Krylov subspace methods, was discussed. Of these, the CGS method was demonstrated to be a powerful method for solving linear systems with nonsymmetric coefficient matrices. The section on linear algebraic equation solvers concluded with an overview of a few other state-of-the-art Krylov subspace methods. The final section of the chapter dealt with treatment of differential equations with nonlinear source terms. One of the popular methods for solving linear systems – the multigrid method – was not discussed in this chapter. The development of the multigrid method requires a thorough understanding of the mathematical properties of the coefficient matrix and convergence errors. These important issues are discussed in Chapter 4, followed by a detailed presentation of the multigrid method.

REFERENCES

[1] Patankar SV. Numerical heat transfer and fluid flow. Washington: Hemisphere; 1980.

[2] Peaceman DW, Rachford HH Jr. The numerical solution of parabolic and elliptic differential equations. J Soc Ind Appl Math 1955;3:28–41.

[3] Douglas J Jr, Gunn JE. A general formulation of alternating direction methods. Numerische Mathematik 1964;6(1):428–53.

[4] Stone HL. Iterative solution of implicit approximations of multi-dimensional partial differential equations. SIAM J Numer Anal 1968;5:530–58.

[5] Lanczos C. Solution of systems of linear equations by minimized iteration. J Res Nat Bur Standards 1952;49:33–53.

[6] Saad Y. Iterative methods for sparse linear systems. 2nd ed. SIAM: Philadelphia; 2003.

[7] Sonneveld P. CGS: a fast Lanczos-type solver for nonsymmetric linear systems. SIAM J Sci Stat Comput 1989;10(1):36–52.

[8] Van der Vorst HA. Bi-CGSTAB: a fast and smoothly converging variant of Bi-CG for the solution of nonsymmetric linear systems. SIAM J Sci Stat Comput 1992;13(2):631–44.

[9] Saad Y, Schultz MH. GMRES: a generalized minimal residual algorithm for solving nonsymmetric linear systems. SIAM J Sci Stat Comput 1986;(7):856–69.

[10] Paige CC, Saunders MA. Solution of sparse indefinite systems of linear equations. SIAM J Numer Anal 1975;12:617–29.

[11] Saad Y, SPASEKIT, Available from: http://www-users.cs.umn.edu/~saad/software/SPARSKIT/.

[12] Kumar A, Mazumder S. Coupled solution of the species conservation equations using unstructured finite-volume method. Int J Numer Methods Fluids 2010;64(4):409–42.

[13] Mazumder S. A new numerical procedure for coupling radiation in participating media with other modes of heat transfer. J Heat Transf 2005;127(9):1037–45.

[14] Whitaker S. The Forchheimer Equation: a theoretical development. Transport Porous Med 1996;25(1):27–61.

EXERCISES

3.1 Derive the formulas for the pentadiagonal solver shown in Code Snippet 3.4. Also, count the number of long operations in the whole algorithm.

3.2 Solve the problem presented in Example 3.2 using the 4th order central difference scheme, and both tridiagonal and pentadiagonal alternating direction implicit methods. Use a 81×81 mesh. Plot the convergence curves on the same graph, and comment on your results.

3.3 Repeat Example 3.2 with the second-order central difference scheme and the Gauss–Seidel method. However, instead of using the Gauss–Seidel method "as is," use the following update formula for Step 2:

$$\phi_{i,j}^{(n+1)} = \omega \left[\frac{S_{i,j} - a_E \phi_{i+1,j}^{(n)} - a_W \phi_{i-1,j}^{(n+1)} - a_N \phi_{i,j+1}^{(n)} - a_S \phi_{i,j-1}^{(n+1)}}{a_O} \right] + (1-\omega)\phi_{i,j}^{(n)},$$

where ω is the so-called *linear relaxation factor*. When $\omega = 1$, the update formula, presented in Section 3.2.3, is recovered. Use a 81×81 mesh. Plot convergence curves for $\omega = 0.5$, 1, and 1.5 on the same graph. Comment on your results.

3.4 Consider the following second-order PDE and boundary conditions:

$$\frac{\partial^2 \phi}{\partial x^2} + \frac{\partial^2 \phi}{\partial y^2} = S_\phi = 50000 \cdot \exp[-50\{(1-x)^2 + y^2\}] \cdot [100\{(1-x)^2 + y^2\} - 2]$$

$$\phi(1, y) = 100(1-y) + 500 \exp(-50y^2)$$
$$\phi(0, y) = 500 \exp(-50\{1+y^2\})$$
$$\phi(x, 0) = 100x + 500 \exp(-50(1-x)^2)$$
$$\phi(x, 1) = 500 \exp(-50\{(1-x)^2 + 1\})$$

Note that this is a linear system of equations. The above system has an analytical solution, given by

$$\phi(x, y) = 500 \exp(-50\{(1-x)^2 + y^2\}) + 100x(1-y)$$

In Exercise 2.6, you have already derived the finite difference equations for this problem. Solve those equations using the following methods: (a) Jacobi, (b) Gauss–Seidel, (c) ADI, (d) Stone's method, (e) method of steepest descent, and (f) CG methods. Use 41×41, 81×81, and 161×161 meshes. For each mesh size, plot the residuals of all 6 methods on a single graph. Comment on your results. Also, make a table showing iterations for convergence and CPU times taken for each method and mesh. Comment on your results.

3.5 Repeat Exercise 3.4 with the conjugate gradient squared method and a nonuniform mesh in both x and y directions. For creating this nonuniform mesh, use the formula shown in Exercise 2.5 with a stretching factor of 1.001. Plot the residuals and comment on your findings.

3.6 Consider the following second-order PDE:

$$\frac{\partial^2 \phi}{\partial x^2} + \frac{\partial^2 \phi}{\partial y^2} = 1, \text{ subject to the following boundary conditions:}$$

$$\frac{\partial \phi}{\partial x}(1, y) = 1$$
$$\phi(0, y) = 0$$
$$\phi(x, 0) = 0$$
$$\phi(x, 1) = 0$$

Apply the central difference scheme for the interior nodes and a first-order scheme for the nodes on the right boundary. Solve the resulting equations using (a) Stone's method, and (b) CGS method. Use an 81×81 mesh. Make contour plots of the solution, and line plots of the residuals. Comment on your solution, as well as ease of implementation of the two solvers compared to Exercises 3.4 and 3.5. What difficulties might you encounter if the right boundary nodes were treated using a second-order scheme?

3.7   Consider the following second-order nonlinear ordinary differential equation and boundary conditions:

$$\frac{d^2\phi}{dx^2} = \exp(C\phi), \qquad \phi(0) = 0, \phi(1) = 1 \text{, where } C \text{ is a constant.}$$

a.   Discretize the equation using the central difference scheme. Use equal node spacing with 41 nodes. Do not linearize the source. Solve the resulting discrete equations using TDMA. Since the equation is nonlinear, you will have to iterate. Set the tolerance for the residual to 10^{-10}. Plot the solution ($\phi(x)$ versus x) for $C = 2$. Also, plot the residuals ($R2$ versus the number of iterations) on a semilog plot. Repeat the calculation for $C = 4$, and plot the solution and residuals again.

b.   Repeat (a), but now linearize the source term using the Taylor series expansion method discussed in Section 3.3. Discuss your findings from (a) and (b).

Stability and Convergence of Iterative Solvers

Several iterative solvers were presented in Chapter 3 and a comparative study was conducted, in which the same boundary value problem was solved to assess the *pros* and *cons* of each method. Three critical questions were not addressed in the preceding chapter. (a) For a given iterative method, under what condition may convergence be elusive? (b) Why is the rate of convergence different for different iterative methods? (c) Irrespective of the method, why does the convergence deteriorate with mesh refinement? This chapter aims to answer these three critical questions.

At the start of Section 3.2, it was demonstrated that convergence is not always guaranteed. Lack of diagonal dominance (violation of the Scarborough criterion) was suggested as a possible cause for divergence. Further, it was insinuated that some mathematical property of the coefficient matrix is really the root cause of convergence or divergence, and also the rate of convergence in cases when the convergence is attainable. In this chapter, aside from diagonal dominance, we will develop a set of mathematical criteria, based on the properties of the coefficient matrix and the iterative scheme, that will enable us to determine *a priori*, i.e., without actually solving the linear system of equations, whether a given partial differential equation (PDE) with a given discretization scheme and a given iterative method of solution will converge or not.

4.1 EIGENVALUES AND CONDITION NUMBER

The stability and convergence of the iterative solution of a linear system is deeply rooted in the eigenvalues of the linear system under consideration. Therefore, we first discuss calculation of the eigenvalues and the implication of their magnitudes. The eigenvalues of a matrix $[A]$ can be computed using the equation

$$[A][q] = \lambda[q], \tag{4.1}$$

where the scalar, λ, is the so-called *eigenvalue*, and $[q]$ is the so-called *eigenvector*. The product of a square matrix and a vector in hyperdimensional space (or column matrix), as in the left-hand side of Eq. (4.1), will result in another vector. In general, this new vector may have no relation to the original vector. If, however, the result of the product is the original vector times some scalar quantity, then, the vector is the so-called eigenvector of the matrix $[A]$, and the scalar premultiplier is known as the eigenvalue of $[A]$. It is easy to see that the eigenvalue represents a stretching factor. The nontrivial solution to Eq. (4.1) can be found if and only if

$$\det\left([A] - \lambda[I]\right) = 0,$$ (4.2)

where $[I]$ is the identity matrix. Equation (4.2) represents a polynomial equation of degree K (i.e., the number of equations or unknowns), and is also known as the *characteristic equation*. The polynomial resulting from the left-hand side of Eq. (4.2) is also known as the *characteristic polynomial* of matrix $[A]$. In general, the roots – K of them – resulting from the solution of Eq. (4.2) may be complex numbers. In summary, a square matrix $[A]$ of size $K \times K$ will have K eigenvalues, which may be real or complex.

The condition number of a square matrix $[A]$, comprised of elements that are real only, is defined as the ratio of its largest to its smallest eigenvalue (by moduli), i.e.,

$$\kappa(A) = \left|\frac{\lambda_{\max}(A)}{\lambda_{\min}(A)}\right|.$$ (4.3)

Since the eigenvalues may be real or complex even for a matrix comprised of real numbers as its elements, as is always the case for coefficient matrices arising out of discretization of PDEs, the moduli of the eigenvalues must be used in the case when they are complex. Also, whether the eigenvalues are positive or negative does not affect the condition number, since the moduli are used in the definition.

Using the definitions provided by Eqs. (4.2) and (4.3), it follows that an identity matrix has a condition number equal to unity since all its eigenvalues are also equal to unity. In terms of iterative solution of a linear system, this is the best-case scenario, because if $[A] = [I]$, no iterations would be necessary to solve the linear system. Hence, a matrix with a condition number close to unity is known as a *well-conditioned matrix*. Such matrices are amenable to efficient iterative solution, as we shall see shortly. On the other hand, a matrix with a large condition number is known as an *ill-conditioned matrix*, and convergence for such a linear system may be difficult or elusive. A matrix with a condition number equal to infinity is known as a *singular matrix*. If the coefficient matrix is singular, the matrix is not invertible. For the particular scenario under consideration, i.e., solution of PDEs, the coefficient matrix is rarely singular. For practical problems, singular matrices can only arise due to programming errors, whereby one of the diagonal elements has been incorrectly assigned a zero value. From Eq. (4.3), a small condition number implies that the maximum and minimum eigenvalues must be fairly close to each other. In the numerical methods literature, this is often referred to as *clustering of eigenvalues*. A matrix whose eigenvalues are clustered is preferable for iterative solution over one whose eigenvalues are scattered. An example, illustrating the use of Eqs. (4.2) and (4.3), is presented next.

EXAMPLE 4.1

In this example, we calculate the eigenvalues and condition numbers of two matrices considered at the beginning of Section 3.2, namely

$$[A] = \begin{bmatrix} 5 & 2 & 2 \\ 2 & -6 & 3 \\ 1 & 2 & 7 \end{bmatrix} \text{ and } [C] = \begin{bmatrix} 5 & 2 & 2 \\ 2 & -6 & 3 \\ 7 & 2 & 1 \end{bmatrix}$$

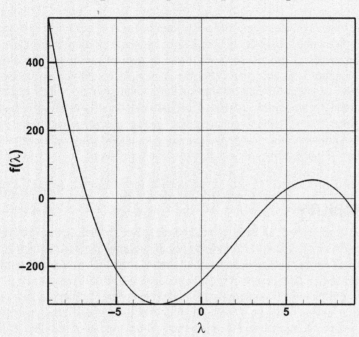

Using Eq. (4.2), we get

$$\begin{vmatrix} 5-\lambda & 2 & 2 \\ 2 & -6-\lambda & 3 \\ 1 & 2 & 7-\lambda \end{vmatrix} = 0,$$

or $f(\lambda) = -\lambda^3 + 6\lambda^2 + 49\lambda - 242 = 0$. A plot of the cubic function is shown in the above figure. The three roots of the cubic equation are -6.701, 4.297, and 8.404, as indicated by the function changing sign three times. The condition number of $[A]$ is $\kappa(A) = |8.404 / 4.297| = 1.96$. In a similar manner, the eigenvalues of the matrix $[C]$ can be determined, and they are -6.671, -1.314, and 7.985. The resulting condition number of $[C]$ is $\kappa(C) = |7.985 / (-1.314)| = 6.07$.

In Section 3.2, the Jacobi method was used to solve the system $[A][\phi] = [\,9 \quad -1 \quad 10\,]^T$, and convergence was attained. However, the same method resulted in divergence when attempting to solve $[C][\phi] = [\,9 \quad -1 \quad 10\,]^T$, where the matrices $[A]$ and $[C]$ are as shown in Example 4.1. Earlier discussions insinuated that this change from convergence to divergence was caused by a change is some property of the coefficient matrix. Example 4.1 clearly shows that interchanging the two coefficients in the last row of the coefficient matrix drastically alters the condition number of the matrix. The second case, which has a much larger condition number, is the case where convergence could not be attained. Thus, it is fair to conclude that the condition number of the coefficient matrix has some relation to the convergence of an iterative solver used to solve the linear system of equations. The closer the condition number is to unity, the better the convergence, and vice versa. However, it is not clear what cut-off value of the condition number, if any, might cause divergence. Further, the condition number of the coefficient matrix is not sufficient to explain why the same system of equations may reach convergence with some iterative scheme and not with others since the condition number of the coefficient matrix is independent of the iterative scheme used to solve the system.

4.2 STABILITY

Stability is a property of an iterative method (or solver), which states whether it will be convergent or divergent. It is a property of the iterative method, and not the coefficient matrix *per se*, although, as we shall find shortly, the two are interconnected. Also, stability is not a measure of the rate of convergence or divergence. It is simply a forecast of whether a given iterative method will converge or diverge. According to the Lax equivalence theorem, a consistent numerical scheme will be convergent if it is stable. In order to understand the mathematical foundation of stability, let us consider the Gauss–Seidel iteration scheme, discussed in Section 3.2.2, for the simple 3×3 linear system, as follows:

$$
\begin{aligned}
a_{11}\phi_1 + a_{12}\phi_2 + a_{13}\phi_3 &= Q_1 \\
a_{21}\phi_1 + a_{22}\phi_2 + a_{23}\phi_3 &= Q_2 . \\
a_{31}\phi_1 + a_{32}\phi_2 + a_{33}\phi_3 &= Q_3
\end{aligned}
\tag{4.4}
$$

Using the Gauss–Seidel iteration scheme, Eq. (4.4) may be rewritten as

$$
\begin{aligned}
a_{11}\phi_1^{(n+1)} &= Q_1 - a_{12}\phi_2^{(n)} - a_{13}\phi_3^{(n)} \\
a_{22}\phi_2^{(n+1)} &= Q_2 - a_{21}\phi_1^{(n+1)} - a_{23}\phi_3^{(n)} , \\
a_{33}\phi_3^{(n+1)} &= Q_3 - a_{31}\phi_1^{(n+1)} - a_{32}\phi_2^{(n+1)}
\end{aligned}
\tag{4.5}
$$

which may be further rearranged to write

$$
\begin{aligned}
a_{11}\phi_1^{(n+1)} &= Q_1 - a_{12}\phi_2^{(n)} - a_{13}\phi_3^{(n)} \\
a_{21}\phi_1^{(n+1)} + a_{22}\phi_2^{(n+1)} &= Q_2 - a_{23}\phi_3^{(n)} . \\
a_{31}\phi_1^{(n+1)} + a_{32}\phi_2^{(n+1)} + a_{33}\phi_3^{(n+1)} &= Q_3
\end{aligned}
\tag{4.6}
$$

Equation (4.6) may be written in matrix form as

$$\left[\begin{bmatrix} 1 & 0 & 0 \\ 0 & 1 & 0 \\ 0 & 0 & 1 \end{bmatrix} + \begin{bmatrix} 0 & 0 & 0 \\ \bar{a}_{21} & 0 & 0 \\ \bar{a}_{31} & \bar{a}_{32} & 0 \end{bmatrix}\right] \begin{bmatrix} \phi_1^{(n+1)} \\ \phi_2^{(n+1)} \\ \phi_3^{(n+1)} \end{bmatrix} = \begin{bmatrix} \bar{Q}_1 \\ \bar{Q}_2 \\ \bar{Q}_3 \end{bmatrix} - \begin{bmatrix} 0 & \bar{a}_{12} & \bar{a}_{13} \\ 0 & 0 & \bar{a}_{23} \\ 0 & 0 & 0 \end{bmatrix} \begin{bmatrix} \phi_1^{(n)} \\ \phi_2^{(n)} \\ \phi_3^{(n)} \end{bmatrix},$$

(4.7)

where $\bar{a}_{ij} = a_{ij}/a_{ii}$ and $\bar{Q}_i = Q_i/a_{ii}$ has been used in deriving Eq. (4.7) from Eq. (4.6). Defining an upper and a lower triangular matrix as

$$[U] = -\begin{bmatrix} 0 & \bar{a}_{12} & \bar{a}_{13} \\ 0 & 0 & \bar{a}_{23} \\ 0 & 0 & 0 \end{bmatrix}; \quad [L] = -\begin{bmatrix} 0 & 0 & 0 \\ \bar{a}_{21} & 0 & 0 \\ \bar{a}_{31} & \bar{a}_{32} & 0 \end{bmatrix},$$

(4.8)

Eq. (4.7) may be written as

$$\left[[I]-[L]\right]\left[\phi^{(n+1)}\right]=\left[\bar{Q}\right]+[U]\left[\phi^{(n)}\right],$$

(4.9)

which, upon further rearrangement, yields

$$\left[\phi^{(n+1)}\right]=\left[[I]-[L]\right]^{-1}[U]\left[\phi^{(n)}\right]+\left[[I]-[L]\right]^{-1}\left[\bar{Q}\right]=[B]\left[\phi^{(n)}\right]+\left[[I]-[L]\right]^{-1}\left[\bar{Q}\right].$$

(4.10)

   Although derived using a 3×3 linear system as an example, it is clear that Eq. (4.10) is the Gauss–Seidel scheme written in matrix form for a general $K \times K$ linear system. The matrix $[B]$, which, in this particular case, is equal to $\left[[I]-[L]\right]^{-1}[U]$, is also known as the *iteration matrix*. It is termed so because it essentially operates on the solution vector at the previous iteration, $[\phi^{(n)}]$, to yield the solution at the current iteration, $[\phi^{(n+1)}]$, as is evident from Eq. (4.10). Two important points must be noted at this juncture. First, the $[L]$ and $[U]$ matrices are upper and lower triangles of the normalized $[A]$ matrix. Therefore, the $[B]$ matrix is intimately tied to the $[A]$ matrix. Second, depending on the iteration scheme or iterative method used, the exact relation between $[A]$ and $[B]$ will change. Since, by definition, the exact numerical solution is obtained when the residual is zero (see Section 3.1) and $[\phi^{(n+1)}] = [\phi^{(n)}]$, it follows that

$$\left[\phi^{(E)}\right]=[B]\left[\phi^{(E)}\right]+\left[[I]-[L]\right]^{-1}\left[\bar{Q}\right],$$

(4.11)

where $[\phi^{(E)}]$ denotes the exact numerical solution. Subtracting Eq. (4.10) from Eq. (4.11) yields

$$\left[\phi^{(E)}\right]-\left[\phi^{(n+1)}\right]=[B]\left\{\left[\phi^{(E)}\right]-\left[\phi^{(n)}\right]\right\}.$$

(4.12)

The quantity $\lfloor\phi^{(E)}\rfloor-\lfloor\phi^{(n)}\rfloor$ denotes the convergence error at the nth iteration, discussed earlier in Section 3.2. Denoting it by $[\varepsilon^{(n)}]$, Eq. (4.12) may be written as

$$\left[\varepsilon^{(n+1)}\right]=[B]\left[\varepsilon^{(n)}\right].$$

(4.13)

Equation (4.13) shows that the matrix $[B]$ operates on the error at the nth iteration to produce the error at the $(n+1)$th iteration. In other words, it acts as an amplifier for the error. This is the reason matrix $[B]$ is also known as the *amplification matrix*. If the amplification matrix is such that the errors get amplified from iteration to iteration, the iterative scheme is unstable, by definition. It is easy to imagine what characteristics the matrix $[B]$ must possess in order for the error to decay from the nth iteration to the $(n+1)$th iteration. For example, if Eq. (4.13) were a scalar equation, the scalar B must be less than unity for the error to decay. In the vector case, the largest eigenvalue, by magnitude or modulus, of matrix $[B]$ must be less than unity. The largest eigenvalue of the iteration or amplification matrix is also known as the *spectral radius of convergence*. Hence, the stability criterion of any iterative method for solving a system of linear algebraic equations may be written as

$$\lambda_{SR} = \max\left(\left|\lambda_i(B)\right|\right) < 1, \tag{4.14}$$

where λ_{SR} denotes the spectral radius of convergence. It is clear from the stability criterion shown in Eq. (4.14) that the stability of an iterative solver depends not only on the $[A]$ matrix (since $[A]$ is related to $[B]$), but also on the iterative scheme, which dictates the exact relation between $[A]$ and $[B]$. This explains two previous findings: (1) the reason why with the same iterative method, some problems converge, while others do not, and (2) the reason why the same problem converges with some iterative methods but not with others. An example is considered next to illustrate the effect of the coefficient matrix on stability.

EXAMPLE 4.2

In this example, we calculate the spectral radius of convergence of the iteration matrices resulting from Jacobi iterations, with the two matrices considered in Example 4.1, namely

$$[A] = \begin{bmatrix} 5 & 2 & 2 \\ 2 & -6 & 3 \\ 1 & 2 & 7 \end{bmatrix} \text{ and } [C] = \begin{bmatrix} 5 & 2 & 2 \\ 2 & -6 & 3 \\ 7 & 2 & 1 \end{bmatrix}$$

Using Eq. (4.8), we first calculate the normalized upper and lower triangular matrices:

$$[L] = \begin{bmatrix} 0 & 0 & 0 \\ 0.333 & 0 & 0 \\ -0.142 & -0.285 & 0 \end{bmatrix} \text{ and } [U] = \begin{bmatrix} 0 & -0.4 & -0.4 \\ 0 & 0 & 0.5 \\ 0 & 0 & 0 \end{bmatrix}$$

Following the steps described earlier for the Gauss–Seidel method, the Jacobi iteration may be written as $\left[\varepsilon^{(n+1)}\right] = [B]\left[\varepsilon^{(n)}\right] = \{[L] + [U]\}\left[\varepsilon^{(n)}\right]$, where the $[L]$ and $[U]$ matrices are as shown previously. The amplification matrix is thus cal-

culated as $[B] = \begin{bmatrix} 0 & -0.4 & -0.4 \\ 0.333 & 0 & 0.5 \\ -0.142 & -0.285 & 0 \end{bmatrix}$. The eigenvalues of the $[B]$ matrix

may be computed using Eq. (4.2), resulting in
$$
\begin{vmatrix}
-\lambda & -0.4 & -0.4 \\
0.333 & -\lambda & 0.5 \\
-0.142 & -0.285 & -\lambda
\end{vmatrix} = 0 ,
$$

which yields the following characteristic equation: $\lambda^3 + 0.219\lambda - 0.066 = 0$.

Solution of this yields the following roots: $\lambda_1 = 0.24$, $\lambda_2 = -0.12 + 0.512i$, and $\lambda_3 = -0.12 - 0.512i$. Thus, $|\lambda_1| = 0.24$, and $|\lambda_2| = |\lambda_3| = 0.525$. The spectral radius is the largest of the three eigenvalues, and is equal to 0.525, indicating that the Jacobi method is stable for coefficient matrix $[A]$.

The same procedure is repeated next for matrix $[C]$, and yields the following three eigenvalues: $\lambda_1 = 1.638$, $\lambda_2 = -0.819 + 0.585i$, and $\lambda_2 = -0.819 - 0.585i$. Thus, $|\lambda_1| = 1.638$ and $|\lambda_2| = |\lambda_3| = 1.006$. The spectral radius is the largest of the three eigenvalues and is equal to 1.638, indicating that the Jacobi method is unstable for the coefficient matrix $[C]$. This finding is consistent with what has already been shown in Section 3.2. While Example 3.1 showed that the linear system with coefficient matrix $[C]$ is less likely to converge than the one with coefficient matrix $[A]$, it failed to demarcate between convergence and divergence. The stability criterion provides a clear mathematical demarcation between the two scenarios, as shown in this example.

The preceding example clearly shows that the spectral radius of convergence is indeed a foolproof measure of stability. Unfortunately, computing the eigenvalues of a $K \times K$ matrix using Eq. (4.2) is a daunting task. It implies finding the roots of a polynomial equation of Kth degree. Although several well-established methods are available to solve the characteristic polynomial equation, all of them require application of a complicated numerical method. Hence, finding the eigenvalues is, arguably, more difficult than solving the set of linear equations itself. It is, therefore, fair to conclude that although we have established the theory for determining the stability of an iterative method, putting it to practice is still elusive. One observation worth making at this point is that, in practice, only the largest eigenvalue of $[A]$ is needed, rather than all the eigenvalues. In the section to follow, a method for computing the spectral radius of convergence, based on Fourier decomposition of the errors, is presented. This stability analysis method, based on the Fourier decomposition of the errors, is referred to in the literature as von Neumann stability analysis, named after John von Neumann*.

---

*__John von Neumann__ (1903–1957) was an applied mathematician and physicist, who was born in Budapest, Hungary, and later migrated to the United States and became a naturalized citizen. He is known for his contributions to mathematics (functional analysis, game theory), physics (quantum mechanics), and computing (cellular automata, linear programming). He spent the major part of his academic career at Princeton University and later worked at the Los Alamos National Laboratory. He is one of the few American scientists recognized on a postal stamp (in 2005).

4.2.1 FOURIER DECOMPOSITION OF ERRORS

As discussed in the preceding section, the error between the exact numerical solution and the partially converged solution is known as the convergence error. In this section, this error is decomposed into its Fourier or spectral components. If we consider a one-dimensional problem in the interval $[0,L]$, then, within that interval, the error will be a continuous function, and may appear, for example, like the function shown in Fig. 4.1(a). In this case, Dirichlet boundary conditions have been assumed at the two ends, as indicated by zero error values at the two ends. In this continuous representation, the error can have oscillations of any frequency (or wavelength). In reality, however, the error can be computed only at a finite number of spatial locations corresponding to the nodes. Thus, any wavelength of oscillation below the grid size

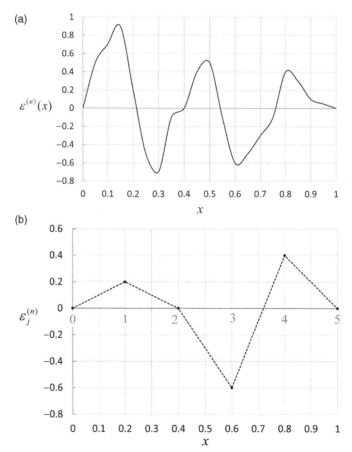

FIGURE 4.1 Example of (a) Continuous and (b) Discrete Representations of the Convergence Error

In the case of the discrete representation, the errors are only available at the nodes numbered 0 through $M-1$. In this example, $L=1$ and $M=6$.

cannot be resolved, as is evident in Fig. 4.1(b). In this example, only 6 nodes ($=M$) have been used. Clearly, the error distribution shown in Fig. 4.1(b) is missing some of the high-frequency oscillations depicted in the continuous representation in Fig. 4.1(a).

In general, the error distribution, either continuous or discrete, may be written as a Fourier series. In the case of the continuous representation, using the Fourier series in half range, the error may be expressed as

$$\varepsilon^{(n)}(x) = \sum_{m=0}^{\infty} A_m^{(n)} \cos\left(m\pi \frac{x}{L} \right) + B_m^{(n)} \sin\left(m\pi \frac{x}{L} \right), \qquad (4.15)$$

where $\varepsilon^{(n)}(x)$ denotes the convergence error at the nth iteration. In general, both the sine and the cosine series must be retained so that the boundary conditions can be satisfied. For example, if Dirichlet boundary conditions are used, the error must be zero at the two boundaries, i.e., at $x = 0$ and at $x = L$. This condition cannot be satisfied with a cosine series alone unless all the coefficients, $A_m^{(n)}$, are set to zero. Similarly, if Neumann or Robin boundary conditions are used, the error is nonzero at the boundaries, and this condition cannot be satisfied with a sine series alone. Since both sine and cosine terms must be retained, for the sake of generality, it is more convenient to use the so-called Euler form of the Fourier series, written as

$$\varepsilon^{(n)}(x) = \sum_{m=0}^{\infty} C_m^{(n)} \exp\left(i\, m\pi \frac{x}{L} \right), \qquad (4.16)$$

where $i = \sqrt{-1}$. The continuous representation has infinite terms. Thus, oscillations of the error of all conceivable frequencies can be captured. In contrast, the discrete representation, which is more suitable for our purposes, may be written as

$$\varepsilon^{(n)}(x_j) = \varepsilon_j^{(n)} = \sum_{m=0}^{M-1} C_m^{(n)} \exp\left(i\, m\pi \frac{x_j}{L} \right), \qquad (4.17)$$

where $x_j = j\Delta x$ are the nodal locations, with $j = 0,1,2,...M-1$ being the nodal indices. $\varepsilon_j^{(n)}$ denotes the convergence error at the jth node and nth iteration. Here, the total number of nodes is denoted by M. Since there are M equally spaced nodes, it follows that $\Delta x = L/(M-1)$, and therefore, $x_j/L = j/(M-1)$. With these simplifications, Eq. (4.17) may be written as

$$\varepsilon_j^{(n)} = \sum_{m=0}^{M-1} C_m^{(n)} \exp\left(i\, m\pi \frac{j}{M-1} \right) = \sum_{m=0}^{M-1} C_m^{(n)} \exp\left(i\, j\theta_m \right), \qquad (4.18)$$

where the phase angle, θ_m, is given by

$$\theta_m = \frac{m\pi}{M-1}. \qquad (4.19)$$

To understand the Fourier decomposition in further depth, let us go back to the example shown in Fig. 4.1(b). Here, the error is zero at the two boundaries, and

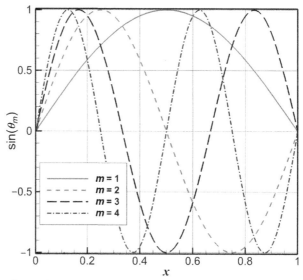

FIGURE 4.2 Fourier (or Spectral) Components of the Error Distribution Shown in Fig. 4.1(b)

therefore, the Fourier series reduces to only the sine series. Further, since $M = 6$, the phase angle becomes $\theta_m = m\pi / 5$. Therefore, the error in this particular case may be written as

$$\varepsilon_j^{(n)} = \sum_{m=0}^{5} C_m^{(n)} \sin\left(j\theta_m \right). \tag{4.20}$$

When $m = 0$, the phase angle is equal to zero, and the corresponding term in the Fourier series vanishes. Similarly, when $m = 5$, the phase angle becomes $\theta_5 = \pi$, and again, the corresponding term vanishes. Thus, in this particular case, the error is essentially a linear combination of four sine waves, which are shown in Fig. 4.2.

As convergence is approached, each of these Fourier components (sine waves) must get damped out, and their amplitudes, $C_m^{(n)}$, must decrease with increasing values of n. If the amplitude of even one of the four waves increases with iterations, the error will grow, and finally result in divergence. In the next section, we use this fundamental premise to compute the spectral radius of convergence.

4.2.2 SPECTRAL RADIUS OF CONVERGENCE

As discussed earlier, for a classical iterative method, the spectral radius of convergence not only depends on the properties of the coefficient matrix, but also on the iterative method being used to solve the linear system. To illustrate its calculation procedure using Fourier decomposition, we consider solution of the 1D Poisson equation, Eq. (2.1), subject to Dirichlet boundary conditions, Eq. (2.2). The finite difference equations for the interior nodes, resulting from discretization of Eq. (2.1) on a uniform grid, may be written as

$$\phi_j - \frac{1}{2}\phi_{j+1} - \frac{1}{2}\phi_{j-1} = -S_j \frac{(\Delta x)^2}{2}. \tag{4.21}$$

Here, the nodes have been denoted by the index j, rather than i, so as not to confuse with complex number notations. Using the Gauss–Seidel method with left to right sweep, Eq. (4.21) may be written in the following iterative form:

$$\phi_j^{(n+1)} - \frac{1}{2}\phi_{j-1}^{(n+1)} = -S_j \frac{(\Delta x)^2}{2} + \frac{1}{2}\phi_{j+1}^{(n)}. \tag{4.22}$$

The exact numerical solution must also satisfy Eq. (4.21). Therefore,

$$\phi_j^{(E)} - \frac{1}{2}\phi_{j-1}^{(E)} = -S_j \frac{(\Delta x)^2}{2} + \frac{1}{2}\phi_{j+1}^{(E)}. \tag{4.23}$$

Subtracting Eq. (4.22) from Eq. (4.23) yields an equation for the convergence error, namely

$$\varepsilon_j^{(n+1)} - \frac{1}{2}\varepsilon_{j-1}^{(n+1)} = \frac{1}{2}\varepsilon_{j+1}^{(n)}. \tag{4.24}$$

Equation (4.24) shows that the convergence error at the current iteration depends on the convergence error, both current and previous, at the neighboring nodes. The exact dependence is governed by the iteration scheme being used.

The next step is to substitute the Fourier representation of the errors [Eq. (4.18)] into the error equation given by Eq. (4.24). This yields

$$\sum_{m=0}^{M-1} C_m^{(n+1)} \exp\{i\,j\theta_m\} - \frac{1}{2}\sum_{m=0}^{M-1} C_m^{(n+1)} \exp\{i(j-1)\theta_m\} = \frac{1}{2}\sum_{m=0}^{M-1} C_m^{(n)} \exp\{i(j+1)\theta_m\}. \tag{4.25}$$

Since the Fourier components (or modes) are linearly independent, Eq. (4.25) must also be satisfied for each Fourier mode (value of m). It follows then that

$$C_m^{(n+1)} \exp\{i\,j\theta_m\} - \frac{1}{2}C_m^{(n+1)} \exp\{i(j-1)\theta_m\}$$
$$= \frac{1}{2}C_m^{(n)} \exp\{i(j+1)\theta_m\} \quad \forall m = 0,1,...,M-1. \tag{4.26}$$

Simplifying, we obtain

$$C_m^{(n+1)} - \frac{1}{2}C_m^{(n+1)} \exp\{-i\theta_m\} = \frac{1}{2}C_m^{(n)} \exp\{i\theta_m\} \quad \forall m = 0,1,...,M-1, \tag{4.27}$$

which, upon rearrangement, yields

$$\lambda_m = \frac{C_m^{(n+1)}}{C_m^{(n)}} = \frac{\frac{1}{2}\exp\{i\theta_m\}}{1 - \frac{1}{2}\exp\{-i\theta_m\}} = \frac{\exp\{i\theta_m\}}{2 - \exp\{-i\theta_m\}} \quad \forall m = 0,1,...,M-1. \tag{4.28}$$

The left-hand side of Eq. (4.28) has the ratio of the amplitudes of the mth Fourier mode between two successive iterations. For a stable iterative scheme, this ratio must be less than unity for each value of m. As a matter of fact, this ratio is nothing but the mth eigenvalue of the iteration matrix [1,2], and is denoted by λ_m. Of particular interest is the largest eigenvalue, whose magnitude, as stated earlier, is the spectral radius of convergence. The magnitudes of the eigenvalues, shown in Eq. (4.28), can be computed using

$$
|\lambda_m| = \sqrt{\lambda_m \lambda_m^*} = \sqrt{\frac{\exp\{i\theta_m\}}{2-\exp\{-i\theta_m\}}\frac{\exp\{-i\theta_m\}}{2-\exp\{i\theta_m\}}}
$$
$$
= \sqrt{\frac{1}{5-4\cos\theta_m}} \qquad \forall m = 0,1,...,M-1,
$$

(4.29)

where λ_m^* is the complex conjugate of λ_m. Recalling that the phase angle is given by $\theta_m = m\pi/(M-1)$, it can vary between 0 and π. Since $\cos\theta_m$ decreases monotonically from +1 to −1 as θ_m increases from 0 to π, it is clear that $|\lambda_m|$ decreases monotonically with increasing values of m. In other words, small values of m will result in large eigenvalues, and vice versa. The smallest value of m, i.e., $m=0$, is not of relevance since it results in the constant term of the Fourier series, as evident from Eq. (4.17). Therefore, we inspect the case when $m=1$, which will result is the largest eigenvalue. The phase angle that produces the largest eigenvalue is $\theta_1 = \pi/(M-1)$. Table 4.1 shows the largest eigenvalue (or spectral radius) computed using Eq. (4.29) for various node counts.

Although not shown in Table 4.1, it is easy to infer from Eq. (4.29) that the spectral radii of convergence tend to unity with larger numbers of nodes. Thus, one can conclude that the Gauss–Seidel method is stable when used to solve the 1D Poisson equation with Dirichlet boundary conditions, irrespective of the mesh size. This result is not surprising, and has already been illustrated in Example 3.2 for the 2D Poisson equation.

One of the critical findings from the preceding discussion is that the spectral components with large wavelengths (small frequencies) are responsible for instability or divergence. While this result has been shown here for the Gauss–Seidel method, it is true irrespective of the iterative method used. In Fig. 4.2, the largest wavelength component corresponds to the $m=1$ sine wave. This particular spectral component is only anchored at the two end points, and is the most difficult component to damp

Table 4.1 Spectral Radii of Convergence for the Solution of the 1D Poisson Equation with Dirichlet Boundary Conditions at Both Ends, and Using Gauss–Seidel Sweeps From Left to Right

Number of Nodes, M	θ_1 (Radians)	Spectral Radius, λ_{SR}
6	$\pi/5$	0.752
11	$\pi/10$	0.914
21	$\pi/20$	0.976
41	$\pi/40$	0.994
81	$\pi/80$	0.998

out. Next, we consider an example to apply the method, just outlined, to explore the stability of a given equation–solver combination.

EXAMPLE 4.3

In this example, we examine the stability of the 1D advection–diffusion equation subject to two Dirichlet boundary conditions. The central difference scheme is used for discretization of both the advection and the diffusion term, and the Gauss–Seidel method is used for solution of the resulting finite difference equations.

The 1D advection–diffusion equation is written as $u\dfrac{d\phi}{dx} - \Gamma\dfrac{d^2\phi}{dx^2} = 0$. In this example, we will assume both u and Γ to be constants. For a physical problem, u would represent the velocity component in the x direction, and could be either positive or negative, while Γ would represent a positive transport coefficient. The boundary conditions are as follows: $\phi(0) = 1$ and $\phi(1) = 0$. Using the central difference scheme, and following the derivation procedures described in Chapter 2, the finite difference equation for the interior nodes, resulting from the above advection–diffusion equation, may be written as

$$u\frac{\phi_{j+1} - \phi_{j-1}}{2\Delta x} - \Gamma\frac{\phi_{j+1} + \phi_{j-1} - 2\phi_j}{(\Delta x)^2} = 0,$$

which may be rearranged to write

$$\left(\frac{1}{2} - P\right)\phi_{j+1} - \left(\frac{1}{2} + P\right)\phi_{j-1} + 2P\phi_j = 0,$$

where $P = \Gamma/(u\,\Delta x)$. Using Gauss–Seidel iterations, and the procedure described earlier in this section, the error equation is derived as

$$\left(\frac{1}{2} - P\right)\varepsilon_{j+1}^{(n)} - \left(\frac{1}{2} + P\right)\varepsilon_{j-1}^{(n+1)} + 2P\varepsilon_j^{(n+1)} = 0.$$ Next, substituting the Fourier representation of the errors, and by considering linear independence of the Fourier modes, we obtain

$$2PC_m^{(n+1)} - \left(\frac{1}{2} + P\right)C_m^{(n+1)}\exp\{-i\theta_m\} = -\left(\frac{1}{2} - P\right)C_m^{(n)}\exp\{i\theta_m\} \quad \forall m = 0,1,...,M-1,$$

which, upon rearrangement, yields the eigenvalues:

$$\lambda_m = \frac{C_m^{(n+1)}}{C_m^{(n)}} = \frac{-\left(\frac{1}{2} - P\right)\exp\{i\theta_m\}}{2P - \left(\frac{1}{2} + P\right)\exp\{-i\theta_m\}} \quad \forall m = 0,1,...,M-1,$$

where $\theta_m = m\pi/(M-1)$. As discussed earlier, the largest eigenvalue is produced by the $m=1$ component. For this, the eigenvalue is written as

$$\lambda_1 = \frac{-\left(\frac{1}{2}-P\right)\exp\{i\theta_1\}}{2P-\left(\frac{1}{2}+P\right)\exp\{-i\theta_1\}} = \frac{-\left(\frac{1}{2}-P\right)\cos\theta_1 - i\left(\frac{1}{2}-P\right)\sin\theta_1}{2P-\left(\frac{1}{2}+P\right)\cos\theta_1 + i\left(\frac{1}{2}+P\right)\sin\theta_1} = \frac{a+bi}{c+di},$$

where

$$a = -\left(\frac{1}{2}-P\right)\cos\theta_1,\; b = -\left(\frac{1}{2}-P\right)\sin\theta_1,\; c = 2P-\left(\frac{1}{2}+P\right)\cos\theta_1,$$

and $d = \left(\frac{1}{2}+P\right)\sin\theta_1.$

The spectral radius of convergence is calculated as

$$\lambda_{SR} = \sqrt{\lambda_1 \cdot \lambda_1^*} = \frac{\sqrt{(ac+bd)^2 + (bc-ad)^2}}{c^2+d^2}.$$

The table below shows the computed eigenvalues for various mesh counts and various values of $P = \Gamma/(u\,\Delta x)$. Since, u could be either positive or negative, both positive and negative values of P are considered.

	Spectral radius of convergence		
P	$M = 11$	$M = 51$	$M = 101$
−0.6	0.995181	0.999804	0.999951
−0.5	1	1	1
−0.4	1.004869	1.000195	1.000049
−0.3	1.009305	1.00037	1.000093
−0.2	1.012206	1.000484	1.000121
−0.1	1.011057	1.000439	1.00011
0	1	1	1
0.1	0.965197	0.998523	0.99963
0.2	0.875531	0.993917	0.998468
0.3	0.678119	0.97713	0.994131
0.4	0.352501	0.882454	0.96626
0.5	0	0	0
0.6	0.268018	0.810855	0.940609

The above table shows that the central difference scheme combined with the Gauss–Seidel method is unstable for small negative values of P in the range −0.5 to 0. It is stable for all other values of P. This finding is true irrespective of mesh size. Physically, P represents a competition between diffusion and advection. It is the reciprocal of the so-called grid Peclet number (see Chapter 6). When P is large, diffusion dominates, and the solution method is stable. When P is small, advection dominates. However, the solution method is still stable as long as the flow direction is positive. It is only when the flow direction reverses (negative P) that instability sets in. It is because of this unstable behavior that the central difference scheme is usually not used to treat advection in fluid flow calculations. It is worth also noting that when $-0.5 \leq P \leq 0$, the Scarborough criterion is violated, although as discussed earlier, this does not immediately imply divergence.

Whether or not the aforementioned predictions of stability analysis hold water is tested next by actually attempting to solve the system of linear algebraic equations using the Gauss–Seidel method. The figure below shows the residuals for the coarsest mesh ($M = 11$) for three different values of P.

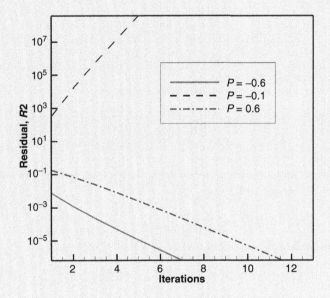

The residual plots clearly show that, as predicted by stability analysis, the solution method is stable for both negative and positive values of P as long as the value is sufficiently large. The Scarborough criterion for this particular case is $|2P| \geq 1$. Therefore, it is not surprising that the scheme is stable for $|P| \geq \frac{1}{2}$. For small negative values of P, the solution method is unstable and the residuals blow up. The results shown here are somewhat counterintuitive because one would generally expect the convergence behavior of $P = -0.1$ to be bounded by the convergence behavior of $P = -0.6$ and $P = 0.6$. This example highlights the value of stability analysis.

In Section 3.2, we remarked that violation of the Scarborough criterion (or lack of diagonal dominance) does not automatically imply instability and divergence. The preceding example further illustrates this point. For both $-\frac{1}{2} \leq P \leq 0$ and $0 \leq P \leq \frac{1}{2}$, the magnitude of the diagonal, and the sum of the magnitudes of the off-diagonals are the same, and the Scarborough criterion is violated. Yet, in the latter case, the iteration scheme is stable. This implies that the stability of an iterative method, while related to diagonal dominance and the Scarborough criterion, cannot be predicted by inspecting the satisfaction of that criterion alone. One must perform a rigorous stability analysis to understand the complete picture.

The procedure for stability analysis, just described, can be applied to any classical (semi-implicit) iterative method and linear system combination, including those arising out of multidimensional problems. To illustrate this, let us consider the stability of the Gauss–Seidel method for the 2D Poisson equation subject to Dirichlet boundary conditions. The finite-difference equations for this particular case, following Eq. (2.48), are

$$\frac{\phi_{p+1,q} + \phi_{p-1,q} - 2\phi_{p,q}}{(\Delta x)^2} + \frac{\phi_{p,q+1} + \phi_{p,q-1} - 2\phi_{p,q}}{(\Delta y)^2} = S_{p,q}, \qquad (4.30)$$

which may be rearranged to write

$$\left(\frac{2}{(\Delta x)^2} + \frac{2}{(\Delta y)^2}\right)\phi_{p,q} - \frac{1}{(\Delta x)^2}\phi_{p+1,q} - \frac{1}{(\Delta x)^2}\phi_{p-1,q} - \frac{1}{(\Delta y)^2}\phi_{p,q+1} - \frac{1}{(\Delta y)^2}\phi_{p,q-1}$$
$$= a_O\phi_{p,q} + a_E\phi_{p+1,q} + a_W\phi_{p-1,q} + a_N\phi_{p,q+1} + a_S\phi_{p,q-1} = -S_{p,q} \qquad (4.31)$$

Using the Gauss–Seidel iteration scheme with sweeps from left to right and bottom to top, Eq. (4.31) may be written as

$$a_O\phi_{p,q}^{(n+1)} + a_W\phi_{p-1,q}^{(n+1)} + a_S\phi_{p,q-1}^{(n+1)} = -S_{p,q} - a_E\phi_{p+1,q}^{(n)} - a_N\phi_{p,q+1}^{(n)}. \qquad (4.32)$$

The exact numerical solution must also satisfy Eq. (4.32). Therefore,

$$a_O\phi_{p,q}^{(E)} + a_W\phi_{p-1,q}^{(E)} + a_S\phi_{p,q-1}^{(E)} = -S_{p,q} - a_E\phi_{p+1,q}^{(E)} - a_N\phi_{p,q+1}^{(E)}. \qquad (4.33)$$

Subtracting Eq. (4.32) from Eq. (4.33) results in the error equation:

$$a_O\varepsilon_{p,q}^{(n+1)} + a_W\varepsilon_{p-1,q}^{(n+1)} + a_S\varepsilon_{p,q-1}^{(n+1)} = -a_E\varepsilon_{p+1,q}^{(n)} - a_N\varepsilon_{p,q+1}^{(n)}. \qquad (4.34)$$

In this case, the error must be expressed as 2D Fourier series of the following form:

$$\varepsilon_{p,q}^{(n)} = \sum_{l=0}^{N-1}\sum_{m=0}^{M-1} C_{l,m}^{(n)} \exp(i[p\theta_l + q\theta_m]), \qquad (4.35)$$

where the two phase angles are given by

$$\theta_l = \frac{l\pi}{(N-1)}, \quad \theta_m = \frac{m\pi}{(M-1)}, \qquad (4.36)$$

where N and M are the number of nodes in the x and y directions, respectively. Substitution of Eq. (4.36) into Eq. (4.34) yields

$$a_O \sum_{l=0}^{N-1} \sum_{m=0}^{M-1} C_{l,m}^{(n+1)} \exp(i[p\theta_l + q\theta_m]) + a_W \sum_{l=0}^{N-1} \sum_{m=0}^{M-1} C_{l,m}^{(n+1)} \exp(i[(p-1)\theta_l + q\theta_m])$$

$$+ a_S \sum_{l=0}^{N-1} \sum_{m=0}^{M-1} C_{l,m}^{(n+1)} \exp(i[p\theta_l + (q-1)\theta_m]) = -a_E \sum_{l=0}^{N-1} \sum_{m=0}^{M-1} C_{l,m}^{(n)} \exp(i[(p+1)\theta_l + q\theta_m]) \cdot$$

$$- a_N \sum_{l=0}^{N-1} \sum_{m=0}^{M-1} C_{l,m}^{(n)} \exp(i[p\theta_l + (q+1)\theta_m])$$

$$(4.37)$$

Once again, due to the linear independence of the Fourier modes, it follows that

$$a_O C_{l,m}^{(n+1)} \exp(i[p\theta_l + q\theta_m]) + a_W C_{l,m}^{(n+1)} \exp(i[(p-1)\theta_l + q\theta_m])$$
$$+ a_S C_{l,m}^{(n+1)} \exp(i[p\theta_l + (q-1)\theta_m]) = -a_E C_{l,m}^{(n)} \exp(i[(p+1)\theta_l + q\theta_m]) \cdot \quad (4.38)$$
$$- a_N C_{l,m}^{(n)} \exp(i[p\theta_l + (q+1)\theta_m])$$

Simplification yields an expression for the eigenvalues, written as

$$\lambda_{l,m} = \frac{C_{l,m}^{(n+1)}}{C_{l,m}^{(n)}} = \frac{-a_E \exp(i\theta_l) - a_N \exp(i\theta_m)}{a_O + a_W \exp(-i\theta_l) + a_S \exp(-i\theta_m)}. \quad (4.39)$$

As in the 1D case, the largest eigenvalue is manifested in the case when $l = m = 1$, resulting in $\pi/(N-1)$ and $\pi/(M-1)$ as the two phase angles. Thus,

$$\lambda_{SR} = \sqrt{\lambda_{1,1} \lambda_{1,1}^*}, \quad \lambda_{1,1} = \frac{-a_E \exp(i\pi/(N-1)) - a_N \exp(i\pi/(M-1))}{a_O + a_W \exp(-i\pi/(N-1)) + a_S \exp(-i\pi/(M-1))}. \quad (4.40)$$

An example that demonstrates stability analysis for a 2D problem is considered next.

EXAMPLE 4.4

In this example, we examine the stability of the 2D Poisson equation subject to Dirichlet boundary conditions on all sides. The computational domain is considered to be a square of unit length. The central difference scheme is used for discretization. Stability analyses of both the Gauss–Seidel method and the line-by-line method with row-wise sweeps (LBL-Row) are performed.

Following Eq. (4.31), the coefficients of the matrix $[A]$ are derived as

$$a_O = \frac{2}{(\Delta x)^2} + \frac{2}{(\Delta y)^2}, \; a_E = a_W = -\frac{1}{(\Delta x)^2}, \; a_N = a_S = -\frac{1}{(\Delta y)^2}. \; \text{Using} \; \Delta x = \frac{L}{N-1} \; \text{and}$$

$\Delta y = \dfrac{L}{M-1}$, Eq. (4.40) may be rewritten as

$$[\lambda_{1,1}]_{GS} = \frac{(N-1)^2 \exp(i\pi/(N-1)) + (M-1)^2 \exp(i\pi/(M-1))}{2[(N-1)^2 + (M-1)^2] - (N-1)^2 \exp(-i\pi/(N-1)) - (M-1)^2 \exp(-i\pi/(M-1))}.$$

The eigenvalues for the LBL-Row can be easily derived following the procedure described above. In this case, only the northern node is treated explicitly, and is written as

$$[\lambda_{1,1}]_{Row} = \frac{(M-1)^2 \exp(i\pi/(M-1))}{\left\{ \begin{array}{c} 2[(N-1)^2+(M-1)^2]-(N-1)^2[\exp(-i\pi/(N-1))+\exp(i\pi/(N-1))] \\ -(M-1)^2\exp(-i\pi/(M-1)) \end{array} \right\}}$$

The spectral radii of convergence using the two iterative methods are listed in the table below for various mesh counts. As an example, an equal number of nodes are used in the x and y directions.

$N(= M)$	Spectral radius for Gauss–Seidel	Spectral radius for LBL-Row
11	0.9144	0.8419
21	0.9762	0.9535
51	0.9961	0.9921
101	0.9990	0.9980

The results show that both the Gauss–Seidel and the LBL-Row are stable for any mesh size. This finding is consistent with the convergence plots shown in Examples 3.2 and 3.3.

Strictly speaking, the Fourier decomposition based von Neumann stability analysis, just presented, is applicable only when Dirichlet boundary conditions or periodic boundary conditions are applied. In the case of Dirichlet boundary conditions, the error at the boundaries is zero. Therefore, any instability that may be responsible for divergence must have been generated by the interior nodes. In the case of periodic boundary conditions, the solution repeats itself beyond the boundaries, and therefore, the nodal equations for the interior nodes are also applicable at the boundaries, implying that the same stability analysis is valid also for boundary nodes. If the boundary conditions are either of Neumann or Robin type, the stability analysis performed on the interior node equations is not valid for the boundary nodes. Depending on the exact boundary condition and its discrete treatment, the onset of instability may be at the boundary. In such a case, the actual convergence behavior may be somewhat different from the predictions of stability analysis. This may be cited as one of the limitations of the Fourier decomposition–based method for finding the spectral radius of convergence. The alternative – a far more tedious one – is to find the eigenvalues of the amplification matrix directly by writing a computer program that will make use of a numerical method for finding eigenvalues. The full amplification matrix will also contain the coefficients arising out of the nodal equations at the boundaries. Therefore, any instability arising out of the boundary nodes will also be captured in this general approach.

4.2.3 USE OF INERTIAL DAMPING FACTOR

An unstable method may be made stable by altering the linear system in a manner that decreases the eigenvalues of the iteration matrix. The action of altering the linear system to change its eigenvalues is generally referred to as preconditioning, which we will discuss in detail in Section 4.4. Preconditioning usually involves multiplying the coefficient matrix with a preconditioning matrix, and then solving the resulting system. It is generally done to accelerate convergence rather than to render an unstable method stable, although that could also be a goal. Here, a simple strategy that may be used to favorably alter the eigenvalues, and consequently, the spectral radius of convergence, is discussed.

To demonstrate the strategy, let us start with the algebraic equations given by Eq. (4.31). Although Eq. (4.31) was derived from the Poisson equation, it is, in fact, the general five-band form arising from finite difference discretization of a 2D PDE on a structured mesh. First, we rewrite this equation in iterative form, using the LBL-Row, as an example iterative method. This yields

$$a_O\phi_{p,q}^{(n+1)} + a_W\phi_{p-1,q}^{(n+1)} + a_S\phi_{p,q-1}^{(n+1)} + a_E\phi_{p+1,q}^{(n)} = -S_{p,q} - a_N\phi_{p,q+1}^{(n)}. \tag{4.41}$$

Next, we modify Eq. (4.41) by adding a term to both sides of Eq. (4.41), as follows:

$$\alpha a_O\phi_{p,q}^{(n+1)} + a_O\phi_{p,q}^{(n+1)} + a_W\phi_{p-1,q}^{(n+1)} + a_S\phi_{p,q-1}^{(n+1)} + a_E\phi_{p+1,q}^{(n)} = -S_{p,q} - a_N\phi_{p,q+1}^{(n)} + \alpha a_O\phi_{p,q}^{(n)}. \tag{4.42}$$

At convergence $\phi_{p,q}^{(n+1)} = \phi_{p,q}^{(n)}$. Therefore, adding $\alpha a_O\phi_{p,q}^{(n+1)}$ to the left-hand side of equation (4.41) while adding $\alpha a_O\phi_{p,q}^{(n)}$ to the right-hand side of the same equation, does not alter the final solution to Eq. (4.31). This modification only alters the path to convergence. The quantity, α, which is generally a positive constant, is known as the *inertial damping factor*. The name stems from the fact that such damping was first introduced to damp out instabilities due to the inertial (or advection) terms in general advection–diffusion equations. Now, following the procedure discussed in the preceding section, the eigenvalues of the iteration matrix, resulting from Eq. (4.42), are derived as

$$\lambda_{l,m} = \frac{-a_N\exp(i\theta_m) + \alpha a_O}{(1+\alpha)a_O + a_E\exp(i\theta_l) + a_W\exp(-i\theta_l) + a_S\exp(-i\theta_m)}. \tag{4.43}$$

It can be deduced easily from the derivation of Eq. (4.43) that, irrespective of the iteration scheme used, or the dimensionality of the problem at hand, the expression for the eigenvalues with inertial damping is the expression without it but with αa_O added to both the numerator and the denominator. In general, α may assume any positive value. The exact value depends largely on the problem at hand, as we shall see later. Inertial damping will always enhance the stability characteristics of any iterative method, i.e., unstable methods may become less unstable or completely stable. Whether or not inertial damping will result in an enhanced rate of convergence, is yet another issue. That depends on the problem (PDE) at hand, and consequently, the relative magnitudes of the link coefficients in Eq. (4.43). This is best demonstrated by the example problem presented next.

EXAMPLE 4.5

In this example, we consider the 1D advection–diffusion equation that was considered in Example 4.3. Here, we investigate the stability of the Gauss–Seidel method for this equation with and without inertial damping.

Following Example 4.3, the spectral radius of convergence is given by

$$\lambda_{SR} = \sqrt{\lambda_1 \cdot \lambda_1^*} \text{ , where } \lambda_1 = \frac{-\left(\frac{1}{2}-P\right)\exp\{i\theta_1\}+2P\alpha}{(1+\alpha)2P-\left(\frac{1}{2}+P\right)\exp\{-i\theta_1\}} \text{ , and } \theta_1 = \pi/(M-1).$$

The table below shows the spectral radius of convergence with $M = 11$ (11 nodes) for different values of the inertial damping factor, α.

Spectral radius of convergence

P	$\alpha = 0$	$\alpha = 1$	$\alpha = 2$	$\alpha = 5$
−0.6	0.995181	0.985526	0.987976	0.992878
−0.5	1	0.987688	0.989064	0.993179
−0.4	1.004869	0.990419	0.990522	0.993607
−0.3	1.009305	0.993916	0.992566	0.994261
−0.2	1.012206	0.998349	0.99557	0.995379
−0.1	1.011057	1.003131	1	0.99766
0	1	1	1	1
0.1	0.965197	0.867945	0.666667	0.96293
0.2	0.875531	0.579229	0.907312	0.982542
0.3	0.678119	0.82234	0.947126	0.986165
0.4	0.352501	0.888262	0.959408	0.987651
0.5	0	0.914483	0.965197	0.988457
0.6	0.268018	0.928084	0.968536	0.988961

The above table shows that as the inertial damping factor (α) is increased, the spectral radii of convergence that were above unity, for P between −0.5 and 0, decreases to values below unity. In other words, stability is attained with an inertial damping factor of 2 or greater, in this particular case. The case when $P = 0$ may be ignored because when $P = 0$, the equation becomes a pure advection equation. One important point to note is that in the regions where the spectral radii were low to begin with, e.g., for values of P between 0 and 0.6, they increase significantly with the introduction of inertial damping. While the spectral radii are still in the stable range, this increase in the spectral radii has serious negative implications on the rate of convergence, as we shall see in the section to follow. The predictions of stability theory are, once again, tested by solving the linear system using the Gauss–Seidel method for $P = -0.1$.

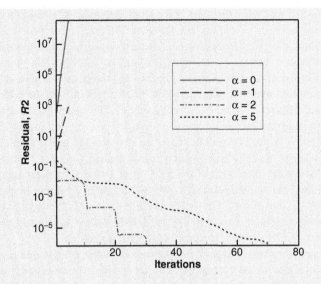

The figure above shows the residuals obtained for various values of the inertial damping factor. For values of α less than 2, divergence occurs, as predicted by stability theory. For $\alpha = 2$, although convergence is finally attained, the residuals exhibit odd behavior. This is because $\alpha = 2$ is right on the cusp of stability (spectral radius of convergence is exactly equal to 1), and numerical round-off plays a role in determining whether the system will finally converge or diverge. The fact that it finally attains convergence in this case is somewhat fortuitous. For $\alpha = 5$, as predicted by stability theory, the method is stable, and consequently, the residuals decrease monotonically.

4.3 RATE OF CONVERGENCE

Thus far, the focus of the discussion has been the determination of whether a particular iterative method is stable or not. The Lax equivalence theorem relates stability to convergence but not to the rate of convergence. Beyond the stability of the method comes the issue of the rate of convergence. In Chapter 3, we considered several iterative methods for solving linear systems. All of them were stable, for example, for the Poisson equation. However, their rates of convergence differed significantly. In this section, we focus on establishing a criterion that may provide an estimate of the rate of convergence of an iterative method. In Section 4.1, we alluded to the fact that the rate of convergence has some relation to the condition number of the coefficient matrix $[A]$. However, the condition number of $[A]$ alone cannot explain why the alternating direction implicit (ADI) method, for example, has a faster rate of convergence than the Gauss–Seidel method for the same problem. In Section 4.2, it was established that the stability of the method depends on whether the largest eigenvalue of the iteration or amplification matrix, $[B]$, exceeds unity. For any classical iterative method, the rate of convergence, in fact, depends on the condition number of the amplification matrix. As discussed earlier, the amplification matrix, $[B]$, is a combination of the properties of the $[A]$ matrix and the iterative method being used. Hence, its condition number reflects the combined effect of both the system of equations being solved as well as how it is being solved. The closer the condition number of the amplification matrix is to unity, the better the convergence, and vice versa.

In Section 4.2, it was shown that the largest eigenvalue of the amplification matrix corresponds to the largest wavelength or smallest frequency Fourier component, i.e., when $m = 1$, and $\theta = \pi/(M-1)$. Conversely, the smallest eigenvalue corresponds to the smallest wavelength or largest frequency Fourier component, i.e., when $m = M-1$, and $\theta = \pi$. The only exception is the Jacobi method. Thus, the condition number, which is the ratio of the largest to the smallest eigenvalue [Eq. (4.3)] can be easily computed from the Fourier decomposition.

The fact that the smallest eigenvalue depends on a phase angle that is independent of mesh count or size, i.e., $\theta = \pi$, implies that the condition number for a given iterative method will be solely dependent on the largest eigenvalue. In other words, for a given iterative scheme, the spectral radius of convergence is sufficient to predict not only stability (i.e., whether the method will converge or not), but also the rate of convergence. For example, Table 4.1 shows that the spectral radius of convergence of the Gauss–Seidel method increases with mesh refinement. This implies that as the mesh is refined, the rate of convergence will deteriorate, as has already been noted in Example 3.2. Similarly, Example 4.5 shows that the spectral radius of convergence increases when inertial damping is introduced for values of P in the range 0–0.6. This implies that the convergence of the Gauss–Seidel method will be slower in this range with inertial damping as opposed to without. When it comes to comparison of the rates of convergence of two different iterative methods, it is, of course, necessary to compute the condition numbers of the amplification matrices corresponding to the two methods separately. Even though the smallest eigenvalue is independent of mesh size in both methods, they are different for the two methods. Thus, comparison of the spectral radii of the two methods alone is not sufficient to determine which of the two methods will converge faster. An example is considered next to highlight some of the issues just discussed.

EXAMPLE 4.6

The problem considered here is the same as that in Example 3.3. In Example 3.3, the Poisson equation was solved on a unit square with Dirichlet boundary conditions using the line-by-line method with both row-wise (LBL-Row) and column-wise (LBL-Column) sweeps. In this example, we compute the condition number of the iteration matrices in the two cases in an effort to correlate the findings with the residuals (convergence plots) already computed in Example 3.3.

Using Eq. (4.43), the eigenvalues of the iteration matrix for LBL-Row is written as

$$\left[\lambda_{l,m}\right]_{LBL-Row} = \frac{-a_N \exp(i\theta_m)}{a_O + a_E \exp(i\theta_l) + a_W \exp(-i\theta_l) + a_S \exp(-i\theta_m)}$$

Similarly, the eigenvalues of the iteration matrix for LBL-Column is written as

$$\left[\lambda_{l,m}\right]_{LBL-Column} = \frac{-a_E \exp(i\theta_l)}{a_O + a_W \exp(-i\theta_l) + a_S \exp(-i\theta_m) + a_N \exp(i\theta_m)},$$

where the coefficients are given by $a_O = \dfrac{2}{(\Delta x)^2} + \dfrac{2}{(\Delta y)^2}$, $a_E = a_W = -\dfrac{1}{(\Delta x)^2}$, $a_N = a_S = -\dfrac{1}{(\Delta y)^2}$. The grid spacings are defined as $\Delta x = \dfrac{L}{N-1}$ and $\Delta y = \dfrac{L}{M-1}$. The smallest eigenvalues are computed using $\theta_l = \theta_m = \pi$, while the largest eigenvalues are computed using $\theta_l = \pi/(N-1)$, and $\theta_m = \pi/(M-1)$. The computed condition numbers of the iteration matrix, $\kappa(B)$, are listed in the table below for the case when the number of nodes in the x and y directions are the same, i.e., $N = M$.

$N(= M)$	$\kappa(B)$ for LBL-Row	$\kappa(B)$ for LBL-Column
11	5.89	5.89
21	6.67	6.67
41	6.91	6.91
81	6.97	6.97

Since the grid spacing is equal in the x and y directions, the condition numbers for LBL-Row and LBL-Column methods are identical, implying the row-wise versus column-wise sweeps will produce identical rates of convergence, as already confirmed by the residual plots shown in Example 3.3.

Next, we consider the case when the grid spacing in the x and y directions are not identical. In Example 3.4, the same problem was also solved using the ADI method. In the case of the ADI method, the eigenvalues of the row-wise and column-wise sweeps must be combined. Noting that one row-wise sweep and one column-wise sweep counts together as two iterations, one may write

$$\left[\lambda_{l,m}\right]_{ADI}^2 = \frac{C_{l,m}^{(n+1)}}{C_{l,m}^{(n)}} \frac{C_{l,m}^{(n+2)}}{C_{l,m}^{(n+1)}} = \left[\lambda_{l,m}\right]_{LBL-Row} \times \left[\lambda_{l,m}\right]_{LBL-Column},$$

which results in $\left[\lambda_{l,m}\right]_{ADI} = \sqrt{\left[\lambda_{l,m}\right]_{LBL-Row} \times \left[\lambda_{l,m}\right]_{LBL-Column}}$.

The computed condition numbers in this case are shown in the table below for two combinations of N and M for the three methods.

$N \times M$	$\kappa(B)$ for LBL-Row	$\kappa(B)$ for LBL-Column	$\kappa(B)$ for ADI
41×81	3.98	18.76	8.64
81×41	18.76	3.98	8.64

In this case, the condition numbers for LBL-Row and LBL-Column are significantly different. For the 41×81 mesh, row-wise sweeps are strongly favorable over column-wise sweeps. This finding is consistent with the residual plots shown in Example 3.3 for the unequal grid aspect ratio case: row-wise sweeps required less than 2000 iterations, while column-wise sweeps required about 7000 iterations. In the case of the ADI method, the rate of convergence is expected to be bounded by the rates of convergence of the LBL-Row and LBL-Column methods based on the computed condition numbers. This finding is also confirmed by the residual plot shown in Example 3.4, which showed that the ADI method requires about 2800 iterations.

In Chapter 3 and in earlier sections of the present chapter, diagonal dominance was cited as a favorable property of the coefficient matrix for a stable iterative solution. Inertial damping, which was introduced to artificially enhance stability, strengthens the diagonal. The reason why diagonal dominance enhances the stability of classical iterative methods becomes evident from the eigenvalue expressions derived earlier in this chapter. Each of these eigenvalue expressions has the following general form:

$$\lambda = \frac{\sum \text{explicit terms}}{\text{Diagonal} + \sum \text{implicit terms}}. \tag{4.44}$$

This implies that the larger the diagonal, the smaller the eigenvalues, and greater the chance of the spectral radius of convergence being less than unity. It should be noted that a stronger diagonal is likely to decrease all the eigenvalues. Therefore, whether a stronger diagonal will improve the condition number or not is not obvious, and is problem dependent. However, as far as stability is concerned, it is a general rule of thumb that a stronger diagonal is preferable for iterative solution.

Another revelation about the eigenvalue expression in Eq. (4.44) is that the so-called "explicit" terms will not always increase the eigenvalue. They will do so only if the coefficients of the explicit terms – assuming that there are more than one – are

of the same sign. If they are of opposite signs, the eigenvalues may actually reduce by treating a term explicitly rather than implicitly. For example, when using
a fourth-order central difference scheme [see Eq. (3.36b)] the eastern coefficient is
$4/[3(\Delta x)^2]$, while the east-eastern coefficient is $-1/[12(\Delta x)^2]$. In this case, explicit
treatment of the east-eastern term (and other similar terms) actually aids the rate of
convergence rather than hampering it. The reader is encouraged to solve Exercise 3.2
to corroborate this claim.

When it comes to predicting the rate of convergence of the Krylov subspace
methods and other fully implicit methods, one has to rely upon the condition
number of the coefficient matrix itself to get some estimate of the rate of convergence. For example, the convergence of the method of steepest descent is known
to scale as $\kappa(A)$, while that of the conjugate gradient (CG) method is known to
scale as $\sqrt{\kappa(A)}$[1]. The rate of convergence in such methods is usually enhanced
by a preconditioning step that alters the condition number, thereby enabling faster
convergence. The reader is referred to more advanced texts [1] for a discussion on
the rate of convergence of Krylov subspace methods. The general idea of preconditioning is discussed next.

4.4 PRECONDITIONING

In the preceding sections of the present chapter, it has been discussed and demonstrated through examples that the rate of convergence is dependent, in some
manner, on the condition number of the coefficient or $[A]$ matrix. Although the dependence is quite convoluted and is affected also by the precise iteration scheme
being used, it is fair to conclude that decreasing the condition number of $[A]$
is beneficial for convergence. The term preconditioning refers to the process by
which the original linear system is altered such that the condition number of the
altered coefficient matrix is smaller than the condition number of the original
coefficient matrix – the hope being that such a strategy would result in more
rapid convergence if the altered system is solved instead of the original system.
In order to understand how preconditioning works, let us start with the original
linear system:

$$[A][\phi] = [Q]. \tag{4.45}$$

Next, we multiply both sides of the equation by a matrix $[M]^{-1}$, such that

$$[M]^{-1}[A][\phi] = [M]^{-1}[Q]. \tag{4.46}$$

The matrix $[M]^{-1}$ is known as the preconditioning matrix, or the *preconditioner*,
in short. More specifically, in Eq. (4.46), it serves as a left preconditioner, since
a left multiplication has been employed to alter the original equation. The altera-
tion of Eq. (4.45) to Eq. (4.46) is motivated by the hope that Eq. (4.46) may be

computationally more efficient to solve than Eq. (4.45). However, this objective will come to fruition only if the following criteria are satisfied:

1. The condition number of $[M]^{-1}[A]$ is less than the condition number of $[A]$.
2. The preconditioner $[M]^{-1}$ is readily available, and computable in an efficient manner.
3. Extra matrix–vector multiplications, such as $[M]^{-1}[Q]$, are computable in an efficient manner.

The first of these criteria must be satisfied in order to reduce the iteration count, while the other two must be met to keep the computational time per iteration relatively low. Finding a preconditioner that satisfies all three criteria is quite challenging, and is, in fact, a prolific research area.

It is obvious that $[M]^{-1} = [A]^{-1}$ is the best preconditioner, since in this case, Eq. (4.46) would reduce to $[I][\phi] = [A]^{-1}[Q]$ — an equation that does not require any iterations to solve. However, finding $[A]^{-1}$ is more difficult than solving Eq. (4.45) itself. Therefore, this idea is not practically feasible. Nonetheless, it does lead to the offshoot idea that, perhaps, $[M]^{-1}$ could be an approximation of $[A]^{-1}$, such that the product $[M]^{-1}[A]$ would be fairly close to an identity matrix, and would have a condition number far closer to unity than the condition number of the original coefficient matrix.

Over the past several decades, many preconditioners have been proposed and demonstrated for a wide variety of PDEs. A discussion of all of those preconditioners is beyond the scope of an introductory text, such as this. Here, only a brief summary is presented, followed by a more detailed discussion of how two such popular preconditioners may be implemented. Broadly, the preconditioners used widely in serial (as opposed to parallel) scientific computations may be classified as follows:

- Classical iteration-based preconditioners, which include
  - Jacobi preconditioner, either point- or block-wise;
  - Symmetric or nonsymmetric Gauss–Seidel preconditioners, either point- or block-wise;
- Incomplete lower upper (ILU) decomposition-based preconditioners, which include
  - ILU without fill-in, or ILU(0);
  - ILU with fill-in, or ILU(n);
  - ILU with thresholding, or ILUT;
- Incomplete Cholesky factorization based preconditioners
- Polynomial preconditioners

This classification, is, by no means, all-inclusive, and many different variations of the preconditioners exist. Two of the preconditioners listed, namely the Jacobi preconditioner and the ILU(0) preconditioner, are discussed next. The discussion includes both how the preconditioner is determined, as well as how it is integrated with a solver. Here, the CG method is chosen as the core solver with which the preconditioner is integrated.

Integration of a preconditioner with the CG solver essentially entails replacing the matrix $[A]$ everywhere in the algorithm by $[M]^{-1}[A]$. However, finding $[M]^{-1}$ is far from trivial, and therefore, the entire algorithm has to be rewritten in a manner that does not require computing $[M]^{-1}$ explicitly. Nonetheless, if $[A]$ were to be replaced by $[M]^{-1}[A]$ and $[Q]$ were to be replaced by $[M]^{-1}[Q]$, the main steps of the CG solver would be as follows:

- Compute the residual vector, $[\hat{R}^{(n)}] = [M]^{-1}[Q] - [M]^{-1}[A][\phi^{(n)}]$, and $\hat{R}2^{(n)}$.
- Set the initial search direction vector equal to the residual vector: $[\hat{D}^{(0)}] = [\hat{R}^{(0)}]$.
- Compute $\alpha^{(n+1)}$ using $\alpha^{(n+1)} = \dfrac{[\hat{R}^{(n)}]^T[\hat{R}^{(n)}]}{[\hat{D}^{(n)}]^T[M]^{-1}[A][\hat{D}^{(n)}]}$.
- Update solution using $[\phi^{(n+1)}] = [\phi^{(n)}] + \alpha^{(n+1)}[\hat{D}^{(n)}]$.
- Compute the new residual vector, $[\hat{R}^{(n+1)}] = [M]^{-1}[Q] - [M]^{-1}[A][\phi^{(n+1)}]$, and $\hat{R}2^{(n+1)}$.
- Compute $\beta^{(n+1)} = \dfrac{[\hat{R}^{(n+1)}]^T[\hat{R}^{(n+1)}]}{[\hat{R}^{(n)}]^T[\hat{R}^{(n)}]} = \dfrac{(\hat{R}2^{(n+1)})^2}{(\hat{R}2^{(n)})^2}$.
- Update search direction vector: $[\hat{D}^{(n+1)}] = [\hat{R}^{(n+1)}] + \beta^{(n+1)}[\hat{D}^{(n)}]$.

As discussed earlier, one of the challenges in executing the above-modified algorithm is that $[M]^{-1}$ is generally not known explicitly. Next, we discuss two preconditioning choices where the preconditioner, $[M]^{-1}$, may be determined with relative ease. One important point to note is that, after preconditioning, the matrix $[M]^{-1}[A]$ must have the same properties (other than the condition number) as the $[A]$ matrix. For example, if the $[A]$ matrix is symmetric positive definite, a symmetric preconditioner must also be used such that $[M]^{-1}[A]$ is also symmetric positive definite.

One of the preconditioners often used is the point Jacobi preconditioner, which is given by

$$[M]^{-1} = [D]^{-1}, \tag{4.47}$$

where $[D] = \text{diag}([A])$. Since $[D]$ is a diagonal matrix, its inverse is easily determined. From a practical implementation point of view, since $[M]^{-1}[A] = [D]^{-1}[A]$, using the point Jacobi preconditioner essentially implies scaling each of the linear algebraic equations by its respective diagonal element prior to executing the CG algorithm, i.e., $A_{ij} = A_{ij}/A_{ii}$ and $Q_i = Q_i/A_{ii}$. Since point Jacobi preconditioning only alters the diagonal elements, the symmetry of the coefficient matrix is retained, and therefore, the CG solver can still be used. Next, an example is considered to assess the effect of point Jacobi preconditioning on the rate of convergence.

EXAMPLE 4.7

The solution to the problem considered in Examples 3.2–3.6 is attempted here with the CG method with and without preconditioning using the point Jacobi preconditioner. A grid size of 161×161 is used. The figure below shows the convergence behavior of the CG method with and without point Jacobi preconditioning.

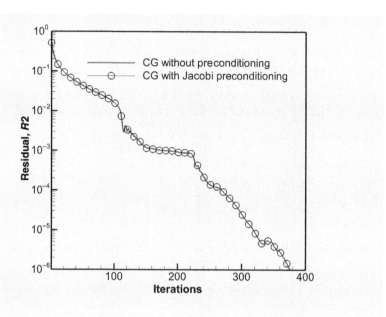

In this case, the scaled residuals (residual normalized by the maximum residual) are considered since the actual residuals with and without preconditioning are different. As shown in the figure above, the residuals, with and without preconditioning, are identical. This implies that preconditioning did not impact the condition number at all. It is actually well known [1] that for the Poisson equation with constant and identical coefficients for all the equations (as in the case of a uniform mesh), although point Jacobi preconditioning alters the eigenvalues of [A], their ratio (the condition number) remains unchanged. However, this behavior is specific to this particular problem. In general, point Jacobi preconditioning does enhance convergence rate and is the easiest to use among all preconditioning techniques.

Another popular preconditioner is the ILU decomposition-based preconditioner. As a matter of fact, Stone, in the process of developing the strongly implicit method presented in Section 3.2.5, started with the basic premise of finding a preconditioner based on approximate or incomplete LU factorization of the coefficient matrix. In Section 3.2.5, it was shown that when a sparse lower triangular, [L], matrix, having the same sparsity pattern as the lower triangle of [A], is multiplied with a sparse upper triangular, [U], matrix, having the same sparsity pattern as the upper triangle of [A], the sparsity pattern of [L][U] is not the same as that of [A]. In fact, the product [L][U] is far less sparse than [A] itself. Since the objective is to find an approximate factorization, namely [A] ≈ [L][U], one can arbitrarily throw away some of the new nonzeroes that arise out of the product [L][U] that did not originally exist in [A]. For example, one can decide to retain the same sparsity pattern in the product [L][U] as in [A]. This results in the so-called ILU(0) decomposition, or ILU decomposition without fill-in. More accurate approximations may be obtained by including some of the new nonzero elements, resulting in the so-called ILU(n) decomposition, where "n" refers to the degree of fill-in [1]. Here, we will discuss only the ILU(0) decomposition-based preconditioning.

In order to find $[A] \approx [L][U]$, let us first consider the sparsity pattern, $NZ([A])$. $NZ([A])$ denotes the set of index pairs (i, j), such that $A_{i,j} \neq 0$. In ILU(0), $[L]$ and $[U]$ are defined such that the sparsity pattern of $[A]$ is identical to that of $[L][U]$, i.e., $NZ([A]) = NZ([L][U])$. In addition, $LU_{i,j} = A_{i,j}$ for all index pairs $(i, j) \in NZ([A])$. The computation of the elements of the matrices $[L]$ and $[U]$ are best depicted by Code Snippet 4.1. Following our earlier notation, the coefficient matrix is assumed to be of size $K \times K$. Also, following standard convention, the diagonal elements of $[L]$ are assumed to be unity, while the diagonal elements of $[U]$ are assumed to be that of $[A]$.

CODE SNIPPET 4.1: ILU(0) DECOMPOSITION. THE ALGORITHM IS SIMILAR TO THE FORWARD ELIMINATION ALGORITHM DESCRIBED IN SECTION 3.1 AND SHOWN IN CODE SNIPPET 3.1.

```
! Temporarily Modify Elements of [A]
For k = 1 : K-1
  For i = k+1 : K
    If (NZA(i,k) == 1) Then
      A(i,k) = A(i,k)/A(k,k) ! Elements of [A] are modified
    End If
    For j = k+1 : K
      If (NZA(i,j) == 1) Then
        A(i,j) = A(i,j) + A(i,k)*A(k,j) ! Elements of [A] are modified
      End If
    End
  End
End
! Fill in Elements of [L] and [U]
For i = 1 : K
  For j = i : K
    U(i,j) = A(i,j) ! Elements of [U] includes diagonal
  End
  For j = 1: i-1
    L(i,j) = A(i,j) ! Elements of [L]
  End
End
```

The determination of $[L]$ and $[U]$ is a preprocessing step, and does not need to be performed within the iteration loop of the solver. In general, storage of the sparsity pattern of $[A]$, namely $NZ([A])$, requires additional memory. This requires storage of integers only, and one can make use of single precision (since the entries are 0 and 1 only), so that the memory requirement is not insignificant. If, however, the computations are performed on a structured mesh, the locations of the nonzeroes are already known, and storage of the sparsity pattern may be avoided completely.

Once the matrices $[L]$ and $[U]$ have been determined, they can be integrated with the solver of choice to generate the algorithm of the preconditioned solver. The algorithm of the preconditioned CG solver with ILU(0) preconditioning is presented below. It assumes that the matrices $[L]$ and $[U]$ are already known.

ALGORITHM: PRECONDITIONED CG METHOD WITH ILU(0) PRECONDITIONING

Step 1: Guess values of ϕ at all nodes, i.e., $\phi_{i,j}$ $\forall i = 1,...,N$ and $\forall j = 1,...,M$. We denote these values as $\phi^{(0)}$. If any of the boundaries have Dirichlet boundary conditions, the guessed values for the boundary nodes corresponding to that boundary must be equal to the prescribed boundary values.

Step 2: Compute the residual vector, $[R^{(0)}] = [Q] - [A][\phi^{(0)}]$, and $R2^{(0)}$.

Step 2p: Compute the modified residual vector, $[\hat{R}^{(0)}] = [M]^{-1}[Q] - [M]^{-1}[A][\phi^{(0)}] = [M]^{-1}[R^{(0)}]$.

Step 3: Set the initial search direction vector equal to the modified residual vector: $[D^{(0)}] = [\hat{R}^{(0)}]$.

Step 4: Compute $\alpha^{(n+1)}$ using $\alpha^{(n+1)} = \dfrac{[R^{(n)}]^T[R^{(n)}]}{[D^{(n)}]^T[A][D^{(n)}]}$.

Step 5: Update solution using $[\phi^{(n+1)}] = [\phi^{(n)}] + \alpha^{(n+1)}[D^{(n)}]$.

Step 6: Compute the new residual vector, $[R^{(n+1)}] = [Q] - [A][\phi^{(n+1)}]$, and $R2^{(n+1)}$.

Step 6p: Compute the modified residual vector, $[\hat{R}^{(n+1)}] = [M]^{-1}[R^{(n+1)}]$.

Step 7: Compute $\beta^{(n+1)} = \dfrac{[R^{(n+1)}]^T[\hat{R}^{(n+1)}]}{[R^{(n)}]^T[\hat{R}^{(n)}]}$.

Step 8: Update search direction vector: $[D^{(n+1)}] = [\hat{R}^{(n+1)}] + \beta^{(n+1)}[D^{(n)}]$.

Step 9: Monitor convergence, i.e., check if $R2^{(n+1)} < \varepsilon_{tol}$? If Yes, then go to Step 10. If No, then go to Step 4 (i.e., proceed to next iteration)

Step 10: Stop iteration and postprocess the results.

In the above algorithm, two new steps, namely Step 2p and Step 6p, have been introduced to integrate preconditioning, in addition to modifications of some of the other steps and equations. Steps 2p and 6p perform the same operation: computation of the modified residual from the original residual. The equation, $[\hat{R}] = [M]^{-1}[R]$, may be alternatively written as

$$[M][\hat{R}] = [R]. \tag{4.48}$$

Further, noting that $[M] = [L][U]$, one can write

$$[L][U][\hat{R}] = [R]. \tag{4.49}$$

Defining a new vector, $[Y]$, such that

$$[U][\hat{R}] = [Y], \tag{4.50}$$

Eq. (4.49) may be written as

$$[L][Y] = [R]. \tag{4.51}$$

Thus, solution of Eq. (4.49) has been split into a two-step process in which Eq. (4.51) is first solved to obtain $[Y]$. Then, Eq. (4.50) is solved to obtain $[\hat{R}]$. The salient point with regard to solution of both Eq. (4.50) and Eq. (4.51) is that both $[L]$ and $[U]$ are triangular matrices, and therefore, solution of both equations requires either forward substitution or backward substitution only. Further, since both $[L]$ and $[U]$ are sparse, the number of operations required for solution of these two equations is not very significant. In summary, Steps 2p and 6p can be executed in a fairly efficient manner. In most cases, the extra computational time required for these two extra steps is offset by the efficiency gains resulting from a reduction in the iteration count. Next, an example that illustrates the CG solver with ILU(0) preconditioning is considered.

EXAMPLE 4.8

The solution to the problem considered in Examples 3.2–3.6, and also in Example 4.7, is attempted here with the CG method and ILU(0) preconditioning. The first task in applying ILU(0) preconditioning is to determine the coefficients of the matrices $[L]$ and $[U]$. Following the procedure followed in deriving Stone's method (Section 3.2.5), we first represent matrix $[A]$ with five diagonals, namely B, D, E, F, and H, as shown in Eq. (3.39). Next, following Stone's derivation, we propose $[L]$ and $[U]$ be comprised of three diagonals each, such that

$$[L] = \begin{bmatrix} 1_1 & & & & \\ c_2 & 1 & & & \\ & & \ddots & & \\ b_{N+1} & & \ddots & \ddots & \\ & & \ddots & & \\ & & b_K & & c_K & 1 \end{bmatrix}, \quad [U] = \begin{bmatrix} g_1 & e_1 & & f_1 & \\ & g_2 & e_2 & & \ddots \\ & & & & f_{K-N} \\ & & & \ddots & \ddots \\ & & & & e_{K-1} \\ & & & & g_K \end{bmatrix}.$$

As shown in Section 3.2.5, with this proposed structure for $[L]$ and $[U]$, the product $[L][U]$ results in seven diagonals instead of five. In other words, the sparsity pattern of $[L][U]$ is different from that of $[A]$. ILU(0), however, stipulates that the sparsity pattern of $[L][U]$ must be the same as that of $[A]$. Therefore, the two extra diagonals are neglected. Setting the remaining five diagonals of $[L][U]$ equal to those in $[A]$, we obtain the following relationships

$E_k = b_k f_{k-N} + c_k e_{k-1} + g_k$, $B_k = b_k g_{k-N}$, $D_k = c_k g_{k-1}$, $F_k = e_k$, and $H_k = f_k$, which may be solved to obtain the necessary elements of $[L]$ and $[U]$. Once the elements of $[L]$ and $[U]$ have been determined, the above algorithm for ILU(0) preconditioned CG method can be executed without additional complexity. For the present computations a grid size of 161×161 is used. The figure below shows the convergence behavior of the CG method with and without ILU(0) preconditioning.

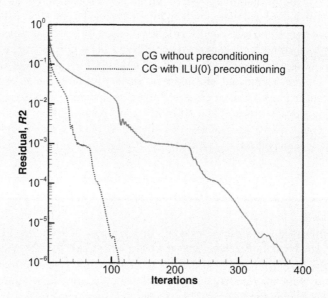

It is evident from the residual plots that unlike the case of point Jacobi preconditioning, ILU(0) preconditioning has significant impact on the convergence of the CG method: the number of iterations required for convergence is reduced by more than a factor of three. For this particular example, the computational times required are less than a second on an Intel core i7 processor. Therefore, the difference in computational time between the two approaches was not reliably recordable. Nonetheless, the dramatic reduction in iteration count is indicative of the power of appropriate preconditioning, especially when the system of equations is large, and the eigenvalues are scattered.

The preceding discussion and example establishes the rationale for using preconditioning as a means to accelerate convergence, and shows the power of preconditioning when performed appropriately. In practice, most Krylov subspace solvers are used in combination with preconditioning to fully capitalize upon their strengths. Generally, preconditioning accelerates convergence by clustering the eigenvalues of the system – in particular, the largest eigenvalue is significantly reduced. Another way to alter the eigenvalues is to use the so-called multigrid method. This is discussed next.

4.5 MULTIGRID METHOD

In Section 4.2.2, it was shown that instability is caused by the large wavelength (small frequency) component of the error, which amplifies with successive iterations. In Section 4.3, it was further demonstrated that for cases that are stable, the convergence rate is dictated by the large wavelength components of the error. These components typically have an amplification factor only slightly less than unity and tend to decay very slowly. Based on these findings, it is fair to conclude that any method that specifically targets and reduces the large wavelength components of the error will exhibit faster rates of convergence. Multigrid methods are designed to do just that.

One of the important findings from Section 4.2 is that the largest eigenvalue is lowered as the mesh is coarsened. This implies that a coarse mesh will result in faster convergence, as evident in all the examples shown in Chapter 3. The multigrid method capitalizes upon the idea of selectively smoothing (reducing) large wavelength errors on a coarse mesh rather than on the original fine mesh because it is more effectively done. The idea led to the so-called geometric multigrid (GMG) method, in which multiple meshes are employed to reduce the errors rapidly. In more recent times, with an increase in geometric complexity of scientific and engineering computations of PDEs, and the increase in the use of unstructured grids, it has become increasingly difficult to use multiple grids. This has led to the development of the so-called algebraic multigrid (AMG) method, in which the general multigrid idea is applied at the linear algebraic equation level without resorting to multiple grids. This is done by developing equivalent coarse-grid approximations from the discrete fine-grid equations. In the next subsection, we discuss the GMG method in significant detail to first explain the working principles of the multigrid idea. This is followed by a brief discussion of the AMG method and how it differs from the GMG method.

4.5.1 GEOMETRIC MULTIGRID (GMG) METHOD

The GMG method is best understood by considering the scenario where only two grids – "coarse" and "fine" – are used. The two-grid algorithm serves as the core framework for a general multigrid algorithm, as will be shown later. Prior to designing any multigrid algorithm, it should be noted that the accuracy of the final solution must be that of the fine mesh. The coarse mesh can only be used to accelerate the convergence; not to compute the final solution. The two-grid algorithm is presented next, with a discussion of the relevant concepts at each step. As an example, we consider the solution of the 2D Poisson equation on a rectangular domain with Dirichlet boundary conditions on all sides.

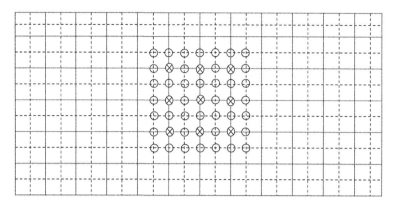

○ Node locations for fine (F) grid

⊗ Node locations for both fine (F) and coarse (C) grid

FIGURE 4.3 Coarse- and Fine-Grid Nodal Arrangements in a Two-Grid Algorithm with Uniform Mesh Spacing

Coarse-grid nodes, denoted by crosses, are at the intersections of solid lines, while fine-grid nodes, denoted by open circles, are at the intersections of both dotted and solid lines.

Step 1: Generating meshes and their relationships.

The first step in the execution of the GMG algorithm is generation and storage of the mesh at various levels. Figure 4.3 shows the "coarse" (C) and "fine" (F) grids for a rectangular domain in the case of a two-grid algorithm. Every coarse-grid node, defined by the pair (*I,J*), has a corresponding fine-grid node sitting atop it, and is defined by the pair (*i,j*). Therefore, the fine-grid indices may be expressed in terms of the coarse-grid indices as follows:

$$(i, j) = (2I - 1, 2J - 1). \tag{4.52}$$

If the total number of fine-grid and coarse-grid nodes in the *i* (or *j*) direction are denoted by N_F (or M_F) and N_C (or M_C), respectively, then the following relationships are also true:

$$\begin{aligned} N_F &= 2N_C - 1 \\ M_F &= 2M_C - 1 \end{aligned}. \tag{4.53}$$

Thus, for example, a 21×21 mesh should be combined with a 41×41 mesh to develop a two-grid algorithm. It follows that the grid spacings for the coarse and fine grids are given by

$$\begin{aligned} \Delta x_C &= \frac{L}{N_C - 1}, \quad \Delta y_C = \frac{H}{M_C - 1} \\ \Delta x_F &= \frac{L}{N_F - 1}, \quad \Delta y_F = \frac{H}{M_F - 1} \end{aligned}. \tag{4.54}$$

The corresponding nodal equations on the coarse and fine grids are written as

$$\left(\frac{2}{(\Delta x_F)^2}+\frac{2}{(\Delta y_F)^2}\right)\phi_{i,j}^F-\frac{1}{(\Delta x_F)^2}\phi_{i+1,j}^F-\frac{1}{(\Delta x_F)^2}\phi_{i-1,j}^F$$
$$-\frac{1}{(\Delta y_F)^2}\phi_{i,j+1}^F-\frac{1}{(\Delta y_F)^2}\phi_{i,j-1}^F=-S_{i,j}, \tag{4.55a}$$

$$\left(\frac{2}{(\Delta x_C)^2}+\frac{2}{(\Delta y_C)^2}\right)\phi_{I,J}^C-\frac{1}{(\Delta x_C)^2}\phi_{I+1,J}^C-\frac{1}{(\Delta x_C)^2}\phi_{I-1,J}^C$$
$$-\frac{1}{(\Delta y_C)^2}\phi_{I,J+1}^C-\frac{1}{(\Delta y_C)^2}\phi_{I,J-1}^C=-S_{I,J}. \tag{4.55b}$$

Step 2: Set initial guess
The next step in the algorithm is to initialize the dependent variable on the fine grid. Since the fine-grid solution is what we are ultimately interested in, it is sufficient to initialize (guess) the solution at the fine-grid nodes. Let this solution be denoted by $\phi_{i,j}^{F(0)}$. If Dirichlet boundary conditions are used, the initial guess at the boundary nodes should be set equal to the prescribed boundary value.

Step 3: Smoothing on fine grid
Next, the algebraic equations on the fine grid are solved using a solver of choice, but only to partial convergence. While any solver, discussed in Chapter 3, may be used for this purpose, it is important to choose one that is easy to implement and whose computational workload per iteration is small. This is because in the context of the GMG algorithm, the solver is not solely responsible for reducing the errors. Rather, the multigrid framework is. In other words, the overall iteration count is not dictated by the solver but rather by the multigrid treatment of the errors. The solver to be used is also known as the *smoother*, and the operation of solving the fine-grid equations to partial convergence is known as *smoothing*. Based on the criterion of low computational workload per iteration, it is customary to use classical iterative solvers for the smoothing operation rather than fully implicit solvers, such as the Krylov subspace solvers. The Gauss–Seidel method is a popular choice as the smoother for multigrid algorithms, primarily because of its extremely low workload per iteration, its ease of implementation, and also the fact that it can be used for both structured and unstructured mesh topologies.

It is clear that solving the fine-grid equations to full convergence in this step would be tantamount to not using the multigrid method at all. Instead, the solution is taken only to partial convergence. To the best of the author's knowledge, there is no reported mathematical analysis that shows the optimum level of partial convergence that is universally applicable to all problems. Generally, iterations are continued until the $R2^F$ decreases by a factor of about two. Often, calculation and monitoring of the scaled residual, as would be required if the residual were to be decreased by a factor two, is completely bypassed, and one or two sweeps of Gauss–Seidel is executed instead. This makes the implementation even easier.

At this point in the algorithm, the solution on the fine grid, $\phi_{i,j}^F$, contains errors due to partial convergence. This error (equal to the difference between the exact numerical solution and the solution at the current iteration) has several different wavelength components. As discussed in Section 4.2.1, of these, the components that have large wavelengths are the most difficult to damp out (or have their amplitudes reduced). Therefore, instead of continuing with regular smoothing (Gauss–Seidel iterations on the fine grid, for example), which would not specifically target the large wavelength components, we transfer this error to a coarse grid and then smooth it on the coarse grid so that they are reduced rapidly. In preparation for these actions, the following step is executed next.

Step 4: Computation of residual on fine grid
The residual and L^2Norm are computed on the fine grid using $[R]^F = [Q] - [A]^F [\phi]^F$, and $R2^F = \sqrt{\left[[R]^F\right]^T [R]^F}$.

Step 5: Transfer of fine-grid residual to coarse grid (restriction)
The residual computed on the fine grid in Step 4 is next transferred to the coarse grid. This process is known as *restriction*. Since every fine-grid node has a coarse-grid node sitting atop it, the restriction operation simply involves copying the residuals on the fine grid to an array (data structure) that stores residuals for the coarse grid. The loops used to perform this transfer should run over the indices of the coarse grid and Eq. (4.52) should be made use of to obtain corresponding fine-grid indices. Henceforth, the residual transferred from the fine to the coarse grid will be denoted by $[R]^{C \leftarrow F}$. For a more complex grid structure, the coarse- and fine-grid nodes may not always overlap. In such a scenario, interpolation will be necessary to execute the transfer process.

Step 6: Smoothing on coarse grid
The transferred residuals are next smoothed on the coarse grid with the specific intent to damp out the large wavelength components of the error rapidly. This operation entails solution of the equation $[A]^C [\phi']^C = [R]^{C \leftarrow F}$ to partial convergence, where $[A]^C$ is the coefficient matrix computed on the coarse mesh [see Eq. (4.55b)] and $[\phi']^C$ is the predicted correction on the coarse mesh. As in Step 3, a solver with a low computational workload and one that is easy to implement must be used. The equation $[A]^C [\phi']^C = [R]^{C \leftarrow F}$ essentially represents the governing linear system in correction form (see Section 3.2.1). Generally, tighter tolerance or more sweeps (typically two or three, as opposed to one) is needed in this step than in Step 3 to obtain an accurate enough prediction of the correction. The Gauss–Seidel method is also a commonly used method for this step.

Step 7: Transfer of coarse-grid correction to fine grid (prolongation)
The correction obtained on the coarse grid, $[\phi']^C$, is next transferred to the fine grid. This process is known as *prolongation*. Execution of the prolongation step involves interpolation since all fine-grid nodes do not have coarse-grid nodes sitting atop them. The transfer of $[\phi']^C$ can be classified into three categories. For the fine-grid nodes that have coarse grid-nodes sitting atop them, the correction is transferred directly. For the remaining nodes, two possibilities exist, as shown in Fig. 4.4.

Based on the two possibilities depicted in Fig. 4.4, either two-point or four-point interpolation is needed to compute fine-grid values from coarse-grid values, as follows:
Two-point:

$$\left(\phi'_{i+1,j}\right)^{F \leftarrow C} = \frac{\left(\phi'_{I,J}\right)^C + \left(\phi'_{I+1,J}\right)^C}{2}, \quad \left(\phi'_{i,j+1}\right)^F = \frac{\left(\phi'_{I,J}\right)^C + \left(\phi'_{I,J+1}\right)^C}{2}, \quad (4.56a)$$

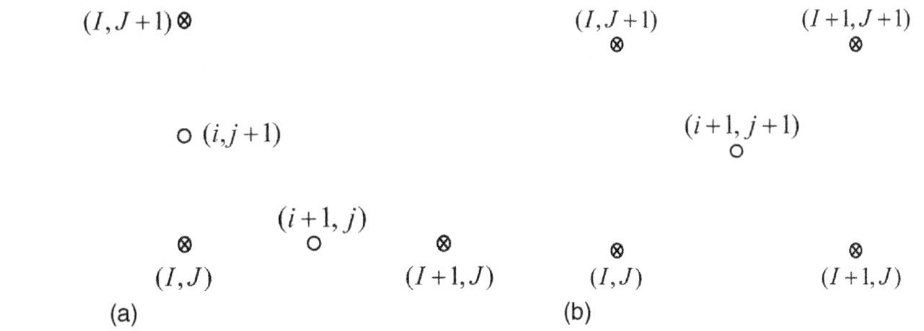

$(I, J+1) \otimes$ $\qquad\qquad$ $(I, J+1)$ $\qquad$ $(I+1, J+1)$
$\qquad\qquad\qquad\qquad\qquad\qquad\qquad\quad \otimes \qquad\qquad\qquad \otimes$

$\circ \; (i, j+1)$ $\qquad\qquad\qquad\qquad\qquad$ $(i+1, j+1)$
$\qquad\qquad\qquad\qquad\qquad\qquad\qquad\qquad\qquad\quad \circ$

$\qquad\qquad\qquad (i+1, j)$
$\qquad \otimes \qquad\quad \circ \qquad\qquad \otimes \qquad\qquad\qquad \otimes \qquad\qquad\qquad \otimes$
$\qquad (I, J) \qquad\qquad\qquad (I+1, J) \qquad\quad (I, J) \qquad\qquad\qquad (I+1, J)$
$\quad$(a) $\qquad\qquad\qquad\qquad\qquad\qquad\quad$ (b)

FIGURE 4.4 Two Interpolation Scenarios in the Two-Grid Algorithm

(a) Two-point interpolation for fine-grid nodes placed along coarse-grid lines, and (b) four-point interpolation for fine-grid nodes offset from coarse-grid lines. Coarse-grid nodes are denoted by crosses, while fine-grid nodes are denoted by open circles.

Four-point:

$$\left(\phi'_{i+1, j+1} \right)^{F \leftarrow C} = \frac{\left(\phi'_{I,J} \right)^{C} + \left(\phi'_{I+1,J} \right)^{C} + \left(\phi'_{I,J+1} \right)^{C} + \left(\phi'_{I+1,J+1} \right)^{C}}{4}. \qquad (4.56b)$$

The transferred correction is denoted by $[\phi']^{F \leftarrow C}$. In the general case of a nonuniform or curvilinear mesh, distance-weighted interpolation must be used. Distance-weighted interpolation is discussed in detail in Chapter 7.

Step 8: Update of fine-grid solution
The fine-grid solution, obtained in Step 3, is next updated by adding to it the correction obtained in Step 7: $[\phi]^{F} = [\phi]^{F} + [\phi']^{F \leftarrow C}$. At this juncture, one complete cycle of the multigrid (two-grid) algorithm has been completed.

Step 9: Check for convergence
Convergence is checked by monitoring the residual computed at Step 4, i.e., is $R2^{F} < \varepsilon_{tol}$? Although the residual computed in Step 4 is lagging behind by one iteration, it is preferable to use it to monitor convergence to avoid computation of the residual twice within the same iteration. If the convergence criterion has not been satisfied, Steps 3–9 must be repeated.

The two-grid algorithm, just described, may be thought of as a detour from the original Gauss–Seidel method (assuming that Gauss–Seidel is the smoother and only one sweep of Gauss–Seidel is performed in Step 3) in which, rather than arrive directly at Step 9 from Step 5, a detour is taken, wherein the errors are smoothed further with particular emphasis on reducing the large-wavelength components of the errors. Whether or not this strategy is really effective will be examined through an example problem shortly. Assuming that the strategy is effective and results in significant reduction in iteration count, it is important at this point to tally the number of extra floating-point operations introduced by the extra steps. For solution of a 2D PDE, Step 3 requires four multiplications and one division per node if one sweep of the Gauss–Seidel method is used, resulting in $5N_F M_F$ long operations. Another five multiplications per node are needed to compute the residual in Step 4, bringing the total to $10N_F M_F$ long operations. Step 5 does not require any arithmetic operation. Step 6 requires five long operations per coarse grid node per sweep. However, the number of coarse grid nodes is approximately one-fourth that of the fine grid nodes. Assuming that three sweeps are used, Step 6 effectively requires $3.75N_F M_F$ long operations, resulting in a total of $13.75N_F M_F$ long operations. The prolongation operation of Step 7 requires approximately $0.75N_F M_F$ long operations to execute [Eq. (4.56)]. Thus, the total number of long operations needed by the two-grid GMG algorithm is approximately $14.5N_F M_F$. In contrast, the core Gauss–Seidel algorithm would require approximately $10N_F M_F$ long operations. Thus, the workload increase per iteration in the two-grid GMG is about 50%. As long as the iteration count is reduced by more than 50%, the two-grid GMG algorithm is expected to be beneficial from the overall computational time standpoint. An example is considered next to assess the *pros* and *cons* of the two-grid algorithm.

EXAMPLE 4.9

The solution to the problem, considered in Examples 3.2–3.6, is attempted here with the two-grid algorithm. The following grid combination is used: 81×81 (coarse) with 161×161 (fine). The residual plot for the 81×81 (coarse)/161×161 (fine) grid combination is shown below along with the residuals of the pure Gauss–Seidel method executed on the 161×161 grid separately.

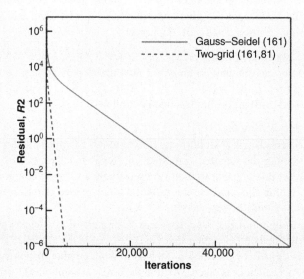

The geometric two-grid algorithm results in tremendous reduction in the iteration count, requiring only 4450 iterations as opposed to the 57,786 iterations required by the pure Gauss–Seidel method – a factor of 13 reduction. As far as computational times are concerned, the two-grid algorithm on the 81/161 grid combination required 5.42 seconds, compared to 24.1 seconds required by Gauss–Seidel – a factor of 4.44 reduction. Clearly, the computational times did not scale exactly as the iterations. As discussed earlier, this is due to the increased workload per iteration. The benefit of using the multigrid method is best understood by closer examination of the errors before and after the coarse-grid smoothing operation. The figures below show the convergence error after one sweep (first figure below) of Gauss–Seidel (Step 3), and also shortly after the prolongation and update (second figure below) step (Step 8).

It is worth recalling that Steps 4–8 represent a detour (the multigrid smoothing of the errors) from the main Gauss–Seidel algorithm.

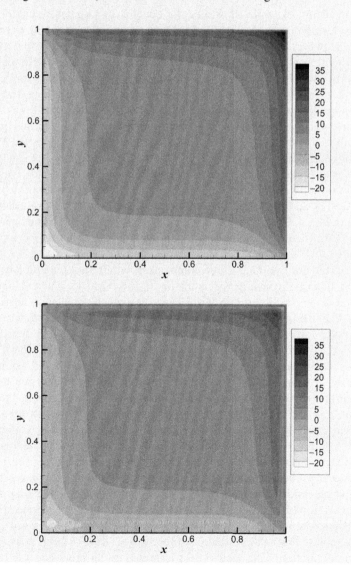

> The error contours exhibit a noticeable reduction in the error after the coarse-grid smoothing operation. In particular, the sharp peaks and valleys have been significantly reduced: the maximum positive error has been reduced from 38.2 to 26.1, while the maximum negative error has been reduced from -25.4 to -16.3.

The preceding example illustrates the benefits of targeted reduction of the large-wavelength components of the error as a means to accelerating overall convergence. However, this remarkable idea would remain largely underutilized if one were to stop at using just two grids. General-purpose multigrid algorithms make use of the basic two-grid idea to construct a hierarchical error reduction framework using many grid levels. The sequence of smoothing–restriction–correction–prolongation steps that are followed in such algorithms is referred to as *multigrid scheduling*. While a large variety of multigrid scheduling algorithms are available, the three most commonly used ones are the V-cycle, the W-cycle, and the full multigrid (FMG) cycle.

In order to understand the role of additional (beyond two) grid levels in the GMG algorithm, we first consider the V-cycle multigrid algorithm, depicted in Fig. 4.5(a). For additional clarity, a three-grid algorithm with all relevant details is shown in Fig. 4.5(b). In the discussion to follow, instead of using superscripts "C" and "F" to denote grid levels, we will use the grid level numbers shown in Fig. 4.5(a).

The V-cycle multigrid algorithm commences with iterative solution of the linear system to partial convergence on the finest grid (Grid 1). The residual is then transferred to the next finest grid (Grid 2) and smoothed using an iterative solver. If, rather than using a few sweeps on Grid 2, iterations were continued, the solution for the correction, $[\phi']^2$, would be a lot more accurate and devoid of the large-wavelength components corresponding to Grid 2. However, as we already know, this would require a large number of iterations on Grid 2, and would defeat the purpose of using a multigrid method. Instead, we could envision using another multigrid algorithm to damp out the errors on Grid 2. Of course, that would require at least one more grid, i.e., Grid 3. Essentially, this leads to the idea of using a multigrid algorithm within the original multigrid algorithm, as depicted in Fig. 4.5(b). If this process were to be continued, we would end up having several multigrid algorithms nested within each other, leading to the concept of *recursion*. If programmed in a modular fashion, with advanced programming languages, it is possible to use recursion with relative ease. When does the process of using additional grid levels stop? Recalling that the objective is to obtain an accurate prediction of the correction on the coarsest mesh level, the process may be stopped when the mesh is so coarse that a direct solution of the system $[A][\phi'] = [R]$ is possible using Gaussian elimination. This scenario for terminating the process is the best-case

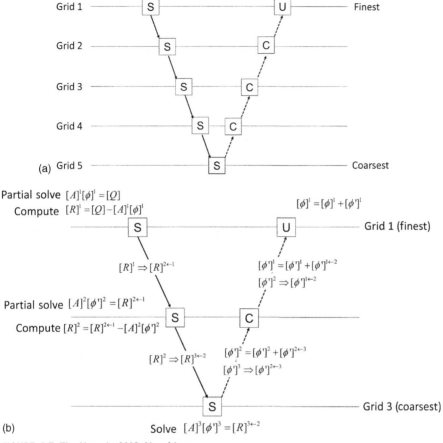

FIGURE 4.5 The V-cycle GMG Algorithm

(a) The scheduling sequence, and (b) detailed work plan in the context of a three-grid algorithm. "S" refers to the smoothing operation, "C" refers to the error correction operation, and "U" refers to the final solution update. The solid arrows represent restriction steps, while the dotted arrows represent prolongation steps.

or ideal scenario. In practice, the number of grid levels is often determined by the overheads incurred in interpolation, storage of multiple grids, and other factors. It is worth remembering that practical problems are rarely solved on a rectangular domain with perfectly orthogonal (Cartesian) meshes, and therefore, one has to actually generate multiple meshes and construct interpolation functions prior to executing the multigrid algorithm. Hence, in practice, the number of grid levels is often not sufficient and the coarsest grid is not coarse enough to enable direct solution of the linear system. In order to highlight some of the aforementioned issues, Example 4.9 is repeated with multiple grid levels and the V-cycle.

EXAMPLE 4.10

The solution to the problem considered in Example 4.9 is repeated here with the V-cycle GMG algorithm. The finest grid considered is a 161×161 mesh, and the grid is progressively coarsened by a factor of two to obtain additional grid levels. The table below summarizes the number of iterations and the computational time required by the overall algorithm as a function of the number of grid levels. The CPU times reported are for computations performed on a 2.2 GHz Intel core i7 processor with 8 GB of RAM. The tolerance used for convergence was 10^{-6}. The residuals for the four cases are also shown in the figure below.

Grid levels	Iterations	CPU time (s)
2 (161,81)	4450	5.42
3 (161,81,41)	947	1.28
4 (161,81,41,21)	219	0.32
5 (161,81,41,21,11)	66	0.10

The results shown in the table above and the figure below clearly illustrate the remarkable power of the multigrid method. With the addition of each grid level the number of iterations decreases by approximately a factor of 4 to 5. Of course, the CPU time does not scale exactly with the number of iterations, as expected, due to increased workload per iteration, especially when the number of grid levels used is relatively large.

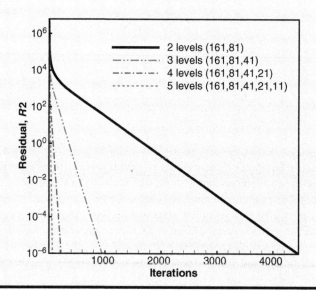

One of the remarkable properties of the multigrid algorithm is its scaling. Of the solvers considered in Chapter 3, the best scaling was produced by the CG solver, where the number of iterations required for convergence increased by a factor of approximately two when the number of nodes was quadrupled (see Example 3.8). Such scaling is typical of the CG or conjugate gradient square (CGS) method, in which the workload (or computational time) scales as $K^{3/2}$ [1]. If the problem considered in Example 4.10 is computed on a 81×81 grid with the V-cycle GMG method with 4 grid levels, convergence is attained in 62 iterations. This is quite remarkable because it implies that the same problem can be solved on a 81×81 grid, as well as a 161×161 grid with roughly the same number of iterations (the 161×161 grid required 66 iterations) by simply using an additional coarse-grid correction. Since the workload on the coarsest mesh is negligible, it implies that the total workload scales approximately as the number of nodes, since the number of iterations remains more or less unchanged between the 81×81 and the 161×161 grid. Thus, the multigrid algorithm is fundamentally an $\mathcal{O}(K)$ algorithm, in which the workload (or computational time) is directly proportional to the number of unknowns, K. As mentioned in Chapter 3, linear scaling with problem size is the best-case scenario as far as the performance of iterative solvers is concerned.

Other multigrid scheduling cycles aim to improve upon the performance of the V-cycle. The W-cycle, and the FMG cycles are depicted schematically in Fig. 4.6. The W-cycle scheduling makes use of the fact that the computational workload at the coarsest grid levels is almost negligible. Therefore, rather than execute the upward prolongation steps all the way to the finest grid, a second set of restriction and smoothing operations are performed on the coarse-grid levels prior to executing the complete prolongation-update branch. Analysis shows that this strategy generally leads to a reduction in the iteration count. The FMG algorithm starts at the coarsest grid level. The solution at the coarsest grid is interpolated (prolongated) to the next fine grid and is used as an initial guess for the smoothing on that grid. The error is then restricted back to the coarsest mesh and in the next step, two successive prolongation steps are executed using two grid levels above the coarsest grid followed by two successive restriction steps, and so on. The FMG algorithm essentially executes a series of inverted two-grid V-cycles with growing sizes of V. It is believed to yield superior performance to either the V- or W-cycles [3,4], and is the best option for applications where adaptive grid refinement is used. Other multigrid schedules, such as the F-cycle (F stands for flexible) and the saw-tooth cycle, are also used. For a description of these, and for further in-depth reading on the multigrid method, the reader is referred to texts focused specifically on multigrid methods [3,4].

4.5.2 ALGEBRAIC MULTIGRID (AMG) METHOD

While the GMG method is one of the most powerful methods for solving a large linear system of equations, unlike the Krylov subspace-based methods, it is not a plugin-type method. The formulation and execution of the GMG method is inherently tied to a grid, thereby significantly hampering its generality as a linear system solver.

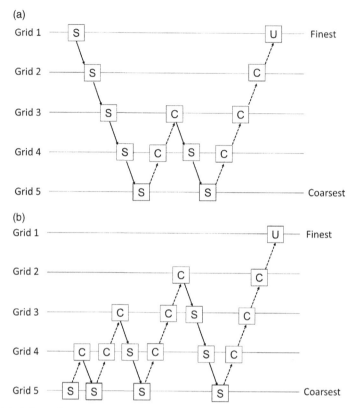

FIGURE 4.6 Schematic Representation of Two Commonly Used Multigrid Cycles

(a) The W-cycle, and (b) the FMG cycle. "S," "C," and "U" refer to smoothing, correction, and updating, respectively.

For example, it cannot be used to solve a general system of linear algebraic equations that did not arise out of discretization of a PDE on a grid. The AMG method was designed with the goal to capitalize upon the powerful multigrid idea of targeted reduction of large-wavelength errors while removing any connection of the method to an underlying grid so that the method can be used as a plugin-type solver. The development of the AMG method was prompted by the increasing popularity of unstructured meshes, in which generating a large number of interconnected grid levels is all but impossible.

The Gauss–Seidel method, which is the most commonly used smoother in multigrid methods, can be used to solve any linear system, and does not require any geometric information if the matrix coefficients are known. Thus, it can be used equally effectively as a smoother in the context of the AMG method, and is, in fact, used as the default smoother in AMG methods. However, computation of the matrix coefficients requires coarse- and fine-grid information. If the very idea of multiple grids is removed, one has to first overcome the challenge of constructing coarse- and fine-grid equivalents of the discrete equations to be able to exercise the multigrid idea. The fine-grid equations are our starting point, and are always available. In AMG, the coarse-grid equivalent equations are constructed from the fine-grid equations at the algebraic equation level by combining fine-grid equations in some fashion – a process known as *agglomeration*. The exact algorithm used to combine the fine-grid equations to form equivalent coarse-grid equations results in variations of the AMG method. Another aspect that differs between various AMG methods is the interpolation scheme used in the restriction and prolongation steps. Since there is no grid any more, the interpolation functions between two equivalent grid levels must be constructed without resorting to a grid. Two commonly used agglomeration and interpolation schemes are discussed next.

The coarse-grid equivalent equations are constructed from the fine-grid equations by agglomerating the fine-grid equations in some fashion. In order to understand this process, let us consider the following set of algebraic equations arising out of finite difference discretization of a 2D PDE on a structured mesh:

$$a_{O,k}^F \phi_k^F + a_{E,k}^F \phi_{k+1}^F + a_{W,k}^F \phi_{k-1}^F + a_{N,k}^F \phi_{k+N}^F + a_{S,k}^F \phi_{k-N}^F = Q_k^F \ \forall k \in F, \qquad (4.57)$$

where $a_{O,k}$ denotes the diagonal coefficient of the nodal equation for node k, and $a_{E,k}$, $a_{W,k}$, $a_{N,k}$, and $a_{S,k}$ denote the off-diagonal coefficients – also known as the link coefficients, since these coefficients link the node k to the surrounding nodes. It is worth reiterating that these equations were derived by discretizing the governing PDE on the finest grid, as indicated by the superscript "F". Equation (4.57) may also be re-written for the surrounding nodes, as follows:

$$a_{O,k+1}^F \phi_{k+1}^F + a_{E,k+1}^F \phi_{k+2}^F + a_{W,k+1}^F \phi_k^F + a_{N,k+1}^F \phi_{k+N+1}^F + a_{S,k+1}^F \phi_{k-N+1}^F = Q_{k+1}^F, \quad (4.58a)$$

$$a_{O,k-1}^F \phi_{k-1}^F + a_{E,k-1}^F \phi_k^F + a_{W,k-1}^F \phi_{k-2}^F + a_{N,k-1}^F \phi_{k+N-1}^F + a_{S,k-1}^F \phi_{k-N-1}^F = Q_{k-1}^F, \quad (4.58b)$$

$$a_{O,k+N}^F \phi_{k+N}^F + a_{E,k+N}^F \phi_{k+N+1}^F + a_{W,k+N}^F \phi_{k+N-1}^F + a_{N,k+N}^F \phi_{k+2N}^F + a_{S,k+N}^F \phi_k^F = Q_{k+N}^F, \quad (4.58c)$$

$$a_{O,k-N}^F \phi_{k-N}^F + a_{E,k-N}^F \phi_{k-N+1}^F + a_{W,k-N}^F \phi_{k-N-1}^F + a_{N,k-N}^F \phi_k^F + a_{S,k-N}^F \phi_{k-2N}^F = Q_{k-N}^F. \quad (4.58d)$$

In order to construct a coarse-grid equivalent equation, Eq. (4.57) may be agglomerated (added) with any of the four equations for the surrounding nodes shown in Eq. (4.58). Which equation is the most logical choice, depends on the link coefficients. When two equations are agglomerated, the physical interpretation is that two nodes have been agglomerated. If the two nodes happen to have the same value of ϕ, this makes logical sense. If we consider the set of nodes O, E, W, N, and S, the two nodes that will have the closest value of ϕ are the two that have the largest link coefficients connecting any two given pairs. Having large link coefficients between a pair of nodes is tantamount to a strong information exchange (or transport) between the pair. For example, if we are solving the heat conduction equation $\nabla \cdot (\kappa \nabla T) = 0$ on a nonuniform structured mesh, the link coefficients will be proportional to κ / δ^2, where κ is the thermal conductivity, and δ is the distance between any two nodes.

Thus, any pair of nodes that has the largest value of κ/δ^2 is the best candidate for agglomeration. Physically, this implies that either the thermal conductivity has to be very large, or the nodes are at close proximity, or both. Both of these conditions will cause the temperature of the two nodes to be close to each other, and therefore, make them suitable candidates for agglomeration. In the case of the solution of the Poisson equation on a uniform mesh, since all four link coefficients have the same value, all four nodes (E, W, N, and S) are equally good candidates for agglomeration with O. Let us assume that we agglomerate O with E. This implies addition of Eq. (4.57) with Eq. (4.58a). The result is

$$
\begin{aligned}
a_{O,k}^F \phi_k^F + a_{E,k}^F \phi_{k+1}^F + a_{W,k}^F \phi_{k-1}^F + a_{N,k}^F \phi_{k+N}^F + a_{S,k}^F \phi_{k-N}^F + a_{O,k+1}^F \phi_{k+1}^F \\
+ a_{E,k+1}^F \phi_{k+2}^F + a_{W,k+1}^F \phi_k^F + a_{N,k+1}^F \phi_{k+N+1}^F + a_{S,k+1}^F \phi_{k-N+1}^F = Q_{k+1}^F + Q_k^F \cdot
\end{aligned}
\tag{4.59}
$$

Assuming that $\phi_k^F = \phi_{k+1}^F = \phi_j^C$, where j is the index of the newly formed coarse-grid node ($j \in C$), Eq. (4.59) may be rewritten as

$$
\begin{aligned}
\left(a_{O,k}^F + a_{O,k+1}^F + a_{E,k}^F + a_{W,k+1}^F \right)\phi_j^C + a_{W,k}^F \phi_{k-1}^F + a_{N,k}^F \phi_{k+N}^F + a_{S,k}^F \phi_{k-N}^F \\
+ a_{E,k+1}^F \phi_{k+2}^F + a_{N,k+1}^F \phi_{k+N+1}^F + a_{S,k+1}^F \phi_{k-N+1}^F = Q_{k+1}^F + Q_k^F \cdot
\end{aligned}
\tag{4.60}
$$

Further, if we assume that none of the other fine-grid nodal equations are combined, the fine-grid index, k, can be replaced by the coarse-grid index, j, resulting in

$$
\begin{aligned}
\left(a_{O,k}^F + a_{O,k+1}^F + a_{E,k}^F + a_{W,k+1}^F \right)\phi_j^C + a_{W,k}^F \phi_{j-1}^C + a_{N,k}^F \phi_{j+N}^C + a_{S,k}^F \phi_{j-N}^C \\
+ a_{E,k+1}^F \phi_{j+2}^C + a_{N,k+1}^F \phi_{j+N+1}^C + a_{S,k+1}^F \phi_{j-N+1}^C = Q_{k+1}^F + Q_k^F \cdot
\end{aligned}
\tag{4.61}
$$

In Eq. (4.61), the quantity within brackets premultiplying ϕ_j^C is the new diagonal of the "coarse node" that was formed by agglomerating the governing nodal equations for k and $k+1$. Based on Eq. (4.61), the following general relations may be deduced for the agglomeration process:

$$
a_{O,j}^C = a_{O,k}^F + a_{O,k+1}^F + a_{E,k}^F + a_{W,k+1}^F,
\tag{4.62a}
$$

$$
Q_j^C = Q_{k+1}^F + Q_k^F.
\tag{4.62b}
$$

Equation (4.62a) states the diagonal element of the new "coarse" level equation is the sum of the diagonal elements of the old "fine" level equations and the sum of the link coefficients that connected the two agglomerated equations or nodes. Equation (4.62b) states that the right-hand side of the new "coarse" equation is the sum of the right-hand sides of the two agglomerated fine equations. The preceding two rules, given by Eq. (4.62), are universal no matter how many equations (nodes) are agglomerated. All of the other link coefficients remain unchanged in this case because no other equations were agglomerated. It should be emphasized that Eqs. (4.61) and (4.62) were derived without resorting to any geometric information. Only the link coefficients were made use of. If other equations are now agglomerated in addition, the

(a)   (b)

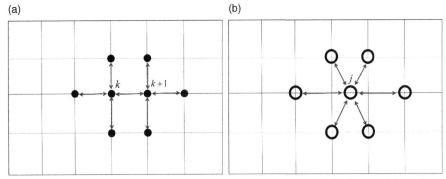

FIGURE 4.7 Illustration of the Agglomeration Process in AMG to Develop "Coarse" Equations

(a) Fine-grid stencil and links (denoted by arrows) and (b) "coarse-grid stencil" and links. The fine-grid nodes are denoted by solid circles while the coarse-grid ones are denoted by open circles.

same procedure and rules, given by Eq. (4.62), must be followed. This makes the process of agglomeration inherently sequential, and the computational algorithm must obey this protocol. Although the "coarse" equation was derived without resorting to geometric information, the agglomeration process does have geometric implications as illustrated in Fig. 4.7. It is clear from Fig. 4.7 that the agglomeration process has the potential to destroy the structured property of the original fine mesh. However, if Gauss–Seidel is used as the smoother, the fact that the coarse equations correspond to a virtual unstructured mesh in physical space has no downside, since the method is a point-wise iterative method. This is one of the main reasons why Gauss–Seidel is the preferred smoother in AMG methods.

The construction of coarse-grid equivalent equations from the fine-grid equations is one important step in the development of the AMG method. Another important step is the development of interpolation functions that will allow transfer of information between the fine and coarse levels, as required in the restriction and prolongation steps. Since the coarse level is now only virtual, geometry-based interpolation can no longer be used, and a different approach must be resorted to.

Equation (4.61) is the coarse-grid equivalent equation that was derived from Eqs (4.57) and (4.58). If the same procedure is followed and applied to the correction form of the equations, rather than to the original equations, the following equation would result:

$$\left(a_{O,k}^F + a_{O,k+1}^F + a_{E,k}^F + a_{W,k+1}^F\right)\phi_j'^C + a_{W,k}^F\phi_{j-1}'^C + a_{N,k}^F\phi_{j+N}'^C + a_{S,k}^F\phi_{j-N}'^C$$
$$+ a_{E,k+1}^F\phi_{j+2}'^C + a_{N,k+1}^F\phi_{j+N+1}'^C + a_{S,k+1}^F\phi_{j-N+1}'^C = R_{k+1}^F + R_k^F .$$

(4.63)

Equation (4.63) represents the coarse-grid correction equation that would have to be solved (using Gauss–Seidel) right after the restriction step to obtain the coarse-grid correction. It is worth noting that the right-hand side of this equation contains the residuals of two fine-grid nodes. In essence, rather than compute the residuals on the fine mesh and then transfer it explicitly to the coarse mesh via geometric interpolation, in this case, the coarse-grid correction equations have been derived directly since geometric interpolation can no longer be employed.

Once the coarse-grid correction, $\phi_j^{\prime C}$, has been computed, it has to be passed back to the fine grid (prolongation) so that the fine-grid solution can be updated. Since the assumption $\phi_k^{\prime F} = \phi_{k+1}^{\prime F} = \phi_j^{\prime C}$ was made in the first place to derive the coarse-grid equation from the fine-grid equations, the prolongation step essentially involves transferring the same correction to both (or more) the agglomerated nodes (or equations).

The preceding discussion provides a high-level outline of the main steps in the AMG method. There are many variations of the method – each differing in the way the coarse-grid correction equations are constructed, as well as the way the coarse-grid correction is transferred back to the fine grid. For a more comprehensive discussion of the AMG method, the reader is referred to journal articles and advanced texts [5–7] on this subject.

In this chapter, the stability and rate of convergence of iterative solvers was investigated. The von Neumann stability analysis procedure was presented and demonstrated for classical semi-implicit iterative solvers. Several examples were shown to demonstrate that the predictions of stability analysis indeed go hand in hand with the convergence behavior of iterative methods. Preconditioning of the coefficient matrix is a commonly used strategy to alter the condition number of the linear system. A general discussion of the philosophy of preconditioning was followed by detailed discussion of Jacobi and ILU preconditioning. An example was shown in which the ILU(0) preconditioner was integrated with the CG solver. One of the major findings that emanated from the Fourier decomposition of the convergence error is that the large wavelength (small frequency) error components are primarily responsible for instability and/or poor rate of convergence. With this important finding as a backdrop, the multigrid method was presented. The geometric two-grid method was discussed in-depth, followed by a presentation of various multigrid scheduling algorithms. Examples showed that the V-cycle multigrid method is an $\mathcal{O}(K)$ method – superior in performance to all other iterative linear system solvers discussed in this text. Other multigrid scheduling algorithms were also discussed briefly. An overview of the algebraic multigrid method was also provided. Thus far, the focus of this text has been on elliptic PDEs. In the next chapter, we switch gears and concentrate on parabolic and hyperbolic PDEs, commonly encountered in the context of the solution of unsteady (or transient) problems in science and engineering.

REFERENCES

[1] Saad Y. Iterative methods for sparse linear systems. 2nd ed. SIAM: Philadelphia; 2003.
[2] Pletcher RH, Tannehill JC, Anderson D. Computational Fluid Mechanics and Heat Transfer. 3rd ed. CRC Press: Boca Raton, FL; 2011.

[3] Wesseling P. An introduction to multigrid methods. R.T. Edwards Inc: Philadelphia; 1992.

[4] Hackbusch W. Multi-grid methods and applications. Springer: Berlin; 2010.

[5] Brandt A, McCormick SF, Ruge J. Algebraic multigrid (AMG) for sparse matrix equations. In: Evans DJ, editor. Sparsity and its applications. Cambridge: Cambridge University Press; 1984. p. 257–84.

[6] Stüben K. Algebraic multigrid (AMG): an introduction with applications. In: Trottenberg U, Oosterlee CW, Sch&uller A, editors. Multigrid. New York: Academic Press; 2000.

[7] Stüben K. "A review of algebraic multigrid," J Comput Appl Math. 2001;128: 281–309.

EXERCISES

4.1   Calculate the condition number of the following matrix:

$$\begin{bmatrix} 5 & -1 & 3 \\ 2 & 6 & -2 \\ 3 & 0 & -4 \end{bmatrix}$$

4.2   The following system of equations is to be solved using the Gauss–Seidel method. Perform a stability analysis to predict if it will converge.

$$\begin{bmatrix} 5 & -1 & -3 \\ -2 & 6 & -2 \\ -1 & 0 & 2 \end{bmatrix} \begin{bmatrix} \phi_1 \\ \phi_2 \\ \phi_3 \end{bmatrix} = \begin{bmatrix} 1 \\ 2 \\ 1 \end{bmatrix}$$

Verify your findings by writing a computer program and solving the linear system using the Gauss–Seidel method.

4.3   For the differential equation $\dfrac{d^3\phi}{dx^3} = S_\phi$, derive a first-order accurate finite difference equation for a generic interior node assuming equal grid spacing. Use nodes O, E, W, and WW in your derivation, where O refers to the central node, E refers to the node to its right, W refers to the node to its left, and WW denotes the second node to its left. Write the final discrete equation in the following form: $A_W\phi_W + A_O\phi_O + A_E\phi_E + A_{WW}\phi_{WW} = S_O$. Now consider solution of the discrete equation, just derived, using an iterative tridiagonal solution technique in which the term that does not belong to the three central diagonals (i.e., $A_{WW}\phi_{WW}$) is treated explicitly. Calculate the spectral radius of convergence for such an iterative scheme. Assume 15 equally spaced nodes in your computational domain.

4.4   Consider the following second-order PDE and boundary conditions:

$$\frac{\partial^2\phi}{\partial x^2} + \frac{\partial^2\phi}{\partial y^2} = S_\phi = 50000 \cdot \exp[-50\{(1-x)^2 + y^2\}] \cdot [100\{(1-x)^2 + y^2\} - 2]$$

$$\phi(1,y) = 100(1-y) + 500\exp(-50y^2)$$
$$\phi(0,y) = 500\exp(-50\{1+y^2\})$$
$$\phi(x,0) = 100x + 500\exp(-50(1-x)^2)$$
$$\phi(x,1) = 500\exp(-50\{(1-x)^2+1\})$$

Note that this is a linear system of equations. The above system has an analytical solution, given by

$$\phi(x,y) = 500\exp(-50\{(1-x)^2+y^2\}) + 100x(1-y)$$

In Exercise 2.6, you have already derived the finite difference equations for this problem. Derive additional equations if necessary, and write a computer program to solve the above problem using the GMG method with just two grids: 81×81 and 41×41. Use one sweep of the Gauss–Seidel method for smoothing on the fine grid. For coarse-grid smoothing, use two different approaches: (a) Gauss–Seidel smoothing with the number of sweeps increasing from one to five (each case is a different computation), and (b) direct solver (Gaussian elimination). Plot the residuals in each case. Comment on your findings.

4.5 Using Example 4.8 as a guide, write a computer program to solve Exercise 4.4 (above) with the CG method using ILU(0) preconditioning. Use different grid sizes ranging from 21×21 to 161×161. Comment on the scaling properties of the solver.

4.6 Consider the following linear system:

$$
\begin{bmatrix}
6 & -2 & 0 & -1 & 0 & 0 \\
-2 & 7 & -1 & 0 & -1 & 0 \\
0 & -1 & 7 & -3 & 0 & -1 \\
-1 & 0 & -3 & 7 & 0 & -2 \\
0 & 0 & -1 & -2 & 7 & -3 \\
0 & -1 & -1 & 0 & -3 & 6
\end{bmatrix}
\begin{bmatrix}
\phi_1 \\ \phi_2 \\ \phi_3 \\ \phi_4 \\ \phi_5 \\ \phi_6
\end{bmatrix}
=
\begin{bmatrix}
3 \\ 3 \\ 2 \\ 1 \\ 1 \\ 1
\end{bmatrix}
$$

The objective is to solve the above linear system using the AMG method with two grid levels. To derive the coarse-grid equivalent equations, agglomerate Equations 1 and 2, 3 and 4, and 5 and 6. Write a computer program that executes the two-grid algorithm with six "fine" and three "coarse" equations. Compare the convergence of the AMG method that you just developed with that of the pure Gauss–Seidel method.

Treatment of the Time Derivative (Parabolic and Hyperbolic PDEs)

From a mathematical perspective, time may be thought of as just another independent variable. However, the independent variable, time, has some attributes that render it special. For one, time usually starts at zero and proceeds toward infinity, making the domain semi-infinite. This also implies that "downstream" information in time is never known, and therefore, the boundary conditions in time are applied at the start of the time domain, and not at the end. Hence, they are usually referred to as initial conditions rather than boundary conditions. Finally, since the vast majority of partial differential equations (PDEs) of interest to a numerical analyst represent mathematical models arising out of physical problems, it is generally worth reconciling them with the natural interpretations of time and space, so that the solution of a PDE can easily be interpreted from a physical perspective. It is to be noted that there is a class of numerical methods for solving PDEs in which space and time are unified in their treatment [1]. In this text, however, we will comply with the vast majority of formulations that retain time and space as separate entities.

Chapter 3 introduced the finite difference method for developing discrete node-based approximations to derivatives. The procedures outlined therein for the treatment of spatial derivatives are, for the most part, directly applicable to the treatment of time derivatives. A few special situations arise when it comes to application of boundary conditions for hyperbolic PDEs. These special cases will become evident in due course. Other than that, this chapter essentially extends the concepts presented in Chapter 3 to the treatment of the time derivative. In addition, the von Neumann stability analysis procedure, introduced in Chapter 4 for analysis of classical iterative solvers, is extended to the prediction of stability of time-marching schemes for parabolic and hyperbolic PDEs.

5.1 STEADY-STATE VERSUS TIME-MARCHING

For practical scientific and engineering analysis, depending on the specific application at hand, either the steady-state solution or the unsteady solution of the problem may be of interest. The strategy of solution and the numerical method to be used is often dictated by which of these two solutions is sought. In order to bring to light these issues, as an example, let us consider the following PDE:

$$\frac{\partial T}{\partial t} = \alpha \nabla^2 T + \dot{Q}_{gen}. \tag{5.1}$$

Equation (5.1) is the so-called heat conduction equation [2], where T denotes temperature, t denotes time, α denotes the thermal diffusivity, and $\dot{Q}_{gen}$ denotes the heat generation rate per unit volume. In an application where the steady-state solution (spatial temperature distribution) to Eq. (5.1) is sought, the solution may be obtained by adopting one of two approaches: (a) Eq. (5.1) may be time-marched to "large" times, in which case the problem is parabolic in time and elliptic in space (see Section 1.2), or (b) the time derivative (left-hand side) may be dropped, and the resulting equation may be solved directly to obtain the steady-state solution. In the latter case, the resulting equation is the Poisson equation, and the resulting problem is an elliptic boundary value problem, which was discussed at length in Chapter 3. Both methods have their respective *pros* and *cons*.

The time-marching approach has the advantage that one does not have to know *a priori* whether a unique steady-state solution to the problem (PDE with certain initial and boundary conditions) exists or not. Through time-marching, the existence or nonexistence of a steady-state solution can be established. For many physical problems, the same governing equation, depending on the parameter range of the computation, the geometry, and the boundary conditions, may or may not have a unique steady-state solution. Fluid flow driven by natural convection is one such example. Hence, this may be cited as a notable advantage of the time-marching method. On the flip side, the time that is required to reach steady state is also not known *a priori*, and therefore, one might have to advance the solution for a large number of time steps before a steady-state solution is reached, rendering the computation time consuming. Another advantage of the time-marching approach is that the resulting coefficient matrix (if an implicit time-marching method is used) is generally better conditioned than the corresponding matrix arising out of the steady-state counterpart. Therefore, iterative solution of the linear system arising out of a time-marching approach with a finite time step size is usually favorable over solution of the linear system arising out of the steady-state equation. This advantage will be discussed further in sections to follow.

The steady-state approach has the advantage that a steady-state solution can be obtained directly without having to monitor whether steady state has been reached. However, as mentioned earlier, it is based on the presumption that a unique steady-state solution exists. If a unique steady-state solution does not exist, the solution will likely not converge. In some cases, the residuals may decrease several orders of magnitude before increasing again, and this behavior may repeat itself. Such behavior is indicative of the fact that the governing equation and boundary conditions cannot be satisfied at all nodes simultaneously, and the solution may flip-flop between multiple solutions. If a unique steady-state solution does indeed exist, the steady-state approach has the advantage that no truncation error is introduced due to discretization of the time derivative. In contrast, in the time-marching approach, because of the infiltration of temporal discretization errors into the solution, the accuracy of the steady-state solution may depend on the time step size used to march to the steady-state solution.

Since both methods have their respective *pros* and *cons*, both steady-state and time-marching approaches have been widely used for obtaining steady-state numerical solutions to PDEs. In the computational fluid dynamics arena, for example, there are two distinct schools of thought [3]: density-based methods, which typically employ time-marching, and pressure-based methods, which typically solve steady-state equations directly. In this chapter, the focus is not on finding a steady-state solution to a time-dependent PDE *per se*, but rather, to find the transient (or time-accurate) solution to the PDE – a solution that accurately captures both the temporal and spatial evolution of the solution. Therefore, this chapter will present time advancement not as just a means to attain steady state, but as a necessary step toward finding the solution at many instants of time along the path to steady state.

5.2 PARABOLIC PARTIAL DIFFERENTIAL EQUATIONS

In general, the class of PDEs discussed in this section is parabolic in time and elliptic in space. This is true irrespective of whether the PDE is 2D or 3D in space. If the PDE being considered is only 1D in space, it is a purely parabolic problem. Equation (5.1) is an example of such a PDE. Examples of the application of this class of PDEs are abundant: heat transfer (energy conservation equation), mass (or species) transport equations, the Navier–Stokes equations for fluid flow at low Reynolds numbers, among others. The general form of such a PDE may assume the following:

$$\frac{\partial \phi}{\partial t} = \alpha \nabla^2 \phi + S_\phi,$$ (5.2)

where α is a positive constant. As mentioned earlier, the result at steady state is the Poisson equation: $\nabla^2 \phi = -S_\phi / \alpha$. Since the spatial operator is elliptic, Eq. (5.2) requires boundary conditions to be specified on all boundaries of the computational domain. Furthermore, since the equation has a first derivative in time, an initial condition is also necessary. The boundary conditions may be of any of the three types discussed in Chapter 3. In order to anchor the solution (eliminate the possibility of multiple solutions), the initial condition must be of Dirichlet type and is written in general form as

$$\phi(0, \mathbf{r}) = \phi_0(\mathbf{r}),$$ (5.3)

where $\mathbf{r}$ represents the position vector of any location within the computational domain.

Prior to discretizing the time derivative, it is necessary to discretize the time axis into a set of temporal nodes, much like spatial nodes used for spatial discretization. The location of the temporal nodes and the size of the time step depend on factors similar to those encountered for spatial discretization: (a) truncation error associated with discretization of the time derivative, and (b) the need to resolve the necessary physics within the prescribed time step. These issues will be revisited in due course. For now, let us consider a set of equally spaced temporal nodes such that

$$t_n = n\Delta t, \tag{5.4}$$

where Δt is the time step size, henceforth referred to simply as time step, and n is the time index. The value of n represents the number of time steps that have already been advanced. When $n = 0$, not a single time step has been advanced – this represents the initial condition. t_n represents the time elapsed after n time steps. In general, one may also use unequal time steps to advance in time. In such a scenario, one must compute time cumulatively yielding

$$t_{n+1} = t_n + (\Delta t)_n, \tag{5.5}$$

where $(\Delta t)_n$ is the time step taken between time index n and $n + 1$.

5.2.1 EXPLICIT OR FORWARD EULER METHOD

In order to derive a finite difference approximation for the first derivative in time, shown in Eq. (5.2), as in the case of spatial discretization, we resort to Taylor series expansions, but now, in time, yielding:

$$\phi_{i,j,n+1} = \phi_{i,j,n} + \Delta t \left.\frac{\partial \phi}{\partial t}\right|_{i,j,n} + \frac{(\Delta t)^2}{2}\left.\frac{\partial^2 \phi}{\partial t^2}\right|_{i,j,n} + \frac{(\Delta t)^3}{6}\left.\frac{\partial^3 \phi}{\partial t^3}\right|_{i,j,n} + \dots, \tag{5.6}$$

where the first two subscripts, i and j, are reserved for spatial node indices, while the third subscript, n, is used to denote the time index. Rearranging Eq. (5.6), we obtain an expression for the first derivative in time:

$$\left.\frac{\partial \phi}{\partial t}\right|_{i,j,n} = \frac{\phi_{i,j,n+1} - \phi_{i,j,n}}{\Delta t} - \frac{\Delta t}{2}\left.\frac{\partial^2 \phi}{\partial t^2}\right|_{i,j,n} + \dots. \tag{5.7}$$

In Eq. (5.6), the Taylor series has been written such that the solution at the next (or forward) time index, $n + 1$, is obtained using the solution at the one before (i.e., time index n). Hence, this method of temporal discretization is also known as the *forward Euler method*. It is clear from Eq. (5.7) that the truncation error associated with the difference approximation just derived is first order in time. Prior to substitution of Eq. (5.7) into Eq. (5.2), we note that Eq. (5.7) provides an approximation for the time derivative at time index n. Since the entire governing equation [Eq. (5.2)] must be satisfied at the same instant of time, the right-hand side of Eq. (5.2) must also be expressed at time index n. Using the spatial difference approximations derived in Chapter 2 and Eq. (5.7), Eq. (5.2) may be written as

$$\frac{\phi_{i,j,n+1} - \phi_{i,j,n}}{\Delta t} = \alpha \left[\frac{\phi_{i+1,j,n} - 2\phi_{i,j,n} + \phi_{i-1,j,n}}{(\Delta x)^2} + \frac{\phi_{i,j+1,n} - 2\phi_{i,j,n} + \phi_{i,j-1,n}}{(\Delta y)^2}\right] + S_{i,j}.$$
$$+ \varepsilon\left(\Delta t, (\Delta x)^2, (\Delta y)^2\right) \tag{5.8}$$

As indicated, the truncation error associated with derivation of Eq. (5.8) from Eq. (5.2) is first order in time and second order in space. In Eq. (5.8), all quantities at n denote values at the previous time index, and are known. Essentially, they represent initial conditions for the current time step. The only unknown in the entire equation is $\phi_{i,j,n+1}$. Thus, Eq. (5.8) may be rearranged to express the only unknown quantity in terms of all the known quantities, yielding

$$\phi_{i,j,n+1} = \phi_{i,j,n} + \alpha \Delta t \left[\frac{\phi_{i+1,j,n} - 2\phi_{i,j,n} + \phi_{i-1,j,n}}{(\Delta x)^2} + \frac{\phi_{i,j+1,n} - 2\phi_{i,j,n} + \phi_{i,j-1,n}}{(\Delta y)^2} \right] + S_{i,j} \Delta t. \quad (5.9)$$

Equation (5.9) is an explicit expression that enables computation of ϕ at the current time step, $n+1$, using the values of ϕ at the previous time step, n. Thus, in addition to being known as the forward Euler method, this method of time advancement is also known as the *explicit method*. The philosophy of time advancement in the explicit method is illustrated in Fig. 5.1.

The explicit time advancement scheme has the notable advantage that no matrix inversion (or solution of simultaneous equations) is necessary to determine either a transient or a steady-state solution to the governing equation. Not only does it make the computational algorithm straightforward to implement, but it also results in a tremendous reduction in runtime memory because the coefficient matrix no longer needs to be stored. If the number of grid points is large, storage of the coefficient matrix, even considering its sparseness, may often be a showstopper. For example, when conducting direct numerical simulation (DNS) of turbulent flows [4], the number of grid points may often exceed several billion. For such unsteady computations, explicit time advancement is very attractive and is routinely used.

A critical issue with any time advancement scheme is its underlying stability. If the initial condition is perturbed by some finite error, will this error grow or decay? This is the central question at hand. If the error continues to grow with time

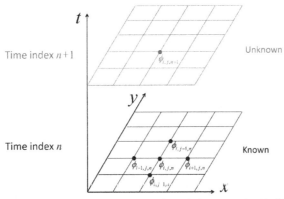

FIGURE 5.1 Schematic Representation of the Relationship Between the Variables at two Successive Time Steps in the Explicit (Forward Euler) Time Advancement Method for Parabolic PDEs

advancement, then the scheme is temporally unstable. If the error does not grow and just oscillates within a finite bound, similar in magnitude to the initial perturbation, then the scheme is temporally stable.

To analyze the stability of the explicit (forward Euler) time advancement scheme, we will once again apply the von Neumann stability analysis procedure presented in Chapter 4. The notable difference between the procedure outlined in Chapter 4 and the one applied here is that in this case, the amplitudes of the individual spectral components of the error may grow or decay with time rather than with iterations. To begin the stability analysis, we rewrite Eq. (5.8) but only for the spatially 1D case since we are primarily interested in stability or instability due to time advancement. For the 1D case, as discussed earlier, the governing PDE is purely parabolic and may be written as

$$\frac{\phi_{j,n+1} - \phi_{j,n}}{\Delta t} = \alpha \left[\frac{\phi_{j+1,n} - 2\phi_{j,n} + \phi_{j-1,n}}{(\Delta x)^2} \right] + S_j, \tag{5.10}$$

where, following the practice adopted in Chapter 4, the subscript j has been used to denote spatial nodes, and the symbol i is reserved for an imaginary number. The exact numerical solution must also obey Eq. (5.10). By exact numerical solution, we mean the solution obtained after the time interval Δt, but using vanishingly small time steps to get to the time interval Δt. This so-called exact solution will have the same spatial truncation error as the solution using a single time step, but negligible temporal truncation error. If subtracted from Eq. (5.10), the resulting error is expected to be free of spatial truncation error and have embedded in it the temporal truncation error associated with the single finite-sized time step, Δt. If the next time step is of equal size, in principle, this error must stay roughly the same order of magnitude since the truncation error at each time step is proportional to time step size. If, on the other hand, the error grows with each successive time step, the growth process can only be attributed to an instability introduced by the time advancement scheme. It is noteworthy that temporal discretization introduces a truncation error into the solution. How this error behaves with time, however, is dictated by how the equations are solved at each time step (i.e., the time advancement algorithm). The basic concept begs analogy with spatial discretization. In that case, the same equations, solved using different iterative solvers, exhibited different stability characteristics. Based on the preceding discussion, the error equation may be written as

$$\frac{\varepsilon_{j,n+1} - \varepsilon_{j,n}}{\Delta t} = \alpha \left[\frac{\varepsilon_{j+1,n} - 2\varepsilon_{j,n} + \varepsilon_{j-1,n}}{(\Delta x)^2} \right]. \tag{5.11}$$

Next, following Eq. (4.38), we may write the error as

$$\varepsilon_{j,n} = \sum_{m=0}^{M-1} C_{m,n} \exp\left(i m\pi \frac{j}{M-1} \right) = \sum_{m=0}^{M-1} C_{m,n} \exp\left(i j\theta_m \right), \tag{5.12}$$

where $\varepsilon_{j,n}$ represents the error at the jth node and the nth time step. The amplitude of the mth spectral component, $C_{m,n}$, is a function of time, as indicated by the subscript n attached to it. In general, this amplitude could have two components – one that oscillates with time, and one that grows or decays with time. Thus, it is convenient to write this amplitude as

$$C_{m,n} = \exp(\beta_m \, n\Delta t), \qquad (5.13)$$

where β_m is a complex number, such that if written explicitly in terms of its real and imaginary components (i.e., $\beta_m = \beta_{m,r} + i\beta_{m,i}$), the amplitude becomes

$$C_{m,n} = \exp(\beta_m \, n\Delta t) = \exp([\beta_{m,r} + i\beta_{m,i}]n\Delta t) = \exp(\beta_{m,r} \, n\Delta t) \cdot \exp(i\beta_{m,i} \, n\Delta t). \quad (5.14)$$

The term $\exp(\beta_{m,r} \, n\Delta t)$ is nonoscillatory and may grow or decay with time, while the term $\exp(i\beta_{m,i} \, n\Delta t)$ is oscillatory and always bounded. Substituting Eq. (5.13) into Eq. (5.12) yields

$$\varepsilon_{j,n} = \sum_{m=0}^{M-1} \exp(\beta_m \, n\Delta t)\exp\left(i\, j\theta_m\right) = \sum_{m=0}^{M-1} \xi_m^n \exp\left(i\, j\theta_m\right). \qquad (5.15)$$

For simplicity of notation, the symbol $\xi_m = \exp(\beta_m \, \Delta t)$ has been introduced in the second part of Eq. (5.15). From Eq. (5.15), it is clear that at time $t = 0$,

$$\varepsilon_{j,0} = \sum_{m=0}^{M-1} \xi_m^0 \exp\left(i\, j\theta_m\right) = \sum_{m=0}^{M-1} 1 \cdot \exp\left(i\, j\theta_m\right). \qquad (5.16)$$

In other words, the time-varying part of the error is unity, and the error only has a spatially varying component. If the quantity $\xi_m = \exp(\beta_m \, \Delta t)$ is greater than unity, the temporal component of the error will grow steadily since $\xi_m^n \to \infty$ if $\left|\xi_m\right| > 1$ and $n \to \infty$. Thus, the stability criterion for any time advancement scheme, with the error given by Eq. (5.15), may be written as

$$\left|\xi_m\right| \le 1. \qquad (5.17)$$

Equation (5.17) states that the stability criterion must be satisfied for each and every spectral component (or Fourier mode) m.

Returning to our error equation, we next substitute Eq. (5.11) into Eq. (5.15). This yields

$$\sum_{m=0}^{M-1} \xi_m^{n+1} \exp\left(i\, j\theta_m\right) - \sum_{m=0}^{M-1} \xi_m^n \exp\left(i\, j\theta_m\right)$$
$$= \frac{\alpha\Delta t}{(\Delta x)^2}\left[\sum_{m=0}^{M-1} \xi_m^n \exp\left(i[j+1]\theta_m\right) - 2\sum_{m=0}^{M-1} \xi_m^n \exp\left(i\, j\theta_m\right) + \sum_{m=0}^{M-1} \xi_m^n \exp\left(i[j-1]\theta_m\right)\right].$$
$$(5.18)$$

Noting again that Fourier modes are independent, Eq. (5.18) must, therefore, be satisfied for all values of m. Using this fact, and simplifying Eq. (5.18) further, we obtain

$$\xi_m \exp\left(i\, j\theta_m\right) - \exp\left(i\, j\theta_m\right)$$
$$= \frac{\alpha\Delta t}{(\Delta x)^2}\left[\exp\left(i[j+1]\theta_m\right) - 2\exp\left(i\, j\theta_m\right) + \exp\left(i[j-1]\theta_m\right)\right]. \qquad (5.19)$$

Dividing Eq. (5.19) by $\exp(i\,j\theta_m)$ yields

$$\xi_m - 1 = \frac{\alpha\,\Delta t}{(\Delta x)^2}\left[\exp(i\theta_m) - 2 + \exp(-i\theta_m)\right] = \frac{\alpha\,\Delta t}{(\Delta x)^2}\left[2\cos\theta_m - 2\right]. \quad (5.20)$$

Using the trigonometric identity $1 - \cos\theta = 2\sin^2(\theta/2)$, Eq. (5.20) may be further simplified to yield

$$\xi_m = 1 - \frac{\alpha\,\Delta t}{(\Delta x)^2}\left[4\sin^2\left(\frac{\theta_m}{2}\right)\right]. \quad (5.21)$$

Using Eq. (5.21) in combination with the constraint posed by Eq. (5.17), the stability criterion becomes

$$-1 \le 1 - \frac{\alpha\,\Delta t}{(\Delta x)^2}\left[4\sin^2\left(\frac{\theta_m}{2}\right)\right] \le 1. \quad (5.22)$$

The upper limit of this inequality is always satisfied since $(\alpha\,\Delta t/(\Delta x)^2)\left[4\sin^2(\theta_m/2)\right]$ is always positive, on account of the fact that α is a positive constant. Hence, we turn our attention to the lower limit of the inequality, which may be written as

$$\frac{\alpha\,\Delta t}{(\Delta x)^2}\left[4\sin^2\left(\frac{\theta_m}{2}\right)\right] \le 2, \quad (5.23)$$

which may be further simplified to write

$$\frac{\alpha\,\Delta t}{(\Delta x)^2} \le \frac{1}{2\sin^2\left(\dfrac{\theta_m}{2}\right)}. \quad (5.24)$$

The strictest constraint (worst-case scenario) is manifested by the particular Fourier mode for which $\sin^2(\theta_m/2) = 1$. For smaller values of $\sin^2(\theta_m/2)$, the constraint defined by the inequality is not as strict, and therefore, is not of interest. Using $\sin^2(\theta_m/2) = 1$, the stability criterion of the explicit (forward Euler) method for a 1D parabolic PDE may be written as

$$\frac{\alpha\,\Delta t}{(\Delta x)^2} \le \frac{1}{2}. \quad (5.25)$$

The quantity, $\alpha\,\Delta t/(\Delta x)^2$, is a nondimensional quantity. For example, in the case of the heat equation, α denotes thermal diffusivity and $\alpha\,\Delta t/(\Delta x)^2$ is the so-called *grid Fourier number*. It is known as the grid Fourier number (rather than Fourier number) because it is based on grid spacing rather than a global length scale.

For fixed grid spacing, Δx, Eq. (5.25) is equivalent to placing a restriction on the time step that may be used to attain a stable solution with explicit time advancement:

$$\Delta t \le \frac{1}{2}\frac{(\Delta x)^2}{\alpha}. \quad (5.26)$$

One of the important implications of Eq. (5.26) is that the time step size must also be correspondingly adjusted to a smaller value if the grid is refined. This requirement places a severe burden on computational efficiency. For example, if the grid size is halved, the time step size must be reduced by a factor of 4. This implies that, even for a 1D problem, the computational time needed to reach a certain instant of real time will go up by a factor of 8 if the grid size is halved. Thus, even though the explicit time advancement method is advantageous from the point of view of memory and ease of implementation, it can become computationally quite expensive when used for fine grids because of the restriction on the time step size that can be used.

If such stability analysis were to be performed on a spatially 2D PDE, the following stability criterion would result for the explicit method:

$$\alpha \Delta t \left[\frac{1}{(\Delta x)^2} + \frac{1}{(\Delta y)^2} \right] \le \frac{1}{2}. \tag{5.27}$$

For equal grid spacing in the two Cartesian directions (i.e., $\Delta x = \Delta y$), Eq. (5.27) reduces to

$$\Delta t \le \frac{1}{4} \frac{(\Delta x)^2}{\alpha}, \tag{5.28}$$

which implies that the restriction on the time step is even more severe for multidimensional problems. If the grid spacing is halved in both directions in this case, the time step size has to be reduced by a factor of 4 and the total workload to advance to a certain instant in real time will increase by a factor of 16. It is worth noting that this behavior bears a strong resemblance to that of the classical iterative schemes discussed in Chapter 3 for iterative solution of steady-state equations. In that case, we noted that the workload increased approximately by a factor of 16 when grid spacing was halved in both directions because of the number of iterations increasing by a factor of 4. Thus, explicit time advancement and direct solution of steady-state equations are comparable methods for use in proceeding to steady state in terms of scaling the computational workload with mesh size.

Thus far, the stability criterion has been portrayed as a purely mathematical constraint that must be obeyed to ensure a stable solution. It is also worth exploring its implications by applying it to a physical problem and examining the resulting solution from a physical perspective. In order to do so, we consider solution of the 1D unsteady heat conduction equation without any heat source. In this particular case the dependent variable, ϕ, represents temperature. Just three nodes are considered, as shown in Fig. 5.2. Of these three nodes, only one (Node 2) is an interior node, and the other two are boundary nodes with prescribed fixed temperatures. Initial conditions are also shown in Fig. 5.2. Using the explicit time advancement method, the discrete equation for Node 2 may be written as

$$\phi_{2,1} = \phi_{2,0} + \frac{\alpha \Delta t}{(\Delta x)^2} \left[\phi_{3,0} - 2\phi_{2,0} + \phi_{1,0} \right]. \tag{5.29}$$

Using the initial and boundary conditions shown in Fig. 5.2, we obtain $\phi_{2,1} = 400 - \left(\alpha \Delta t / (\Delta x)^2 \right) \left[200 \right]$. If the stability criterion is violated and $\left(\alpha \Delta t / (\Delta x)^2 \right)$ is chosen to be equal to unity, for example, we obtain $\phi_{2,1} = 200$ K. This result violates the second law of thermodynamics. Physically, in the absence of any heat

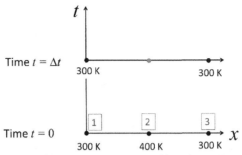

FIGURE 5.2 Grid, Geometry, Initial, and Boundary Conditions for an Example 1D Heat Conduction Problem

sink ($S_2 = 0$), Node 2 can cool to a minimum temperature of 300 K. Any temperature below that is physically impossible. If, on the other hand, $\alpha \Delta t / (\Delta x)^2 \leq 1/2$, we obtain $300\,\text{K} \leq \phi_{2,1} \leq 400$ K, which is physically meaningful. The preceding example shows that when the stability criterion is violated, the resulting solution may also violate physical principles. The overall algorithm for execution of the explicit (forward Euler) time advancement method for a parabolic PDE is presented next.

ALGORITHM: EXPLICIT (FORWARD EULER) METHOD FOR PARABOLIC PDEs

Step 1: Set up the initial conditions for ϕ at all nodes (i.e., $\phi_{i,j,0}$ $\forall i = 1,...,N$ and $\forall j = 1,...,M$).
Step 2: Choose an appropriate time step size (Δt) that obeys the stability criterion, such as that given by Eq. (5.26) (for problems that are spatially 1D) or that given by Eq. (5.27) (for problems that are spatially 2D).
Step 3: Compute all nodal values of ϕ at the next time step (i.e., compute $\phi_{i,j,1}$ $\forall i = 1,...,N$ and $\forall j = 1,...,M$) using an explicit update formula similar to Eq. (5.29). For boundary nodes with non-Dirichlet-type boundary conditions, equations similar to Eq. (5.29) must be derived following the procedures outlined in Chapter 3 and applied.
Step 4: Reset the initial conditions (i.e., $\phi_{i,j,0} = \phi_{i,j,1}$ $\forall i = 1,...,N$ and $\forall j = 1,...,M$).
Step 5: Proceed to the next time step and repeat Steps 3 and 4 until the desired instant of time or prescribed number of time steps has been reached.

An example is next considered to demonstrate the explicit (forward Euler) method and the implications the stability criterion has on the solution procedure.

EXAMPLE 5.1

In this example we consider solution of the following 1D unsteady PDE in $[0,\pi]$:

$$\frac{\partial \phi}{\partial t} = \frac{\partial^2 \phi}{\partial x^2},$$ subject to the following boundary and initial conditions:

Boundary conditions: $\phi(t,0) = e^{-t}$, $\phi(t,\pi) = -e^{-t}$
Initial condition: $\phi(0,x) = \cos x$.

The closed-form analytical solution to the problem is given by $\phi(t,x) = e^{-t}\cos x$, and the steady-state solution is zero everywhere. For numerical solution using the explicit (forward Euler) method, we consider 21 nodes in the spatial domain, resulting in $\Delta x = \pi/20$. In this case $\alpha = 1$. Therefore, the maximum-allowable time step based on the stability criterion is $(\Delta t)_{max} = (\Delta x)^2/2 = 0.0123\,\text{s}$. Two different time step sizes are considered: $0.8(\Delta t)_{max}$ and $1.6(\Delta t)_{max}$. The solutions in both cases are shown in the figures below after the same instants of time.

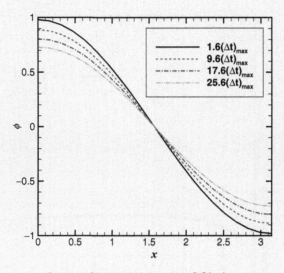

Solution Obtained Using $\Delta t = 0.8(\Delta t)_{max}$

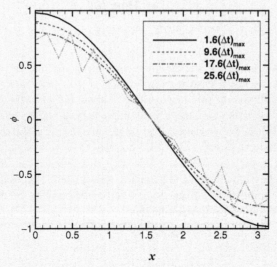

Solution Obtained Using $\Delta t = 1.6(\Delta t)_{max}$

It is clear from the solutions shown in the figure immediately above that instabilities set in after some time period and errors blow up when time step constraint is violated. As to exactly how quickly they blow up depends on the problem at hand. To understand the behavior of errors better, the error (defined here as the difference between the analytical and numerical solution) is plotted in the figure below as a function of time at Node 5 of the solution domain.

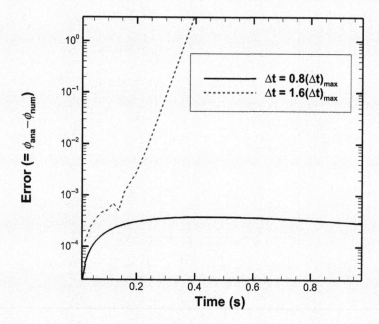

As discussed earlier, if the stability criterion is obeyed the error stays within bounds, as in the case with $\Delta t = 0.8(\Delta t)_{max}$. In fact, closer examination reveals that the error first increases up to about 0.4 s and then decreases with time in this particular case. This is because temporal truncation errors vanish as the solution approaches steady state. In contrast, if the stability criterion is violated, the initial error, which happens to be quite small, continues to amplify, as in the case with $\Delta t = 1.6(\Delta t)_{max}$. It oscillates for some period before completely blowing up and becoming unbounded. This example demonstrates the implications of time step size on the stability of the explicit (forward Euler) time advancement method.

5.2.2 IMPLICIT OR BACKWARD EULER METHOD

As discussed in the preceding section, although the explicit (forward Euler) method is easy to implement and requires little memory, time step restriction can often result in a huge penalty on computational time. In particular, if a nonuniform mesh is used, the maximum allowable time step is dictated by the smallest grid spacing, which can often be an order of magnitude less than that computed for the average grid spacing.

In the backward Euler method, the difference approximation to the time derivative is derived by employing a backward Taylor series expansion in time, as suggested by the name of the method. This yields

$$\phi_{i,j,n} = \phi_{i,j,n+1} - \Delta t \left.\frac{\partial \phi}{\partial t}\right|_{i,j,n+1} + \left.\frac{(\Delta t)^2}{2}\frac{\partial^2 \phi}{\partial t^2}\right|_{i,j,n+1} - \left.\frac{(\Delta t)^3}{6}\frac{\partial^3 \phi}{\partial t^3}\right|_{i,j,n+1} + \dots \quad (5.30)$$

Rearranging to express the first derivative in time in terms of other terms, Eq. (5.30) yields

$$\left.\frac{\partial \phi}{\partial t}\right|_{i,j,n+1} = \frac{\phi_{i,j,n+1} - \phi_{i,j,n}}{\Delta t} + \left.\frac{\Delta t}{2}\frac{\partial^2 \phi}{\partial t^2}\right|_{i,j,n+1} + \dots \quad (5.31)$$

Comparing Eq. (5.31) with Eq. (5.7) (a similar equation for the explicit method) reveals that the expression for the time derivative in both cases is identical. The leading order error term in both cases is also first order, but the exact expressions are slightly different. Since the time derivative in this case is expressed at the current time step, $n+1$, the right-hand side of the governing equation [Eq. (5.2)] must also be expressed at the same instant of time. Consequently, the finite difference equation for the implicit (backward Euler) method becomes

$$\frac{\phi_{i,j,n+1} - \phi_{i,j,n}}{\Delta t} = \alpha\left[\frac{\phi_{i+1,j,n+1} - 2\phi_{i,j,n+1} + \phi_{i-1,j,n+1}}{(\Delta x)^2} + \frac{\phi_{i,j+1,n+1} - 2\phi_{i,j,n+1} + \phi_{i,j-1,n+1}}{(\Delta y)^2}\right] + S_{i,j}$$
$$+ \varepsilon\left(\Delta t, (\Delta x)^2, (\Delta y)^2\right) \quad (5.32)$$

The scaling behavior of the truncation error of the implicit method, as indicated in Eq. (5.32), is exactly the same as that of the explicit method, indicating that for a given grid spacing and time step size, both methods are expected, magnitude-wise, to yield similar errors. As opposed to the explicit method [Eq. (5.8)], Eq. (5.32) has more than one unknown. All quantities at time index $n+1$ are unknown. Thus, Eq. (5.32) represents a set of implicit (or simultaneous) equations, as suggested by the name of the method. The dependency between known and unknown quantities is depicted in Fig. 5.3. Rearranging Eq. (5.32) such that known and unknown quantities are on different sides of the equation, we may write

$$\left(\frac{1}{\Delta t} + \frac{2\alpha}{(\Delta x)^2} + \frac{2\alpha}{(\Delta y)^2}\right)\phi_{i,j,n+1} - \frac{\alpha}{(\Delta x)^2}\phi_{i+1,j,n+1} - \frac{\alpha}{(\Delta x)^2}\phi_{i-1,j,n+1}$$
$$- \frac{\alpha}{(\Delta y)^2}\phi_{i,j+1,n+1} - \frac{\alpha}{(\Delta y)^2}\phi_{i,j-1,n+1} = \frac{1}{\Delta t}\phi_{i,j,n} + S_{i,j} \quad (5.33)$$

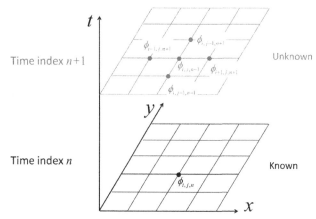

FIGURE 5.3 Schematic Representation of the Relationship Between the Variables at two Successive Time Steps in the Implicit (Backward Euler) Time Advancement Method for Parabolic PDEs

As in the steady-state case, Eq. (5.33) represents a five-banded system of equations. This system of equations may be solved using any of the iterative solvers discussed in Chapters 3 and 4. It is easy to see from the limit $\Delta t \rightarrow \infty$ that Eq. (5.33) reduces to Eq. (2.48), which is the Poisson equation discretized using the finite difference procedure.

Since the implicit method involves solution of a linear system of equations, it is obviously much more complicated to implement than the explicit method. Hence, the implicit method would only have value as a time advancement method if it offered better stability characteristics than the explicit method, since both methods also offer similar (first-order in time) accuracy. Next, we perform von Neumann analysis of the implicit time advancement method for a problem that is spatially 1D. Problems that are spatially 1D result in tridiagonal matrix systems that can be solved using a direct solver. Thus, any instability in the solution procedure can be attributed solely to the time advancement procedure. For problems that are spatially multidimensional, on the other hand, an iterative solver is necessary to solve the resulting linear system. In such a scenario, instabilities may result from spatial iterations (as seen in Chapter 4), from the time advancement process, or from a combination of both, thereby making isolation of instabilities due to time advancement difficult.

Following the derivation of Eq. (5.11), we first write the corresponding error equation for the implicit time advancement method for a 1D parabolic PDE:

$$\frac{\varepsilon_{j,n+1} - \varepsilon_{j,n}}{\Delta t} = \alpha \left[\frac{\varepsilon_{j+1,n+1} - 2\varepsilon_{j,n+1} + \varepsilon_{j-1,n+1}}{(\Delta x)^2} \right]. \tag{5.34}$$

Next, we substitute the Fourier representation of the error [Eq. (5.15)] into Eq. (5.34), to yield

$$
\sum_{m=0}^{M-1} \xi_m^{n+1} \exp\left(i\, j\theta_m\right) - \sum_{m=0}^{M-1} \xi_m^{n} \exp\left(i\, j\theta_m\right)
$$
$$
= \frac{\alpha \Delta t}{(\Delta x)^2} \left[\sum_{m=0}^{M-1} \xi_m^{n+1} \exp\left(i[j+1]\theta_m\right) - 2\sum_{m=0}^{M-1} \xi_m^{n+1} \exp\left(i\, j\theta_m\right) + \sum_{m=0}^{M-1} \xi_m^{n+1} \exp\left(i[j-1]\theta_m\right) \right].
$$

$$(5.35)$$

Noting, as before, that the Fourier modes are independent, Eq. (5.35) must be satisfied for all values of m. Using this fact, and simplifying Eq. (5.35) further, we obtain

$$
\xi_m \exp\left(i\, j\theta_m\right) - \exp\left(i\, j\theta_m\right)
$$
$$
= \frac{\alpha \Delta t}{(\Delta x)^2} \xi_m \left[\exp\left(i[j+1]\theta_m\right) - 2\exp\left(i\, j\theta_m\right) + \exp\left(i[j-1]\theta_m\right) \right]. \quad (5.36)
$$

Dividing Eq. (5.36) by $\exp\left(i\, j\theta_m\right)$, and simplifying using the same trigonometric identities used to derive Eq. (5.21), we obtain

$$
\xi_m - 1 = -\frac{4\alpha \Delta t}{(\Delta x)^2} \xi_m \left[\sin^2\left(\frac{\theta_m}{2}\right) \right], \quad (5.37)
$$

which, upon further simplification, yields

$$
\xi_m = \frac{1}{1 + \dfrac{4\alpha \Delta t}{(\Delta x)^2} \sin^2\left(\dfrac{\theta_m}{2}\right)}. \quad (5.38)
$$

It is self-evident from Eq. (5.38) that the criterion $-1 \le \xi_m \le 1$ (or $|\xi_m| \le 1$) is always obeyed irrespective of the values of θ_m, Δt, or Δx. This implies that the implicit or backward Euler method is unconditionally stable. In practice, this means that an arbitrarily large time step size may be used with the backward Euler method.

Even though the implicit (backward Euler) method allows use of arbitrarily large time steps from a stability perspective, one must keep in mind that a large time step implies large temporal truncation error and inaccurate solution. Thus, although large time step size is acceptable from a stability perspective, it is not desirable from an accuracy perspective. Nonetheless, one of the notable advantages of the implicit time advancement method is its unconditional stability. On account of this favorable attribute, it is almost always the default method of time advancement in general-purpose unsteady PDE solvers, such as those used in commercially available finite element or computational fluid dynamics codes. The overall solution algorithm for the implicit (backward Euler) time advancement method for a parabolic PDE is described next.

ALGORITHM: IMPLICIT (BACKWARD EULER) METHOD FOR PARABOLIC PDEs

Step 1: Set up the initial conditions for ϕ at all nodes (i.e., $\phi_{i,j,0}$ $\forall i = 1,...,N$ and $\forall j = 1,...,M$).

Step 2: Choose an appropriate time step size (Δt) and begin time advancement.

Step 3: Set up a linear system $[A][\phi]=[Q]$ in which matrix elements are derived from nodal equations, such as Eq. (5.33).

Step 4: Compute $[\phi]$ at the next time step (i.e., compute $\phi_{i,j,1}$ $\forall i = 1,...,N$ and $\forall j = 1,...,M$) by solving the linear system set up in Step 3 using any of the solvers discussed in Chapters 3 and 4, or any other solver of choice. Convergence must be monitored at this step if an iterative solver is used, and iterations must be continued until the prescribed convergence criterion has been satisfied.

Step 5: Reset the initial conditions (i.e., $\phi_{i,j,0}=\phi_{i,j,1}$ $\forall i = 1,...,N$ and $\forall j = 1,...,M$).

Step 6: Proceed to the next time step and repeat Steps 3–5 until the desired instant of time or prescribed number of time steps has been reached.

It was remarked in Section 5.1 that when steady-state solution to a problem is desired, direct solution of the steady-state equation is preferable to time-marching the unsteady-state equation to steady state since discretization of the time derivative introduces additional errors. On the other hand, it was stated immediately after Eq. (5.33) that the limit $\Delta t \rightarrow \infty$ demonstrates that Eq. (5.33) reduces to the steady-state equation, which would suggest that time-marching the solution to a steady state and direct solution of the steady-state equation are one and the same. The confusion arises from the fact that Eq. (5.33) has temporal discretization error, $(\Delta t/2)(\partial^2 \phi/\partial t^2)\big|_{i,j,n+1}$, built into it [see Eq. (5.31)]. Theoretically, when steady state is reached, the solution, by definition, is independent of time, and all its derivatives with respect to time are zero. In this scenario, temporal discretization error will vanish even if $\Delta t \rightarrow \infty$. In practice, however, the solution is time-marched to steady state only within a certain prescribed tolerance. Therefore, temporal truncation error remains nonzero and could be substantial if large time step sizes were used to reach steady state. However, this is highly problem dependent and depends on the tolerances used. When steady-state equations are solved directly, such a situation and associated ambiguities do not arise, and spatial discretization error is the only error in the solution. Next, we consider an example that highlights the implications of time step size on the stability and accuracy of the implicit time advancement method.

EXAMPLE 5.2

In this example, we repeat the problem considered in Example 5.1, but do so with the implicit (backward Euler) method. For numerical solution, we again consider 21 nodes in the spatial domain, resulting in $\Delta x = \pi/20$. The same two time step sizes are considered: $0.8(\Delta t)_{max}$ and $1.6(\Delta t)_{max}$. Here $(\Delta t)_{max}$ is the maximum allowable time step of the explicit method (see Example 5.1). The solutions using both time steps are shown in the figures below after the same instants of time.

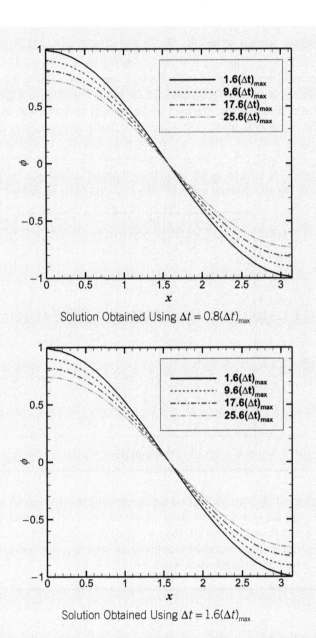

Solution Obtained Using $\Delta t = 0.8(\Delta t)_{max}$

Solution Obtained Using $\Delta t = 1.6(\Delta t)_{max}$

Unlike the explicit method, use of a time step larger than the maximum allowable time step in the implicit method does not cause instabilities in the solution, confirming the findings of von Neumann stability analysis. Although not visually discernible from the figures immediately above, the error in the solutions obtained with $1.6(\Delta t)_{max}$ is approximately twice as large as that obtained using $0.8(\Delta t)_{max}$, confirming that truncation errors are first order. This is clearer in the figure below, in which the error at Node 5 is plotted as a function of time for three different time step sizes. Like in the explicit method, the error grows in magnitude in the implicit method and then starts to decay after about 0.4 s. The sign of the error is opposite that of the explicit method, as is evident from Eqs. (5.7) and (5.31). Finally, the error scales linearly with time step size. When time step size is increased from $0.8(\Delta t)_{max}$ to $1.6(\Delta t)_{max}$, the error approximately doubles, whereas it increases approximately by a factor of 10 when time step size is increased to $8(\Delta t)_{max}$, indicative of the fact that the implicit method is first-order accurate in time.

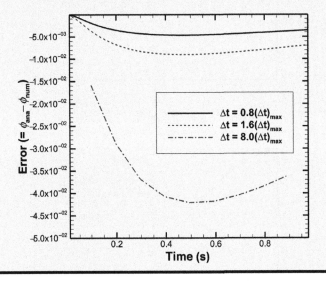

It was remarked in Section 5.1 that time-marching the solution to steady state is often advantageous over direct solution of the steady-state equations because the matrix is often better conditioned. To lend credibility to this remark, we next consider an example in which a time-marched solution using the implicit method is compared with direct solution of the steady-state equation.

EXAMPLE 5.3

In this example we consider solution of a 2D unsteady parabolic PDE using two different approaches: (1) implicit time-marching of the solution to steady state, and (2) obtaining the steady-state solution directly. Space is discretized using a 41×41 mesh. The Gauss–Seidel method is used for iterative solution of the linear system of equations in both approaches. The governing equation being solved is as follows:

$$\frac{\partial \phi}{\partial t} = \nabla^2 \phi - S_\phi,$$

such that the steady-state form of the equation is $\nabla^2 \phi = S_\phi$, where

$$S_\phi = 1000\left[2\sinh(x-\tfrac{1}{2})+4(x-\tfrac{1}{2})\cosh(x-\tfrac{1}{2})+(x-\tfrac{1}{2})^2\sinh(x-\tfrac{1}{2})\right]$$
$$+1000\left[2\sinh(y-\tfrac{1}{2})+4(y-\tfrac{1}{2})\cosh(y-\tfrac{1}{2})+(y-\tfrac{1}{2})^2\sinh(y-\tfrac{1}{2})\right]$$

The boundary conditions are as follows:

$$\phi(t,0,y)=1000\left[\tfrac{1}{4}\sinh(-\tfrac{1}{2})+(y-\tfrac{1}{2})^2\sinh(y-\tfrac{1}{2})\right],$$
$$\phi(t,1,y)=1000\left[\tfrac{1}{4}\sinh(\tfrac{1}{2})+(y-\tfrac{1}{2})^2\sinh(y-\tfrac{1}{2})\right],$$
$$\phi(t,x,0)=1000\left[(x-\tfrac{1}{2})^2\sinh(x-\tfrac{1}{2})+\tfrac{1}{4}\sinh(-\tfrac{1}{2})\right]$$
$$\phi(t,x,1)=1000\left[(x-\tfrac{1}{2})^2\sinh(x-\tfrac{1}{2})+\tfrac{1}{4}\sinh(\tfrac{1}{2})\right].$$

The initial condition used is $\phi(0,x,y)=0$.
The closed-form analytical solution to the steady-state problem is given by:

$$\phi(\infty,x,y)=1000\left[(x-\tfrac{1}{2})^2\sinh(x-\tfrac{1}{2})+(y-\tfrac{1}{2})^2\sinh(y-\tfrac{1}{2})\right].$$

The convergence tolerance of the Gauss–Seidel iteration-based solver was set to 10^{-9} for the scaled residual (i.e., a residual drop by 9 orders of magnitude both for the steady-state solution and for the unsteady implicit solution which had iterations in each time step). For the implicit time-marching scheme, a time step equal to 10 times that of the maximum-allowable time step for the corresponding explicit method was used. The residual of the steady-state solution and that at two different time steps of the implicit time-marching method are shown in the figure below.

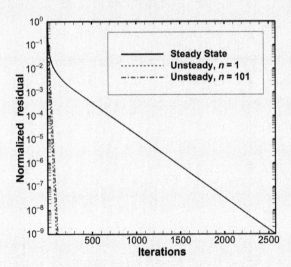

The residual plots clearly confirm the previously made remark that the coefficient matrix is better conditioned for the unsteady implicit system of equations than the corresponding steady-state equations. Steady-state convergence was attained after 2576 iterations, whereas convergence was attained only after an average of 112 iterations within each time step. With the implicit time-marching method, the steady-state solution was attained after 322 time steps.

Steady state was detected by monitoring the magnitude of the L^2Norm of the difference in solution between previous and current time steps. A tolerance of 10^{-4} was used for this purpose. The accuracy of the solution obtained using the two methods is assessed next by examining the error between numerical and analytical solutions at steady state. Error distributions are shown in the figures below. Solution errors using the two different approaches are identical, showing that the solution is not contaminated at all by temporal discretization errors in this case and that $(\Delta t/2)(\partial^2\phi/\partial t^2)\big|_{i,j,n+1}$ is indeed very small.

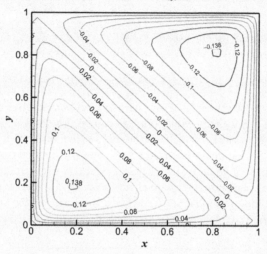

Error in Solution Using Steady State Approach

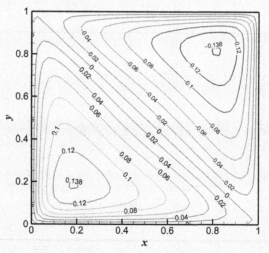

Error in Solution Using Implicit Time Marching Approach

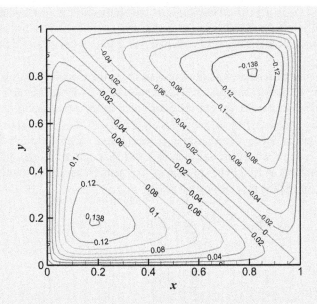

The time-marching approach required 322 time steps and approximately 112 iterations per time step. This implies a total of $322 \times 112 = 36{,}064$ iterations (or sweeps through the computational domain). In contrast, the steady-state approach required only 2576 sweeps. Thus, the time-marching approach proved computationally much more expensive. If, however, only the steady-state solution is of interest, and the time-accurate solution is not of interest, it is not necessary to use a very tight tolerance (such as 10^{-9} used here) for convergence within each time step. To test this hypothesis, the solution was recomputed with a tolerance of 10^{-3} (the error is shown in the figure above). The error in the steady-state solution did not change at all when a more relaxed convergence criterion at each time step was used. On the positive side, however, the number of iterations per time step reduced from 110 to 38, implying a total of $322 \times 38 = 12{,}236$ iterations. In practice, it is possible to use an even more relaxed convergence criterion to make total iteration counts in the two methods comparable. This example shows that both the steady-state and time-marching approaches to obtaining a steady-state solution to a PDE are viable alternatives, with each approach having its own respective *pros* and *cons*.

5.2.3 CRANK–NICOLSON METHOD

Both the explicit (forward Euler) and implicit (backward Euler) methods have temporal truncation errors that are first order. This means that small time steps must be used to obtain an accurate solution. Further, the explicit method has additional restrictions due to its stability. For many practical applications, using excessively small time steps is not affordable from a computational efficiency standpoint. The Crank–Nicolson method [5] was proposed in 1947 to address this critical shortcoming of the forward and backward Euler methods. It is a higher order (in time) implicit method.

To derive the difference equations necessary for the Crank–Nicolson method, we first conceive of a time step size $\Delta t/2$. Correspondingly, we use the time index $n + \frac{1}{2}$ for the time instant $t + \Delta t/2$. This time instant and time index will not actually be used. They are only introduced for the purposes of derivation. As we shall shortly find, they will be eliminated during the derivation process. Next, we perform two Taylor series expansions: one in which the solution at $t + \Delta t$ is expanded about the solution at $t + \Delta t/2$, and another in which the solution at t is expanded about the solution at $t + \Delta t/2$. These two expansions may be written as

$$\phi_{i,j,n+1} = \phi_{i,j,n+\frac{1}{2}} + \frac{\Delta t}{2}\frac{\partial \phi}{\partial t}\bigg|_{i,j,n+\frac{1}{2}} + \frac{1}{2}\left(\frac{\Delta t}{2}\right)^2\frac{\partial^2 \phi}{\partial t^2}\bigg|_{i,j,n+\frac{1}{2}} + \frac{1}{6}\left(\frac{\Delta t}{2}\right)^3\frac{\partial^3 \phi}{\partial t^3}\bigg|_{i,j,n+\frac{1}{2}} + \cdots, \quad (5.39a)$$

$$\phi_{i,j,n} = \phi_{i,j,n+\frac{1}{2}} - \frac{\Delta t}{2}\frac{\partial \phi}{\partial t}\bigg|_{i,j,n+\frac{1}{2}} + \frac{1}{2}\left(\frac{\Delta t}{2}\right)^2\frac{\partial^2 \phi}{\partial t^2}\bigg|_{i,j,n+\frac{1}{2}} - \frac{1}{6}\left(\frac{\Delta t}{2}\right)^3\frac{\partial^3 \phi}{\partial t^3}\bigg|_{i,j,n+\frac{1}{2}} + \cdots. \quad (5.39b)$$

Subtracting Eq. (5.39b) from Eq. (5.39a), and rearranging, we obtain:

$$\frac{\partial \phi}{\partial t}\bigg|_{i,j,n+\frac{1}{2}} = \frac{\phi_{i,j,n+1} - \phi_{i,j,n}}{\Delta t} - \frac{1}{24}(\Delta t)^2\frac{\partial^3 \phi}{\partial t^3}\bigg|_{i,j,n+\frac{1}{2}} + \cdots. \quad (5.40)$$

It is evident from Eq. (5.40) that the truncation error incurred in approximating the first derivative is second order in time. However, the time derivative in this case has been derived at $t + \Delta t/2$, rather than at t (explicit method) or at $t + \Delta t$ (implicit method). Therefore, when substituted into the governing PDE, the right-hand side of the PDE must also be considered at the same instant of time (i.e., at $t + \Delta t/2$). Using the spatial difference approximations derived in Chapter 2, and substituting Eq. (5.40) into Eq. (5.2), we obtain:

$$\frac{\phi_{i,j,n+1} - \phi_{i,j,n}}{\Delta t} = \alpha\left[\frac{\phi_{i+1,j,n+\frac{1}{2}} - 2\phi_{i,j,n+\frac{1}{2}} + \phi_{i-1,j,n+\frac{1}{2}}}{(\Delta x)^2} + \frac{\phi_{i,j+1,n+\frac{1}{2}} - 2\phi_{i,j,n+\frac{1}{2}} + \phi_{i,j-1,n+\frac{1}{2}}}{(\Delta y)^2}\right]$$
$$+ S_{i,j} + \varepsilon\left((\Delta t)^2, (\Delta x)^2, (\Delta y)^2\right) \quad (5.41)$$

Since there is no temporal node at $n + \frac{1}{2}$, in practice, all quantities on the right-hand side of Eq. (5.8) must be approximated further. This may also be accomplished using Taylor series expansions. Addition of Eqs. (5.39a) and (5.39b), followed by rearrangement, yields

$$\phi_{i,j,n+\frac{1}{2}} = \frac{\phi_{i,j,n} + \phi_{i,j,n+1}}{2} - \frac{1}{8}(\Delta t)^2\frac{\partial^2 \phi}{\partial t^2}\bigg|_{i,j,n+\frac{1}{2}} + \cdots. \quad (5.42)$$

Substitution of Eq. (5.42) and similar approximations into Eq. (5.41) yields

$$
\frac{\phi_{i,j,n+1} - \phi_{i,j,n}}{\Delta t} = \frac{\alpha}{2}\left[\frac{\phi_{i+1,j,n} - 2\phi_{i,j,n} + \phi_{i-1,j,n}}{(\Delta x)^2} + \frac{\phi_{i,j+1,n} - 2\phi_{i,j,n} + \phi_{i,j-1,n}}{(\Delta y)^2} \right]
$$
$$
+ \frac{\alpha}{2}\left[\frac{\phi_{i+1,j,n+1} - 2\phi_{i,j,n+1} + \phi_{i-1,j,n+1}}{(\Delta x)^2} + \frac{\phi_{i,j+1,n+1} - 2\phi_{i,j,n+1} + \phi_{i,j-1,n+1}}{(\Delta y)^2} \right] + S_{i,j}.
$$
$$
+ \varepsilon\left((\Delta t)^2, (\Delta x)^2, (\Delta y)^2 \right) \tag{5.43}
$$

The approximation made in arriving at Eq. (5.43) from Eq. (5.41) introduced a second-order temporal truncation error, as shown in Eq. (5.42). Thus, the overall temporal accuracy of the Crank–Nicolson method (second order) remains unaltered by the approximation. Rearranging Eq. (5.43) to place known and unknown quantities on different sides of the equation, we obtain

$$
\left(\frac{1}{\Delta t} + \frac{\alpha}{(\Delta x)^2} + \frac{\alpha}{(\Delta y)^2} \right)\phi_{i,j,n+1} - \frac{\alpha}{2(\Delta x)^2}\phi_{i+1,j,n+1} - \frac{\alpha}{2(\Delta x)^2}\phi_{i-1,j,n+1}
$$
$$
- \frac{\alpha}{2(\Delta y)^2}\phi_{i,j+1,n+1} - \frac{\alpha}{2(\Delta y)^2}\phi_{i,j-1,n+1} \qquad \cdot \tag{5.44}
$$
$$
= \frac{\phi_{i,j,n+1}}{\Delta t} + \frac{\alpha}{2}\left[\frac{\phi_{i+1,j,n} - 2\phi_{i,j,n} + \phi_{i-1,j,n}}{(\Delta x)^2} + \frac{\phi_{i,j+1,n} - 2\phi_{i,j,n} + \phi_{i,j-1,n}}{(\Delta y)^2} \right] + S_{i,j}
$$

It is worth noting that the artificial temporal node, $n + \frac{1}{2}$, was eliminated while deriving Eq. (5.44). As in the case of the implicit (backward Euler) method, Eq. (5.44) represents an implicit set of equations in standard five-band form, implying that the Crank–Nicolson method is also implicit. The algorithm for the Crank–Nicolson method is identical to that of the backward Euler method, described in Section 5.2.2, with the notable exception that the coefficient matrix, $[A]$, and the right-hand-side vector, $[Q]$, used in Step 3 of the algorithm, will be different. They are to be derived using Eq. (5.44), rather than Eq. (5.33). The dependencies between the known and unknown quantities in the Crank–Nicolson method are depicted pictorially in Fig. 5.4.

Important questions that need to be asked here are whether the Crank–Nicolson method is stable and whether there are any stability constraints associated with it. To answer these questions, we once again perform von Neumann stability analysis of the method. Following the same procedure as in Sections 5.2.1 and 5.2.2, the error equation corresponding to the 1D version of Eq. (5.43) [cf. Eq. (5.34)] may be written as

$$
\frac{\varepsilon_{j,n+1} - \varepsilon_{j,n}}{\Delta t} = \frac{\alpha}{2}\left[\frac{\varepsilon_{j+1,n} - 2\varepsilon_{j,n} + \varepsilon_{j-1,n}}{(\Delta x)^2} \right] + \frac{\alpha}{2}\left[\frac{\varepsilon_{j+1,n+1} - 2\varepsilon_{j,n+1} + \varepsilon_{j-1,n+1}}{(\Delta x)^2} \right], \tag{5.45}
$$

where $\varepsilon_{j,n}$ carries the usual meaning. Substituting the Fourier representation of the error [Eq. (5.15)] into Eq. (5.45), we obtain

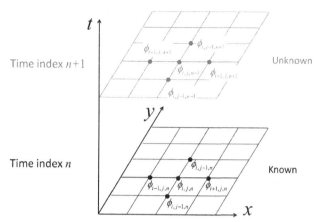

FIGURE 5.4 Schematic Representation of the Relationship Between the Variables at two Successive Time Steps in the Crank–Nicolson Time Advancement Method for Parabolic PDEs

$$\sum_{m=0}^{M-1} \xi_m^{n+1} \exp(i\,j\theta_m) - \sum_{m=0}^{M-1} \xi_m^{n} \exp(i\,j\theta_m)$$

$$= \frac{\alpha \Delta t}{2(\Delta x)^2} \left[\sum_{m=0}^{M-1} \xi_m^{n} \exp(i[j+1]\theta_m) - 2\sum_{m=0}^{M-1} \xi_m^{n} \exp(i\,j\theta_m) + \sum_{m=0}^{M-1} \xi_m^{n} \exp(i[j-1]\theta_m) \right]$$

$$+ \frac{\alpha \Delta t}{2(\Delta x)^2} \left[\sum_{m=0}^{M-1} \xi_m^{n+1} \exp(i[j+1]\theta_m) - 2\sum_{m=0}^{M-1} \xi_m^{n+1} \exp(i\,j\theta_m) + \sum_{m=0}^{M-1} \xi_m^{n+1} \exp(i[j-1]\theta_m) \right]$$

$$(5.46)$$

Noting the independence of the Fourier modes, and dividing by $\xi_m^{n} \exp(i\,j\theta_m)$, we obtain

$$\xi_m - 1 = \frac{\alpha \Delta t}{2(\Delta x)^2}\left[\exp(i\theta_m) - 2 + \exp(-i\theta_m)\right]$$

$$+ \frac{\alpha \Delta t}{2(\Delta x)^2}\xi_m\left[\exp(i\theta_m) - 2 + \exp(-i\theta_m)\right]$$

$$= \frac{\alpha \Delta t}{(\Delta x)^2}\left[\cos(\theta_m) - 1\right] + \frac{\alpha \Delta t}{(\Delta x)^2}\xi_m\left[\cos(\theta_m) - 1\right] = -\frac{2\alpha \Delta t}{(\Delta x)^2}[\xi_m + 1]\sin^2\left(\frac{\theta_m}{2}\right)$$

$$(5.47)$$

Rearranging Eq. (5.47) yields

$$\xi_m\left[1 + \frac{2\alpha \Delta t}{(\Delta x)^2}\sin^2\left(\frac{\theta_m}{2}\right)\right] = 1 - \frac{2\alpha \Delta t}{(\Delta x)^2}\sin^2\left(\frac{\theta_m}{2}\right). \tag{5.48}$$

Denoting the quantity $\left[2\alpha \Delta t/(\Delta x)^2\right]\sin^2(\theta_m/2)$ by ζ_m, Eq. (5.48) may be rearranged to yield

$$\xi_m = \frac{1 - \zeta_m}{1 + \zeta_m}. \tag{5.49}$$

Stability hinges on the criterion $-1 \le \xi_m \le 1$ being satisfied. Since the quantity, $\zeta_m = \frac{2\alpha\Delta t}{(\Delta x)^2}\sin^2(\theta_m/2)$, is always positive, it follows that the upper bound of the inequality is always true. The lower bound of the right-hand side of Eq. (5.49) will occur for large positive values of ζ_m and will tend to -1 in the limit $\zeta_m \to \infty$. For any finite value of ζ_m, the right-hand side is greater than -1. In summary, both bounds of the stability criterion are satisfied irrespective of the value of ζ_m. Thus, the Crank–Nicolson method is unconditionally stable for the unsteady diffusion equation. This makes it an attractive choice for computing unsteady problems since accuracy can be enhanced without loss of stability at almost the same computational cost per time step. Another way of interpreting the increased order of accuracy of the method is that larger time steps may be used to get comparable accuracy with the forward or backward Euler methods. This is a significant advantage for problems where a large number of time steps have to be executed to reach the desired instant of time. These issues are highlighted in Example 5.4.

EXAMPLE 5.4

In this example we repeat the problem considered in Example 5.1, but use the Crank–Nicolson method instead. For comparison, the same problem is also solved using the backward Euler method (Example 5.2). As a result of 21 nodes being considered in the spatial domain, we get $\Delta x = \pi/20$. A time step size equal to $4(\Delta t)_{max}$ is considered in which $(\Delta t)_{max}$ is the maximum allowable time step of the explicit method (see Example 5.1). Solutions using both methods are shown in the figures below after the same instants of time. The first observation is that the Crank–Nicolson method is unconditionally stable, as predicted by stability analysis.

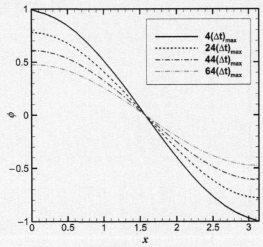

Solutions Obtained at Different Instances of Time Using the Implicit (Backward Euler) Method With $\Delta t = 4(\Delta t)_{max}$

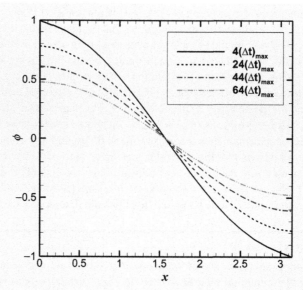

Solutions Obtained at Different Instances of Time Using the Crank–Nicolson Method With $\Delta t = 4(\Delta t)_{max}$

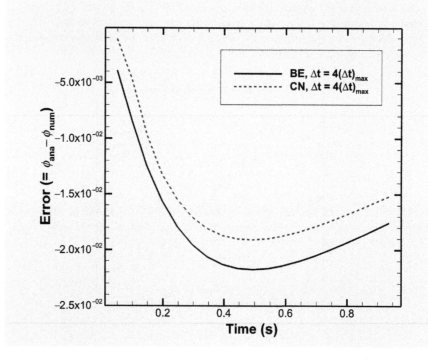

Although the solutions look identical, the errors in them are somewhat different. The figure above shows the error at Node 5 as a function of time using both methods. Since the Crank–Nicolson method is second-order accurate in time, the errors produced by it are clearly smaller than those produced by the backward Euler method at all instants of time. It is worth noting that the error shown here is not just the temporal truncation error, but total error, since it also includes spatial truncation error. Temporal truncation error may be isolated by solving the same problem with an extremely fine mesh. This can then be used to examine the temporal scaling behavior of both methods. This is left as an exercise for the reader.

5.2.4 TIME-SPLITTING ALTERNATE DIRECTION IMPLICIT METHOD

The ADI method was introduced in Section 3.2.4, in which it was demonstrated for iterative solution of the linear system of equations arising out of elliptic PDEs. The same method is discussed here in the context of time-advancing schemes for parabolic PDEs. The method has also been used for hyperbolic PDEs [6]. The ADI method for time-dependent PDEs was first proposed in 1955 by Peaceman and Rachford [7] in the context of solution of the unsteady heat conduction equation. Douglas and Rachford [8] extended it in 1956 to solution of 3D problems and proposed an algorithm that differed slightly from the original Peaceman–Rachford algorithm. The most general version of the method was finally proposed in 1964 by Douglas and Gunn [9]. It is popularly known as the Douglas–Gunn time-splitting procedure. In the most general form of the algorithm, an arbitrary number of time splits may be used.

The time-splitting ADI method is an implicit method in time. The general idea of the two-step time-splitting process is that the general implicit equation arising out of discretization of a 2D PDE, such as Eq. (5.32), can be solved by solution of two spatially 1D unsteady problems: one in the x-direction and the other in the y-direction. In general, since the procedure is applied to discretized equations, the splitting does not have to be strictly direction based, but can be nodal index based (i.e., solution along i lines followed by solution along j lines). Thus, the procedure is applicable as long as the mesh is structured. A minimum of three such splits are necessary for 3D geometry. The motivation behind such splitting is that implicit equations arising from spatially 1D problems can be solved using direct tridiagonal solvers that are computationally efficient and do not require any iterations to couple spatial nodes. Consequently, no additional instabilities are generated as a result of the combined effect of time advancement and spatial iterations, as may occur during iterative solution of implicit equations arising from spatially multidimensional problems. We present only the Douglas–Rachford variation of the ADI method in this section. This is fundamentally the same as the Douglas–Gunn method when no more than two time splits are used.

In the Douglas–Rachford method, the difference equation derived using the backward Euler method [Eq. (5.32)] is split into two equations as follows:

$$\frac{\phi^*_{i,j} - \phi_{i,j,n}}{\Delta t} = \alpha \left[\frac{\phi^*_{i+1,j} - 2\phi^*_{i,j} + \phi^*_{i-1,j}}{(\Delta x)^2} + \frac{\phi_{i,j+1,n} - 2\phi_{i,j,n} + \phi_{i,j-1,n}}{(\Delta y)^2} \right] + S_{i,j}, \qquad (5.50a)$$

$$\frac{\phi_{i,j,n+1} - \phi^*_{i,j}}{\Delta t} = \alpha \left[\frac{\phi_{i,j+1,n+1} - 2\phi_{i,j,n+1} + \phi_{i,j-1,n+1}}{(\Delta y)^2} - \frac{\phi_{i,j+1,n} - 2\phi_{i,j,n} + \phi_{i,j-1,n}}{(\Delta y)^2} \right], \quad (5.50b)$$

where ϕ^* denotes an intermediate (approximate) solution. It may be interpreted as a solution either at time Δt before the current time step or at time Δt after the previous time step. Each of the above two equations can be rearranged to yield a tridiagonal system of equations along constant j and constant i lines, respectively, as follows:

$$\left(\frac{1}{\Delta t} + \frac{2\alpha}{(\Delta x)^2} \right) \phi^*_{i,j} - \frac{\alpha}{(\Delta x)^2} \phi^*_{i+1,j} - \frac{\alpha}{(\Delta x)^2} \phi^*_{i-1,j}$$
$$= \frac{\phi_{i,j,n}}{\Delta t} + \alpha \left[\frac{\phi_{i,j+1,n} - 2\phi_{i,j,n} + \phi_{i,j-1,n}}{(\Delta y)^2} \right] + S_{i,j}, \quad (5.51a)$$

$$\left(\frac{1}{\Delta t} + \frac{2\alpha}{(\Delta y)^2} \right) \phi_{i,j,n+1} - \frac{\alpha}{(\Delta y)^2} \phi_{i,j+1,n+1} - \frac{\alpha}{(\Delta y)^2} \phi_{i,j-1,n+1}$$
$$= \frac{\phi^*_{i,j}}{\Delta t} - \alpha \left[\frac{\phi_{i,j+1,n} - 2\phi_{i,j,n} + \phi_{i,j-1,n}}{(\Delta y)^2} \right]. \quad (5.51b)$$

When Eq. (5.50a) is added to Eq. (5.50b), the result is not Eq. (5.32), but rather the following equation:

$$\frac{\phi_{i,j,n+1} - \phi_{i,j,n}}{\Delta t} = \alpha \left[\frac{\phi^*_{i+1,j} - 2\phi^*_{i,j} + \phi^*_{i-1,j}}{(\Delta x)^2} + \frac{\phi_{i,j+1,n+1} - 2\phi_{i,j,n+1} + \phi_{i,j-1,n+1}}{(\Delta y)^2} \right] + S_{i,j}. \quad (5.52)$$

This implies that an additional error is incurred over and above the truncation error already present in the original backward Euler algorithm [Eq. (5.32)] when the time-splitting algorithm is used. To gain insight into this additional error, we once again employ Taylor series expansions such that

$$\phi^*_{i,j} = \phi_{i,j,n+1} - \Delta t \frac{\partial \phi}{\partial t}\bigg|_{i,j,n+1} + \frac{(\Delta t)^2}{2} \frac{\partial^2 \phi}{\partial t^2}\bigg|_{i,j,n+1} - \frac{(\Delta t)^3}{6} \frac{\partial^3 \phi}{\partial t^3}\bigg|_{i,j,n+1} + \dots \quad (5.53)$$

Similar expansions may be written for nodes $i+1, j$ and $i-1, j$. Substituting Eq. (5.53) and other similar expansions into Eq. (5.52), we obtain

$$\frac{\phi_{i,j,n+1} - \phi_{i,j,n}}{\Delta t} = \alpha \left[\frac{\phi_{i+1,j,n+1} - 2\phi_{i,j,n+1} + \phi_{i-1,j,n+1}}{(\Delta x)^2} + \frac{\phi_{i,j+1,n+1} - 2\phi_{i,j,n+1} + \phi_{i,j-1,n+1}}{(\Delta y)^2} \right] + S_{i,j}$$
$$- \frac{\alpha \Delta t}{(\Delta x)^2} \left[\frac{\partial \phi}{\partial t}\bigg|_{i+1,j,n+1} - 2\frac{\partial \phi}{\partial t}\bigg|_{i,j,n+1} + \frac{\partial \phi}{\partial t}\bigg|_{i-1,j,n+1} \right] + \dots \quad (5.54)$$

The last term in Eq. (5.54) represents the additional error that results from splitting the original backward Euler method [Eq. (5.32)] into the two-step algorithm, given by Eq. (5.50). Using the central difference approximation [cf. Eq. (2.8)], the last term in Eq. (5.54) may also be written as

$$\frac{\partial\phi}{\partial t}\bigg|_{i+1,j,n+1} - 2\frac{\partial\phi}{\partial t}\bigg|_{i,j,n+1} + \frac{\partial\phi}{\partial t}\bigg|_{i-1,j,n+1} = (\Delta x)^2 \frac{\partial^2}{\partial x^2}\left(\frac{\partial\phi}{\partial t}\right)\bigg|_{i,j,n+1} + \frac{(\Delta x)^4}{12}\frac{\partial^4}{\partial x^4}\left(\frac{\partial\phi}{\partial t}\right)\bigg|_{i,j,n+1} + \dots .$$

(5.55)

Substituting Eq. (5.55) into Eq. (5.54) yields:

$$\frac{\phi_{i,j,n+1} - \phi_{i,j,n}}{\Delta t} = \alpha\left[\frac{\phi_{i+1,j,n+1} - 2\phi_{i,j,n+1} + \phi_{i-1,j,n+1}}{(\Delta x)^2} + \frac{\phi_{i,j+1,n+1} - 2\phi_{i,j,n+1} + \phi_{i,j-1,n+1}}{(\Delta y)^2}\right] + S_{i,j}$$

$$- \alpha\Delta t\left[\frac{\partial^2}{\partial x^2}\left(\frac{\partial\phi}{\partial t}\right)\bigg|_{i,j,n+1} + \frac{(\Delta x)^2}{12}\frac{\partial^4}{\partial x^4}\left(\frac{\partial\phi}{\partial t}\right)\bigg|_{i,j,n+1}\right] + \dots$$

$$= \alpha\left[\frac{\phi_{i+1,j,n+1} - 2\phi_{i,j,n+1} + \phi_{i-1,j,n+1}}{(\Delta x)^2} + \frac{\phi_{i,j+1,n+1} - 2\phi_{i,j,n+1} + \phi_{i,j-1,n+1}}{(\Delta y)^2}\right] + S_{i,j}$$

$$- \alpha^2\Delta t\left[\frac{\partial^2}{\partial x^2}\left(\frac{\partial^2\phi}{\partial x^2} + \frac{\partial^2\phi}{\partial y^2}\right)\bigg|_{i,j,n+1} + \frac{(\Delta x)^2}{12}\frac{\partial^4}{\partial x^4}\left(\frac{\partial^2\phi}{\partial x^2} + \frac{\partial^2\phi}{\partial y^2}\right)\bigg|_{i,j,n+1}\right] + \dots$$

(5.56)

The final expression in Eq. (5.56) was obtained by employing the governing equation [Eq. (5.2)] for error expression. It is clear from Eq. (5.56) that the leading order error term is first order in time. This implies that while the time-splitting procedure has introduced a new error over and above the truncation errors in the backward Euler method, this error is also first order in time, and therefore, the overall fidelity of the scheme – first order in time and second order in space – remains unaltered.

The stability of the Douglas–Rachford method is unconditional since each of the two split steps essentially constitutes solution of a 1D problem using the backward Euler method. Formal stability analyses of the method may be found in the original references [7–9], which also present the three-step process for 3D problems. We conclude this section by presenting the algorithm of the two-step time-splitting ADI method (Douglas–Rachford method).

ALGORITHM: DOUGLAS–RACHFORD ADI METHOD FOR PARABOLIC PDEs

Step 1: Set up the initial conditions for ϕ at all nodes (i.e., $\phi_{i,j,0}$ $\forall i = 1,\dots,N$ and $\forall j = 1,\dots,M$).

Step 2: Choose an appropriate time step size (Δt) and begin time advancement.

Step 3a: Set up and solve the tridiagonal system of equations given by Eq. (5.51a). This predicts $\phi^*_{i,j}$ $\forall i = 1,\dots,N$ and $\forall j = 1,\dots,M$.

Step 3b: Set up and solve the tridiagonal system of equations given by Eq. (5.51b). This yields the solution at the next time step (i.e., $\phi_{i,j,1}$ $\forall i = 1,\dots,N$ and $\forall j = 1,\dots,M$).

Step 4: Reset the initial conditions (i.e., $\phi_{i,j,0} = \phi_{i,j,1}$ $\forall i = 1,\dots,N$ and $\forall j = 1,\dots,M$).

Step 5: Proceed to the next time step and repeat Steps 3–4 until the desired instant of time or prescribed number of time steps has been reached.

5.3 HYPERBOLIC PARTIAL DIFFERENTIAL EQUATIONS

A hyperbolic PDE arises from mathematical modeling of physical phenomena that involve wave propagation. Consequently, it is often referred to as the *wave equation*. Examples include vibration of strings, membranes, rods, beams, and plates, as well as inviscid fluid flow. As discussed in Section 1.1, the commonly used governing equation for fluid flow, in the absence of viscosity, is a hyperbolic equation, known as the Euler equation [6]. It is a reduced form of the Navier–Stokes equations, and is used extensively to model high-speed fluid flow. The canonical form of the hyperbolic wave equation may be written as

$$\frac{\partial^2 \phi}{\partial t^2} = c^2 \, \nabla^2 \phi, \tag{5.57}$$

where c^2 is a positive constant. Physically, c represents the speed at which a wave propagates. Equation (5.57) is purely hyperbolic for spatially 1D cases, whereas it is elliptic in space and hyperbolic in time for spatially multidimensional cases. At steady state, Eq. (5.57) reduces to the Laplace equation.

Since the spatial operator is elliptic, boundary conditions must be prescribed on all spatial boundaries of the computational domain. Furthermore, since Eq. (5.57) contains a second derivative in time, at least two boundary conditions are required in the time domain. As discussed in Chapter 1 and Section 5.2, both conditions in time must be specified at the same initial instant of time (i.e., at $t = 0$) since the time domain is semi-infinite. In other words, Eq. (5.57) requires two initial conditions. In order to have a well-posed problem and a unique solution to Eq. (5.57), one of the initial conditions must be of Dirichlet type [6, 10]. The other may be either of Neumann type or of Robin type (see Chapter 2 for a discussion of these types of conditions). If the second initial condition is of Neumann type – the most common scenario by far – the two initial conditions (Dirichlet type and Neumann type) in conjunction are often referred to as the Cauchy* condition or the Cauchy initial condition. The Cauchy condition is different from the Robin condition in that a linear combination of the dependent variable and its derivative is prescribed in the latter, whereas the dependent variable and its derivative are both prescribed in the former, but separately. The canonical form of the Cauchy initial condition may be written as

$$\phi(0, \mathbf{r}) = \alpha(\mathbf{r}), \tag{5.58a}$$

$$\left. \frac{\partial \phi}{\partial t} \right|_{(0, \mathbf{r})} = \beta(\mathbf{r}), \tag{5.58b}$$

---

*Augustin–Louis Cauchy (1789–1857) was a French mathematician whose contribution to mathematics and mathematical physics laid the foundation for much of modern applied mathematics and physics. His contributions can be found in topics ranging from elasticity theory to astronomy. He is regarded as the pioneer of complex analysis and, to this day, has more theorems and physical principles named after him than any other individual.

where $\alpha(\mathbf{r})$ and $\beta(\mathbf{r})$ are arbitrary functions of space. The explicit and implicit methods for time-advancing hyperbolic PDEs are discussed in the following two subsections.

5.3.1 EXPLICIT METHOD

As in the case of parabolic problems, a difference approximation to the second derivative in time can be derived by beginning with a forward Taylor series expansion in time, yielding

$$\phi_{i,j,n+1} = \phi_{i,j,n} + \Delta t \left.\frac{\partial \phi}{\partial t}\right|_{i,j,n} + \frac{(\Delta t)^2}{2} \left.\frac{\partial^2 \phi}{\partial t^2}\right|_{i,j,n} + \frac{(\Delta t)^3}{6} \left.\frac{\partial^3 \phi}{\partial t^3}\right|_{i,j,n} + \frac{(\Delta t)^4}{24} \left.\frac{\partial^4 \phi}{\partial t^4}\right|_{i,j,n} + \cdots \quad (5.59)$$

Substituting the two initial conditions in Eq. (5.58) into Eq. (5.59), we obtain

$$\phi_{i,j,n+1} = \alpha_{i,j} + \Delta t \beta_{i,j} + \frac{(\Delta t)^2}{2} \left.\frac{\partial^2 \phi}{\partial t^2}\right|_{i,j,n} + \frac{(\Delta t)^3}{6} \left.\frac{\partial^3 \phi}{\partial t^3}\right|_{i,j,n} + + \frac{(\Delta t)^4}{24} \left.\frac{\partial^4 \phi}{\partial t^4}\right|_{i,j,n} + \cdots \quad (5.60)$$

Rearranging Eq. (5.60) yields:

$$\left.\frac{\partial^2 \phi}{\partial t^2}\right|_{i,j,n} = \frac{2}{(\Delta t)^2} \left[\phi_{i,j,n+1} - \alpha_{i,j} - \Delta t \beta_{i,j} \right] - \frac{\Delta t}{3} \left.\frac{\partial^3 \phi}{\partial t^3}\right|_{i,j,n} + \cdots \quad (5.61)$$

It is evident that the approximation given by Eq. (5.61) is only first-order accurate in time. However, this is not too much of an inconvenience since Eq. (5.61) is to be used only in the first time step (i.e., $n = 0$) and can be replaced by a better (second-order) approximation from the second time step ($n > 0$) onwards when solutions at more than two time instants become available. Another point to note is that the explicit method is likely to encounter a stability constraint (as will be confirmed later) that may restrict its time step in any event. The truncation error will be small as a result of the use of small time steps even though the approximation is only first order.

In order to derive a higher (second) order approximation for subsequent time steps, we employ an additional Taylor series expansion:

$$\phi_{i,j,n-1} = \phi_{i,j,n} - \Delta t \left.\frac{\partial \phi}{\partial t}\right|_{i,j,n} + \frac{(\Delta t)^2}{2} \left.\frac{\partial^2 \phi}{\partial t^2}\right|_{i,j,n} - \frac{(\Delta t)^3}{6} \left.\frac{\partial^3 \phi}{\partial t^3}\right|_{i,j,n} + \frac{(\Delta t)^4}{24} \left.\frac{\partial^4 \phi}{\partial t^4}\right|_{i,j,n} + \cdots \quad (5.62)$$

Addition of Eqs. (5.59) and (5.62) yields

$$\left.\frac{\partial^2 \phi}{\partial t^2}\right|_{i,j,n} = \frac{\phi_{i,j,n+1} - 2\phi_{i,j,n} + \phi_{i,j,n-1}}{(\Delta t)^2} - \frac{1}{12}(\Delta t)^2 \left.\frac{\partial^4 \phi}{\partial t^4}\right|_{i,j,n} + \cdots \quad (5.63)$$

Eq. (5.63) represents the central difference approximation in time and has the same truncation error in time as its spatial counterpart [cf. Eq. (2.8)]. Now, employing the central difference approximation for spatial derivatives at the same instant of time at which the time derivative has been formulated (at time index n), we may write

$$n=0: \frac{2}{(\Delta t)^2}\left[\phi_{i,j,n+1} - \alpha_{i,j} - \Delta t \beta_{i,j}\right] = c^2\left[\frac{\alpha_{i+1,j} - 2\alpha_{i,j} + \alpha_{i-1,j}}{(\Delta x)^2} + \frac{\alpha_{i,j+1} - 2\alpha_{i,j} + \alpha_{i,j-1}}{(\Delta y)^2}\right],$$

(5.64a)

$$n>0: \frac{1}{(\Delta t)^2}\left[\phi_{i,j,n+1} - 2\phi_{i,j,n} + \phi_{i,j,n-1}\right]$$

$$= c^2\left[\frac{\phi_{i+1,j,n} - 2\phi_{i,j,n} + \phi_{i-1,j,n}}{(\Delta x)^2} + \frac{\phi_{i,j+1,n} - 2\phi_{i,j,n} + \phi_{i,j-1,n}}{(\Delta y)^2}\right].$$

(5.64b)

Rearranging in explicit form, Eq. (5.64) may be written as

$$n=0: \phi_{i,j,n+1} = \alpha_{i,j} + \Delta t \beta_{i,j} + \frac{c^2 (\Delta t)^2}{2}\left[\frac{\alpha_{i+1,j} - 2\alpha_{i,j} + \alpha_{i-1,j}}{(\Delta x)^2} + \frac{\alpha_{i,j+1} - 2\alpha_{i,j} + \alpha_{i,j-1}}{(\Delta y)^2}\right],$$

(5.65a)

$$n>0: \phi_{i,j,n+1} = 2\phi_{i,j,n} - \phi_{i,j,n-1}$$

$$+ c^2 (\Delta t)^2\left[\frac{\phi_{i+1,j,n} - 2\phi_{i,j,n} + \phi_{i-1,j,n}}{(\Delta x)^2} + \frac{\phi_{i,j+1,n} - 2\phi_{i,j,n} + \phi_{i,j-1,n}}{(\Delta y)^2}\right].$$

(5.65b)

The dependency between unknown and known quantities in the explicit time advancement method for hyperbolic PDEs is depicted schematically in Fig. 5.5.

As with all the methods described in Section 5.2, stability is an important consideration when it comes to time advancement methods. As before, we will apply the von Neumann stability analysis procedure to the spatially 1D difference equation. This means the 1D version of Eq. (5.64b) in this particular case. This equation is

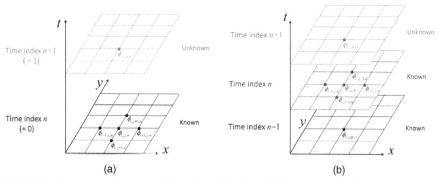

FIGURE 5.5 Schematic Representation of the Relationship Between the Variables at Successive Time Steps in the Explicit Method for Hyperbolic PDEs

(a) First time step ($n = 0$), and (b) subsequent time steps ($n > 0$).

chosen over Eq. (5.64a) because it can be applied to all time steps except the first. Following the procedure described in Section 5.2.1, the error equation corresponding to the 1D version of Eq. (5.64b) may be written as

$$\frac{\varepsilon_{i,j,n+1} - 2\varepsilon_{i,j,n} + \varepsilon_{i,j,n-1}}{(\Delta t)^2} = c^2 \left[\frac{\varepsilon_{i+1,j,n} - 2\varepsilon_{i,j,n} + \varepsilon_{i-1,j,n}}{(\Delta x)^2} \right]. \tag{5.66}$$

Next, we follow the same procedure as used several times in Section 5.2. This entails substituting the Fourier representation of the error [Eq. (5.15)] into Eq. (5.66), considering the linear independence of Fourier modes, and finally dividing the resulting equation by $\xi_m^n \exp(i\, j\theta_m)$. The details are omitted here for the sake of brevity. Execution of these steps finally yields

$$\xi_m - 2 + \xi_m^{-1} = \frac{c^2 (\Delta t)^2}{(\Delta x)^2} \left[\exp(i\theta_m) - 2 + \exp(-i\theta_m) \right]. \tag{5.67}$$

Simplifying further and using the trigonometric identity $1 - \cos\theta = 2\sin^2(\theta/2)$, we obtain

$$\xi_m + \frac{1}{\xi_m} = 2 - \frac{c^2 (\Delta t)^2}{(\Delta x)^2} \left[4\sin^2 \left(\frac{\theta_m}{2} \right) \right] = G_m, \tag{5.68}$$

where the right-hand side of Eq. (5.68) has been denoted by G_m for simplicity of notation. In addition, we introduce a new notation as follows:

$$R = \frac{c(\Delta t)}{(\Delta x)}, \tag{5.69}$$

such that

$$G_m = 2 - R^2 \left[4\sin^2 \left(\frac{\theta_m}{2} \right) \right]. \tag{5.70}$$

Since c denotes speed, it follows that the quantity, R, is nondimensional. This is known as the *Courant*[†] *number* or the *CFL number*, named after Courant, Friedrichs, and Lewy, who first investigated the stability of the hyperbolic wave equation in 1928 in a seminal publication [11]. Using the new notations, Eq. (5.68) becomes

$$\xi_m^2 - G_m \xi_m + 1 = 0. \tag{5.71}$$

[†]**Richard Courant** (1888–1972) was an applied mathematician from Germany. He migrated to the United States in 1936 and held a professorship at New York University for much of his life. He founded the Courant Institute of Mathematical Sciences, which is one of the eminent schools for graduate studies in applied mathematics. Along with his advisor – the famous mathematician, David Hilbert – Courant coauthored a book in 1924 entitled *Methods of Mathematical Physics*, which continues to be a popular text even today.

The two roots of Eq. (5.71) are

$$\xi_{m,1} = \frac{G_m + \sqrt{G_m^2 - 4}}{2}, \quad \xi_{m,2} = \frac{G_m - \sqrt{G_m^2 - 4}}{2}. \tag{5.72}$$

When $\sin^2(\theta_m/2) = 0$, it follows from Eq. (5.70) that $G_m = 2$, and consequently, Eq. (5.72) yields $\xi_{m,1} = \xi_{m,2} = 1$, which satisfies the stability criterion $-1 \le \xi_m \le 1$. On the other hand, when $\sin^2(\theta_m/2) = 1$, $G_m = 2 - 4R^2$. Substituting in Eq. (5.72) and simplifying, we obtain

$$\xi_{m,1} = 1 - 2R^2 + 2R\sqrt{R^2 - 1}, \quad \xi_{m,2} = 1 - 2R^2 - 2R\sqrt{R^2 - 1}. \tag{5.73}$$

Next, we examine the implications of these two roots satisfying the stability criterion. We start with the first root satisfying the lower limit (i.e., $1 - 2R^2 + 2R\sqrt{R^2 - 1} \ge -1$), which, upon simplification, yields $R^2 \ge 1$ or $\left(\frac{c\Delta t}{\Delta x}\right)^2 \ge 1$. This condition is rather strange because it almost implies that the method will be stable only with large time steps, which begs the question as to what happens if $R^2 < 1$. Is the method unstable if $R^2 < 1$? For $0 < R^2 < 1$, the two roots $\xi_{m,1}$ and $\xi_{m,2}$ are both complex and complex conjugates of each other with magnitudes less than unity. Therefore, inspection of the lower limit for the first root does not yield any realistic stability constraint and can be satisfied with any time step size. Therefore, we move on to the upper limit for the first root (i.e., $1 - 2R^2 + 2R\sqrt{R^2 - 1} \le 1$). Upon simplification, this inequality yields $R\sqrt{R^2 - 1} \le R^2$. Further simplification yields $R^2 \ge 0$, which is of course always true. In summary, inspection of the first root provides no clues as to the stability of the method.

Imposing the lower bound for the second root yields $1 - 2R^2 - 2R\sqrt{R^2 - 1} \ge -1$. Upon simplification, this yields $R^2 \le 1$, which is a meaningful stability criterion. Thus, the criterion for stability of the explicit time advancement method for the spatially 1D hyperbolic wave equation is that the CFL or Courant number must be less than unity. This criterion is often referred to as the *CFL criterion*, and may be mathematically written as

$$\frac{c\Delta t}{\Delta x} \le 1. \tag{5.74}$$

Physically, $c\Delta t$ represents the distance traveled by a wave over a time period Δt. The CFL criterion states that if the distance traveled by a wave in a single time step is greater than the grid spacing, instability sets in and the solution may blow up. It is easy to show that imposing the upper stability bound for the second root (i.e., $1 - 2R^2 - 2R\sqrt{R^2 - 1} \le 1$) yields $R^2 \le 0$, which can only be satisfied by an imaginary value of R. Therefore, it is not meaningful from a physical standpoint. Next, we consider an example that explores the implications of the CFL criterion by applying it to a physical problem and examining the resulting solution from a physical perspective.

EXAMPLE 5.5

In this example, we consider the vibration of a string held fixed at its two ends. The governing equation to be solved is $\dfrac{\partial^2 \phi}{\partial t^2} = \dfrac{\partial^2 \phi}{\partial x^2}$, and the boundary conditions are $\phi(t,0) = \phi(t,1) = 0$. The initial conditions are $\phi(0,x) = 0$ everywhere except $\phi(0,0.5) = 1$, and $(\partial \phi / \partial t)\big|_{0,x} = 0$. In this particular case, the dependent variable, ϕ, represents the displacement of the string. Just three nodes are considered, as shown in the figure below. Only one of the three nodes (Node 2) is an interior node. The other two are boundary nodes, whose displacements are zero at all times (i.e., they are held fixed). The center of the string is initially displaced (as given by $\phi(0,0.5) = 1$) and then released (also shown in the figure).

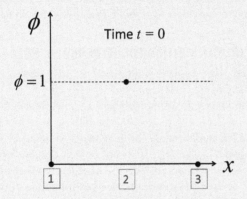

In this case the governing equation represents an undamped oscillation, and the string will continue to vibrate with displacement of the middle node ranging between +1 and −1. Using the explicit time advancement method, the difference equation for Node 2 in the first time step may be written as

$$\phi_{2,1} = \alpha_2 + \Delta t \beta_2 + \frac{R^2}{2}\left[\alpha_3 - 2\alpha_2 + \alpha_1\right]$$

$$= 1 + 0 + \frac{R^2}{2}[0 - 2 + 0] = 1 - R^2$$

It is clear from the above equation that $-1 \le \phi_{2,1} \le 1$ if $R^2 \le 2$. In other words, a CFL number (R) greater than unity would still produce a physically meaningful solution. Clearly, this is not what stability analysis predicted. However, stability analysis was only conducted on those equations used from the second time step onward (i.e., for $n > 0$). Therefore, in order to test the predictions of stability analysis, we must perform a calculation for the second time step. Let us assume that in the first time step we use $R^2 = 0.5$, such that the solution after the first time step becomes $\phi_{2,1} = 0.5$. For the second time step, Eq. (5.65b) reduces to $\phi_{2,2} = 2\phi_{2,1} - \alpha_2 + R^2\left[\phi_{3,1} - 2\phi_{2,1} + \phi_{1,1}\right] = 1 - 1 + R^2[0 - 1 + 0] = -R^2$. Clearly, $-1 \le \phi_{2,2} \le 1$ as long as $R^2 \le 1$. Thus, the computed results are consistent with stability predictions and the CFL criterion.

The CFL criterion manifests itself in many application areas. Examples include the computation of fluid flow [3,6], front (or interface) tracking algorithms using the volume-of-fluid method [12] or the level set method [13], and solution of the Boltzmann transport equation [14]. Essentially, whenever an advection-type operator is integrated explicitly in time, the CFL criterion must be satisfied because the wave nature of transport processes is borne out by the advection term of a PDE, as shown in Section 1.2. Having established the stability bounds of the explicit method, we turn now to the overall solution algorithm for solving a hyperbolic PDE using the explicit method.

ALGORITHM: EXPLICIT METHOD FOR HYPERBOLIC PDEs

Step 1: Set up the initial conditions at all nodes (i.e., $\alpha_{i,j}$ and $\beta_{i,j}$ $\forall i = 1,...,N$ and $\forall j = 1,...,M$).
Step 2: Choose an appropriate time step size (Δt) that obeys the CFL stability criterion.
Step 3: Compute all nodal values of ϕ at the next time step (i.e., compute $\phi_{i,j,1}$ $\forall i = 1,...,N$ and $\forall j = 1,...,M$ using an explicit update formula similar to Eq. (5.64a)). For boundary nodes with non-Dirichlet–type boundary conditions, equations similar to Eq. (5.64a) must be derived following the procedures outlined in Chapter 3 and applied.
Step 4: Reset the initial conditions (i.e., $\alpha_{i,j} = \phi_{i,j,1}$ $\forall i = 1,...,N$ and $\forall j = 1,...,M$).
Step 5: Repeat Step 3 using Eq. (5.64b).
Step 6: Reset the initial conditions (i.e., $\alpha_{i,j} = \phi_{i,j,2}$ $\forall i = 1,...,N$ and $\forall j = 1,...,M$).
Step 7: Proceed to the next time step and repeat Steps 5 and 6 until the desired instant of time or prescribed number of time steps has been reached.

5.3.2 IMPLICIT METHOD

As in the case of parabolic PDEs, the explicit method is easy to implement because it does not involve solution of a linear system of equations. However, the CFL stability criterion can often pose a severe constraint on time step size for practical problems, especially if a nonuniform mesh is used, in which case, the maximum allowable time step is the smallest of the allowable time steps computed at each node. The implicit method was devised to combat the time step restriction of the explicit method. The second derivative in time can be derived here by beginning with a backward Taylor series expansion as follows:

$$\phi_{i,j,n} = \phi_{i,j,n+1} - \Delta t \left.\frac{\partial \phi}{\partial t}\right|_{i,j,n+1} + \frac{(\Delta t)^2}{2}\left.\frac{\partial^2 \phi}{\partial t^2}\right|_{i,j,n+1} - \frac{(\Delta t)^3}{6}\left.\frac{\partial^3 \phi}{\partial t^3}\right|_{i,j,n+1} + ... \quad (5.75)$$

Similarly, the first derivative at time index n may be expressed as

$$\left.\frac{\partial\phi}{\partial t}\right|_{i,j,n} = \left.\frac{\partial\phi}{\partial t}\right|_{i,j,n+1} - \Delta t \left.\frac{\partial^2\phi}{\partial t^2}\right|_{i,j,n+1} + \frac{(\Delta t)^2}{2}\left.\frac{\partial^3\phi}{\partial t^3}\right|_{i,j,n+1} - \frac{(\Delta t)^3}{6}\left.\frac{\partial^4\phi}{\partial t^4}\right|_{i,j,n+1} +\ldots, \quad (5.76)$$

Rearranging yields

$$\left.\frac{\partial\phi}{\partial t}\right|_{i,j,n+1} = \left.\frac{\partial\phi}{\partial t}\right|_{i,j,n} + \Delta t \left.\frac{\partial^2\phi}{\partial t^2}\right|_{i,j,n+1} - \frac{(\Delta t)^2}{2}\left.\frac{\partial^3\phi}{\partial t^3}\right|_{i,j,n+1} + \frac{(\Delta t)^3}{6}\left.\frac{\partial^4\phi}{\partial t^4}\right|_{i,j,n+1} +\ldots. \quad (5.77)$$

Substituting Eq. (5.77) into Eq. (5.75) and simplifying yields

$$\phi_{i,j,n} = \phi_{i,j,n+1} - \Delta t \left.\frac{\partial\phi}{\partial t}\right|_{i,j,n} - \frac{(\Delta t)^2}{2}\left.\frac{\partial^2\phi}{\partial t^2}\right|_{i,j,n+1} + \frac{(\Delta t)^3}{3}\left.\frac{\partial^3\phi}{\partial t^3}\right|_{i,j,n+1} +\ldots. \quad (5.78)$$

Substituting the two initial conditions [Eq. (5.58)] into Eq. (5.78), we obtain

$$\alpha_{i,j} = \phi_{i,j,n+1} - \Delta t \beta_{i,j} - \frac{(\Delta t)^2}{2}\left.\frac{\partial^2\phi}{\partial t^2}\right|_{i,j,n+1} + \frac{(\Delta t)^3}{3}\left.\frac{\partial^3\phi}{\partial t^3}\right|_{i,j,n+1} +\ldots, \quad (5.79)$$

Rearranging yields the desired expression for the second derivative, which can only be used for the first time step ($n = 0$):

$$\left.\frac{\partial^2\phi}{\partial t^2}\right|_{i,j,n+1} = \frac{2}{(\Delta t)^2}\left[\phi_{i,j,n+1} - \alpha_{i,j} - \Delta t \beta_{i,j}\right] + \frac{2\Delta t}{3}\left.\frac{\partial^3\phi}{\partial t^3}\right|_{i,j,n+1} +\ldots. \quad (5.80)$$

As in the explicit formulation [Eq. (5.61)], the expression for the second derivative in time is first-order accurate for the first time step. For subsequent time steps, another Taylor series expansion, in addition to Eq. (5.75), may be used. It may be written as

$$\phi_{i,j,n-1} = \phi_{i,j,n+1} - 2\Delta t \left.\frac{\partial\phi}{\partial t}\right|_{i,j,n+1} + \frac{(2\Delta t)^2}{2}\left.\frac{\partial^2\phi}{\partial t^2}\right|_{i,j,n+1} - \frac{(2\Delta t)^3}{6}\left.\frac{\partial^3\phi}{\partial t^3}\right|_{i,j,n+1} +\ldots. \quad (5.81)$$

Multiplying Eq. (5.75) by 2, subtracting the result from Eq. (5.81), and simplifying yields

$$\left.\frac{\partial^2\phi}{\partial t^2}\right|_{i,j,n+1} = \frac{\phi_{i,j,n-1} - 2\phi_{i,j,n} + \phi_{i,j,n+1}}{(\Delta t)^2} + \Delta t \left.\frac{\partial^3\phi}{\partial t^3}\right|_{i,j,n+1} +\ldots. \quad (5.82)$$

In contrast with the explicit method [Eq. (5.63)], the difference approximation for the second derivative in time for subsequent time steps is only first-order accurate, as shown by Eq. (5.82). This reduction in order may be cited as a major disadvantage of the implicit method compared with the explicit method when it comes to hyperbolic PDEs.

Using Eqs. (5.80) and (5.82), in conjunction with the central difference approximation for the spatial derivatives at the very instant of time the time derivative was formulated (at time index $n+1$), we may write

$$n=0:\frac{2}{(\Delta t)^2}\left[\phi_{i,j,n+1}-\alpha_{i,j}-\Delta t\beta_{i,j}\right]$$

$$=c^2\left[\frac{\phi_{i+1,j,n+1}-2\phi_{i,j,n+1}+\phi_{i-1,j,n+1}}{(\Delta x)^2}+\frac{\phi_{i,j+1,n+1}-2\phi_{i,j,n+1}+\phi_{i,j-1,n+1}}{(\Delta y)^2}\right],$$

(5.83a)

$$n>0:\frac{1}{(\Delta t)^2}\left[\phi_{i,j,n+1}-2\phi_{i,j,n}+\phi_{i,j,n-1}\right]$$

$$=c^2\left[\frac{\phi_{i+1,j,n+1}-2\phi_{i,j,n+1}+\phi_{i-1,j,n+1}}{(\Delta x)^2}+\frac{\phi_{i,j+1,n+1}-2\phi_{i,j,n+1}+\phi_{i,j-1,n+1}}{(\Delta y)^2}\right].$$

(5.83b)

Equation (5.83) may be rearranged such that it can be written in implicit form, with known and unknown quantities on different sides of the equation, as follows:

$$n=0:\left(\frac{2}{(\Delta t)^2}+\frac{2c^2}{(\Delta x)^2}+\frac{2c^2}{(\Delta y)^2}\right)\phi_{i,j,n+1}-\frac{c^2}{(\Delta x)^2}\phi_{i+1,j,n+1}-\frac{c^2}{(\Delta x)^2}\phi_{i-1,j,n+1}$$

$$-\frac{c^2}{(\Delta y)^2}\phi_{i,j+1,n+1}-\frac{c^2}{(\Delta y)^2}\phi_{i,j-1,n+1}=\frac{2}{(\Delta t)^2}\left[\alpha_{i,j}+\Delta t\beta_{i,j}\right],$$

(5.84a)

$$n>0:\left(\frac{1}{(\Delta t)^2}+\frac{2c^2}{(\Delta x)^2}+\frac{2c^2}{(\Delta y)^2}\right)\phi_{i,j,n+1}-\frac{c^2}{(\Delta x)^2}\phi_{i+1,j,n+1}-\frac{c^2}{(\Delta x)^2}\phi_{i-1,j,n+1}$$

$$-\frac{c^2}{(\Delta y)^2}\phi_{i,j+1,n+1}-\frac{c^2}{(\Delta y)^2}\phi_{i,j-1,n+1}=\frac{1}{(\Delta t)^2}\left[2\phi_{i,j,n}-\phi_{i,j,n-1}\right].$$

(5.84b)

As in the case of parabolic PDEs, Eq. (5.84) may be solved using any of the iterative methods discussed in Chapters 3 and 4. The dependency between unknown and known quantities in the implicit time advancement method for hyperbolic PDEs is depicted schematically in Fig. 5.6.

Next, we examine the stability of the implicit method for hyperbolic PDEs. Following the same steps undertaken for the explicit method [cf. Eq. (5.67)] we obtain

$$\xi_m-2+\xi_m^{-1}=\frac{c^2(\Delta t)^2}{(\Delta x)^2}\xi_m\left[\exp(i\theta_m)-2+\exp(-i\theta_m)\right].$$

(5.85)

Using the necessary trigonometric identity, and simplifying yields

$$\left[1+4R^2\sin^2\left(\frac{\theta_m}{2}\right)\right]\xi_m^2-2\xi_m+1=0,$$

(5.86)

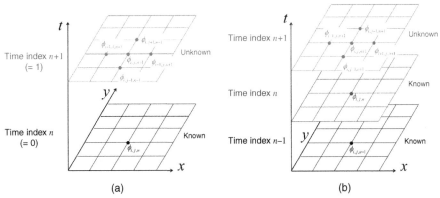

FIGURE 5.6 Schematic Representation of the Relationship Between the Variables at Successive Time Steps in the Implicit Method for Hyperbolic PDEs

(a) First time step ($n = 0$), and (b) subsequent time steps ($n > 0$).

where, as before, R is the CFL number defined by Eq. (5.69). The two roots of Eq. (5.86) are both complex and may be written as

$$\xi_{m,1} = \frac{1 + 2R\sin\left(\dfrac{\theta_m}{2}\right)i}{1 + 4R^2\sin^2\left(\dfrac{\theta_m}{2}\right)}, \quad \xi_{m,2} = \frac{1 - 2R\sin\left(\dfrac{\theta_m}{2}\right)i}{1 + 4R^2\sin^2\left(\dfrac{\theta_m}{2}\right)} \tag{5.87}$$

Since the two roots shown in Eq. (5.87) are complex conjugates of each other, their magnitudes are the same and equal to:

$$|\xi_m| = \sqrt{\xi_{m,1}\,\xi_{m,2}} = \sqrt{\frac{1}{1 + 4R^2\sin^2\left(\dfrac{\theta_m}{2}\right)}}, \tag{5.88}$$

which is less than or equal to unity for any realizable values of R^2 and θ_m. Thus, the implicit method for hyperbolic PDEs, like the implicit method for parabolic PDEs, is unconditionally stable. The overall solution algorithm of the implicit method is presented below.

ALGORITHM: IMPLICIT METHOD FOR HYPERBOLIC PDEs

Step 1: Set up the initial conditions for all nodes (i.e., $\alpha_{i,j}$ and $\beta_{i,j}$, $\forall i = 1,...,N$ and $\forall j = 1,...,M$).

Step 2: Choose an appropriate time step size (Δt) and begin time advancement.

Step 3: Set up a linear system $[A][\phi] = [Q]$, such as Eq. (5.84a), in which matrix elements are derived from nodal equations.

Step 4: Compute $[\phi]$ at the first time step (i.e., compute $\phi_{i,j,1}$ $\forall i = 1,...,N$ and $\forall j = 1,...,M$) by solving the linear system set up in Step 3 using any of the solvers discussed in Chapters 3 and 4 or any other solver of choice. If an iterative solver is used, convergence must be monitored at this step and iterations must be continued until the prescribed convergence criterion has been satisfied.

Step 5: Reset the initial conditions (i.e., $\phi_{i,j,n} = \phi_{i,j,n+1}$ $\forall i = 1,...,N$ and $\forall j = 1,...,M$) and proceed to the next time step.

Step 6: Set up a linear system $[A][\phi] = [Q]$, such as Eq. (5.84b), in which matrix elements are derived from nodal equations.

Step 7: Compute $[\phi]$ at the next time step (i.e., compute $\phi_{i,j,n+1}$ $\forall i = 1,...,N$ and $\forall j = 1,...,M$) by solving the linear system set up in Step 6.

Step 8: Repeat Steps 5–7 until the desired instant of time or prescribed number of time steps has been reached.

Example 5.6 compares the *pros* and *cons* of the explicit and implicit methods for hyperbolic PDEs.

EXAMPLE 5.6

In this example we assess the *pros* and *cons* of the explicit and implicit time advancement methods for hyperbolic PDEs. The physical problem under consideration is the vibration of a string held fixed at its two ends. The governing equation to be solved is $(\partial^2 \phi / \partial t^2) = (\partial^2 \phi / \partial x^2)$ and the boundary conditions are $\phi(t,0) = \phi(t,1) = 0$. The initial condition is a triangular displacement profile given by $\phi(0,x) = \begin{cases} 2x, & 0 \le x \le 0.5 \\ 2 - 2x, & 1 \ge x \ge 0.5 \end{cases}$ and an initial velocity of $(\partial \phi / \partial t)|_{0,x} = 0$. In this particular case the dependent variable, ϕ, represents the displacement of the string. The numerical solution uses 41 spatial nodes. The closed-form analytical solution can be obtained by solving the resulting eigenvalue problem using Separation of Variables. It is written as

$$\phi(t,x) = \frac{8}{\pi^2} \sum_{n=1}^{\infty} \frac{\sin(n\pi/2)}{n^2} \cos(n\pi t) \sin(n\pi x).$$

First, the problem is solved using the explicit method with a CFL number (R) equal to 0.5 (i.e., $\Delta t = \Delta x / 2$). Exact analytical and the numerical results are shown in the figure below at three different instances of time. The error between the analytical and numerical solution is also shown. Even though the errors are quite small (less than 1%), the solutions exhibit spurious high-frequency oscillations that are not present in the exact analytical solution. However, since the CFL number used obeys the stability constraint, the error does not grow monotonically as time increases. This is depicted in the figure by the almost equal magnitude of the maximum error at all three instances of time. Moreover, solution symmetry is retained at all instances of time.

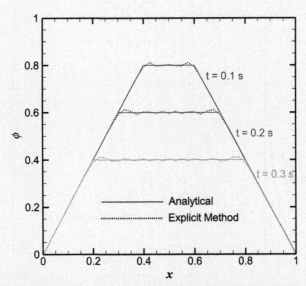

Solution Using the Explicit Method and CFL Number Equal to 0.5

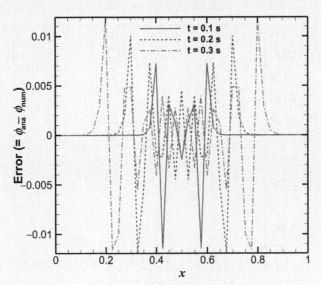

Error Between Analytical and the Explicit Numerical Solution
with CFL Number Equal to 0.5

Next, the problem is solved with the same CFL number using the implicit method. Moreover, the same plots are generated such that the implicit method can be compared with the explicit method. These are shown in the figures below.

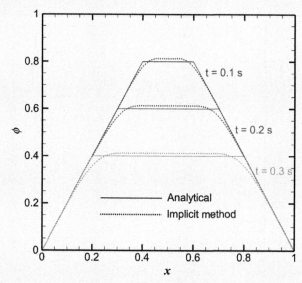

Solution Using the Implicit Method and CFL Number Equal to 0.5

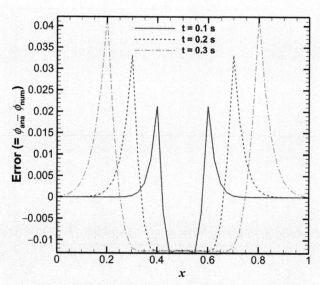

Error Between Analytical and the Implicit Numerical Solution
with CFL Number Equal to 0.5

In this case, no high-frequency oscillations are observed in the solution. Instead, the solution is smeared in regions where sharp spatial gradients exist. Maximum errors are significantly larger (about 2–4%) than those (below 1%)

in the explicit method. This is because the implicit method is first-order accurate in time [Eq. (5.82)] while the explicit method is fortuitously second-order accurate [Eq. (5.63)] except in the very first time step.

Next, to confirm the findings of von Neumann stability analysis, numerical solutions are attempted with a CFL number equal to 2. The results are shown in the figure below.

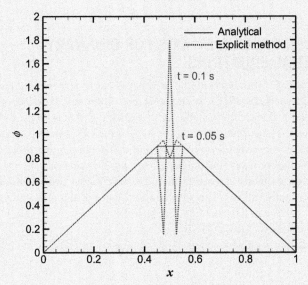

Solution Using the Explicit Method and CFL Number Equal to 2

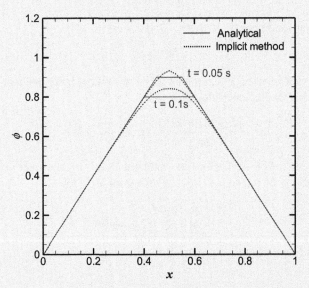

Solution Using the Implicit Method and CFL Number Equal to 2

The results clearly show that the explicit method is unstable with a CFL number equal to 2, as predicted by stability analysis, whereas the implicit method is stable. Moreover, the errors incurred in the implicit solution are now significantly larger as a result of a larger time step size being used. This example conclusively demonstrates the value of stability analysis and its implications.

5.4 HIGHER ORDER METHODS FOR ORDINARY DIFFERENTIAL EQUATIONS

Sections 5.2 and 5.3 discussed the basic time advancement methods used for both parabolic and hyperbolic PDEs. With the exception of the Crank–Nicolson method, which is second-order accurate, all the methods discussed thus far are only first-order accurate in time. The explicit method for hyperbolic PDEs is also second-order accurate, but that is somewhat fortuitous and not by design. Higher order temporal accuracy is desirable for many practical applications. This section caters to this need by discussing a few methods that are explicit but have higher order temporal accuracy. The discussion in this section is restricted to ordinary differential equations (ODEs) and is meant to serve as a foundation for the following section in which the methods discussed here are extended to PDEs.

Our discussion begins by considering solution of an ODE of the following general form:

$$\frac{d\phi}{dt} = f(t,\phi), \tag{5.89}$$

with the initial condition:

$$\phi(0) = \phi_0. \tag{5.90}$$

It is worth recalling at this juncture that with the forward Euler method, we obtained a difference approximation of the form:

$$\frac{\phi(t+\Delta t)-\phi(t)}{\Delta t} = f(t,\phi(t)) + \varepsilon(\Delta t). \tag{5.91}$$

Our objective here is to derive an approximation that is higher order in accuracy. To accomplish this goal, we define two new quantities as follows:

$$\begin{aligned} F_1 &= \Delta t \ f(t,\phi) \\ F_2 &= \Delta t \ f(t+\alpha \Delta t, \phi + \beta F_1) \end{aligned}, \tag{5.92}$$

where α and β are arbitrary constants that will be determined later. Let us also propose that

$$\phi(t+\Delta t) = \phi(t) + w_1 F_1 + w_2 F_2, \tag{5.93}$$

where w_1 and w_2 are constant weights. They may be chosen with the goal of making the approximation shown in Eq. (5.93) as close as possible to an exact Taylor series expansion, namely

$$\phi(t + \Delta t) = \phi(t) + \Delta t \frac{d\phi}{dt}\bigg|_t + \frac{1}{2}(\Delta t)^2 \frac{d^2\phi}{dt^2}\bigg|_t + \frac{1}{6}(\Delta t)^3 \frac{d^3\phi}{dt^3}\bigg|_t + \dots \qquad (5.94)$$

Before comparing Eqs. (5.93) and (5.94) to derive the weights, we substitute Eq. (5.92) into Eq. (5.93). This yields

$$\phi(t + \Delta t) = \phi(t) + w_1 \Delta t \, f(t,\phi) + w_2 \Delta t \, f(t + \alpha \Delta t, \phi + \beta \Delta t \, f(t,\phi)), \qquad (5.95)$$

As a consequence of performing a Taylor series expansion of the last term in Eq. (5.95), the equation may be written as

$$\phi(t + \Delta t) = \phi(t) + w_1 \Delta t \, f(t,\phi) + w_2 \Delta t \left[f(t,\phi) + \alpha \Delta t \frac{\partial f}{\partial t}\bigg|_{t,\phi} + \beta \Delta t \, f(t,\phi) \frac{\partial f}{\partial \phi}\bigg|_{t,\phi} \right]$$

$$+ w_2 \Delta t \left[\frac{(\alpha \Delta t)^2}{2!} \frac{\partial^2 f}{\partial t^2}\bigg|_{t,\phi} + \frac{(\beta \Delta t \, f(t,\phi))^2}{2!} \frac{\partial^2 f}{\partial \phi^2}\bigg|_{t,\phi} + \frac{\alpha \Delta t \beta \Delta t \, f(t,\phi)}{2!} \frac{\partial^2 f}{\partial t \partial \phi}\bigg|_{t,\phi} \right] + \dots$$

$$\qquad (5.96)$$

Gathering similar terms on the right-hand side, Eq. (5.96) may be rewritten as

$$\phi(t + \Delta t) = \phi(t) + (w_1 + w_2)\Delta t \, f(t,\phi) + w_2 \alpha (\Delta t)^2 \frac{\partial f}{\partial t}\bigg|_{t,\phi} + w_2 \beta (\Delta t)^2 f(t,\phi) \frac{\partial f}{\partial \phi}\bigg|_{t,\phi}$$

$$+ w_2 (\Delta t)^3 \left[\frac{\alpha^2}{2!} \frac{\partial^2 f}{\partial t^2}\bigg|_{t,\phi} + \frac{(\beta f(t,\phi))^2}{2!} \frac{\partial^2 f}{\partial \phi^2}\bigg|_{t,\phi} + \frac{\alpha \beta f(t,\phi)}{2!} \frac{\partial^2 f}{\partial t \partial \phi}\bigg|_{t,\phi} \right] + \dots \qquad (5.97)$$

Since Eq. (5.97) contains partial derivatives of f, it cannot be compared directly with Eq. (5.94). Direct comparison between Eqs. (5.94) and (5.97) can be made by differentiating Eq. (5.89) with respect to time, yielding

$$\frac{d^2\phi}{dt^2} = \frac{d}{dt}[f(t,\phi)] = \frac{\partial f}{\partial t} + \frac{\partial f}{\partial \phi} \frac{d\phi}{dt} = \frac{\partial f}{\partial t} + f(t,\phi) \frac{\partial f}{\partial \phi}. \qquad (5.98)$$

Substituting Eqs. (5.89) and (5.98) into Eq. (5.94) and rearranging yields

$$\phi(t + \Delta t) = \phi(t) + \Delta t \, f(t,\phi) + \frac{1}{2}(\Delta t)^2 \left[\frac{\partial f}{\partial t}\bigg|_{t,\phi} + f(t,\phi) \frac{\partial f}{\partial \phi}\bigg|_{t,\phi} \right] + \frac{1}{6}(\Delta t)^3 \frac{d^3\phi}{dt^3}\bigg|_t + \dots \quad (5.99)$$

As a consequence of making Eq. (5.99) equivalent to Eq. (5.97) by enforcing equality of terms up to second order, we obtain

$$w_1 + w_2 = 1, \; \alpha w_2 = \frac{1}{2}, \; \beta w_2 = \frac{1}{2}. \qquad (5.100)$$

Eq. (5.100) represents three equations with four unknowns. Therefore, a unique solution does not exist. However, one can choose a value for one of the four constants and calculate the other three. A combination that is commonly used [15,16] is

$$w_1 = w_2 = \frac{1}{2}, \ \alpha = \beta = 1. \qquad (5.101)$$

Substituting Eq. (5.101) into Eq. (5.95) and rearranging yields

$$\frac{\phi(t+\Delta t)-\phi(t)}{\Delta t} = \frac{1}{2\Delta t}[F_1+F_2]+\varepsilon[(\Delta t)^2] = \frac{1}{2}[f(t,\phi)+f(t+\Delta t,\phi+F_1)]+\varepsilon[(\Delta t)^2], \qquad (5.102)$$

where F_1 and F_2 are given by Eq. (5.92). The difference approximation for the time derivative shown in Eq. (5.102) is second-order accurate in time. Consequently, this method of time advancement is known as the Runge–Kutta method of order 2 (RK2).

The Runge–Kutta method is arguably one of the most popular methods of time advancement because it can easily be extended to higher order. The fourth-order version of the method, RK4, may be written as [15,16]

$$\frac{\phi(t+\Delta t)-\phi(t)}{\Delta t} = \frac{1}{6\Delta t}[F_1+2F_2+2F_3+F_4]+\varepsilon[(\Delta t)^4], \qquad (5.103)$$

where

$$\begin{aligned} F_1 &= \Delta t \ f(t,\phi) \\ F_2 &= \Delta t \ f(t+\frac{\Delta t}{2},\phi+\frac{1}{2}F_1) \\ F_3 &= \Delta t \ f(t+\frac{\Delta t}{2},\phi+\frac{1}{2}F_2) \\ F_4 &= \Delta t \ f(t+\frac{\Delta t}{2},\phi+F_3) \end{aligned} \qquad (5.104)$$

Example 5.7 assesses the accuracy of these two Runge–Kutta schemes.

EXAMPLE 5.7

In this example we assess the accuracy of RK2, RK4, and the forward Euler (FE) method. The ODE to be solved is $\frac{d\phi}{dt}=2-2\phi$, with the initial condition $\phi(0)=0$. The exact analytical solution to the system is written as $\phi(t)=1-e^{-2t}$. The numerical solution is computed with a time step equal to 0.1 s. The maximum allowable time step for the forward Euler method for this particular problem is 0.5 s. The table below shows solutions obtained by the three methods alongside the exact analytical solution.

Time(s)	Analytical	FE	RK2	RK4
0.1	0.181269	0.200000	0.180000	0.181267
0.2	0.329680	0.360000	0.327600	0.329676
0.3	0.451188	0.488000	0.448632	0.451183
0.4	0.550671	0.590400	0.547878	0.550665
0.5	0.632121	0.672320	0.629260	0.632115
0.6	0.698806	0.737856	0.695993	0.698800
0.7	0.753403	0.790285	0.750715	0.753398
0.8	0.798103	0.832228	0.795586	0.798098
0.9	0.834701	0.865782	0.832380	0.834696
1	0.864665	0.892626	0.862552	0.864660

The data in the table immediately above demonstrate that as the order of temporal accuracy is increased, the solution approaches the exact analytical solution.

As with all explicit methods, the stability of the Runge-Kutta method is a concern. For an ODE of the form shown in Eq. (5.89), a universal statement cannot be made regarding its stability. The stability depends on the forcing function (i.e., the right-hand side of Eq. (5.89)). Since the focus of the present text is PDEs, a detailed discussion of the stability of Runge–Kutta methods in the context of solution of ODEs is not included and the reader is referred to other texts [15, 16] that provide detailed stability analysis.

The Runge–Kutta method, along with its early predecessors for solving ODEs, such as the Taylor series method [15], belongs to a class of methods known as *single time step methods*. These methods do not require solutions at multiple previous time steps to construct solutions for future instances of time. Therefore, they can be elegantly extended to arbitrarily higher orders of temporal accuracy. In practice, the solution is desired within a certain error tolerance, and consequently, the order of the method can be adapted to the problem at hand and the accuracy sought. This has led to the development of adaptive Runge–Kutta methods, based on the so-called Runge–Kutta–Fehlberg algorithm [15], in which the temporal truncation error is monitored at each time step and the order of the method is adjusted on the fly to decrease the error if deemed necessary.

Although single time step methods are elegantly extendible to higher-order accuracy, increased accuracy comes at increased computational cost. For example, the explicit (forward Euler) method for an ODE requires only one functional evaluation per time step, while RK2 requires two [Eq. (5.102)] and RK4 requires four [Eq. (5.103)]. The computational cost of the higher-order renditions of the method can often be prohibitive when it comes to using these methods in the context of PDEs in which time advancement has to be executed for thousands of individual spatial nodes (as will be seen in the following section). In an effort to circumvent this practical problem, so-called *multi time step methods* have been developed. In such methods, the higher derivatives are eliminated using Lagrange polynomial approximations (see Chapter 8). One of the most popular multistep high-order explicit methods is the *Adams–Bashforth method*. It can be extended to fifth-order accuracy by using the following hierarchical relationships [15, 16]:

$$\phi(t + \Delta t) = \phi(t) + \Delta t \, f(t, \phi(t)), \tag{5.105a}$$

$$\phi(t + 2\Delta t) = \phi(t + \Delta t) + \Delta t \left[\frac{3}{2} f(t + \Delta t, \phi(t + \Delta t)) - \frac{1}{2} f(t, \phi(t)) \right], \tag{5.105b}$$

$$\phi(t + 3\Delta t) = \phi(t + 2\Delta t)$$
$$+ \Delta t \left[\frac{23}{12} f(t + 2\Delta t, \phi(t + 2\Delta t)) - \frac{4}{3} f(t + \Delta t, \phi(t + \Delta t)) + \frac{5}{12} f(t, \phi(t)) \right], \tag{5.105c}$$

$$\phi(t + 4\Delta t) = \phi(t + 3\Delta t)$$
$$+ \Delta t \left[\frac{55}{24} f(t + 3\Delta t, \phi(t + 3\Delta t)) - \frac{59}{24} f(t + 2\Delta t, \phi(t + 2\Delta t)) + \frac{37}{24} f(t + \Delta t, \phi(t + \Delta t)) - \frac{3}{8} f(t, \phi(t)) \right], \tag{5.105d}$$

$$\phi(t + 5\Delta t) = \phi(t + 4\Delta t)$$
$$+ \Delta t \left[\frac{1901}{720} f(t + 3\Delta t, \phi(t + 3\Delta t)) - \frac{1387}{360} f(t + 3\Delta t, \phi(t + 3\Delta t)) + \frac{109}{30} f(t + 2\Delta t, \phi(t + 2\Delta t)) - \frac{637}{360} f(t + \Delta t, \phi(t + \Delta t)) + \frac{251}{720} f(t, \phi(t)) \right]. \tag{5.105e}$$

The algorithm for the Adams–Bashforth method is executed as follows: first, starting with the initial condition at $t = 0$, Eq. (5.105a) is used to compute the solution at $t + \Delta t$. Next, Eq. (5.105b) is used to compute the solution at $t + 2\Delta t$. The process is continued until the solution at $t + 5\Delta t$ has been computed using Eq. (5.105e). From this point on, the solution at the five previous time steps is already available and Eq. (5.105e) is repeatedly used to advance the solution in time. This implies that the method is less than fifth-order accurate up to the fifth time step and fifth-order accurate beyond the fifth time step. The accuracy of the solution in the first few time steps is often a cause for concern. In such a case, it is customary to use the RK4 method up to the fourth time step and the Adams–Bashforth method from the fifth time step onward. One final point to note is that, although the Adams–Bashforth method was designed to be used as a fifth-order accurate method, it can be used as a lower order method if so desired.

Another popular variation of the Adams–Bashforth explicit time advancement method is the *Adams–Moulton method*. This method is semi-implicit in time and usually used in predictor–corrector mode. The solution at every new time step is first predicted using the Adams–Bashforth method and then corrected using the

Adams–Moulton predictor step, which increases the accuracy of the method by an order. Therein lies the advantage of the two-step predictor–corrector process. The corrector steps are written as

$$\phi(t+\Delta t)=\phi(t)+\frac{1}{2}\Delta t\left[f(t+\Delta t,\phi(t+\Delta t))+f(t,\phi(t))\right],\qquad(5.106a)$$

$$\phi(t+2\Delta t)=\phi(t+\Delta t)+\Delta t\left[\begin{array}{l}\dfrac{5}{12}f(t+2\Delta t,\phi(t+2\Delta t))+\dfrac{2}{3}f(t+\Delta t,\phi(t+\Delta t))\\[2mm]-\dfrac{1}{12}f(t,\phi(t))\end{array}\right],$$
$$(5.106b)$$

$$\phi(t+3\Delta t)=\phi(t+2\Delta t)+\Delta t\left[\begin{array}{l}\dfrac{3}{8}f(t+3\Delta t,\phi(t+3\Delta t))+\dfrac{19}{24}f(t+2\Delta t,\phi(t+2\Delta t))\\[2mm]-\dfrac{5}{24}f(t+\Delta t,\phi(t+\Delta t))\\[2mm]+\dfrac{1}{24}f(t,\phi(t))\end{array}\right],$$
$$(5.106c)$$

$$\phi(t+4\Delta t)=\phi(t+3\Delta t)+\Delta t\left[\begin{array}{l}\dfrac{251}{720}f(t+4\Delta t,\phi(t+4\Delta t))+\dfrac{646}{720}f(t+3\Delta t,\phi(t+3\Delta t))\\[2mm]-\dfrac{264}{720}f(t+2\Delta t,\phi(t+2\Delta t))\\[2mm]+\dfrac{106}{720}f(t+\Delta t,\phi(t+\Delta t))\\[2mm]-\dfrac{19}{720}f(t,\phi(t))\end{array}\right].$$
$$(5.106d)$$

The solution algorithm of the Adams–Moulton method can be precisely clarified by considering a case where only second-order temporal accuracy is desired. In such a case, Eq. (5.105a) is first solved to predict $\phi(t+\Delta t)$. This solution is then substituted into the right-hand side of Eq. (5.106a) to correct the solution. The resulting solution is second-order accurate. In fact, Eq. (5.106a) represents the so-called trapezoidal rule for numerical integration (see Chapter 8) and is known to be second-order accurate [15].

The methods discussed thus far for time advancement of a time-dependent ODE (scalar case) can also be extended to a system of ODEs (vector case). Let us consider a system of N ODEs of the following form:

$$\begin{aligned}\frac{d\phi_1}{dt}&=f_1(t,\phi_1,\phi_2,\phi_3,...,\phi_N)\\[1mm]\frac{d\psi_2}{dt}&=f_2(t,\phi_1,\phi_2,\phi_3,...,\phi_N)\\[1mm]&\vdots\\[1mm]\frac{d\phi_N}{dt}&=f_N(t,\phi_1,\phi_2,\phi_3,...,\phi_N)\end{aligned}\qquad,\qquad(5.107a)$$

with initial conditions:

$$\begin{aligned}\phi_1(0) &= \phi_{0,1}\\ \phi_2(0) &= \phi_{0,2}\\ &\vdots\\ \phi_N(0) &= \phi_{0,N}\end{aligned} \qquad \text{(5.107b)}$$

In compact matrix form, Eq. (5.107) may be written as

$$\frac{d\Phi}{dt} = f(t, \phi_1, \phi_2, \phi_3, ..., \phi_N),$$

$$\Phi(0) = \Phi_0, \qquad \text{(5.108)}$$

where

$$\begin{aligned}\Phi &= [\phi_1 \quad \phi_2 \quad \phi_3 \quad \cdots \quad \phi_N]^T\\ f &= [f_1 \quad f_2 \quad f_3 \quad \cdots \quad f_N]^T\\ \Phi_0 &= [\phi_{0,1} \quad \phi_{0,2} \quad \phi_{0,3} \quad \cdots \quad \phi_{0,N}]^T\end{aligned} \qquad \text{(5.109)}$$

For example, the Runge–Kutta method can be directly applied to Eq. (5.108) to yield formulas similar to Eqs (5.102) and (5.103):

$$\text{RK2}: \quad \begin{aligned}&\Phi(t+\Delta t) = \Phi(t) + \frac{\Delta t}{2}(\mathbf{F}_1 + \mathbf{F}_2),\\ &\mathbf{F}_1 = f(t, \Phi),\\ &\mathbf{F}_2 = f(t+\Delta t, \Phi + \Delta t\, \mathbf{F}_1),\end{aligned} \qquad \text{(5.110a)}$$

$$\text{RK4}: \quad \begin{aligned}&\Phi(t+\Delta t) = \Phi(t) + \frac{\Delta t}{6}(\mathbf{F}_1 + 2\mathbf{F}_2 + 2\mathbf{F}_3 + \mathbf{F}_4),\\ &\mathbf{F}_1 = f(t, \Phi),\\ &\mathbf{F}_2 = f(t+\frac{\Delta t}{2}, \Phi + \frac{1}{2}\Delta t\, \mathbf{F}_1),\\ &\mathbf{F}_3 = f(t+\frac{\Delta t}{2}, \Phi + \frac{1}{2}\Delta t\, \mathbf{F}_2),\\ &\mathbf{F}_4 = f(t+\frac{\Delta t}{2}, \Phi + \Delta t\, \mathbf{F}_3).\end{aligned} \qquad \text{(5.110b)}$$

Since integration in the RK method is explicit in time, even though Eq. (5.110) is a vector equation, it can be treated as a set of scalar equations that are decoupled from one another and solved (evaluated) sequentially, thereby making the method easy to implement.

The subject of integration (time advancement) of time-dependent ODEs has been heavily researched over the past century. The literature on this area is rich with many explicit and implicit time advancement schemes of varying orders of accuracy

[3,6,14,15]. Discussion of all these schemes is beyond the scope and intended coverage of this chapter. The brief description provided here serves only to lay the foundation for our topic of interest: the solution of PDEs.

5.5 METHOD OF LINES

The Method of Lines is a method that enables reduction of a PDE to a set of ODEs. This reduction is motivated by the realization that powerful higher order algorithms (and codes) for time integration, as discussed in the preceding section, already exist and may be used conveniently for the solution of PDEs. In essence, the method of lines is based upon the general idea of transformation from an Eulerian reference frame to a Lagrangian reference frame. This transformation can be understood by considering a spatially 1D parabolic PDE:

$$\frac{\partial \phi}{\partial t} = \frac{\partial^2 \phi}{\partial x^2}. \tag{5.111}$$

This equation must be satisfied at each node, $x = x_i$, within the computational domain for numerical solution. Therefore, we may write

$$\left.\frac{\partial \phi}{\partial t}\right|_{x=x_i} = \left.\frac{\partial^2 \phi}{\partial x^2}\right|_{x=x_i}. \tag{5.112}$$

Applying the central difference formula to the right-hand side of Eq. (5.112), we may further write

$$\left.\frac{\partial \phi}{\partial t}\right|_{x=x_i} = \frac{\phi_{i+1} - 2\phi_i + \phi_{i-1}}{(\Delta x)^2}. \tag{5.113}$$

Equation (5.113) essentially states that the rate of change of ϕ with respect to time at $x = x_i$ depends on three quantities: the value of ϕ at $x = x_i$ (i.e., ϕ_i) and two other quantities (i.e., ϕ_{i-1} and ϕ_{i+1}). Therefore, we may interpret Eq. (5.113) from a strictly mathematical perspective as

$$\left.\frac{\partial \phi}{\partial t}\right|_{x=x_i} = \frac{d\phi_i}{dt} = f(\phi_i, \phi_{i-1}, \phi_{i+1}). \tag{5.114}$$

In this interpretation, the rate of change of ϕ with respect to time at $x = x_i$ is no longer a function of space, but rather a function of other variables that are unknown or of dependent variables. These dependent variables, in turn, are also functions of their counterpart dependent variables. The first equality of Eq. (5.114) stems from the fact that each node in our system is stationary (i.e., has no motion). This equality will not hold if, for example, we consider a problem with a moving mesh. In general, equations like Eq. (5.114) may be written for all N nodes, ultimately resulting in N ordinary differential equations. This system of ODEs fits the framework described by Eq. (5.108). Hence, this system of ODEs can be solved using any of the methods discussed in the preceding section. Example 5.8 demonstrates the method of lines for solution of a PDE.

EXAMPLE 5.8

In this example we repeat Example 5.1 using the method of lines. Following the discussion above, this can be done by first writing down the difference equations as a set of ODEs:

$$\frac{d\phi_i}{dt} = f_i(\phi_1,\phi_2,...,\phi_N) = \frac{\phi_{i-1} - 2\phi_i + \phi_{i+1}}{(\Delta x)^2} \quad \forall i = 2, 3,..., N-1$$

Next, we use RK2 to integrate the final set of ODEs. This can be done by first determining $\mathbf{F}_1$ and $\mathbf{F}_2$. From their definitions [Eq. (5.110a)] we have

$$F_{1,i} = f_i = \frac{\phi_{i-1} - 2\phi_i + \phi_{i+1}}{(\Delta x)^2},$$

and

$$F_{2,i} = f_i(\phi_1 + \Delta t\, F_{1,1}, \phi_2 + \Delta t\, F_{1,2},..., \phi_N + \Delta t\, F_{1,N})$$
$$= \frac{\phi_{i-1} + \Delta t\, F_{1,i-1} - 2(\phi_i + \Delta t\, F_{1,i}) + \phi_{i+1} + \Delta t\, F_{1,i+1}}{(\Delta x)^2}$$

When $i = 2$, the above formula for $F_{2,2}$ requires $F_{1,1}$. However, $F_{1,1}$ cannot be computed using the above formula for $F_{1,i}$ because there is no node with index 0. Instead, the governing equation, in conjunction with the boundary condition, can be used to derive $F_{1,1} = (d\phi_1/dt) = -e^{-t}$ since $F_{1,1} = f_1$. Similarly, when $i = N-1$, $F_{1,N} = (d\phi_N/dt) = e^{-t}$. With these expressions, the RK2 method can now be applied in a straightforward manner. A total of 21 nodes ($= N$) are considered for spatial discretization, resulting in $\Delta x = \pi/20$. Therefore, the maximum allowable time step, based on the stability criterion of the forward Euler method, is $(\Delta t)_{max} = (\Delta x)^2/2 = 0.0123\,\text{s}$. The time step size used is $0.8(\Delta t)_{max}$. Temporal evolution of the solution and the temporal behavior of the error at Node 5 is shown in the two figures below. For comparison, the error in the solution using the forward Euler method is also shown. The solutions produced are similar, at least visually, to those produced by the other methods (see Examples 5.1 and 5.2). As far as the error is concerned, the solution is expected to be second-order accurate temporally since the RK2 method is used for final integration of the set of ODEs. Indeed, the magnitude of the error incurred by the method of lines is smaller than the error incurred by the forward Euler method. It is worth recalling that the error shown in the figure below is the sum of spatial and temporal discretization errors. Hence, the errors incurred by the two methods do not exactly scale as the ratio of the temporal discretization errors of the two methods.

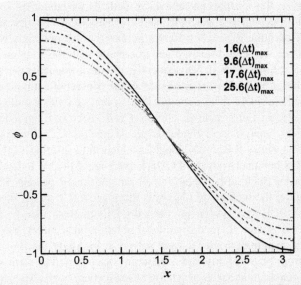

Solution Obtained Using the Method of Lines with RK2 Integration

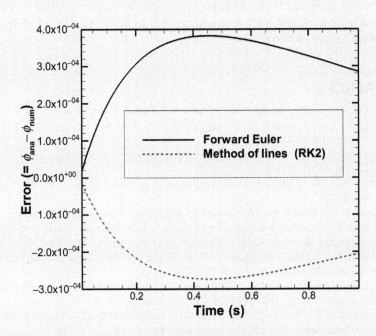

Error in Solution at the node 5 Using the Forward Euler Method
and the Method of Lines (RK2)

To summarize, this Chapter presented the methods for numerical solution to unsteady problems encountered in science and engineering, which typically results in parabolic or hyperbolic PDEs. A different perspective for obtaining a steady-state solution to a PDE was first discussed and compared with the solution of boundary value problems (elliptic PDEs). Subsequently, the most popular time advancement (or time-marching) methods were presented in the context of solution of both parabolic and hyperbolic PDEs. These include the explicit or forward Euler method, the implicit or backward Euler method, and the Crank–Nicolson method. Formal von Neumann stability analysis was conducted to establish the stability bounds of each method. The Peaceman–Rachford–Douglass–Gunn time-splitting ADI method was also presented. Since many powerful ODE solvers are available today in the public domain, it is beneficial to utilize them for solution of time-dependent PDEs. In light of this realization, the method of lines was introduced as a means of converting a time-dependent PDE to a system of coupled ODEs that can then be integrated in time with an ODE solver of any desired level of temporal accuracy. Throughout this chapter, several example problems were solved to highlight the *pros* and *cons* of each method from both stability and accuracy perspectives. The methods presented in the chapter constitute the proverbial "tip of the iceberg" only. Many popular time advancement methods, such as the leapfrog and Lax–Wendroff methods [6], used extensively for the solution of the Euler equation, were not covered. Indeed, an entire text could be dedicated to the discussion of time advancement algorithms. For further reading on these and other time advancement methods, the reader is referred to application-specific advanced texts [3,6,17].

REFERENCES

[1] Chang S-C. The method of space-time conservation element and solution element – a new approach for solving the Navier-Stokes and Euler equations. J Comput Phys 1995;119(2):295–324.

[2] Incropera FP, Dewitt DP, Bergman TL, Lavine AS. Introduction to heat transfer. 6th ed. New York: Wiley; 2011.

[3] Ferziger JH, Perić M. Computational methods for fluid dynamics. 3rd ed. Springer: Berlin; 2002.

[4] Pope SB. Turbulent flows. Cambridge University Press: Cambridge, UK; 2000.

[5] Crank J, Nicolson P. A practical method for numerical evaluation of solutions of partial differential equations of the heat conduction type. Proc Camb Phil Soc 1947;43(1):50–67.

[6] Pletcher RH, Tannehill JC, Anderson D. Computational fluid mechanics and heat transfer. 3rd ed. CRC Press: Boca Raton, FL; 2011.

[7] Peaceman DW, Rachford HH Jr. The numerical solution of parabolic and elliptic differential equations. J Soc Ind Appl Math 1955;3(1):28–41.

[8] Douglas J Jr, Rachford HH Jr. On the numerical solution of heat conduction problems in two and three space variables. Trans Am Math Soc 1956;82(2):421–39.

[9] Douglas J Jr, Gunn JE. A general formulation of alternating direction methods. Numerische Mathematik 1964;6(1):428–53.

[10] Kreyzig E. Advanced engineering mathematics. 10th ed. New York: John Wiley and Sons; 2010.

[11] Courant R, Friedrichs KO, Lewy H. Über die partiellen Differenzengleichungen der Mathematischen Physik. Math Ann 1928;100:32–74. English translation, with commentaries by Lax, P.B., Widlund, O.B., Parter, S.V., in IBM J. Res. Develop. 11, 1967.

[12] Hirt CW, Nichols BD. Volume of fluid (VOF) method for the dynamics of free boundaries. J Comput Phys 1981;39(1):201–25.

[13] Osher SJ, Fedkiw RP. Level set methods and dynamic implicit surfaces. Springer-Verlag: Berlin; 2002.

[14] Ali SA, Kollu G, Mazumder S, Sadayappan P, Mittal A. Large-scale parallel computation of the phonon Boltzmann transport equation. Intl J Therm Sci 2014;86:341–51.

[15] Ward WE, Kincaid DR. Numerical mathematics and computing. 7th ed. Cengage Learning: Boston, MA; 2012.

[16] Butcher JC. Numerical methods for ordinary differential equations. New York: John Wiley and Sons; 2003.

[17] Sheng X, Song W. Essentials of computational electromagnetics. Wiley-IEEE Press: Singapore; 2012.

EXERCISES

5.1   Consider the following PDE:

$$\frac{\partial \phi}{\partial t} = \frac{\partial^2 \phi}{\partial x^2}$$

The initial condition is $\phi(0,x) = 1$, and the boundary conditions are as follows:

$$\frac{\partial \phi}{\partial x}(t,0) = 0$$

$$\frac{\partial \phi}{\partial x}(t,1) = -\phi(t,1)$$

The analytical solution to the problem is given by:

$$\phi(t,x) = \sum_{n=1}^{\infty} C_n \exp(-\lambda_n^2 t)\cos(\lambda_n x), \text{ where } C_n = \frac{4\sin \lambda_n}{2\lambda_n + \sin(2\lambda_n)}$$

The first four values of λ are as follows: $\lambda_1 = 0.8603$, $\lambda_2 = 3.4256$, $\lambda_3 = 6.4373$, and $\lambda_4 = 9.5293$. Note that it is sufficient to use just the first four terms of the series for an accurate enough answer for $t > 0.01$.

Use the explicit (forward Euler) method to solve the above problem numerically. Use a time step equal to half of the maximum allowable time step (as dictated by the stability criterion). Use 26 nodes in the x direction. It is sufficient to use a first-order accurate scheme (spatially) for the boundary nodes. Use second-order scheme for interior nodes.

a.   Make a plot of the solutions [$\phi(t,x)$ vs. x] at $t = 0.1$, 0.2, 0.4, and 0.8 on a single graph.

b. Make a plot of the error in the solutions [difference between analytical and numerical solutions] at $t = 0.1, 0.2, 0.4$, and 0.8 on a single graph.

c. Comment on the results.

5.2 Repeat Exercise 5.1 (including all plots) but now using the implicit (backward Euler) method. In addition to using $\Delta t_{max}/2$ (as in Exercise 5.1), also compute one other case using Δt_{max}. Comment on your results.

5.3 Repeat Exercise 5.2 (including all plots) but now using the Crank–Nicolson method. Discuss the results of Exercises 5.1–5.3 by comparing and contrasting your solutions in terms of accuracy, stability, efficiency, and ease of use.

5.4 Derive a second-order accurate implicit scheme in both space and time for the spatially one-dimensional hyperbolic wave equation. It is sufficient to derive this scheme for $n > 2$ (third time step onward). Perform a von Neumann stability analysis on the resulting discrete equation (for the second time step onward) to determine if the method is stable.

5.5 Consider the following PDE:

$$\frac{\partial^2 \phi}{\partial t^2} = \frac{\partial^2 \phi}{\partial x^2}$$

The initial conditions are as follows: $\phi(0,x) = \sin(\pi x)$ and $(\partial \phi / \partial t)(0,x) = (1/4)\sin(2\pi x)$
The boundary conditions are as follows: $\phi(t,0) = 0$, and $\phi(t,1) = 0$
Obtain the analytical solution to the above system using separation of variables.

Use the explicit (forward Euler) method to solve the above problem numerically. Use a time step equal to half of the maximum allowable time step (as dictated by the CFL stability criterion). Use 26 nodes in the x direction.

a. Make a plot of the solutions [$\phi(t,x)$ vs. x] at $t = 0.4, 0.8$, and 1.2 on a single graph.

b. Make a plot of the errors (analytical minus numerical) solutions at the same instances of time as Part (a) on a single graph.

c. Use a CFL number equal to 2 and compute the solution at 0.4, 0.8, and 1.2 s. Plot these solutions. On the same plot show the solution at the same instances of time but with the time step used in Part (a) (i.e., half of the maximum-allowable time step). Make sure you label each curve.

d. Comment on the results from Parts (a) through (c).

5.6 Repeat Exercise 5.5 (including all plots) but now using the implicit (backward Euler) method. Comment on your results.

5.7 Repeat Exercise 5.1 using the method of lines and the RK2 method.

5.8 A numerical analyst has proposed the following second-order accurate (in time) scheme for solution of the 1D unsteady diffusion equation as a substitute for the Crank–Nicolson method:

$$\frac{\phi_{j,n+1}-\phi_{j,n-1}}{2\Delta t} = \frac{\alpha}{\Delta x^2}\Big[\phi_{j+1,n} - 2\phi_{j,n} + \phi_{j-1,n}\Big],$$

where the notations used are the same as those used throughout this chapter. Perform a von Neumann stability analysis to determine whether this time advancement scheme is stable. If a stability criterion exists, what is it?

The Finite Volume Method (FVM)

6

Thus far, the focus of this text has been the finite difference method (FDM). As discussed in Chapter 1, the FDM solves the governing partial differential equation (PDE) directly, and the solution is referred to as the strong form solution. The FDM is based strictly on mathematical principles, the contention being that if the governing PDE is solved accurately enough, the underlying physical law upon which it is constructed, will be automatically satisfied. In contrast, the finite-volume method (FVM) is founded upon an entirely different philosophy – one that emphasizes the satisfaction of the physical law underlying the governing PDE on finite-sized control volumes, rather than the mathematical preciseness of the solution of the PDE itself. It yields the so-called weak form solution.

The FVM enforces the conservation laws underlying the governing PDE on control volumes that constitute the overall computational domain. In principle, it is merely an extension of the basic conservation concepts and associated analysis we perform in our undergraduate curriculum. For example, in mechanics, the free body diagram is the foundation of all analysis. The so-called free body is nothing but a closed control volume to which we apply conservation of momentum (force balance), either linear or angular. Similarly, in thermodynamics, we apply mass and energy conservation laws to control volumes representing devices such as nozzles, pumps, and turbines. If one now imagines splitting (discretizing) the large single control volume into a set of smaller control volumes, and the same conservation laws applied to all the smaller control volumes, the end result is a set of coupled algebraic equations that may be solved conveniently on a computer, and the resulting method is the so-called FVM. The equations are coupled because adjacent control volumes communicate with each other through mass, momentum, or energy exchange. Since the smallest control volume is still of finite size, and not infinitesimally small, the method is referred to as the FVM.

The FVM has its roots in fluid dynamics. The earliest example of using the FVM, at least in a formal sense, can be traced back to the works of Harlow and Welch [1] at the Los Alamos National Laboratory in 1965. They developed the so-called marker-and-cell method for analyzing incompressible fluid flow problems with free surfaces. Later, with a growing realization that conservation of basic physical quantities such as mass, momentum, and energy is at the heart of the vast majority of terrestrial processes, the FVM became the method of choice for analysis of problems involving fluid flow and other transport phenomena. Its widespread adoption may also be attributed to the fact that basic conservation laws are easily understood by scientists and engineers who may not have in-depth knowledge of PDEs and associated numerical methods for their solution. To this day, the FVM continues to be the method of choice for the analysis of problems involving fluid flow and transport phenomena.

This chapter, and the next, provides formal in-depth coverage of the FVM. The present chapter introduces the method to the reader in the context of a structured mesh, while the next chapter extends it to an unstructured mesh. Although the concepts introduced in Chapters 3–5 are applicable to any discretization method, they do require minor modifications when applied to the FVM. These modifications are also highlighted in this chapter.

6.1 DERIVATION OF FINITE VOLUME EQUATIONS

As already described in Chapter 1, in the FVM, the computational domain or total volume of matter to be analyzed is first split or discretized into a set of smaller control volumes. These control volumes are referred to as *cells*. The surfaces bounding the cells are known as *faces*. The vertices of the cells are known as *nodes* or are sometimes referred to as *vertices* for clarity. Figure 6.1 shows a schematic of the finite volume discretization pattern. The nodal locations, denoted by solid circles in Fig. 6.1, are, in fact, the same locations where the nodes in the FDM would reside. In the FVM, starting from the governing differential equation, algebraic equations are constructed that provide spatially (or volume) averaged values of the dependent variable at each cell. The volume-averaged value is defined as

$$\bar{\phi}_O = \frac{1}{V_O} \int_{V_O} \phi \, dV, \tag{6.1}$$

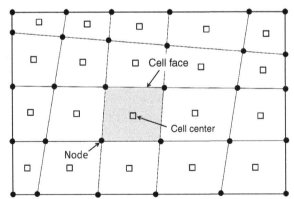

FIGURE 6.1 Schematic Representation of Finite Volume Discretization of a Rectangular Domain Showing Cell Centers, Cell Faces, and Cell Vertices (Nodes)

The node locations (solid black circles) are the same as in the FDM.

where V_o denotes the volume of the control volume over which the average is being computed, and $\bar{\phi}_o$ denotes the average value of the dependent variable, ϕ, over that control volume. For simplicity of notation, henceforth we will denote this average value as ϕ_o, i.e., without using the bar on top. It is generally assumed that the average value of ϕ over the control volume is the same as its value at the geometric centroid of the control volume. While there may be only a handful of functions for which this assumption is preempted, it is always justifiable in the limiting case of the control volume shrinking to a small size. This is because any nonlinear continuous function can be approximated accurately by a linear function if the interval of approximation is sufficiently small (it follows from the fact that the higher order terms of the Taylor series expansion are negligible). The average of a linear function of space is, of course, the same as its value at the geometric centroid of the interval over which the function is being averaged. Hence, even though this approximation may produce errors on a coarse mesh, the error decreases as the mesh is refined, which, as we already know, is in compliance with the general behavior of truncation errors. The average value of the dependent variable in a control volume is generally referred to as the cell-center value or the cell value, while the average value at a face is referred to as the face-center or face value. Like its volumetric counterpart, the average value at the face is also stored at its geometric centroid. In summary, the FVM yields cell-center values. In contrast, as is evident from Fig. 6.1, the FDM yields vertex or nodal values.

As was done in Chapter 2 for the FDM, we start our derivation of the finite volume equations by considering a spatially 1D ordinary differential equation, written as

$$\frac{d}{dx}\left[\Gamma\frac{d\phi}{dx}\right] = -S_\phi,\tag{6.2}$$

where Γ is a coefficient that, in general, may be a function of space or of the dependent variable itself, or both, i.e., $\Gamma = \Gamma(x,\phi)$. If Γ were a constant, Eq. (6.2) would reduce to the 1D version of the Poisson equation. Physically, equations such as Eq. (6.2) are used to model diffusive transport. For example, if ϕ represented temperature, Γ represented the thermal conductivity, and S_ϕ represented the heat generation rate per unit volume, Eq. (6.2) would be used to model heat transport by conduction, and is the so-called steady state heat conduction equation [2]. The rate of transfer of physical quantities, such as heat, per unit area is known as flux, and is written as

$$J = -\Gamma\frac{d\phi}{dx}.\tag{6.3}$$

A much more detailed discussion of the physical significance of such equations is presented in Section 7.1. Prior to continuing further, the reader is encouraged to peruse that section for a deeper understanding of the governing equations discussed thus far in this text and their physical connection.

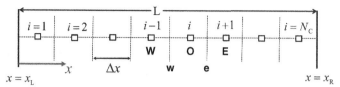

FIGURE 6.2 Schematic Representation Showing Arrangement of Cell Centers (Hollow Squares) and Cell Faces (Dotted Lines) in A 1D Computational Domain

The 1D computational domain over which the governing equation [Eq. (6.2)] is to be solved is illustrated in Fig. 6.2. As in the case of the FDM, we will continue to use directional O, E, W, N, and S notations, except that these notations will now be reserved for cell centers rather than nodes. In addition, the faces of the cell are denoted by e, w, n, and s. The first step in derivation of the finite volume equation for a given control volume is to integrate the governing equation analytically over the control volume or cell. For the cell i shown in Fig. 6.2, this yields

$$\int_{w}^{e} \frac{d}{dx}\left[\Gamma \frac{d\phi}{dx}\right] dx = -\int_{w}^{e} S_\phi \, dx, \tag{6.4}$$

where the integration has been performed from the western face of the cell to the eastern face since the cell is 1D in this particular case. Simplification of Eq. (6.4) yields

$$\left[\Gamma \frac{d\phi}{dx}\right]_{e,i} - \left[\Gamma \frac{d\phi}{dx}\right]_{w,i} = -S_i \Delta x_i, \tag{6.5}$$

where the subscript e,i denotes the eastern face of cell i, and so on. S_i denotes the average value of S_ϕ over the control volume of size Δx_i. Making use of Eq. (6.3), and then rearranging, Eq. (6.5) may rewritten as

$$J_{e,i} - J_{w,i} = S_i \Delta x_i. \tag{6.6}$$

Equation (6.6) represents a conservation equation, which states that the difference in flux between the two faces of the control volume is equal to what is produced within the cell. In essence, the governing PDE [Eq. (6.2)] has been recast into a form that has a clear physical interpretation. The starting point of any finite volume formulation is a flux balance equation similar to Eq. (6.6). This equation is eventually converted to its algebraic equivalent and solved. Since the discrete approximations are applied to this modified form of the governing equation rather than the original governing PDE, and the ensuing solution is that of the modified governing equation, such a solution is referred to as the weak form solution.

Equation (6.6) is applicable to any cell i. If explicitly written for $i = 1, 2, \ldots N_C$, we obtain the following set of equations:

$$
\begin{aligned}
J_{e,1} - J_{w,1} &= S_1 \Delta x_1, \\
J_{e,2} - J_{w,2} &= S_2 \Delta x_2, \\
&\vdots \\
J_{e,N_C} - J_{w,N_C} &= S_{N_C} \Delta x_{N_C}.
\end{aligned}
\tag{6.7}
$$

Since the governing differential equation contains a derivative of the quantity $\Gamma \, d\phi/dx$, it follows then that $\Gamma \, d\phi/dx$ must be continuous. Therefore, by Eq. (6.3), the flux across any cell face must be continuous, implying that $J_{e,i} = J_{w,i+1}$ for $i = 1, 2, \ldots N_C - 1$. Enforcing flux continuity in Eq. (6.7) and adding all N_C equations, we obtain

$$
J_{e,N_C} - J_{w,1} = S_1 \Delta x_1 + S_2 \Delta x_2 + \ldots + S_{N_C} \Delta x_{N_C} = \int_{x_L}^{x_R} S \, dx.
\tag{6.8}
$$

Equation (6.8) is similar to Eq. (6.6), except that it represents conservation or flux balance over the entire computational domain. While Eq. (6.6) shows that the conservation law is valid locally (in individual control volumes), Eq. (6.8) shows that it is also valid globally. Although not shown here, an equation similar to Eq. (6.8) can easily be derived for a multidimensional computational domain. Several salient features of the FVM are delineated by Eq. (6.8). First, it is worth noting that Eq. (6.8) or its multidimensional counterpart is true no matter what the shape or size of the computational domain. This is because the fluxes on the interior faces will always cancel out due to continuity (what goes into one cell through a face comes out of another cell through the same face). Second, Eq. (6.8) or its multidimensional counterpart is true no matter how many cells the computational domain is split into or what the size and shape of the constituent cells are. In other words, the global conservation property of the FVM is independent of the mesh.

Global conservation is, arguably, the most important property when it comes to the solution of PDEs that represents conservation laws. In the FDM, the governing PDE, which only satisfies the underlying conservation law on an infinitesimally small control volume, is enforced at nodes. That does not necessarily translate to enforcement of global conservation. As we shall see in due course, in the FDM, the accuracy of global conservation is strongly dependent on the mesh, and the boundary conditions being used. In contrast, in the FVM, as shown by Eqs (6.6) and (6.8), both local and global conservation are built into the method since conservation at both levels is satisfied independent of the mesh.

Returning to Eq. (6.5), the next task is to obtain approximations to the derivatives. For this, as before, we employ Taylor* series expansions. To enable comparison with

---

*Brook Taylor (1685–1731) was an English mathematician best known for the so-called Taylor's theorem and the Taylor series. The Taylor series was originally proposed by the Scottish mathematician James Gregory but formalized by Taylor in 1715. Taylor's work laid the foundation for what is known today as the calculus of finite differences. He was elected fellow of the Royal Society in 1712.

the finite difference equations derived in Chapter 2, we will consider a uniform mesh, as shown in Fig. 6.2. In order to derive an expression for the first derivative at the eastern face, we perform the following two Taylor series expansions:

$$\phi_{i+1} = \phi_e + \frac{\Delta x}{2}\frac{d\phi}{dx}\bigg|_e + \frac{1}{2}\left(\frac{\Delta x}{2}\right)^2\frac{d^2\phi}{dx^2}\bigg|_e + \frac{1}{6}\left(\frac{\Delta x}{2}\right)^3\frac{d^3\phi}{dx^3}\bigg|_e + ..., \tag{6.9a}$$

$$\phi_i = \phi_e - \frac{\Delta x}{2}\frac{d\phi}{dx}\bigg|_e + \frac{1}{2}\left(\frac{\Delta x}{2}\right)^2\frac{d^2\phi}{dx^2}\bigg|_e - \frac{1}{6}\left(\frac{\Delta x}{2}\right)^3\frac{d^3\phi}{dx^3}\bigg|_e + ..., \tag{6.9b}$$

where the grid spacing is defined as $\Delta x = L/N_C$, where $N_C = N - 1$ is now the number of cells. Subtracting Eq. (6.9b) from (6.9a), and rearranging, we obtain

$$\frac{d\phi}{dx}\bigg|_e = \frac{\phi_{i+1} - \phi_i}{\Delta x} - \frac{(\Delta x)^2}{24}\frac{d^3\phi}{dx^3}\bigg|_e + ... \tag{6.10}$$

Equation (6.10) is the central difference approximation to the first derivative. On a uniform mesh, as shown by Eq. (6.10), the approximation is second-order accurate. The error expression for the leading order error term is similar to the error expression derived earlier for approximation of the second derivative [see Eq. (2.10)]. In this case, however, it is proportional to the third derivative rather than the fourth, which stems from the fact that the governing equation has already been integrated once. A similar expression may also be derived for the first derivative at the western face. Substitution of Eq. (6.10) and similar expressions into Eq. (6.5) yields

$$\Gamma_{e,i}\frac{\phi_{i+1} - \phi_i}{\Delta x} - \Gamma_{w,i}\frac{\phi_i - \phi_{i-1}}{\Delta x} = -S_i\Delta x \qquad \forall i = 2,...,N_C - 1. \tag{6.11}$$

Equation (6.11) is valid only for the interior cells, and its range of validity excludes the two cells adjacent to the left and right boundaries. This is because the central difference approximation [Eq. (6.10)] cannot be used for the eastern face of the rightmost cell (cell N_C) and the western face of the first control volume (cell 1). The derivation of these two boundary flux terms require application of boundary conditions, which is deferred until the next section. If the transport coefficient, Γ, is assumed to be a constant and equal to unity to make the left-hand sides of Eqs (6.2) and (2.1) equivalent, Eq. (6.11), after dividing through by Δx, would reduce to

$$\frac{\phi_{i+1} - 2\phi_i - \phi_{i-1}}{(\Delta x)^2} = -S_i \qquad \forall i = 2,...,N_C \tag{6.12}$$

Equation (6.12) is identical to Eq. (2.11) with the notable exception of the minus sign on the right-hand side of Eq. (6.12), which was simply introduced to lend a physical meaning to the governing equation (it would not be necessary if S_ϕ were interpreted as the destruction rate of ϕ per unit volume instead of the production rate per unit volume). In summary, for a uniform mesh, the discrete algebraic equations for the interior nodes or cells obtained using either the FDM or the FVM are identical. It is this finding that often prompts researchers and authors to contend that the two methods are synonymous. In order to rebuff this contention, one must answer the question: "what makes the two methods different?" First of all, the algebraic equations are identical only for a uniform mesh. Most importantly, as we shall see shortly, the treatment of boundary conditions is entirely different in the two methods, and therein lies their major difference.

Returning to the discrete form of the finite volume equation, namely Eq. (6.11), we note that the values of Γ at the cell faces still need to be determined. In the FVM, information is only stored at cell centers. Therefore, the face values of Γ are not readily available, and are traditionally determined via interpolation. In the case of the uniform mesh used here, the face values may be written as

$$\Gamma_{e,i} = \frac{\Gamma_i + \Gamma_{i+1}}{2}, \quad \Gamma_{w,i} = \frac{\Gamma_i + \Gamma_{i-1}}{2}. \tag{6.13}$$

In the case of a nonuniform and/or nonorthogonal mesh, inverse distance–weighted interpolation must be used to obtain face values from surrounding cell values. This type of interpolation is discussed in detail in Section 7.3.2. Furthermore, if Γ is discontinuous at the face, as may happen in reality at material interfaces, special treatment is necessary to obtain the face value of Γ so that flux continuity is not violated. This special scenario and the associated derivation are also presented in the latter half of Section 7.3.2. Substitution of Eq. (6.13) into Eq. (6.11), followed by rearrangement, yields

$$\left(\frac{\Gamma_i + \Gamma_{i+1}}{2\Delta x} + \frac{\Gamma_i + \Gamma_{i-1}}{2\Delta x} \right) \phi_i - \left(\frac{\Gamma_i + \Gamma_{i+1}}{2\Delta x} \right) \phi_{i+1}$$
$$- \left(\frac{\Gamma_i + \Gamma_{i-1}}{2\Delta x} \right) \phi_{i-1} = S_i \Delta x \quad \forall i = 2, ..., N_C - 1. \tag{6.14}$$

The treatment of boundary conditions in the context of the FVM is discussed next.

6.2 APPLICATION OF BOUNDARY CONDITIONS

In this section, we discuss the treatment of boundary conditions in the context of the FVM. As in Section 2.3, the coverage will include the three main canonical forms, namely Dirichlet, Neumann, and Robin boundary conditions.

One of the salient features of the FVM is that in this method, there is no cell center located at the boundary. The mean value theorem stipulates that the average value of a function must correspond to a value of the function corresponding to a point inside the interval of averaging. Therefore, storing averaged information at a location that is not inside the control volume is invalid. What this also implies is that the boundary condition is never applied directly to a cell center (or to an unknown), which is in stark contrast with the FDM where Dirichlet boundary conditions are directly applied to the boundary nodes (see Section 2.3.1).

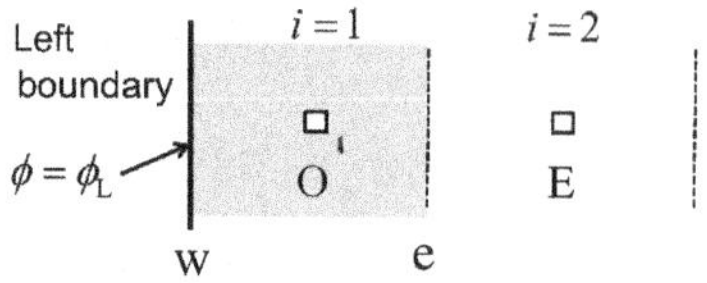

FIGURE 6.3 Schematic Representation of a Boundary Cell with the Dirichlet Boundary Condition Applied to the Left Boundary

Another important feature of the FVM is that the governing equation is satisfied in the true open interval that constitutes the computational domain – (x_L, x_R) in the 1D case. In Section 2.3.1, it was pointed out that in the FDM, if Dirichlet boundary conditions are applied to the boundary nodes, the governing equation cannot be satisfied simultaneously at the boundary nodes. In the 1D case, for example, the governing equation is actually satisfied in the interval $[x_L + \Delta x, x_R - \Delta x]$, and is, therefore, mesh dependent. In the FVM, on the other hand, since integration is performed over each control volume that constitutes the computational domain, the governing equation is, in fact, applied, albeit in an indirect manner, to the extreme edges of the computational domain, i.e., in the open interval in question. This will become clear in the discussion to follow.

We begin our discussion with Dirichlet boundary conditions. In the FVM, since no cell centers exist on the boundaries, the boundary condition, irrespective of type, is applied to the cell faces. Let us consider the leftmost cell ($i = 1$) shown in Fig. 6.2, the stencil of which is shown in Fig. 6.3. In order to derive the discrete equation for cell O, we begin with the general finite volume equation given by Eq. (6.5). The specific objective, in this case, is to derive a discrete approximation for the flux term $\left[\Gamma d\phi/dx\right]_{w,1}$, which may be written as $\Gamma_{w,1} d\phi/dx\big|_{w,1}$. For simplicity of notation, we will drop the subscript "1" in subsequent equations pertaining to this derivation.

In order to obtain an expression for the derivative at the western face, we first perform a Taylor series expansion of O about w, yielding

$$\phi_O = \phi_w + \frac{\Delta x}{2} \frac{d\phi}{dx}\bigg|_w + \frac{1}{2}\left(\frac{\Delta x}{2}\right)^2 \frac{d^2\phi}{dx^2}\bigg|_w + \frac{1}{6}\left(\frac{\Delta x}{2}\right)^3 \frac{d^3\phi}{dx^3}\bigg|_w + ..., \qquad (6.15)$$

which may be rearranged to write

$$\frac{d\phi}{dx}\bigg|_w = \frac{2}{\Delta x}(\phi_O - \phi_w) - \frac{\Delta x}{4}\frac{d^2\phi}{dx^2}\bigg|_w + ... \qquad (6.16)$$

Unfortunately, the approximation shown in Eq. (6.16) is only first-order accurate, and is not suitable for use in conjunction with the approximation of the derivative on the eastern face [Eq. (6.10)] which is second-order accurate. In order to derive a second-order accurate approximation for the derivative, we employ another Taylor series expansion:

$$\phi_E = \phi_w + \frac{3\Delta x}{2}\frac{d\phi}{dx}\bigg|_w + \frac{1}{2}\left(\frac{3\Delta x}{2}\right)^2\frac{d^2\phi}{dx^2}\bigg|_w + \frac{1}{6}\left(\frac{3\Delta x}{2}\right)^3\frac{d^3\phi}{dx^3}\bigg|_w + ... \qquad (6.17)$$

Multiplying Eq. (6.15) by 9, and subtracting Eq. (6.17) from the result yields

$$9\phi_O - \phi_E = 8\phi_w + 3\Delta x\frac{d\phi}{dx}\bigg|_w - \frac{3}{8}(\Delta x)^3\frac{d^3\phi}{dx^3}\bigg|_w + ..., \qquad (6.18)$$

which, upon rearrangement, yields

$$\frac{d\phi}{dx}\bigg|_w = \frac{9\phi_O - \phi_E - 8\phi_w}{3\Delta x} + \frac{1}{8}(\Delta x)^2\frac{d^3\phi}{dx^3}\bigg|_w + ... = \frac{9\phi_O - \phi_E - 8\phi_L}{3\Delta x} + \frac{1}{8}(\Delta x)^2\frac{d^3\phi}{dx^3}\bigg|_w + ...$$
$$(6.19)$$

In the final expression in Eq. (6.19), the Dirichlet boundary condition at the western (left) face, namely $\phi_w = \phi_L$, has been used. As expected, with the use of two Taylor series expansions, the approximation for the derivative is second-order accurate.

The approximation of the transport coefficient, Γ, at the western boundary face requires a slightly different approach since there is no boundary condition for it. A first-order approximation (extrapolation formula) may be derived using a single Taylor series expansion, as follows:

$$\Gamma_O = \Gamma_w + \frac{\Delta x}{2}\frac{d\Gamma}{dx}\bigg|_w + ..., \qquad (6.20)$$

yielding $\Gamma_w \approx \Gamma_O$ with an error that scales linearly (first-order) with the grid spacing. This represents a simple extrapolation of the cell-center value of Γ to the boundary. Likewise, a second-order approximation may be derived using two Taylor series expansions (using O and E), yielding

$$\Gamma_w \approx \frac{3\Gamma_O - \Gamma_E}{2}. \qquad (6.21)$$

It is common practice to use a first-order approximation for Γ since properties are generally weakly varying functions of space or ϕ. Of course, this is quite problem dependent, and the reader is advised to assess the nature of the functional dependence of Γ prior to deciding on an extrapolation order. It is worth noting that the interpolation formula, given by Eq. (6.13), for the interior faces, is actually second-order accurate. For the sake of consistency, we will continue to use second-order accurate approximations in our derivation. Substitution of Eqs (6.10), (6.13), (6.19), and (6.21) into Eq. (6.5) for the cell $i = 1$ yields

$$\frac{\Gamma_1 + \Gamma_2}{2}\left(\frac{\phi_2 - \phi_1}{\Delta x}\right) - \frac{3\Gamma_1 - \Gamma_2}{2}\left(\frac{9\phi_1 - \phi_2 - 8\phi_L}{3\Delta x}\right) = -S_1 \Delta x. \tag{6.22}$$

Eq. (6.22) may be rearranged to place known and unknown quantities on different sides of the equation to write

$$\frac{\Gamma_1 + \Gamma_2}{2}\left(\frac{\phi_2 - \phi_1}{\Delta x}\right) - \frac{3\Gamma_1 - \Gamma_2}{2}\left(\frac{9\phi_1 - \phi_2}{3\Delta x}\right) = -S_1 \Delta x - \frac{3\Gamma_1 - \Gamma_2}{2}\left(\frac{8\phi_L}{3\Delta x}\right). \tag{6.23}$$

A similar procedure may be used to derive the discrete equation for the rightmost cell $(i = N_C)$, and is written as

$$\frac{3\Gamma_{N_C} - \Gamma_{N_C - 1}}{2}\left(\frac{-9\phi_{N_C} + \phi_{N_C - 1}}{3\Delta x}\right) - \frac{\Gamma_{N_C} + \Gamma_{N_C - 1}}{2}\left(\frac{\phi_{N_C} - \phi_{N_C - 1}}{\Delta x}\right)$$
$$= -S_{N_C} \Delta x - \frac{3\Gamma_{N_C} - \Gamma_{N_C - 1}}{2}\left(\frac{8\phi_R}{3\Delta x}\right), \tag{6.24}$$

where ϕ_R denotes the prescribed value of ϕ at the right boundary. Equations (6.14), (6.23), and (6.24) represent the complete set of N_C finite volume equations necessary to determine the N_C cell-center values of ϕ. Prior to writing these equations in tridiagonal matrix form, we will invoke the assumption that $\Gamma = 1$ everywhere so that the resulting linear system can be compared to the same system yielded by the finite difference formulation. For the finite difference formulation, when two Dirichlet boundary conditions are used, the resulting tridiagonal matrix may be written as [following Eq. (2.39); with Dirichlet boundary conditions at both ends and changing the sign of S_ϕ, for reasons mentioned earlier]:

$$\begin{bmatrix} 1 & 0 & \cdots & \cdots & \cdots & & 0 \\ \frac{1}{(\Delta x)^2} & \frac{-2}{(\Delta x)^2} & \frac{1}{(\Delta x)^2} & 0 & \cdots & & 0 \\ & \ddots & \ddots & \ddots & & & \\ \cdots & 0 & \frac{1}{(\Delta x)^2} & \frac{-2}{(\Delta x)^2} & \frac{1}{(\Delta x)^2} & 0 & \cdots & 0 \\ & & 0 & \frac{1}{(\Delta x)^2} & \frac{-2}{(\Delta x)^2} & \frac{1}{(\Delta x)^2} & 0 \\ & & \cdots & 0 & \frac{1}{(\Delta x)^2} & \frac{-2}{(\Delta x)^2} & \frac{1}{(\Delta x)^2} & 0 & \cdots \\ & & & & \ddots & \ddots & \ddots & \\ 0 & & & \cdots & 0 & \frac{1}{(\Delta x)^2} & \frac{-2}{(\Delta x)^2} & \frac{1}{(\Delta x)^2} \\ 0 & & & & & \cdots & 0 & 1 \end{bmatrix} \begin{bmatrix} \phi_1 \\ \phi_2 \\ \vdots \\ \phi_{i-1} \\ \phi_i \\ \phi_{i+1} \\ \vdots \\ \phi_{N-1} \\ \phi_N \end{bmatrix} = \begin{bmatrix} \phi_L \\ -S_2 \\ \vdots \\ -S_{i-1} \\ -S_i \\ -S_{i+1} \\ \vdots \\ -S_{N-1} \\ \phi_R \end{bmatrix}.$$
$$\tag{6.25}$$

In contrast, for the FVM, the resulting tridiagonal matrix, after dividing Eqs (6.14), (6.23), and (6.24) by Δx, may be written as

$$
\begin{bmatrix}
\frac{-4}{(\Delta x)^2} & \frac{4}{3(\Delta x)^2} & 0 & \cdots & \cdots & & & & 0 \\
\frac{1}{(\Delta x)^2} & \frac{-2}{(\Delta x)^2} & \frac{1}{(\Delta x)^2} & 0 & \cdots & & & & 0 \\
& \ddots & \ddots & \ddots & & & & & \\
\cdots & 0 & \frac{1}{(\Delta x)^2} & \frac{-2}{(\Delta x)^2} & \frac{1}{(\Delta x)^2} & 0 & \cdots & & 0 \\
& & 0 & \frac{1}{(\Delta x)^2} & \frac{-2}{(\Delta x)^2} & \frac{1}{(\Delta x)^2} & 0 & & \\
& \cdots & & 0 & \frac{1}{(\Delta x)^2} & \frac{-2}{(\Delta x)^2} & \frac{1}{(\Delta x)^2} & 0 & \cdots \\
& & & & \ddots & \ddots & \ddots & & \\
0 & & & \cdots & 0 & \frac{1}{(\Delta x)^2} & \frac{-2}{(\Delta x)^2} & \frac{1}{(\Delta x)^2} & \\
0 & & & & \cdots & 0 & \frac{4}{3(\Delta x)^2} & \frac{-4}{(\Delta x)^2} &
\end{bmatrix}
\begin{bmatrix}
\phi_1 \\ \phi_2 \\ \vdots \\ \phi_{i-1} \\ \phi_i \\ \phi_{i+1} \\ \vdots \\ \phi_{N_C-1} \\ \phi_{N_C}
\end{bmatrix}
=
\begin{bmatrix}
-S_1 - \frac{8}{3(\Delta x)^2}\phi_L \\
-S_2 \\
\vdots \\
-S_{i-1} \\
-S_i \\
-S_{i+1} \\
\vdots \\
-S_{N_C-1} \\
-S_{N_C} - \frac{8}{3(\Delta x)^2}\phi_R
\end{bmatrix}.
$$
(6.26)

The only difference between the coefficient matrices shown in Eq. (6.25) and Eq. (6.26) is in the first and last rows, and as discussed earlier, this is due to the difference in the way the boundary conditions in the two methods are applied. However, as we shall see in Section 6.4, this relatively minor difference has serious implications on the solution. It is worth emphasizing that both methods are second-order accurate, and therefore, the solution yielded by the two methods are expected to be of comparable accuracy, at least in principle. Next, we discuss the implementation of Neumann boundary conditions in the context of the FVM.

In the case of the Neumann boundary condition, the derivative normal to the boundary is known. If applied to the left end in the setup shown in Fig. 6.3, we may write

$$
\left.\frac{d\phi}{dx}\right|_{x=x_L} = \left.\frac{d\phi}{dx}\right|_{w,1} = J_L.
$$
(6.26)

Substituting this boundary condition, along with Eqs (6.10), (6.13), and (6.21) into Eq. (6.5), for cell $i = 1$, we obtain

$$
\frac{\Gamma_1 + \Gamma_2}{2}\left(\frac{\phi_2 - \phi_1}{\Delta x}\right) - \frac{3\Gamma_1 - \Gamma_2}{2}J_L = -S_1 \Delta x,
$$
(6.27)

which upon rearrangement yields

$$
\frac{\Gamma_1 + \Gamma_2}{2}\left(\frac{\phi_2 - \phi_1}{\Delta x}\right) = -S_1 \Delta x + \frac{3\Gamma_1 - \Gamma_2}{2}J_L.
$$
(6.28)

Physically, the Neumann boundary condition represents a prescribed flux boundary condition in which $-\Gamma\, d\phi/dx$ is prescribed at the boundary. For example, in heat transfer applications, the heat flux, written as $-\kappa\, dT/dx$ may be prescribed at the boundaries, where κ is the thermal conductivity, and T is the temperature. In such scenarios, the temperature at the boundary (ϕ in the general case) is an unknown. In many applications, the value of ϕ at the boundary may be desirable, as in the case of heat transfer where temperature of the boundary is a quantity of practical interest.

The computation of ϕ at the boundary constitutes a postprocessing step that is performed after the solution of ϕ at all cell centers has been obtained by solving the system of algebraic equations. For example, to compute ϕ at the left boundary, we make use of the expression for the first derivative given by Eq. (6.19) to write

$$\left.\frac{d\phi}{dx}\right|_w = \frac{9\phi_1 - \phi_2 - 8\phi_L}{3\Delta x} = J_L, \tag{6.29}$$

which may be rearranged to write

$$\phi_L = \frac{9\phi_1 - \phi_2 - 3J_L\Delta x}{8}. \tag{6.30}$$

Since this is a postprocessing step, the values of ϕ_1 and ϕ_2 are already known at this juncture, and the value of ϕ_L can, therefore, be computed using Eq. (6.30).

The Robin boundary condition may be implemented in a manner similar to Neumann boundary conditions. If applied to the western face of the leftmost control volume, the Robin boundary condition may be written as

$$\alpha\phi_{w,1} + \beta\left.\frac{d\phi}{dx}\right|_{w,1} = \gamma. \tag{6.31}$$

Rearrangement of Eq. (6.31), followed by substitution into Eqs (6.15) and (6.17) yields

$$\phi_O = \frac{\gamma}{\alpha} - \frac{\beta}{\alpha}\left.\frac{d\phi}{dx}\right|_w + \frac{\Delta x}{2}\left.\frac{d\phi}{dx}\right|_w + \frac{1}{2}\left(\frac{\Delta x}{2}\right)^2\left.\frac{d^2\phi}{dx^2}\right|_w + \frac{1}{6}\left(\frac{\Delta x}{2}\right)^3\left.\frac{d^3\phi}{dx^3}\right|_w + \cdots, \tag{6.32a}$$

$$\phi_E = \frac{\gamma}{\alpha} - \frac{\beta}{\alpha}\left.\frac{d\phi}{dx}\right|_w + \frac{3\Delta x}{2}\left.\frac{d\phi}{dx}\right|_w + \frac{1}{2}\left(\frac{3\Delta x}{2}\right)^2\left.\frac{d^2\phi}{dx^2}\right|_w + \frac{1}{6}\left(\frac{3\Delta x}{2}\right)^3\left.\frac{d^3\phi}{dx^3}\right|_w + \cdots \tag{6.32b}$$

Multiplying Eq. (6.32a) by 9, and subtracting Eq. (6.32b) from the result, yields

$$9\phi_O - \phi_E = \frac{8\gamma}{\alpha} + \left(3\Delta x - \frac{8\beta}{\alpha}\right)\left.\frac{d\phi}{dx}\right|_w - \frac{3}{8}(\Delta x)^3\left.\frac{d^3\phi}{dx^3}\right|_w + \cdots, \tag{6.33}$$

which upon rearrangement results in

$$\left.\frac{d\phi}{dx}\right|_w = \frac{9\phi_O - \phi_E - \dfrac{8\gamma}{\alpha}}{3\Delta x - \dfrac{8\beta}{\alpha}} + \frac{\dfrac{3}{8}(\Delta x)^3}{\left(3\Delta x - \dfrac{8\beta}{\alpha}\right)}\left.\frac{d^3\phi}{dx^3}\right|_w + \cdots \tag{6.34}$$

Although not obvious, the leading error term shown in Eq. (6.34) is of second order since it contains $(\Delta x)^3$ in the numerator and (Δx) in the denominator. Substitution of Eq. (6.34), along with Eqs (6.10), (6.13), and (6.21) into Eq. (6.5), for cell $i = 1$, yields

$$\frac{\Gamma_1 + \Gamma_2}{2}\left(\frac{\phi_2 - \phi_1}{\Delta x}\right) - \frac{3\Gamma_1 - \Gamma_2}{2}\left(\frac{9\phi_1 - \phi_2 - \dfrac{8\gamma}{\alpha}}{3\Delta x - \dfrac{8\beta}{\alpha}}\right) = -S_1 \Delta x. \tag{6.35}$$

Rearrangement of Eq. (6.35) results in the desired discrete finite volume equation for the leftmost node:

$$\frac{\Gamma_1 + \Gamma_2}{2}\left(\frac{\phi_2 - \phi_1}{\Delta x}\right) - \frac{3\Gamma_1 - \Gamma_2}{2}\left(\frac{9\alpha\phi_1 - \alpha\phi_2}{3\alpha\Delta x - 8\beta}\right) = -S_1 \Delta x - \frac{3\Gamma_1 - \Gamma_2}{2}\left(\frac{8\gamma}{3\alpha\Delta x - 8\beta}\right). \tag{6.36}$$

In this case, the value of ϕ at the left boundary can be computed by substituting Eq. (6.19) into Eq. (6.31) to obtain

$$\alpha\phi_L + \beta\frac{9\phi_1 - \phi_2 - 8\phi_L}{3\Delta x} = \gamma, \tag{6.37}$$

which upon rearrangement yields

$$\phi_L = \left(\gamma - \beta\frac{9\phi_1 - \phi_2}{3\Delta x}\right) \bigg/ \left(\alpha - \frac{8\beta}{3\Delta x}\right). \tag{6.38}$$

At this juncture, the implementation of the three canonical types of boundary conditions in the context of the FVM has been discussed. Prior to extending the finite volume formulation to multidimensional problems, we consider a simple 1D example in an effort to highlight the concepts discussed in the preceding two sections, especially from the perspective of conservation.

EXAMPLE 6.1

In this example, we solve the following ordinary differential equation:

$\dfrac{d}{dx}\left[\Gamma\dfrac{d\phi}{dx}\right] = -S_\phi$ with $\Gamma = 1$ and $S_\phi = -\cos x$, and the following boundary conditions: $\phi(0) = \phi(1) = 0$. Both the FDM and the FVM are considered, and the results obtained using the two methods are compared and contrasted.

The analytical solution to the above system is given by $\phi(x) = 1 - \cos x + [\cos(1) - 1]x$. The fluxes at the two boundaries are given by $J_L = -\Gamma\,d\phi/dx\big|_{x=0} = 1 - \cos(1)$, and $J_R = -\Gamma\,d\phi/dx\big|_{x=1} = -\sin(1) - [\cos(1) - 1]$, where the fluxes, as given by the above definitions, are in the positive x direction. This means that if the flux value is positive, it is going from left to right, and if it is negative, it is going from right to left. The net production inside the computational domain is given by $S = \int_0^1 S_\phi\,dx = -\sin(1)$. The net imbalance in the computational domain by fundamental conservation principle at steady state is given by imbalance = in – out + produced. If the imbalance is positive, it means that more is coming into the computational domain than is leaving and vice versa. Mathematically, the imbalance is written as $I = J_L - J_R + S$. Using the expressions derived above, the imbalance is $I = 1 - \cos(1) - (-\sin(1) - [\cos(1) - 1]) - \sin(1) = 0$, a result that must be satisfied to ensure global conservation.

First, we consider computations using the FDM. The system of tridiagonal equations to be solved is shown in Eq. (6.25). The fluxes are computed using Eq. (2.20) and its counterpart equation for the right boundary. The net production is computed using the trapezoidal rule of integration [see Eq. (8.33)], which yields $S \approx (\Delta x / 2)[S_1 + S_N] + \Delta x \sum_{i=2}^{N-1} S_i$, where N is the total number of nodes. The two tables below show the solution, $\phi(x)$, with $N = 5$ and $N = 9$, and also the flux summaries in the two cases. The analytical results are also shown for comparison in both tables.

FDM solution with $N = 5$			
x	$\phi_{analytical}$	ϕ_{FDM}	$\phi_{analytical} - \phi_{FDM}$
0.25	−0.083837	−0.084275	0.000438
0.50	−0.107431	−0.107993	0.000561
0.75	−0.076462	−0.076862	0.000399

FDM fluxes with $N = 5$		
	Analytical	FDM
J_L	0.459698	0.458214
J_R	−0.381773	−0.398908
S	−0.841471	−0.837084
I	0.000000	0.020037

FDM solution with $N = 9$			
x	$\phi_{analytical}$	ϕ_{FDM}	$\phi_{analytical} - \phi_{FDM}$
0.25	−0.083837	−0.083946	0.000109
0.50	−0.107431	−0.107571	0.000140
0.75	−0.076462	−0.076562	0.000100

FDM fluxes with $N = 9$		
	Analytical	FDM
J_L	0.459698	0.459809
J_R	−0.381773	−0.386372
S	−0.841471	−0.840375
I	0.000000	0.005806

From the data presented in the tables above, two important points need to be brought to light: (a) the FDM solution, $\phi(x)$, is quite accurate even with just five nodes, and the error reduces exactly by a factor of four when the grid spacing is halved, as would be expected of a second-order central difference scheme, and (b) for both mesh sizes, there is a net flux imbalance, and the imbalance decreases as the mesh is refined.

Next, we consider computations using the FVM. The system of tridiagonal equations to be solved is shown in Eq. (6.26). The fluxes are computed using Eq. (6.19) and its counterpart equation for the right boundary. The net production is computed using $S \approx \sum_{i=1}^{N_C} S_i \Delta x$, where $N_C = N-1$ is the total number of cells. The two tables below show the solution, $\phi(x)$, with $N_C = 4$ and $N_C = 8$, and also the flux summaries in the two cases. The analytical results are also shown for comparison in both tables. In this case, when the mesh is halved, the cell centers of the coarse and fine mesh are not colocated, and thus, the $\phi(x)$ shown for $N_C = 4$ and $N_C = 8$ are at different locations.

FVM solution with $N_c = 4$			
x	$\phi_{analytical}$	ϕ_{FVM}	$\phi_{analytical} - \phi_{FVM}$
0.125	−0.049660	−0.049903	0.000243
0.375	−0.102894	−0.103200	0.000306
0.625	−0.098274	−0.098341	0.000066
0.875	−0.043232	−0.042796	−0.000437

FVM fluxes with $N_c = 4$		
	Analytical	**FVM**
J_L	0.459698	0.461238
J_R	−0.381773	−0.382429
S	−0.841471	−0.843666
I	0.000000	0.000000

FVM solution with $N_c = 8$			
x	$\phi_{analytical}$	ϕ_{FVM}	$\phi_{analytical} - \phi_{FVM}$
0.0625	−0.026779	−0.026813	0.000034
0.3125	−0.095223	−0.095320	0.000097
0.5625	−0.104504	−0.104587	0.000083
0.8125	−0.061190	−0.061190	0.000000

FVM fluxes with $N_c = 8$		
	Analytical	**FVM**
J_L	0.459698	0.460191
J_R	−0.381773	−0.381828
S	−0.841471	−0.842019
I	0.000000	0.000000

Inspection of the FVM results shown in the tables above, along with comparison with the corresponding FDM results, illustrates several important points. First, as in the case of the FDM, the error is small for the FVM even with just four cells (five nodes), and the error scales approximately by a factor of four when the grid size is halved, in keeping with the second-order central difference scheme. The largest error for the FDM with five nodes is 0.000561, while for the FVM, it is 0.000437, which makes the two methods comparable when it comes to the accuracy of $\phi(x)$. The important difference in the two methods is exhibited by the flux values. While the individual fluxes in both cases (FDM versus FVM) have comparable accuracy when compared with the analytical solution, in the FVM, the fluxes are always conserved globally irrespective of the mesh. In contrast, as seen earlier, for the FDM, they are not conserved, and the imbalance (error) is mesh dependent – the error decreases with mesh refinement. While this extremely important conservation property of the FVM was discussed earlier from a theoretical standpoint, this example clearly illustrates that, indeed, the conservation property is built into the FVM.

6.3 FLUX SCHEMES FOR ADVECTION–DIFFUSION

Thus far, our discussion has been limited to Poisson-type equations that are representative of conservation laws in which diffusion is the only mechanism of transport. One of the most prolific uses of the FVM are for problems that involve fluid flow, in which, in addition to diffusive transport, there is also transport due to bulk fluid motion – a phenomenon known as *advection* [2–4].

If the advective flux is also considered, the net flux is a sum of the diffusive and advective fluxes, and Eq. (6.3) assumes the following form in the 1D case:

$$J = J_A + J_D = \rho u \phi - \Gamma \frac{d\phi}{dx}, \tag{6.39}$$

where J_A and J_D are advective and diffusive components of the flux, respectively. Since advection is caused by the bulk motion of the medium, it is written as the product of the mass flow rate per unit area, ρu, and the advected scalar ϕ, where ρ is the density of the medium, and u is its velocity component in the x direction. The governing equation, the so-called advection–diffusion equation, becomes [*cf.* Eq. (6.2)]

$$\frac{d}{dx}\left[\rho u \phi - \Gamma \frac{d\phi}{dx}\right] = S_\phi, \tag{6.40}$$

where the source term S_ϕ is to be interpreted as the production rate of ϕ per unit volume. The corresponding finite volume equation is derived, as before, by integrating the governing equation over a 1D control volume of size $\Delta x_i = x_{e,i} - x_{w,i}$, such that

$$\int_{w,i}^{e,i} \frac{d}{dx}\left[\rho u \phi - \Gamma \frac{d\phi}{dx}\right] dx = \int_{w,i}^{e,i} S_\phi \, dx. \tag{6.41}$$

Performing the integration, we obtain

$$\left[\rho u \phi - \Gamma \frac{d\phi}{dx}\right]_{e,i} - \left[\rho u \phi - \Gamma \frac{d\phi}{dx}\right]_{w,i} = J_{e,i} - J_{w,i} = S_i \Delta x_i, \tag{6.42}$$

which is essentially the same equation as Eq. (6.6) with the exception that in this case, the flux includes both advective and diffusive fluxes. Nonetheless, the fact that Eqs (6.6) and (6.42) are identical conveys the important message that the FVM, in essence, results in a flux conservation equation. The treatment of the terms $\Gamma d\phi/dx|_{e,i}$ and $\Gamma d\phi/dx|_{w,i}$ have already been discussed in Section 6.2. Here, we only discuss treatment of the advective flux terms, such as $\rho u \phi|_{e,i}$. First, we rewrite the advective flux as $\rho u|_{e,i} \phi_{e,i}$. The mass flux at the face, $\rho u|_{e,i}$, is computed by an accompanying flow solver. If a staggered mesh [3] is used, no interpolation is necessary, and the mass flux at the face is directly available from the flow solver. If, on the other hand, a colocated mesh is used, the mass flux at the face is computed by the flow solver from cell-center velocities and densities using a complex interpolation procedure, such as pressure-weighted interpolation [5,6] or momentum-weighted interpolation [7]. A discussion of the exact procedure to do this falls in the realm of computational fluid dynamics, and is well beyond the scope of this text. For the present discussion, we will assume, without loss of generalization, that the mass flux at the face, $\rho u|_{e,i}$, is known. The problem of expressing the advective flux of ϕ in terms of cell-center values of ϕ then boils down to expressing $\phi_{e,i}$ in terms of the cell-center values of ϕ.

To this end, the central difference approximation may be used, which may be derived by adding Eqs (6.9a) and (6.9b) to yield

$$\phi_{e,i} = \frac{\phi_{i+1} + \phi_i}{2} - \frac{1}{2}\left(\frac{\Delta x}{2}\right)^2 \frac{d^2\phi}{dx^2}\bigg|_e + \dots \tag{6.43}$$

for a uniform mesh with grid spacing Δx. Substituting Eq. (6.43) and a similar approximation for $\phi_{w,i}$, along with Eq. (6.10), into Eq. (6.42), we obtain

$$\rho u|_{e,i}\left(\frac{\phi_{i+1}+\phi_i}{2}\right) - \Gamma_{e,i}\left(\frac{\phi_{i+1}-\phi_i}{\Delta x}\right) - \rho u|_{w,i}\left(\frac{\phi_{i-1}+\phi_i}{2}\right) + \Gamma_{w,i}\left(\frac{\phi_i-\phi_{i-1}}{\Delta x}\right) = S_i\Delta x. \tag{6.44}$$

Rearrangement of Eq. (6.44) yields

$$\left(\frac{\Gamma_{e,i}}{\Delta x} + \frac{\Gamma_{w,i}}{\Delta x} + \frac{1}{2}(\rho u|_{e,i} - \rho u|_{w,i})\right)\phi_i - \left(\frac{\Gamma_{e,i}}{\Delta x} - \frac{1}{2}\rho u|_{e,i}\right)\phi_{i+1} \\ - \left(\frac{\Gamma_{w,i}}{\Delta x} + \frac{1}{2}\rho u|_{w,i}\right)\phi_{i-1} = S_i\Delta x \tag{6.45}$$

In this 1D case, mass conservation stipulates that $\rho u|_{e,i} = \rho u|_{w,i}$, which we will, henceforth, denote by ρu. Further, for simplicity, we will assume that Γ is a constant. As we shall see shortly, the assumption of constant Γ does not alter the conclusions of the derivation being conducted. Under the aforementioned assumptions, Eq. (6.45) reduces to

$$\left(\frac{2\Gamma}{\Delta x}\right)\phi_i - \left(\frac{\Gamma}{\Delta x} - \frac{1}{2}\rho u\right)\phi_{i+1} - \left(\frac{\Gamma}{\Delta x} + \frac{1}{2}\rho u\right)\phi_{i-1} = S_i\Delta x, \tag{6.46}$$

As a final step, we multiply Eq. (6.46) by $\Delta x/\Gamma$, to yield

$$2\phi_i - \left(1 - \frac{Pe_\Delta}{2}\right)\phi_{i+1} - \left(1 + \frac{Pe_\Delta}{2}\right)\phi_{i-1} = S_i\frac{(\Delta x)^2}{\Gamma}, \tag{6.47}$$

where $Pe_\Delta = \rho u\Delta x/\Gamma$ is a nondimensional number known as the *grid Peclet number*. As evident from its definition, it is a measure of the relative strength of advection to diffusion. When $Pe_\Delta \gg 1$, advection dominates over diffusion, and when $Pe_\Delta \ll 1$, diffusion dominates over advection. An important result is manifested when $Pe_\Delta > 2$ or if $Pe_\Delta < -2$. In the former case, the magnitude of the link coefficient for ϕ_{i-1} is greater than 2, while in the latter case, the magnitude of the link coefficient for ϕ_{i+1} is greater than 2. In both cases, the magnitude of the diagonal is less than the magnitude of the sum of the off-diagonals of the coefficient matrix, and the Scarborough criterion, given in Eq. (3.18), is violated. This implies that for $|Pe_\Delta| > 2$, Eq. (6.47) may not converge if an iterative solver is used to solve it. This finding was actually confirmed in Example 4.3 in the context of the FDM. Although the present conclusion is being drawn for the 1D advection–diffusion equation, the same holds true for the multidimensional advection–diffusion equation, as well. This is one reason why the pure central difference approximation is rarely used for the treatment of the advective flux in the context of numerical solution of the advection–diffusion equation [3].

6.3.1 UPWIND DIFFERENCE SCHEME

The reason cited in the preceding section for the unsuitability of the central difference approximation for the treatment of advection is purely based on stability considerations. However, this should not be the only reason to discard the central difference approximation because it has the notable advantage that it is second-order accurate. Stability can always be dealt with by using a more implicit iterative solver or by using artificial means such as by using an inertial damping factor (Section 4.2.3).

Another reason why the central difference approximation is not considered a viable choice for the treatment of advection is based on the consideration of the physics of transport processes. Physically, diffusive fluxes follow the gradient of the dependent variable, e.g., heat conduction occurs from high to low temperatures. In contrast, advective fluxes follow the flow. Let us consider a situation where we have pure advection (information carried by bulk motion of the fluid only) and no diffusion. In such a scenario, if the flow is from left to right (eastward), then, it is likely that the value of ϕ at the eastern face of a cell will be heavily influenced by the value of ϕ at the cell center, i.e., $\phi_{e,i} \approx \phi_i$. Conversely, if the flow is from right to left (westward), $\phi_{e,i} \approx \phi_{i+1}$. In other words, the face value is almost equal to the value at the upstream or upwind cell. This type of a direction-dependent fluxing scheme is known as the upwind or upstream differencing scheme. Mathematically, it may be written as

$$\phi_{e,i} = \begin{cases} \phi_i & \text{if } \rho u|_{e,i} > 0 \\ \phi_{i+1} & \text{if } \rho u|_{e,i} < 0 \end{cases}, \tag{6.48a}$$

$$\phi_{w,i} = \begin{cases} \phi_i & \text{if } \rho u|_{w,i} < 0 \\ \phi_{i-1} & \text{if } \rho u|_{w,i} > 0 \end{cases}. \tag{6.48b}$$

When the face velocity is zero, the advection term is zero, and requires no consideration. The upwind difference scheme shown in Eq. (6.48) is only first-order accurate. This is easily shown by performing a Taylor series expansion of the value of ϕ at the cell center about its value at the face, as shown in Eq. (6.9b). This results in a leading truncation error term equal to $-(\Delta x/2)d\phi/dx|_e$. Hence, such an upwind difference scheme is known as a first-order upwind difference scheme. Compared to the central difference scheme, its accuracy is poor. The first-order accuracy of the scheme manifests itself in smearing out regions with sharp gradients in the solution, as will be demonstrated through an example shortly. Substitution of Eqs (6.48) and (6.10) into Eq. (6.42), followed by application of mass conservation, i.e., $\rho u|_{e,i} = \rho u|_{w,i} = \rho u$, yields

$$\rho u \phi_i - \Gamma_{e,i}\left(\frac{\phi_{i+1} - \phi_i}{\Delta x}\right) - \rho u \phi_{i-1} + \Gamma_{w,i}\left(\frac{\phi_i - \phi_{i-1}}{\Delta x}\right) = S_i \Delta x \quad \text{if } \rho u > 0, \tag{6.49a}$$

$$\rho u\,\phi_{i+1} - \Gamma_{e,i}\left(\frac{\phi_{i+1}-\phi_i}{\Delta x}\right) - \rho u\phi_i + \Gamma_{w,i}\left(\frac{\phi_i-\phi_{i-1}}{\Delta x}\right) = S_i\Delta x \quad \text{if } \rho u < 0. \quad (6.49\text{b})$$

Once again, assuming constant Γ and employing the definition of the grid Peclet number, we obtain

$$(2+Pe_\Delta)\phi_i - \phi_{i+1} - (1+Pe_\Delta)\phi_{i-1} = S_i\frac{(\Delta x)^2}{\Gamma} \quad \text{if } Pe_\Delta > 0, \quad (6.50\text{a})$$

$$(2+Pe_\Delta)\phi_i - \phi_{i-1} - (1+Pe_\Delta)\phi_{i+1} = S_i\frac{(\Delta x)^2}{\Gamma} \quad \text{if } Pe_\Delta < 0. \quad (6.50\text{b})$$

Equation (6.50) clearly satisfies the stability constraint imposed by Eq. (3.18) for all values of the grid Peclet number, making the likelihood of a stable iterative solution far stronger. Despite its poor accuracy, it is the improved stability characteristics, coupled with its simplicity that makes the first-order upwind difference scheme one of the most popular choices for the treatment of the advective flux term in numerical computations of advection–diffusion problems.

In an effort to delineate a physical interpretation to the error incurred by the first-order upwind difference scheme, we substitute Eq. (6.48) into Eq. (6.42) but without throwing away the first-order term from the Taylor series expansion so that the upwind approximation is, in essence, second-order accurate. Considering the case of positive flow ($\rho u|_{e,i} = \rho u|_{w,i} = \rho u > 0$) and constant Γ, this yields

$$\rho u\left(\phi_i + \frac{\Delta x}{2}\frac{d\phi}{dx}\Big|_{e,i}\right) - \Gamma\frac{d\phi}{dx}\Big|_{e,i} - \rho u\left(\phi_{i-1} + \frac{\Delta x}{2}\frac{d\phi}{dx}\Big|_{w,i}\right) + \Gamma\frac{d\phi}{dx}\Big|_{w,i} = S_i\Delta x_i, \quad (6.51)$$

which upon rearrangement yields

$$\left[\rho u\phi_i - \left(\Gamma - \frac{\Delta x}{2}\right)\frac{d\phi}{dx}\Big|_{e,i}\right] - \left[\rho u\phi_{i-1} - \left(\Gamma - \frac{\Delta x}{2}\right)\frac{d\phi}{dx}\Big|_{w,i}\right] = S_i\Delta x_i. \quad (6.52)$$

In contrast, the purely first order upwind approximation, given by Eq. (6.48), yields

$$\left[\rho u\phi_i - (\Gamma)\frac{d\phi}{dx}\Big|_{e,i}\right] - \left[\rho u\phi_{i-1} - (\Gamma)\frac{d\phi}{dx}\Big|_{w,i}\right] = S_i\Delta x_i. \quad (6.53)$$

Comparison of Eq. (6.52) with Eq. (6.53) reveals that in the second-order approximation [Eq. (6.52)] the effective diffusion coefficient is $\Gamma - \Delta x/2$, which is a reduced value compared to Γ. In other words, in the second-order approximation, diffusion is weaker as compared to the first-order approximation. Conversely, the first-order upwind approximation results in enhancement of the diffusion flux, and this phenomenon is known as *numerical diffusion*. Clearly, the extent of numerical diffusion is grid dependent since the enhancement of the diffusion coefficient vanishes as the grid size tends to zero. In summary, the aforementioned "smearing" of the solution in regions of strong gradients can be physically interpreted as being the result of enhanced diffusion due to an increased diffusion coefficient resulting from the use of the first-order upwind approximation for the advective flux.

In order to combat the numerical diffusion caused by the first-order upwind differencing scheme, higher order upwind differencing schemes have been proposed. The second-order upwind differencing scheme makes use of two Taylor series expansions to derive an expression for the face value of ϕ. One of these is Eq. (6.9b), which expands the first upwind cell about the face. The second expansion makes use of the next upwind cell, as follows:

$$\phi_{i-1} = \phi_e - \frac{3\Delta x}{2}\frac{d\phi}{dx}\bigg|_e + \frac{1}{2}\left(\frac{3\Delta x}{2}\right)^2\frac{d^2\phi}{dx^2}\bigg|_e - \frac{1}{6}\left(\frac{3\Delta x}{2}\right)^3\frac{d^3\phi}{dx^3}\bigg|_e + ..., \qquad (6.54)$$

wherein positive flow has been assumed. Multiplying Eq. (6.9b) by 3 and subtracting Eq. (6.54) from the result, followed by some rearrangement, yields

$$\phi_e = \frac{3\phi_i - \phi_{i-1}}{2} + \frac{3}{8}(\Delta x)^2\frac{d^2\phi}{dx^2}\bigg|_e + ... \qquad (6.55)$$

Equation (6.55) clearly shows that this scheme is second-order accurate. Similar expressions may be derived for the western face velocity. Combining both faces and both flow directions, the second-order upwind difference scheme may be written as

$$\phi_{e,i} = \begin{cases} \dfrac{3\phi_i - \phi_{i-1}}{2} & \text{if } \rho u|_{e,i} > 0 \\[2mm] \dfrac{3\phi_{i+1} - \phi_{i+2}}{2} & \text{if } \rho u|_{e,i} < 0 \end{cases}, \qquad (6.56a)$$

$$\phi_{w,i} = \begin{cases} \dfrac{3\phi_{i-1} - \phi_{i-2}}{2} & \text{if } \rho u|_{w,i} > 0 \\[2mm] \dfrac{3\phi_i - \phi_{i+1}}{2} & \text{if } \rho u|_{w,i} < 0 \end{cases}. \qquad (6.56b)$$

Substitution of Eqs (6.56) and (6.10) into Eq. (6.42), followed by application of mass conservation, i.e., $\rho u|_{e,i} = \rho u|_{w,i} = \rho u$, results in

$$\rho u\left(\frac{3\phi_i - \phi_{i-1}}{2}\right) - \Gamma_{e,i}\left(\frac{\phi_{i+1} - \phi_i}{\Delta x}\right) - \rho u\left(\frac{3\phi_{i-1} - \phi_{i-2}}{2}\right)$$
$$+\Gamma_{w,i}\left(\frac{\phi_i - \phi_{i-1}}{\Delta x}\right) = S_i\Delta x \quad \text{if } \rho u > 0, \qquad (6.57a)$$

$$\rho u\left(\frac{3\phi_{i+1} - \phi_{i+2}}{2}\right) - \Gamma_{e,i}\left(\frac{\phi_{i+1} - \phi_i}{\Delta x}\right) - \rho u\left(\frac{3\phi_i - \phi_{i+1}}{2}\right)$$
$$+\Gamma_{w,i}\left(\frac{\phi_i - \phi_{i-1}}{\Delta x}\right) = S_i\Delta x \quad \text{if } \rho u < 0. \qquad (6.57b)$$

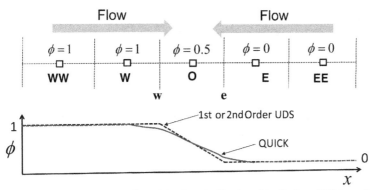

FIGURE 6.4 Schematic Representation of the Results Produced by Various UDS for a Scenario with Opposing Flow on Opposite Faces of the Same Cell

One of the important observations to be made from Eq. (6.57) is that the stencil extends beyond the three cells i, $i-1$, and $i+1$ that would normally materialize with second-order discretization of a 1D problem. For example, for the positive flow case, cell $i-2$ also appears in the equation. The repercussion of stencil extension is that boundary conditions are difficult to apply when the second-order upwind scheme is used, as we shall see shortly. Also, the tridiagonal matrix algorithm (TDMA) cannot be used any more as the core engine of the line-by-line sweeping pattern. The problem may be circumvented either by using a pentadiagonal solver (Section 3.1.2) or by treating the farthest cell (either $i-2$ or $i+2$) explicitly. The former option is computationally more expensive, while the latter option may disrupt convergence [8].

In both first- and second-order upwind difference schemes, information travels solely from the upstream to the downstream side. While this is intuitive, this treatment often results in undesirably sharp changes in the solution. Let us imagine a situation where two flow streams, approaching each other from opposite ends, collide with one another and create a stagnation point, as shown in Fig. 6.4. Additionally, the flow approaching from the left is carrying with it a scalar of unity value, i.e., $\phi = 1$, and the flow approaching from the right is carrying a scalar with $\phi = 0$. If either the first- or the second-order upwind difference schemes are used, the western face of cell O will have a value of $\phi = 1$, and the eastern face will have a value of $\phi = 0$. While these face values may be realistic from the standpoint of advective transport alone, in reality, since advection does not act in isolation but in unison with diffusion, the value of ϕ at the western face of cell O is likely to be somewhere between 0 and 1 – perhaps, closer to 1 than 0 if the transport is advection dominated. If the node downstream to the western face, i.e., node O was also used to formulate a value of ϕ at the western face of cell O, the resulting answer would be between 1 and 0.5, which, based on the preceding discussion, would be more compliant with reality.

In an effort to include some degree of back (or upstream) propagation of information, the quadratic upwind interpolation for convective kinematics (QUICK) scheme was proposed [9]. In the QUICK scheme, three cells are used to estimate the face value – one downstream and two upstream. On an equally spaced grid, this results in a scheme that is third-order accurate. However, for an unequally spaced mesh, the worst-case accuracy is second-order. Hence, the name "quadratic." To derive an expression for ϕ_e, in addition to the two Taylor series expansions about upstream cells, i and $i-1$, as given by Eqs (6.9b) and (6.54), we perform a Taylor series expansion about the downstream cell $i+1$ (for positive flow), leading to

$$\phi_{i+1} = \phi_e + \frac{\Delta x}{2} \frac{d\phi}{dx}\bigg|_e + \frac{1}{2}\left(\frac{\Delta x}{2}\right)^2 \frac{d^2\phi}{dx^2}\bigg|_e + \frac{1}{6}\left(\frac{\Delta x}{2}\right)^3 \frac{d^3\phi}{dx^3}\bigg|_e + ... \tag{6.58}$$

Manipulating these three equations to eliminate both first and second derivatives using the procedure outlined in Section 2.2, we obtain

$$\phi_e = \frac{3}{8}\phi_{i+1} + \frac{3}{4}\phi_i - \frac{1}{8}\phi_{i-1} - \frac{1}{16}(\Delta x)^3 \frac{d^3\phi}{dx^3}\bigg|_e + ... \tag{6.59}$$

The western face velocity may be derived using a similar procedure. In summary, the QUICK scheme may be written as

$$\phi_{e,i} = \begin{cases} \dfrac{3}{8}\phi_{i+1} + \dfrac{3}{4}\phi_i - \dfrac{1}{8}\phi_{i-1} & \text{if } \rho u\big|_{e,i} > 0 \\[2mm] \dfrac{3}{8}\phi_i + \dfrac{3}{4}\phi_{i+1} - \dfrac{1}{8}\phi_{i+2} & \text{if } \rho u\big|_{e,i} < 0 \end{cases}, \tag{6.60a}$$

$$\phi_{w,i} = \begin{cases} \dfrac{3}{8}\phi_i + \dfrac{3}{4}\phi_{i-1} - \dfrac{1}{8}\phi_{i-2} & \text{if } \rho u\big|_{w,i} > 0 \\[2mm] \dfrac{3}{8}\phi_{i-1} + \dfrac{3}{4}\phi_i - \dfrac{1}{8}\phi_{i+1} & \text{if } \rho u\big|_{w,i} < 0 \end{cases}. \tag{6.60b}$$

As mentioned earlier, both for the second-order upwind difference scheme as well as for the QUICK scheme, treatment of boundary conditions requires special attention since the stencil extends beyond the neighboring cells. To highlight this point, let us consider the left boundary of a 1D computational domain. Let us also consider the case where a Dirichlet boundary condition is applied to the leftmost face, as shown in Fig. 6.3. The advective flux at the western face of $i = 1$ is given by $\rho u \phi_L$, and requires no further consideration. If the first-order upwind difference scheme is used, for a positive flow, the advective flux at the eastern face of $i = 1$ is given by $\rho u \phi_1$. However, if the second-order upwind or the QUICK scheme is used, two upstream cell values are needed, which, in this case, are not available. Therefore, expressions given by Eqs (6.56a) and (6.60a) cannot be used. Hence, in the case of the second-order upwind difference scheme, we rederive an expression for the face value $\phi_{e,1}$ using Eq. (9.9b) and a new Taylor series expansion of the boundary value about the eastern face, written as

$$\phi_L = \phi_{e,1} - \Delta x \frac{d\phi}{dx}\bigg|_{e,1} + \frac{1}{2}(\Delta x)^2 \frac{d^2\phi}{dx^2}\bigg|_{e,1} - \frac{1}{6}(\Delta x)^3 \frac{d^3\phi}{dx^3}\bigg|_{e,1} + ... \tag{6.61}$$

Multiplying Eq. (6.9b) by 2 and subtracting Eq. (6.61) from the result, for the second-order upwind difference scheme, we obtain

$$\phi_{e,1} = 2\phi_1 - \phi_L + \frac{1}{4}(\Delta x)^2 \frac{d^2\phi}{dx^2}\bigg|_{e,1} + \dots \qquad (6.62)$$

In order to derive $\phi_{e,1}$ for the QUICK scheme, in addition to Eqs (6.9b) and (6.62), we also make use of Eq. (6.58). Upon manipulation of the three equations to cancel first and second derivatives, for the QUICK scheme, we obtain

$$\phi_{e,1} = \phi_1 - \frac{1}{3}\phi_L + \frac{1}{3}\phi_2 - \frac{1}{24}(\Delta x)^3 \frac{d^3\phi}{dx^3}\bigg|_{e,1} + \dots \qquad (6.63)$$

In order to derive the discrete finite volume equation for the leftmost cell ($i = 1$), either Eq. (6.62) or Eq. (6.63) must be substituted into Eq. (6.42), in conjunction with Eqs. (6.10) and either Eq. (6.57) or Eq. (6.60), to yield the necessary equations for the second-order upwind scheme or the QUICK scheme, respectively. Further details are provided in the example to follow.

EXAMPLE 6.2

In this example, we solve the 1D advection–diffusion equation, written as

$\dfrac{d}{dx}\left(Pe\phi - \dfrac{d\phi}{dx}\right) = 0$, with boundary conditions $\phi(0) = 1$, and $\phi(1) = 0$. The global Peclet number, denoted by $Pe = \rho u L/\Gamma$, is a constant parameter in this study. Positive values of Pe imply positive flow. Here, we will consider the following values of Pe: 0.1, 1, and 10. The values of Pe that are greater than unity imply that advection is dominant over diffusion, while the values of Pe that are less than unity imply that diffusion is dominant over advection. The analytical solution is easily obtained by integrating the above equation twice and applying the two boundary conditions, and is written as

$$\phi(x) = \frac{\exp(Pe) - \exp(Pe \cdot x)}{\exp(Pe) - 1}.$$

For numerical calculations, first-order upwind, second-order upwind, and QUICK schemes are considered for the treatment of the advective flux. In each case, the central difference scheme is used for the treatment of diffusion. Two different mesh sizes are considered: $N_C = 10$ and 20 cells. A summary of the discrete finite volume equations for each of the three fluxing schemes is provided next.

First-order upwind scheme

$$i - 1.(12 + 3Pe\Delta x)\phi_i - 4\phi_{i+1} - (8 + 3Pe\Delta x)\phi_L$$

$$i = 2,\dots, N_C - 1 : (2 + Pe\Delta x)\phi_i - \phi_{i+1} - (1 + Pe\Delta x)\phi_{i-1} = 0$$

$$i = N_C : 12\phi_i - (4 + 3Pe\Delta x)\phi_{i-1} = (8 - 3Pe\Delta x)\phi_R$$

Second-order upwind scheme

$$i = 1 : (12 + 6Pe\Delta x)\phi_i - 4\phi_{i+1} = (8 + 6Pe\Delta x)\phi_L$$

$$i = 2 : (4 + 3Pe\Delta x)\phi_i - 2\phi_{i+1} - (2 + 5Pe\Delta x)\phi_{i-1} = -2Pe\Delta x\phi_L$$

$$i = 3,...,N_C - 1 : (4 + 3Pe\Delta x)\phi_i - 2\phi_{i+1} - (2 + 4Pe\Delta x)\phi_{i-1} + Pe\Delta x\phi_{i-2} = 0$$

$$i = N_C : 24\phi_i - (8 + 9Pe\Delta x)\phi_{i-1} + 3Pe\Delta x\phi_{i-2} = (16 - 6Pe\Delta x)\phi_R$$

QUICK scheme

$$i = 1 : (12 + 3Pe\Delta x)\phi_i + (Pe\Delta x - 4)\phi_{i+1} = (8 + 4Pe\Delta x)\phi_L$$

$$i = 2 : (48 + 10Pe\Delta x)\phi_i + (9Pe\Delta x - 24)\phi_{i+1} - (24 + 27Pe\Delta x)\phi_{i-1} = -8Pe\Delta x\phi_L$$

$$i = 3,...,N_C - 1 : (16 + 3Pe\Delta x)\phi_i + (3Pe\Delta x - 8)\phi_{i+1} - (8 + 7Pe\Delta x)\phi_{i-1} + Pe\Delta x\phi_{i-2} = 0$$

$$i = N_C : (96 - 9Pe\Delta x)\phi_i - (32 + 18Pe\Delta x)\phi_{i-1} + 3Pe\Delta x\phi_{i-2} = (64 - 24Pe\Delta x)\phi_R$$

The three figures, shown below, quantify the solutions and the errors in the solutions obtained using the three different schemes for advection (central difference is used for diffusion in each case) for the three different Peclet numbers using just 10 cells. For solving the system of equations arising out of the use of the second-order upwind difference scheme and the QUICK scheme, Gaussian elimination was used since, as mentioned before, the system of equations is no longer tridiagonal.

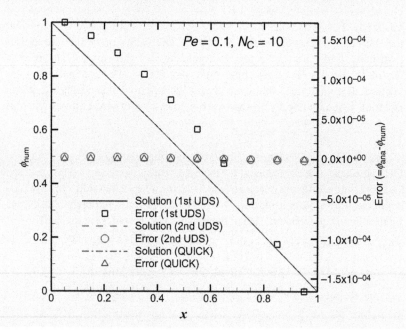

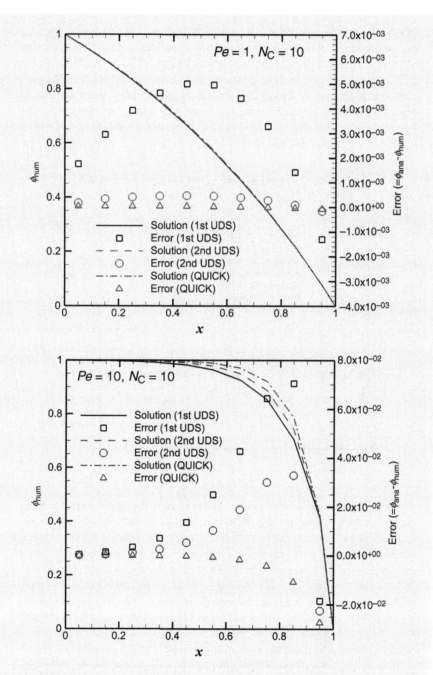

The results show that irrespective of the Peclet number, the second-order upwind difference scheme (2nd UDS) and the QUICK schemes produce significantly lower errors than the first-order upwind difference scheme (1st

UDS). As the Peclet number is increased to increase the effect of advection, the choice of scheme becomes more critical. At the highest Peclet number ($Pe = 10$), 1st UDS shows large errors – as large as 7%. At this high Pe, the QUICK scheme is, on average, more accurate than the 2nd UDS, although it may have higher errors locally. This is because the truncation error also contains the higher order derivatives, which are different in the two cases. The "smearing effect" of the 1st UDS is also evident in the solution. The QUICK scheme has the least smearing effect.

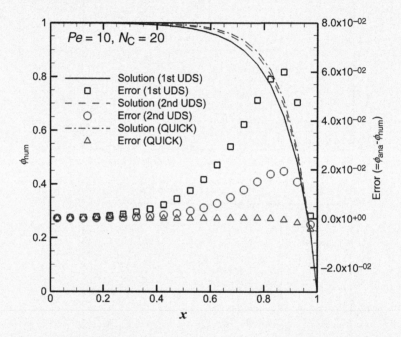

The solution with 20 cells for $Pe = 10$ is shown in the figure above. With reduction in the grid spacing, the errors produced by all three schemes decrease, as evident by comparing the figure with the one above. The 1st UDS is expected to scale linearly with grid spacing. However, that does not appear to be the case in practice. This is because in the FVM, the cell centers for the two meshes do not coincide with each other, the errors shown in the two figures are actually at different locations. Since the second derivative is changing rapidly in this region of the domain, the computed errors also change rapidly from point to point. In regions where the derivatives are not as large, linear scaling is indeed observed. Similarly, quadratic scaling is observed in the case of 2nd UDS.

In summary, this example demonstrates how to apply upwind difference schemes of various orders to the solution of the advection–diffusion equation, and the implications on accuracy of using each scheme.

6.3.2 **EXPONENTIAL SCHEME**

The flux treatment strategy discussed thus far – central difference scheme for diffusion and UDS for advection – is founded on the fundamental premise that advection and diffusion are independent unrelated mechanisms of transport. In mathematical terms, this is often referred to as *operator splitting*. In reality, these two mechanisms are simultaneously at play, and the direction of the net flux is a manifestation of the combined effect of these two mechanisms. Hence, treating them in unison is more in line with the fundamental physics, as is done in the exponential scheme. To derive the exponential scheme, we start with our governing 1D advection–diffusion equation, namely Eq. (6.40), but without a source term, namely

$$\frac{d}{dx}\left[\frac{Pe_\Delta}{\Delta x}\phi - \frac{d\phi}{dx}\right] = 0,\tag{6.64}$$

where Pe_Δ is the grid Peclet number, and for a uniform mesh with grid spacing Δx, is defined as $Pe_\Delta = \rho u \Delta x / \Gamma$. Recalling that the objective is to derive the face value of ϕ from adjacent cell-center values, we solve Eq. (6.64) in the interval (x_i, x_{i+1}), as shown in Fig. 6.5. Since the interval of solution is a small part of the full computational domain, and only spans a length equal to Δx, it is justifiable to assume that ρu and Γ are both constants in that range. Consequently, Pe_Δ may also be assumed to be constant in the interval (x_i, x_{i+1}), and may be denoted as $Pe|_e$, i.e., the value of Pe_Δ at the eastern face. Integrating Eq. (6.64) twice, we obtain

$$\phi(x) = A + B\exp\left(\frac{Pe|_e}{\Delta x}x\right),\tag{6.65}$$

where A and B are integration constants, which are to be determined by applying boundary conditions. The boundary conditions are $\phi(x_i) = \phi_i$ and $\phi(x_{i+1}) = \phi_{i+1}$. Alternatively, by shifting the origin to the cell center i (the governing differential equation remains unchanged by doing this), the boundary conditions may be rewritten as $\phi(0) = \phi_i$, and $\phi(\Delta x) = \phi_{i+1}$. Application of the two boundary conditions to Eq. (6.65) yields

$$A = \frac{\phi_i \exp(Pe|_e) - \phi_{i+1}}{\exp(Pe|_e) - 1}, B = \frac{\phi_{i+1} - \phi_i}{\exp(Pe|_e) - 1}.\tag{6.66}$$

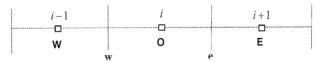

FIGURE 6.5 Schematic Representation of the 1D Stencil Used to Derive the Exponential Scheme

Substituting Eq. (6.66) into Eq. (6.65) and using $x = \Delta x / 2$ in the resulting expression to obtain the eastern face value of ϕ, we obtain

$$\phi_{e,i} = \frac{\phi_{i+1} + \phi_i \exp(Pe|_e/2)}{\exp(Pe|_e/2)+1}. \tag{6.67}$$

It is easy to see that in the limiting case of $Pe|_e \to 0$, i.e., no advection, Eq. (6.67) reduces to $\phi_{e,i} = (\phi_{i+1} + \phi_i)/2$, which is the central difference approximation. The diffusive flux is derived by differentiating Eq. (6.65) with respect to x, and then substituting $x = \Delta x/2$ and Eq. (6.66) into the resulting equation to yield

$$\left.\frac{d\phi}{dx}\right|_{e,i} = B\frac{Pe|_e}{\Delta x}\exp\left(\frac{Pe|_e}{\Delta x}\frac{\Delta x}{2}\right) = \frac{Pe|_e}{\Delta x}\frac{(\phi_{i+1}-\phi_i)}{} \frac{\exp(Pe|_e/2)}{\exp(Pe|_e)-1}. \tag{6.68}$$

Substitution of Eqs (6.67) and (6.68) into Eq. (6.39) yields the expression for the total flux as given by the exponential scheme, and is written as

$$\begin{aligned} J_{e,i} &= \left.\rho u\right|_{e,i}\phi_{e,i} - \Gamma_{e,i}\left.\frac{d\phi}{dx}\right|_{e,i} \\ &= \left.\rho u\right|_{e,i}\frac{\phi_{i+1} + \phi_i \exp(Pe|_e/2)}{\exp(Pe|_e/2)+1} - \Gamma_{e,i}\frac{Pe|_e}{\Delta x}\frac{(\phi_{i+1}-\phi_i)}{}\frac{\exp(Pe|_e/2)}{\exp(Pe|_e)-1}. \end{aligned} \tag{6.69}$$

Noting further that $Pe_{e,i} = \left.\rho u\right|_{e,i}\Delta x/\Gamma_{e,i}$, Eq. (6.69) may be simplified to write

$$J_{e,i} = \frac{\Gamma_{e,i}}{\Delta x}[f_{e,i}^-\phi_i - f_{e,i}^+\phi_{i+1}], \tag{6.70}$$

where

$$f_{e,i}^- = \frac{Pe|_{e,i}\exp(Pe|_{e,i})}{\exp(Pe|_{e,i})-1}, \quad f_{e,i}^+ = \frac{Pe|_{e,i}}{\exp(Pe|_{e,i})-1}. \tag{6.71}$$

Equation (6.71) cannot be used in the limit of pure diffusion, i.e., $Pe|_e \to 0$, since it will give rise to a division by zero. Making use of the fact that $\exp(x) = 1 + x + x^2/2 + ...$, one can write Eq. (6.71) as

$$f_{e,i}^- = \frac{Pe|_{e,i}\left(1 + Pe|_{e,i} + Pe|_{e,i}^2/2 + ...\right)}{Pe|_{e,i} + Pe|_{e,i}^2/2 + ...} = \frac{1 + Pe|_{e,i} + Pe|_{e,i}^2/2 + ...}{1 + Pe|_{e,i}/2 + Pe|_{e,i}^2/6 + ...}. \tag{6.72}$$

In the limit $Pe|_e \to 0$, the higher order terms are negligible, and Eq. (6.71), using the expansions shown in Eq. (6.72), reduces to

$$\lim_{Pe|_{e,i}\to 0} f_{e,i}^- = \frac{1 + Pe|_{e,i}}{1 + \dfrac{Pe|_{e,i}}{2}}, \quad \lim_{Pe|_{e,i}\to 0} f_{e,i}^+ = \frac{1}{1 + \dfrac{Pe|_{e,i}}{2}}. \tag{6.73}$$

Equation (6.73) is convenient for use in situations where the local Peclet number may be small without a loss of accuracy. It is worth pointing out that the low Peclet number limit can be encountered in computations of real-life problems in regions where the local flow velocities may be tiny, e.g., in the eye of a recirculating vortex or at a stagnation point. These scenarios can occur even in flows where the global Peclet number is large.

One of the important aspects of the exponential scheme is that advection and diffusion are treated within the same formulation. Thus, the central difference scheme is not used for the diffusive flux at all. Instead, the diffusive flux is derived from the piecewise analytical solution using Eq. (6.68). This implies that for the sake of consistency, a similar strategy must also be employed for boundary conditions. Let us consider the case where a Dirichlet boundary condition is applied at the left end, as shown in Fig. 6.3. For the eastern face of the leftmost cell ($i = 1$), the flux expression given by Eq. (6.70) may still be used. For the western face, however, a new expression is necessary. To derive such an expression, we follow the same procedure as outlined earlier. The governing equation and its solution, Eq. (6.65), are still valid. The boundary conditions, however, are different: $\phi(0) = \phi_L$, and $\phi(\Delta x/2) = \phi_1$. Using these two boundary conditions and Eq. (6.65), we obtain

$$B = \frac{\phi_1 - \phi_L}{\exp(\dfrac{Pe|_{w,1}}{2}) - 1}, \quad A = \phi_L - B. \tag{6.74}$$

The total flux at the western face of $i = 1$ is then derived as

$$J_{w,1} = \rho u|_{w,1}\,\phi_L - \Gamma_{w,1}\frac{d\phi}{dx}\bigg|_{w,1} = \rho u|_{w,1}\,\phi_L - \Gamma_{w,1}\frac{d\phi}{dx}\bigg|_{x=0} = \rho u|_{w,1}\,\phi_L - \Gamma_{w,1}B\frac{Pe|_{w,1}}{\Delta x}. \tag{6.75}$$

Once again, noting that $Pe_{w,1} = \rho u|_{w,1}\,\Delta x/\Gamma_{w,1}$, and using Eq. (6.74), Eq. (6.75) may be simplified to yield

$$J_{w,1} = \frac{\Gamma_{w,1}}{\Delta x}Pe|_{w,1}\left[\frac{\phi_L \exp(Pe|_{w,1}/2) - \phi_1}{\exp(Pe|_{w,1}/2) - 1}\right]. \tag{6.76}$$

In the limiting case of $Pe|_{w,1} \to 0$, using the same procedure as before, it can be shown that

$$\lim_{Pe|_{w,1} \to 0} J_{w,1} = \frac{\Gamma_{w,1}}{\Delta x}\left[\frac{1 + Pe|_{e,i}/2}{1 + Pe|_{e,i}/8}\phi_L - \frac{1}{1 + Pe|_{e,i}/8}\phi_1\right]. \tag{6.77}$$

Next, we consider an example, which demonstrates use of the exponential scheme.

EXAMPLE 6.3

In this example, we solve the 1D advection–diffusion equation also solved in Example 6.2. As in Example 6.2, we consider the following values of Pe: 0.1, 1, and 10. For numerical calculations, the exponential scheme is used. A coarse mesh with $N_C = 10$ is considered. A summary of the discrete finite volume equations for the exponential scheme is provided below.

$$i=1: \left[\frac{\exp(Pe_\Delta)}{\exp(Pe_\Delta)-1} + \frac{1}{\exp(Pe_\Delta/2)-1} \right] \phi_i - \frac{1}{\exp(Pe_\Delta)-1}\phi_{i+1} = \frac{\exp(Pe_\Delta/2)}{\exp(Pe_\Delta/2)-1}\phi_L$$

$$i=2,...,N_C-1: \left[1+\exp(Pe_\Delta) \right]\phi_i - \phi_{i+1} - \exp(Pe_\Delta)\phi_{i-1} = 0$$

$$i=N_C: \left[\frac{1}{\exp(Pe_\Delta)-1} + \frac{\exp(Pe_\Delta/2)}{\exp(Pe_\Delta/2)-1} \right] \phi_i - \frac{\exp(Pe_\Delta)}{\exp(Pe_\Delta)-1}\phi_{i-1} = \frac{1}{\exp(Pe_\Delta/2)-1}\phi_R$$

where $Pe_\Delta = \rho u \Delta x / \Gamma_\phi$ is the so-called grid Peclet number. Since the global Peclet number, $Pe = \rho u L / \Gamma_\phi$ reduces to $Pe = \rho u / \Gamma_\phi$ by virtue of the domain length being unity, the grid Peclet number may be expressed in terms of the global Peclet number as $Pe_\Delta = Pe\Delta x$.

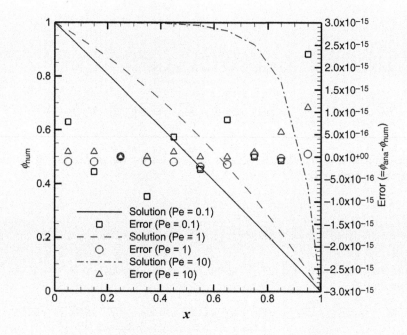

The solution obtained using the exponential scheme and the associated errors are shown in the figure above. It is seen that even with just 10 cells, irrespective of the Peclet number, the errors are within round-off accuracy (zero for all practical purposes). This is to be expected because for this particular case in which the grid Peclet number is constant (independent of x), the exponential scheme essentially recovers the exact solution to the 1D advection–diffusion equation by assimilating piecewise exact solutions.

The findings of Examples 6.2 and 6.3 appear to suggest that the exponential scheme is the scheme of choice for the solution of the advection–diffusion equation. In practice, this is not the case. Most practical problems are multidimensional. Consequently, the locally 1D exact solution that is used to construct the exponential scheme breaks down for such problems, and the solution accuracy is generally comparable to the operator-split differencing schemes discussed in Section 6.3.1. Another not so obvious disadvantage of using the exponential scheme is that it is computationally expensive to implement. The computation of the link coefficients in this scheme requires computation of several exponential functions. On a digital computer, computing exponential functions is far more expensive than multiplication or division operations. This makes the assembly of the linear system far more expensive when the exponential scheme is used compared with other schemes. One notable advantage of the exponential scheme is that the stencil never extends beyond one cell, and thus, boundary conditions are relatively easy to implement. Nonetheless, on account of the disadvantages cited earlier, it is not a very popular scheme in areas such as computational fluid dynamics and heat transfer, except for certain specific applications.

Besides the accuracy and ease of implementation related issues discussed in this section, another important consideration for the treatment of the advective flux is monotonicity of the solution for the dependent variable. Most numerical computations warrant that the solution between two cells or two nodes must be monotonically increasing or decreasing. This demand stems from the requirements posed by the physics of the problem. As an example, if two points A and B have two different temperatures, then at steady state, the temperature anywhere between those two points cannot exceed the bounds of the two end points if a heat source is absent. This constraint is borne out of the second law of thermodynamics. By definition, any curve of degree two or higher has at least one extremum. Therefore, when higher order upwind schemes are used for advection, the solution can have local oscillations and can exceed the bounds of the two end points. This implies, for example, that a face value can be beyond the bounds of the surrounding cell values, which, as discussed earlier, violates the physics. The first-order upwind scheme encounters no such issues, being a simple piecewise linear approximation. This is another reason why, despite its poor accuracy, it is still considered the "bread and butter" scheme for the treatment of advection. Over the past several decades, a large number of monotonicity-preserving schemes have been developed for the treatment of advection. These schemes attempt to use higher order expansions, such as the one derived for the QUICK scheme, but with constraints based on the local Peclet number that prevent overshoots or undershoots in the solution locally. A discussion of these advanced schemes (essentially nonoscillatory (ENO) schemes [10], total variation diminishing (TVD) schemes [11], adaptive blending schemes, etc.) is well beyond an introductory text such as this one, and may be found in advanced texts on computational fluid dynamics [3,4,12] and relevant journal articles [10,11,13,14].

6.4 MULTIDIMENSIONAL PROBLEMS

In the preceding sections, the foundations of the FVM have been presented, and the method has been applied to both pure diffusion and advection–diffusion problems in a single dimension. In this section, we extend the concepts discussed in the preceding section to multidimensional problems but on an orthogonal (Cartesian) grid. In the section to follow, the method is further extended to multidimensional problems on a curvilinear grid.

As in the 1D case, the computational domain is first split into a set of smaller control volumes or cells. A typical structured orthogonal mesh is shown in Fig. 6.6. As mentioned earlier, all information is stored at cell centers, while fluxes are computed at cell faces. As before, the cell centers are denoted by O, E, W, N, and S, while the cell faces are denoted by e, w, n, and s. In 3D geometry, two additional notations are needed. Traditionally, the two cells in the z direction are denoted either by F and B (forward and backward) or by U and D (upstream and downstream).

We begin our derivation of the finite volume equations for a 2D geometry by considering integration of the Poisson (diffusion) equation on an orthogonal Cartesian mesh. The governing equation is the 2D rendition of Eq. (6.2), and is written as

$$\frac{\partial}{\partial x}\left[\Gamma\frac{\partial\phi}{\partial x}\right]+\frac{\partial}{\partial y}\left[\Gamma\frac{\partial\phi}{\partial y}\right]=-S_\phi. \tag{6.78}$$

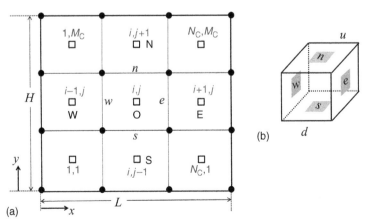

FIGURE 6.6 Schematic Representation of the Orthogonal Cell Arrangement in a 2D Cartesian Geometry (a) and a 3D Hexahedral Cell with the Six Faces (b)

In this case, the flux of ϕ is a vector, written as $\mathbf{J} = J_x \hat{\mathbf{i}} + J_y \hat{\mathbf{j}}$, where, following standard notations, $\hat{\mathbf{i}}$ and $\hat{\mathbf{j}}$ denote unit vectors in the x and y directions, respectively, while the components of the fluxes in the two directions are written as

$$J_x = -\Gamma \frac{\partial \phi}{\partial x}, J_y = -\Gamma \frac{\partial \phi}{\partial y}. \tag{6.79}$$

The extension of Eqs (6.78) and (6.79) to three dimensions is straightforward. Next, we integrate Eq. (6.78) over the control volume O, shown in Fig. 6.6. This yields

$$\int_s^n \int_w^e \frac{\partial}{\partial x}\left[\Gamma \frac{\partial \phi}{\partial x}\right] dx\, dy + \int_s^n \int_w^e \frac{\partial}{\partial y}\left[\Gamma \frac{\partial \phi}{\partial y}\right] dx\, dy = -\int_s^n \int_w^e S_\phi\, dx\, dy, \tag{6.80}$$

where the face notations have been used to define the integration limits. To simplify Eq. (6.80), the order of integration in the second term on the left-hand side is changed, and both terms on the left-hand side integrated once to yield

$$\int_s^n \left[\Gamma \frac{\partial \phi}{\partial x}\right]_w^e dy + \int_w^e \left[\Gamma \frac{\partial \phi}{\partial y}\right]_s^n dx = -\int_s^n \int_w^e S_\phi\, dx\, dy. \tag{6.81}$$

Using the definition of the average [Eq. (6.1)] and realizing that the quantities within square brackets represent average fluxes at individual faces of the cell, Eq. (6.81) may be further simplified (integrated once more) to write

$$\left(\left[\Gamma \frac{\partial \phi}{\partial x}\right]_e - \left[\Gamma \frac{\partial \phi}{\partial x}\right]_w\right)\Delta y + \left(\left[\Gamma \frac{\partial \phi}{\partial y}\right]_n - \left[\Gamma \frac{\partial \phi}{\partial y}\right]_s\right)\Delta x = -S_O\, \Delta x\, \Delta y. \tag{6.82}$$

In deriving Eq. (6.82) from Eq. (6.81), it has been assumed that the average flux at a face is equivalent to the value of the flux computed at the geometric center of the face. This assumption is commensurate with the assumption that the volume average of a variable is the same as the value of the variable evaluated at the geometric centroid. The implications and repercussions of this assumption have already been discussed in Section 6.1. A uniform mesh, $\Delta x = L/N_C$ and $\Delta y = H/M_C$, has also been assumed in deriving Eq. (6.82). Using the definition of the fluxes given by Eq. (6.79), Eq. (6.82) may be written as

$$\left(J_e - J_w\right)\Delta y + \left(J_n - J_s\right)\Delta x = S_O \Delta x \Delta y. \tag{6.83}$$

Although not obvious, the fluxes, J_e, J_w, J_n, and J_s in Eq. (6.83) actually represent fluxes normal to the cell faces. In the case of the orthogonal (Cartesian) grid considered here, the normal flux essentially reduces to either the x or y component. This will become clear in Chapter 7 when we discuss control volumes of arbitrary shape. The finite volume equation [Eq. (6.83)] is nothing but a flux conservation equation [cf. Eq. (6.6)], which is to be expected in light of the discussion in Section 6.1.

The first derivatives appearing in Eq. (6.82) may be approximated using the Taylor series expansion–based procedure outlined for the derivation of Eq. (6.10), and is not repeated here for the sake of brevity. It is worth mentioning here, though, that such 1D Taylor series expansions are made possible only because a perfectly orthogonal mesh is under consideration. If the mesh is not orthogonal, the Taylor series expansions become truly multidimensional, as shown in Eq. (2.43), and derivation of individual partial derivatives in such a scenario would be next to impossible. Using Eq. (6.10) and other similar expressions for the face fluxes, Eq. (6.82) may be written as

$$\Gamma_e\left(\frac{\phi_E-\phi_O}{\Delta x}\right)\Delta y-\Gamma_w\left(\frac{\phi_O-\phi_W}{\Delta x}\right)\Delta y+\Gamma_n\left(\frac{\phi_N-\phi_O}{\Delta y}\right)\Delta x$$
$$-\Gamma_s\left(\frac{\phi_O-\phi_S}{\Delta y}\right)\Delta x=-S_O\Delta x\Delta y,$$

(6.84)

where the symbols carry their usual meanings, and it is understood that the faces are of cell O. Rearranging, Eq. (6.84) may be written in the following five-band form:

$$\left(\frac{\Gamma_e+\Gamma_w}{\Delta x}\Delta y+\frac{\Gamma_n+\Gamma_s}{\Delta y}\Delta x\right)\phi_O-\frac{\Gamma_e}{\Delta x}\Delta y\,\phi_E-\frac{\Gamma_w}{\Delta x}\Delta y\,\phi_W$$
$$-\frac{\Gamma_n}{\Delta y}\Delta x\,\phi_N-\frac{\Gamma_s}{\Delta y}\Delta x\,\phi_S=S_O\Delta x\Delta y.$$

(6.85)

If $\Gamma=1$, and Eq. (6.85) is divided through by the volume $\Delta x\,\Delta y$, it reduces to Eq. (2.48), which was derived using the finite difference procedure.

The application of boundary conditions follows the exact same procedures described in Section 6.2, with the exception that more than one flux term may have to be approximated and finally substituted into the governing equation to obtain the final discrete equation for a boundary cell. This is illustrated in the example below.

EXAMPLE 6.4

In this example, we show how to derive the discrete finite volume equation from the 2D Poisson equation for a corner cell, shown in the figure below. The boundary conditions on its two faces are also shown in the figure.

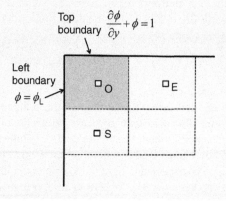

The starting point of the derivation is the finite volume equation, given by Eq. (6.82) with $\Gamma = 1$, since the governing equation in question is the Poisson equation. In this case, expressions for flux need to be derived for the western and northern faces. For the western face flux, following Eq. (6.29), we have

$\left.\dfrac{\partial \phi}{\partial x}\right|_w \approx \dfrac{9\phi_O - \phi_E - 8\phi_L}{3\Delta x}$. To derive the northern face flux, we follow the exact

same procedure used to derive Eq. (6.34). First, we rewrite the boundary condition as

$$\phi_n = 1 - \left.\frac{\partial \phi}{\partial y}\right|_n$$

Next, we perform two Taylor series expansions, as follows:

$$\phi_O = \phi_n - \frac{\Delta y}{2}\left.\frac{\partial \phi}{\partial y}\right|_n + \frac{1}{2}\left(\frac{\Delta y}{2}\right)^2 \left.\frac{\partial^2 \phi}{\partial y^2}\right|_n - \frac{1}{6}\left(\frac{\Delta y}{2}\right)^3 \left.\frac{\partial^3 \phi}{\partial y^3}\right|_n + ..., \text{ and}$$

$$\phi_S = \phi_n - \frac{3\Delta y}{2}\left.\frac{\partial \phi}{\partial y}\right|_n + \frac{1}{2}\left(\frac{3\Delta y}{2}\right)^2 \left.\frac{\partial^2 \phi}{\partial y^2}\right|_n - \frac{1}{6}\left(\frac{3\Delta y}{2}\right)^3 \left.\frac{\partial^3 \phi}{\partial y^3}\right|_n + ...$$

Multiplying the first equation by 9, and subtracting the second equation, we obtain

$$9\phi_O - \phi_S = 8\phi_n - 3\Delta y \left.\frac{\partial \phi}{\partial y}\right|_n + \frac{3}{8}(\Delta y)^3 \left.\frac{\partial^3 \phi}{\partial y^3}\right|_n + ...$$

Substituting the boundary condition into the above equation, we obtain

$$9\phi_O - \phi_S = 8 - (8 + 3\Delta y)\left.\frac{\partial \phi}{\partial y}\right|_n + \frac{3}{8}(\Delta y)^3 \left.\frac{\partial^3 \phi}{\partial y^3}\right|_n + ...,$$

which, upon rearrangement, yields

$$\left.\frac{\partial \phi}{\partial y}\right|_n = \frac{8 + \phi_S - 9\phi_O}{(8 + 3\Delta y)} + \frac{3}{8}\frac{(\Delta y)^3}{(8 + 3\Delta y)}\left.\frac{\partial^3 \phi}{\partial y^3}\right|_n + ...$$

Finally, substituting the two expressions for the derivatives just derived, as well as the central difference formula for the derivatives at the interior faces into Eq. (6.82), we obtain

$$\left(\left[\frac{\phi_E - \phi_O}{\Delta x}\right] - \left[\frac{9\phi_O - \phi_E - 8\phi_L}{3\Delta x}\right]\right)\Delta y + \left(\left[\frac{8 + \phi_S - 9\phi_O}{(8 + 3\Delta y)}\right] - \left[\frac{\phi_O - \phi_S}{\Delta y}\right]\right)\Delta x = -S_O \Delta x \Delta y ,$$

which is the required discrete finite volume equation for the corner cell O.

The treatment of the advective flux in the case of the advection–diffusion equation also remains unaltered in multidimensional geometry. However, one has to make use of the appropriate velocity component and upstream cell in order to apply the upwind difference scheme. This is demonstrated in the example to follow.

EXAMPLE 6.5

In this example, we consider a problem often encountered in the heat transfer area. The problem consists of a planar 2D channel, as shown in the figure below, through which water flows. The water enters the channel from the left with a temperature of 20°C. The channel is heated at the bottom with a heat flux of 10 W/m², while the top is perfectly insulated (zero flux).

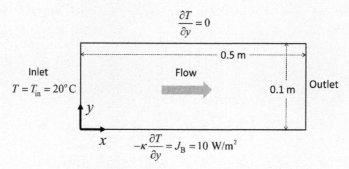

The objective is to find the temperature distribution within the channel. The governing equation is the steady-state energy conservation equation, written as

$$\frac{\partial}{\partial x}(\rho u c_p T) + \frac{\partial}{\partial y}(\rho v c_p T) = \frac{\partial}{\partial x}\left(\kappa \frac{\partial T}{\partial x}\right) + \frac{\partial}{\partial y}\left(\kappa \frac{\partial T}{\partial y}\right),$$

where ρ, c_p, and κ are the density, specific heat capacity, and thermal conductivity of the fluid (water). The x- and y-velocity components are denoted by u and v, respectively. Assuming ρ, c_p, and κ to be constants, and defining $\Gamma = \kappa/\rho c_p$ and $\phi = T$, the governing equation may be rewritten as

$$\frac{\partial}{\partial x}\left(u\phi - \Gamma \frac{\partial \phi}{\partial x}\right) + \frac{\partial}{\partial y}\left(v\phi - \Gamma \frac{\partial \phi}{\partial y}\right) = 0,$$

which is the standard 2D advection–diffusion equation, i.e., the 2D rendition of Eq. (6.39). In this example, we will use the following values for the thermophysical properties: $\rho = 1000$ kg/m³, $c_p = 4200$ J/kg/°C, and $\kappa = 0.613$ W/m/°C, resulting in $\Gamma = 1.45 \times 10^{-7}$ m²/s. Prior to solving the governing equation, the two velocity components u and v must be known. Here, we assume that the flow is laminar and fully developed, such that

$$u(y) = u_{in}(y) = 6 u_{avg}\left[\left(\frac{y}{h}\right) - \left(\frac{y}{h}\right)^2\right],$$ and $v = 0$, where h is the height of the

inlet, and u_{avg} is the average velocity at the inlet in the x direction. The average velocity is adjusted such that the global Peclet number, $Pe = \dfrac{\rho u_{avg} h}{\Gamma}$, is equal to 200. Under these assumptions, the governing discrete finite volume equations, using the 1st UDS for advection and central difference for diffusion may be written as

Interior cells ($2 \leq i \leq N_C - 1$, $2 \leq j \leq M_C - 1$):

$$\left(u_j \Delta y + 2\Gamma \frac{\Delta y}{\Delta x} + 2\Gamma \frac{\Delta x}{\Delta y} \right) \phi_{i,j} - \Gamma \frac{\Delta y}{\Delta x} \phi_{i+1,j} - \left(u_j \Delta y + \Gamma \frac{\Delta y}{\Delta x} \right) \phi_{i-1,j} - \Gamma \frac{\Delta x}{\Delta y} \phi_{i,j+1} - \Gamma \frac{\Delta x}{\Delta y} \phi_{i,j-1} = 0$$

Inlet cells ($i = 1$, $2 \leq j \leq M_C - 1$):

$$\left(u_j \Delta y + 4\Gamma \frac{\Delta y}{\Delta x} + 2\Gamma \frac{\Delta x}{\Delta y} \right) \phi_{i,j} - \frac{4}{3} \Gamma \frac{\Delta y}{\Delta x} \phi_{i+1,j} - \Gamma \frac{\Delta x}{\Delta y} \phi_{i,j+1} - \Gamma \frac{\Delta x}{\Delta y} \phi_{i,j-1} = \left(u_j \Delta y + \frac{8}{3} \Gamma \frac{\Delta y}{\Delta x} \right) \phi_{in}$$

Outlet cells ($i = N_C$, $2 \leq j \leq M_C - 1$); at the outlet, we assume that the diffusive flux is zero:

$$\left(u_j \Delta y + \Gamma \frac{\Delta y}{\Delta x} + 2\Gamma \frac{\Delta x}{\Delta y} \right) \phi_{i,j} - \left(u_j \Delta y + \Gamma \frac{\Delta y}{\Delta x} \right) \phi_{i-1,j} - \Gamma \frac{\Delta x}{\Delta y} \phi_{i,j+1} - \Gamma \frac{\Delta x}{\Delta y} \phi_{i,j-1} = 0$$

Bottom wall cells ($2 \leq i \leq N_C - 1$, $j = 1$):

$$\left(u_j \Delta y + 2\Gamma \frac{\Delta y}{\Delta x} + \Gamma \frac{\Delta x}{\Delta y} \right) \phi_{i,j} - \Gamma \frac{\Delta y}{\Delta x} \phi_{i+1,j} - \left(u_j \Delta y + \Gamma \frac{\Delta y}{\Delta x} \right) \phi_{i-1,j} - \Gamma \frac{\Delta x}{\Delta y} \phi_{i,j+1} = +\Gamma \frac{J_B}{\kappa} \Delta x$$

Top wall cells ($2 \leq i \leq N_C - 1$, $j = M_C$):

$$\left(u_j \Delta y + 2\Gamma \frac{\Delta y}{\Delta x} + \Gamma \frac{\Delta x}{\Delta y} \right) \phi_{i,j} - \Gamma \frac{\Delta y}{\Delta x} \phi_{i+1,j} - \left(u_j \Delta y + \Gamma \frac{\Delta y}{\Delta x} \right) \phi_{i-1,j} - \Gamma \frac{\Delta x}{\Delta y} \phi_{i,j-1} = 0$$

Inlet bottom corner cell ($i = 1$, $j = 1$):

$$\left(u_j \Delta y + 4\Gamma \frac{\Delta y}{\Delta x} + \Gamma \frac{\Delta x}{\Delta y} \right) \phi_{i,j} - \frac{4}{3} \Gamma \frac{\Delta y}{\Delta x} \phi_{i+1,j} - \Gamma \frac{\Delta x}{\Delta y} \phi_{i,j+1} = \left(u_j \Delta y + \frac{8}{3} \Gamma \frac{\Delta y}{\Delta x} \right) \phi_{in} + \Gamma \frac{J_B}{\kappa} \Delta x$$

Inlet top corner cell ($i = 1$, $j = M_C$):

$$\left(u_j \Delta y + 4\Gamma \frac{\Delta y}{\Delta x} + \Gamma \frac{\Delta x}{\Delta y} \right) \phi_{i,j} - \frac{4}{3} \Gamma \frac{\Delta y}{\Delta x} \phi_{i+1,j} - \Gamma \frac{\Delta x}{\Delta y} \phi_{i,j-1} = \left(u_j \Delta y + \frac{8}{3} \Gamma \frac{\Delta y}{\Delta x} \right) \phi_{in}$$

Outlet bottom corner cell $(i = N_C, j = 1)$:

$$\left(u_j \Delta y + \Gamma \frac{\Delta y}{\Delta x} + \Gamma \frac{\Delta x}{\Delta y} \right) \phi_{i,j} - \left(u_j \Delta y + \Gamma \frac{\Delta y}{\Delta x} \right) \phi_{i-1,j} - \Gamma \frac{\Delta x}{\Delta y} \phi_{i,j+1} = +\Gamma \frac{J_B}{\kappa} \Delta x$$

Outlet top corner cell $(i = N_C, j = M_C)$:

$$\left(u_j \Delta y + \Gamma \frac{\Delta y}{\Delta x} + \Gamma \frac{\Delta x}{\Delta y} \right) \phi_{i,j} - \left(u_j \Delta y + \Gamma \frac{\Delta y}{\Delta x} \right) \phi_{i-1,j} - \Gamma \frac{\Delta x}{\Delta y} \phi_{i,j-1} = 0$$

The equations were solved iteratively using the alternating direction implicit (ADI) method on a 200×40 mesh. The convergence criterion was applied to the normalized (by first iteration value) residual, and a tolerance of 10^{-6} was used.

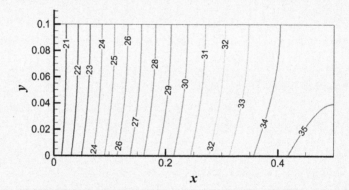

The steady-state temperature distribution (in °C) is shown in the figure above. The fluid is heated from the bottom, and has the highest temperature at the bottom boundary. The fluid flow (advection) from the left to the right forces the isotherms to take a rightward bend. The zero flux at the top boundary manifests itself as contour lines being perpendicular to the top surface.

The convergence behavior is shown in the figure below. Six orders of magnitude convergence requires approximately 137,000 iterations. The convergence exhibits linear behavior on a log scale, as would be expected of a classical iterative solver.

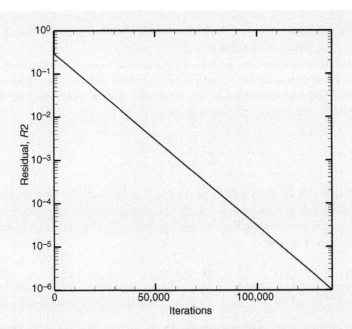

The fluxes at each boundary were computed, and a heat flux summary is presented in the table below.

	Heat flux (W/m) [positive sign = IN]
Inlet	−0.714169
Outlet	−4.285816
Bottom wall	5.000000
Top wall	0.000000
Imbalance	$−1.39×10^{-5}$

At the inlet, heat transfer occurs both by advection and diffusion (conduction). Advection brings in heat, but conduction forces it out. As a result, the net heat flux at the inlet is outward, albeit a small amount. The heat flux imbalance is roughly of the same order of magnitude as the tolerance used for convergence, and is zero for all practical purposes, emphasizing once again that the FVM automatically guarantees conservation – energy conservation in this particular case.

6.5 TWO-DIMENSIONAL AXISYMMETRIC PROBLEMS

As discussed in Section 2.7, many engineering problems are best solved using the cylindrical coordinate system. A subset of these problems is one where variation in the azimuthal direction is absent or negligible. In such a scenario, the problem reduces to the so-called 2D axisymmetric problem, in which the solution is a function of just r and z, as opposed to r, θ, and z. A 2D axisymmetric mesh is shown in Fig. 6.7, along with the coordinate directions and the stencil.

In the discussion to follow, we will apply the finite volume procedure to derive a set of discrete algebraic equations for diffusion (generalized Poisson equation) in the cylindrical coordinate system [*cf.* Eq. (2.71) and Eq. (6.77)]. Under the axisymmetric assumption, the governing equation reduces to

$$\frac{1}{r}\frac{\partial}{\partial r}\left(r\Gamma\frac{\partial\phi}{\partial r} \right) + \frac{\partial}{\partial z}\left(\Gamma\frac{\partial\phi}{\partial z} \right) = -S_\phi . \tag{6.86}$$

The first step in the finite volume procedure is to integrate the governing equation over a control volume. As opposed to the 2D planar Cartesian geometry in which $dV = dx\,dy$, here, the differential volume is written as $dV = 2\pi r dr\,dz$. Thus, Eq. (6.86) may be integrated to write

$$\int_s^n\int_w^e \frac{1}{r}\frac{\partial}{\partial r}\left(r\Gamma\frac{\partial\phi}{\partial r} \right)2\pi r dr\,dz + \int_s^n\int_w^e \frac{\partial}{\partial z}\left(\Gamma\frac{\partial\phi}{\partial z} \right)2\pi r dr\,dz = -\int_s^n\int_w^e S_\phi\, 2\pi r dr\,dz, \tag{6.87}$$

where the symbols carry their usual meaning. Reordering the integrals appropriately so that exact differentials are integrated first, the integrals on the left-hand side may be simplified, and Eq. (6.87) may be written as

$$2\pi\int_w^e\left(\left[r\Gamma\frac{\partial\phi}{\partial r} \right]_n - \left[r\Gamma\frac{\partial\phi}{\partial r} \right]_s \right)dz + 2\pi\int_s^n\left(\left[\Gamma\frac{\partial\phi}{\partial z} \right]_e - \left[\Gamma\frac{\partial\phi}{\partial z} \right]_w \right)r dr = -2\pi\int_s^n\int_w^e S_\phi\, r dr\,dz. \tag{6.88}$$

Integrating once more, and invoking the definition of the average flux at a face, we obtain

$$\left(\left[r\Gamma\frac{\partial\phi}{\partial r} \right]_n - \left[r\Gamma\frac{\partial\phi}{\partial r} \right]_s \right)\Delta z + \left(\left[\Gamma\frac{\partial\phi}{\partial z} \right]_e - \left[\Gamma\frac{\partial\phi}{\partial z} \right]_w \right)\frac{r_n^2 - r_s^2}{2} = -S_O\frac{r_n^2 - r_s^2}{2}\Delta z. \tag{6.89}$$

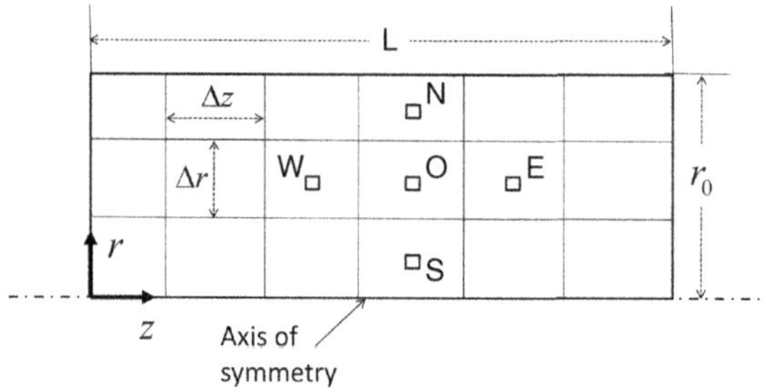

FIGURE 6.7 Schematic Representation of a 2D Axisymmetric Geometry, Stencil, and Mesh

The quantity, $(r_n^2 - r_s^2)/2$, may be written as $(r_n - r_s)(r_n + r_s)/2$, which, for a uniform mesh, reduces to $r_O \Delta r$. For a uniform mesh, using the central difference approximation [as in Eq. (6.10)], Eq. (6.89) may be further simplified to

$$r_n \Gamma_n \left(\frac{\phi_N - \phi_O}{\Delta r} \right) \Delta z - r_s \Gamma_s \left(\frac{\phi_O - \phi_S}{\Delta r} \right) \Delta z$$
$$+ r_O \Gamma_e \left(\frac{\phi_E - \phi_O}{\Delta z} \right) \Delta r - r_O \Gamma_w \left(\frac{\phi_O - \phi_W}{\Delta z} \right) \Delta r = -S_O r_O \Delta r \Delta z \qquad (6.90)$$

The radii at the faces may be expressed in terms of the cell-center radii: $r_n = r_O + \Delta r/2$, and $r_s = r_O - \Delta r/2$. Thus, in five-band form, Eq. (6.90) may be rewritten as

$$\left(\frac{r_O (\Gamma_e + \Gamma_w)}{\Delta z} \Delta r + \frac{r_n \Gamma_n + r_s \Gamma_s}{\Delta r} \Delta z \right) \phi_O - \frac{r_O \Gamma_e}{\Delta z} \Delta r \phi_E - \frac{r_O \Gamma_w}{\Delta z} \Delta r \phi_W$$
$$- \frac{r_n \Gamma_n}{\Delta r} \Delta z \phi_N - \frac{r_s \Gamma_s}{\Delta r} \Delta z \phi_S = S_O \, r_O \Delta r \Delta z \qquad (6.91)$$

The fundamental difference between Eq. (6.91) and the corresponding equation in Cartesian coordinates [Eq. (6.85)] is that the northern and southern coefficients are unequal. This is due to the fact that the surface area increases with radius in the cylindrical coordinate system, and as a result, the areas of the northern and southern faces of the control volume in question are unequal.

Boundary conditions in the cylindrical coordinate system are applied in much the same manner as the Cartesian coordinate system, which has already been discussed in preceding sections. In the FVM, no nodes are located at the axis of symmetry. Therefore, the question of singularity, as was encountered in the finite difference formulation (see Section 2.7), does not arise in the first place, and therefore, no special treatment for the axis is necessary. In order to derive the discrete finite volume equation for a cell adjacent to the axis of symmetry, we use $r_s = 0$ in Eq. (6.90), to yield

$$r_n \Gamma_n \left(\frac{\phi_N - \phi_O}{\Delta r} \right) \Delta z + r_O \Gamma_e \left(\frac{\phi_E - \phi_O}{\Delta z} \right) \Delta r - r_O \Gamma_w \left(\frac{\phi_O - \phi_W}{\Delta z} \right) \Delta r = -S_O \, r_O \Delta r \Delta z. \qquad (6.92)$$

Physically, the contribution of the southern face vanishes because its area is zero. In the example to follow, the FVM is demonstrated for an unsteady problem in cylindrical coordinates.

EXAMPLE 6.6

In this example, we consider an unsteady advection–diffusion problem in cylindrical coordinates – once again, from the heat transfer area. The problem involves unsteady heating of a cylindrical tube by applying a sinusoidally varying (in time) heat flux to the middle one-third of its outer wall. The objective is to find the temperature distribution in the tube as a function of time and space. We will assume that azimuthal symmetry exists, and therefore, 2D axisymmetric computations will be performed. The computational domain, along with boundary conditions, is depicted in the figure below. As in Example 6.5, we will make the assumption of constant thermophysical properties, and axial fully-developed flow (i.e., $v = 0$). Under these assumptions, the governing equation reduces to

$$\frac{\partial \phi}{\partial t} + \frac{1}{r}\frac{\partial}{\partial r}\left(-r\Gamma\frac{\partial \phi}{\partial r}\right) + \frac{\partial}{\partial z}\left(u\phi - \Gamma\frac{\partial \phi}{\partial z}\right) = 0 \,,$$ where ϕ is the temperature. The

thermophysical properties are taken to be that of water: $\rho = 1000$ kg/m^3, $c_p = 4200$ J/kg/°C, and $\kappa = 0.613$ W/m/°C, resulting in $\Gamma = 1.45 \times 10^{-7}$ m^2/s.

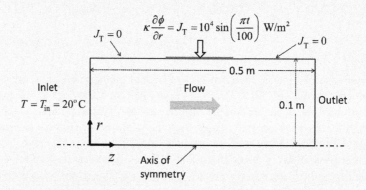

The axial velocity profile is assumed to be that for a fully-developed flow, and is written as

$$u(r) = u_{in}(r) = 2u_{avg}\left[1 - \left(\frac{r}{r_0}\right)^2\right],$$ where r_0 is the outer radius of the tube, and

$u_{avg} = 10^{-3}$ m/s is the average axial velocity of the flow through the tube. The initial conditions are $\phi(0,r,z) = \phi_{in}$.

The governing equation is discretized using the 1st UDS for advection, and backward Euler discretization in time. The resulting discrete finite volume equations are summarized below.

Interior cells ($2 \leq i \leq N_C - 1, \, 2 \leq j \leq M_C - 1$):

$$\left(\frac{r_j \Delta r \Delta z}{\Delta t} + u_j r_j \Delta r + 2\Gamma r_j \frac{\Delta r}{\Delta z} + \Gamma(r_{n,j} + r_{s,j})\frac{\Delta z}{\Delta r}\right)\phi_{i,j,n+1}$$

$$- \Gamma r_j \frac{\Delta r}{\Delta z}\phi_{i+1,j,n+1} - \left(u_j r_j \Delta r + \Gamma r_j \frac{\Delta r}{\Delta z}\right)\phi_{i-1,j,n+1}$$

$$- \Gamma r_{n,j}\frac{\Delta z}{\Delta r}\phi_{i,j+1,n+1} - \Gamma r_{s,j}\frac{\Delta z}{\Delta r}\phi_{i,j-1,n+1} = \frac{r_j \Delta r \Delta z}{\Delta t}\phi_{i,j,n}$$

Inlet cells ($i = 1, 2 \leq j \leq M_C - 1$):

$$\left(\frac{r_j \Delta r \Delta z}{\Delta t} + u_j r_j \Delta r + 4 \Gamma r_j \frac{\Delta r}{\Delta z} + \Gamma (r_{n,j} + r_{s,j}) \frac{\Delta z}{\Delta r} \right) \phi_{i,j,n+1} - \frac{4}{3} \Gamma r_j \frac{\Delta r}{\Delta z} \phi_{i+1,j,n+1}$$

$$- \Gamma r_{n,j} \frac{\Delta z}{\Delta r} \phi_{i,j+1,n+1} - \Gamma r_{s,j} \frac{\Delta z}{\Delta r} \phi_{i,j-1,n+1} = \left(u_j r_j \Delta r + \frac{8}{3} \Gamma r_j \frac{\Delta r}{\Delta z} \right) \phi_{in} + \frac{r_j \Delta r \Delta z}{\Delta t} \phi_{i,j,n}$$

Outlet cells ($i = N_C, 2 \leq j \leq M_C - 1$); at the outlet, we assume that the diffusive flux is zero:

$$\left(\frac{r_j \Delta r \Delta z}{\Delta t} + u_j r_j \Delta r + \Gamma r_j \frac{\Delta r}{\Delta z} + \Gamma (r_{n,j} + r_{s,j}) \frac{\Delta z}{\Delta r} \right) \phi_{i,j,n+1} - \left(u_j r_j \Delta r + \Gamma r_j \frac{\Delta r}{\Delta z} \right) \phi_{i-1,j,n+1}$$

$$- \Gamma r_{n,j} \frac{\Delta z}{\Delta r} \phi_{i,j+1,n+1} - \Gamma r_{s,j} \frac{\Delta z}{\Delta r} \phi_{i,j-1,n+1} = \frac{r_j \Delta r \Delta z}{\Delta t} \phi_{i,j,n}$$

Bottom boundary cells ($2 \leq i \leq N_C - 1, j = 1$):

$$\left(\frac{r_j \Delta r \Delta z}{\Delta t} + u_j r_j \Delta r + 2 \Gamma r_j \frac{\Delta r}{\Delta z} + \Gamma r_{n,j} \frac{\Delta z}{\Delta r} \right) \phi_{i,j,n+1}$$

$$- \Gamma r_j \frac{\Delta r}{\Delta z} \phi_{i+1,j,n+1} - \left(u_j r_j \Delta r + \Gamma r_j \frac{\Delta r}{\Delta z} \right) \phi_{i-1,j,n+1}$$

$$- \Gamma r_{n,j} \frac{\Delta z}{\Delta r} \phi_{i,j+1,n+1} = \frac{r_j \Delta r \Delta z}{\Delta t} \phi_{i,j,n}$$

Top boundary cells ($2 \leq i \leq N_C - 1, j = M_C$)

$$\left(\frac{r_j \Delta r \Delta z}{\Delta t} + u_j r_j \Delta r + 2 \Gamma r_j \frac{\Delta r}{\Delta z} + \Gamma r_{s,j} \frac{\Delta z}{\Delta r} \right) \phi_{i,j,n+1}$$

$$- \Gamma r_j \frac{\Delta r}{\Delta z} \phi_{i+1,j,n+1} - \left(u_j r_j \Delta r + \Gamma r_j \frac{\Delta r}{\Delta z} \right) \phi_{i-1,j,n+1}$$

$$- \Gamma r_{s,j} \frac{\Delta z}{\Delta r} \phi_{i,j-1,n+1} = \Gamma r_{n,j} \Delta z \frac{J_{T,n+1}}{\kappa} + \frac{r_j \Delta r \Delta z}{\Delta t} \phi_{i,j,n}$$

Inlet bottom corner cell ($i = 1, j = 1$):

$$\left(\frac{r_j \Delta r \Delta z}{\Delta t} + u_j r_j \Delta r + 4 \Gamma r_j \frac{\Delta r}{\Delta z} + \Gamma r_{n,j} \frac{\Delta z}{\Delta r} \right) \phi_{i,j,n+1} - \frac{4}{3} \Gamma r_j \frac{\Delta r}{\Delta z} \phi_{i+1,j,n+1}$$

$$- \Gamma r_{n,j} \frac{\Delta z}{\Delta r} \phi_{i,j+1,n+1} = \left(u_j r_j \Delta r + \frac{8}{3} \Gamma r_j \frac{\Delta r}{\Delta z} \right) \phi_{in} + \frac{r_j \Delta r \Delta z}{\Delta t} \phi_{i,j,n}$$

Inlet top corner cell $(i = 1, j = M_C)$:

$$\left(\frac{r_j \Delta r \Delta z}{\Delta t} + u_j r_j \Delta r + 4\Gamma r_j \frac{\Delta r}{\Delta z} + \Gamma r_{s,j} \frac{\Delta z}{\Delta r}\right)\phi_{i,j,n+1}$$

$$-\frac{4}{3}\Gamma r_j \frac{\Delta r}{\Delta z}\phi_{i+1,j,n+1} - \Gamma r_{s,j} \frac{\Delta z}{\Delta r}\phi_{i,j-1,n+1}$$

$$= \left(u_j r_j \Delta r + \frac{8}{3}\Gamma r_j \frac{\Delta r}{\Delta z}\right)\phi_{in} + \Gamma r_{n,j}\Delta z \frac{J_{T,n+1}}{\kappa} + \frac{r_j \Delta r \Delta z}{\Delta t}\phi_{i,j,n}$$

Outlet bottom corner cell $(i = N_C, j = 1)$:

$$\left(\frac{r_j \Delta r \Delta z}{\Delta t} + u_j r_j \Delta r + \Gamma r_j \frac{\Delta r}{\Delta z} + \Gamma r_{n,j} \frac{\Delta z}{\Delta r}\right)\phi_{i,j,n+1}$$

$$-\left(u_j r_j \Delta r + \Gamma r_j \frac{\Delta r}{\Delta z}\right)\phi_{i-1,j,n+1} - \Gamma r_{n,j} \frac{\Delta z}{\Delta r}\phi_{i,j+1,n+1} = \frac{r_j \Delta r \Delta z}{\Delta t}\phi_{i,j,n}$$

Outlet top corner cell $(i = N_C, j = M_C)$:

$$\left(\frac{r_j \Delta r \Delta z}{\Delta t} + u_j r_j \Delta r + \Gamma r_j \frac{\Delta r}{\Delta z} + \Gamma r_{s,j} \frac{\Delta z}{\Delta r}\right)\phi_{i,j,n+1}$$

$$-\left(u_j r_j \Delta r + \Gamma r_j \frac{\Delta r}{\Delta z}\right)\phi_{i-1,j,n+1}$$

$$-\Gamma r_{s,j} \frac{\Delta z}{\Delta r}\phi_{i,j-1,n+1} = \Gamma r_{n,j}\Delta z \frac{J_{T,n+1}}{\kappa} + \frac{r_j \Delta r \Delta z}{\Delta t}\phi_{i,j,n}$$

In the above equations, the notations follow those used in Chapter 5, the third superscript denotes the time index. r_j denotes the radius at cell (i, j), but since the radius only depends on j, a single subscript is used. $r_{n,j}$ and $r_{s,j}$ denote radii at the northern and southern faces of cell (i, j), respectively. $J_{T,n+1}$ denotes the flux at the top wall (outer radius of tube) at the current time step.

The above linear system is solved using row-wise sweeps on a 300 (axial) × 60 (radial) mesh, and four orders of magnitude convergence is enforced at each time step. A time step equal to 10 s is used and 20 time steps are computed. This corresponds to a total time of 200 s, which represents one full sinusoidal pulse of the applied heat flux. During the first half of the pulse (100 s) the heat flux is positive (heating of the tube), while in the second half, the heat flux is negative (cooling of the tube). The temperature distributions (in °C) at four instances are depicted in the snapshots shown below. The time instances chosen correspond to the peak-heating load, zero heating load, peaking cooling load, and zero heating load.

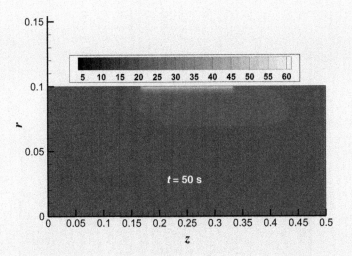

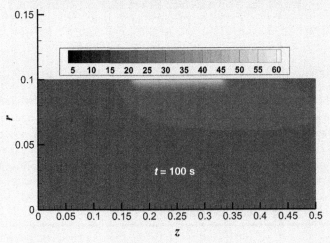

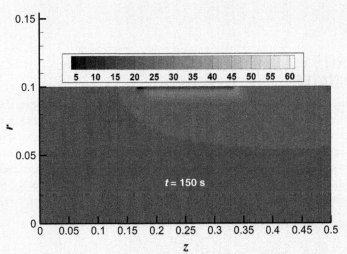

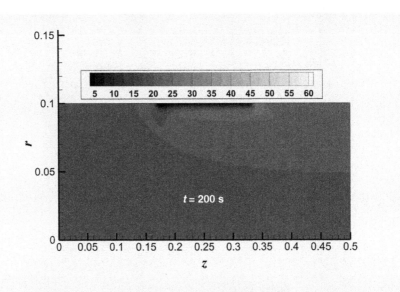

During the first 100 s, heat is added to the system. This is clearly evident in the topmost two figures. At 100 s, cooling commences. Consequently, in the bottom two figures, the wall is observed to be colder than the medium immediately next to it. The effect of advection is clearly evident, as it carries the heat from the left to the right. As far as convergence is concerned, at each time step, approximately 9 iterations were necessary to reach four orders of magnitude convergence. As discussed in Chapter 5, the introduction of the time derivative makes the system of equations better conditioned, enabling rapid convergence within each time step.

6.6 FINITE VOLUME METHOD IN CURVILINEAR COORDINATES

In Section 2.7, transformation to a general curvilinear system was discussed. The reader is encouraged to peruse that section prior to moving ahead, as it contains important transformation relations and the underlying mathematics. The transformation of the multidimensional Poisson (or diffusion) equation to curvilinear coordinates has already been presented in Section 2.7. Here, we extend that derivation to include the advective flux, and subsequently, apply the finite volume integration procedure to derive the relevant discrete algebraic equations.

The decision to include the advective flux is motivated by the fact that the FVM is routinely used for the solution of problems involving fluid flow, in which advection is omnipresent.

We begin our derivation with the generalized advection–diffusion equation, which, following Eq. (6.40), may be written as

$$\frac{\partial}{\partial t}(\rho\phi) + \nabla \bullet (\rho \mathbf{U}\phi - \Gamma \nabla\phi) = \frac{\partial}{\partial t}(\rho\phi) + \frac{\partial}{\partial x_i}\left(\rho u_i\phi - \Gamma\frac{\partial\phi}{\partial x_i}\right) = S_\phi. \qquad (6.93)$$

In the second-half of Eq. (6.93), Cartesian tensor notations have been used. A brief review of tensor notations is provided in Appendix C. The velocity vector is denoted by $\mathbf{U}$, with its components u_i ($i = 1, 2$, and 3). Using the transformation given by Eq. (2.90), it follows that the advective flux may be written as

$$\frac{\partial}{\partial x_i}(\rho u_i\phi) = \frac{\beta_{ij}}{J}\frac{\partial}{\partial \xi_j}(\rho u_i\phi). \qquad (6.94)$$

where, recalling from Section 2.7, J is the determinant of the Jacobian of the forward transformation, and β_{ij} the cofactors of the Jacobian matrix. Following the derivation of Eq. (C.12) (see Appendix C) and Eq. (2.91), the cofactor matrix may be inserted into the derivative to yield an alternative form for the right-hand side of Eq. (6.94), written as

$$\frac{\beta_{ij}}{J}\frac{\partial}{\partial \xi_j}(\rho u_i\phi) = \frac{1}{J}\frac{\partial}{\partial \xi_j}(\rho u_i\beta_{ij}\phi) = \frac{1}{J}\frac{\partial}{\partial \xi_j}(\rho U_j\phi), \qquad (6.95)$$

where the quantity

$$U_j = u_i\beta_{ij} = u_1\beta_{1j} + u_2\beta_{2j} + u_3\beta_{3j} \qquad (6.96)$$

is known as the *contravariant velocity*. Physically, it represents the component of the velocity vector normal to the constant ξ_j surface (see Fig. 2.11). Thus, U_1 is locally normal to the $\xi_2\xi_3$ surface, U_2 is locally normal to the $\xi_1\xi_3$ surface, and so on. The transformation of the diffusive flux has already been performed in Section 2.7, and is given by Eq. (2.95). Combining Eq. (6.95) and Eq. (2.95), and substituting the result into Eq. (6.93), we obtain the transformed form of the general advection–diffusion equation. Using Cartesian tensor notations, it may be written as

$$\frac{\partial}{\partial t}(\rho\phi) + \frac{1}{J}\frac{\partial}{\partial \xi_j}\left(\rho U_j\phi - \frac{\Gamma}{J}B_{kj}\frac{\partial\phi}{\partial \xi_k}\right) = S_\phi. \qquad (6.97)$$

Upon expansion of the repeated indices, Eq. (6.97) yields

$$
\frac{\partial}{\partial t}(\rho\phi) + \frac{1}{J}\frac{\partial}{\partial \xi_1}\left[\rho U_1 \phi - \frac{\Gamma}{J}\left(B_{11}\frac{\partial \phi}{\partial \xi_1} + B_{21}\frac{\partial \phi}{\partial \xi_2} + B_{31}\frac{\partial \phi}{\partial \xi_3}\right)\right]
$$
$$
+ \frac{1}{J}\frac{\partial}{\partial \xi_2}\left[\rho U_2 \phi - \frac{\Gamma}{J}\left(B_{12}\frac{\partial \phi}{\partial \xi_1} + B_{22}\frac{\partial \phi}{\partial \xi_2} + B_{32}\frac{\partial \phi}{\partial \xi_3}\right)\right] \quad , \quad (6.98)
$$
$$
+ \frac{1}{J}\frac{\partial}{\partial \xi_3}\left[\rho U_3 \phi - \frac{\Gamma}{J}\left(B_{13}\frac{\partial \phi}{\partial \xi_1} + B_{23}\frac{\partial \phi}{\partial \xi_2} + B_{33}\frac{\partial \phi}{\partial \xi_3}\right)\right] = S_\phi
$$

where the tensor B_{ij} is given by Eq. (2.94). To derive the discrete finite volume equation, Eq. (6.98) must be integrated over a finite volume. This is discussed next.

The differential volume, dV, is written in Cartesian coordinates as $dV = dx_1 dx_2 dx_3$. Substitution of Eq. (2.83) yields

$$
dV = dx_1 dx_2 dx_3 = \left(\frac{\partial x_1}{\partial \xi_i}d\xi_i\right)\left(\frac{\partial x_2}{\partial \xi_j}d\xi_j\right)\left(\frac{\partial x_3}{\partial \xi_k}d\xi_k\right), \quad (6.99)
$$

which, for any covariant basis set $(\hat{\xi}_1,\hat{\xi}_2,\hat{\xi}_3)$, reduces to [15]

$$
dV = dx_1 dx_2 dx_3 = J d\xi_1 d\xi_2 d\xi_3. \quad (6.100)
$$

Next, using Eq. (6.100), we perform the finite volume integration of Eq. (6.98). For algebraic simplicity, a 2D case is considered. Finite volume integration of the 2D rendition of Eq. (6.98) yields

$$
\int_s^n\int_w^e \frac{\partial}{\partial t}(\rho\phi) J d\xi_1 d\xi_2 + \int_s^n\int_w^e \frac{\partial}{\partial \xi_1}\left[\rho U_1 \phi - \frac{\Gamma}{J}\left(B_{11}\frac{\partial \phi}{\partial \xi_1} + B_{21}\frac{\partial \phi}{\partial \xi_2}\right)\right]d\xi_1 d\xi_2
$$
$$
+ \int_s^n\int_w^e \frac{\partial}{\partial \xi_2}\left[\rho U_2 \phi - \frac{\Gamma}{J}\left(B_{12}\frac{\partial \phi}{\partial \xi_1} + B_{22}\frac{\partial \phi}{\partial \xi_2}\right)\right]d\xi_1 d\xi_2 = \int_s^n\int_w^e S_\phi J d\xi_1 d\xi_2
$$

(6.101)

which, upon simplification for a mesh with equal grid spacing, yields

$$
\frac{\partial}{\partial t}(\rho\phi)\bigg|_o J_o \Delta\xi_1 \Delta\xi_2
$$
$$
+\left[\rho U_1 \phi - \frac{\Gamma}{J}\left(B_{11}\frac{\partial \phi}{\partial \xi_1} + B_{21}\frac{\partial \phi}{\partial \xi_2}\right)\right]_e \Delta\xi_2 - \left[\rho U_1 \phi - \frac{\Gamma}{J}\left(B_{11}\frac{\partial \phi}{\partial \xi_1} + B_{21}\frac{\partial \phi}{\partial \xi_2}\right)\right]_w \Delta\xi_2
$$
$$
+\left[\rho U_2 \phi - \frac{\Gamma}{J}\left(B_{12}\frac{\partial \phi}{\partial \xi_1} + B_{22}\frac{\partial \phi}{\partial \xi_2}\right)\right]_n \Delta\xi_1 - \left[\rho U_2 \phi - \frac{\Gamma}{J}\left(B_{12}\frac{\partial \phi}{\partial \xi_1} + B_{22}\frac{\partial \phi}{\partial \xi_2}\right)\right]_s \Delta\xi_1
$$
$$
= S_o J_o \Delta\xi_1 \Delta\xi_2
$$

(6.102)

The procedures described in the preceding sections and Chapter 5 may be used to reduce Eq. (6.102) systematically. For example, using the Euler time discretization procedure, we obtain

$$\frac{\partial}{\partial t}(\rho \phi)\Big|_{0} J_{0} \Delta \xi_{1} \Delta \xi_{2} \approx \frac{\rho_{0,n+1} \phi_{0,n+1} - \rho_{0,n} \phi_{0,n}}{\Delta t} J_{0} \Delta \xi_{1} \Delta \xi_{2}. \tag{6.103}$$

The advective flux, $\rho U_1 \phi|_e \Delta \xi_2$, may be approximated using the UDS described in Section 6.3.1. Since U_1 is the contravariant velocity component, the quantity $\rho U_1|_e \Delta \xi_2$ represents the mass flux normal to the face e. The sign of this mass flux determines the upwind cell. The quantities, B and J, are computed at cell faces using standard distance-weighted interpolation, described in Section 7.3.2. The derivatives with respect to ξ_1 are easily approximated at eastern and western cell faces using two Taylor series expansions in the ξ_1 direction. Similarly, the derivatives with respect to ξ_2 may be approximated at northern and southern cell faces using two Taylor series expansions in the ξ_2 direction. This results in the following central difference approximations:

$$\frac{\partial \phi}{\partial \xi_1}\Big|_e \approx \frac{\phi_E - \phi_O}{\Delta \xi_1}, \frac{\partial \phi}{\partial \xi_1}\Big|_w \approx \frac{\phi_O - \phi_W}{\Delta \xi_1}$$
$$\frac{\partial \phi}{\partial \xi_2}\Big|_n \approx \frac{\phi_N - \phi_O}{\Delta \xi_2}, \frac{\partial \phi}{\partial \xi_2}\Big|_s \approx \frac{\phi_O - \phi_S}{\Delta \xi_2}. \tag{6.104}$$

For a uniform grid, the approximations shown in Eq. (6.104), like their Cartesian mesh counterparts, are also second-order accurate.

The derivatives with respect to ξ_2 at the eastern and western faces and the derivatives with respect to ξ_1 at northern and southern faces require special consideration. These terms are a manifestation of the fact that the grid lines are now no longer orthogonal to each other. Figure 6.8 shows a typical 2D stencil in curvilinear coordinates.

The derivative of ϕ with respect to ξ_2 at the eastern face may be computed by first computing the value of ϕ at the northeastern and southeastern vertices (denoted by solid circles) of the cell O. In order to compute the vertex (or nodal) values, it is customary to use a four-point interpolation using values of the neighboring cells. Assuming a uniform mesh in both the ξ_1 and ξ_2 directions, the derivatives of ϕ with respect to ξ_2 at the eastern face may then be written as

$$\frac{\partial \phi}{\partial \xi_2}\Big|_e \approx \frac{\frac{\phi_N + \phi_{NE} + \phi_O + \phi_E}{4} - \frac{\phi_S + \phi_{SE} + \phi_O + \phi_E}{4}}{\Delta \xi_2} \approx \frac{\phi_N + \phi_{NE} - \phi_S - \phi_{SE}}{4 \Delta \xi_2}. \tag{6.105}$$

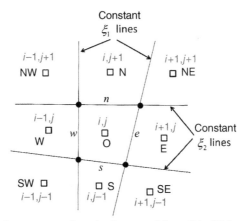

FIGURE 6.8 Schematic Representation of a Structured Stencil in 2D Curvilinear Coordinates

Likewise,

$$
\begin{aligned}
\left.\frac{\partial \phi}{\partial \xi_2}\right|_w &\approx \frac{\phi_N + \phi_{NW} - \phi_S - \phi_{SW}}{4\Delta\xi_2}, \\
\left.\frac{\partial \phi}{\partial \xi_1}\right|_s &\approx \frac{\phi_E + \phi_{SE} - \phi_W - \phi_{SW}}{4\Delta\xi_1}, \\
\left.\frac{\partial \phi}{\partial \xi_1}\right|_n &\approx \frac{\phi_E + \phi_{NE} - \phi_W - \phi_{NW}}{4\Delta\xi_1}.
\end{aligned}
\tag{6.106}
$$

Substitution of Eqs (6.103) through (6.106) into Eq. (6.102) yields the final discrete finite volume equation for cell O. This final step is left as an exercise for the reader. In an example to follow, this final step will be further elaborated upon. At this juncture, it should suffice to point out that the resulting discrete equation will have nine bands (or diagonals) instead of the five that usually arise for a perfectly orthogonal mesh. The extra four bands are due to contributions of the NW, NE, SW, and SE cells to cell O. The structure of the resulting coefficient matrix is depicted in Fig. 6.9. An example that demonstrates the solution of an elliptic PDE in transformed curvilinear coordinates is presented next.

EXAMPLE 6.7

In this example, we consider the solution of the same problem considered in Example 2.4. The objective is to find the solution of the Laplace equation on a rhombus shown in the figure below. The sides of the rhombus have lengths

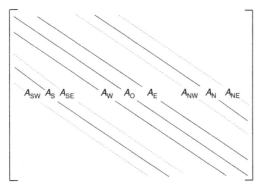

FIGURE 6.9 Structure of the Coefficient Matrix Arising Out of Discretization of the 2D Advection–Diffusion Equation on a Curvilinear Mesh

The bands denoted by dotted lines are borne out of the extra terms arising out of the transformation to curvilinear coordinates.

of 1 unit. In Example 2.4, this problem was solved using the finite difference method. Here, we aim to solve the problem using the FVM. The computational domain, the arrangement of the cells, and the boundary conditions are shown in the figure below.

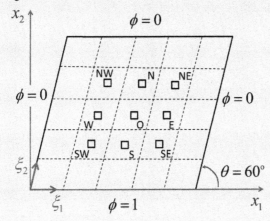

The governing equation, after coordinate transformation, has been already derived in Example 2.4, and is written as

$$\frac{\partial^2 \phi}{\partial \xi_1^2} - 2\cos\theta \frac{\partial^2 \phi}{\partial \xi_1 \partial \xi_2} + \frac{\partial^2 \phi}{\partial \xi_2^2} = 0.$$

To derive the finite volume equation, we integrate the above equation over a finite volume to yield (the reader is referred to the general derivation, presented above, for detailed steps):

$$\left(\left[\frac{\partial\phi}{\partial\xi_1}\right]_e - \left[\frac{\partial\phi}{\partial\xi_1}\right]_w\right)\Delta\xi_2 - \cos\theta\left(\left[\frac{\partial\phi}{\partial\xi_2}\right]_e - \left[\frac{\partial\phi}{\partial\xi_2}\right]_w\right)\Delta\xi_2$$

$$-\cos\theta\left(\left[\frac{\partial\phi}{\partial\xi_1}\right]_n - \left[\frac{\partial\phi}{\partial\xi_1}\right]_s\right)\Delta\xi_1 + \left(\left[\frac{\partial\phi}{\partial\xi_2}\right]_n - \left[\frac{\partial\phi}{\partial\xi_2}\right]_s\right)\Delta\xi_1 = 0$$

The cross-derivative term has been split into two terms and different orders of integration have been used to ensure that there is no directional bias in the discretization. Next, substituting the discrete approximations for the derivatives [Eqs (6.104) through (6.106)], we obtain

$$\left(\frac{\phi_E - \phi_O}{\Delta\xi_1} - \frac{\phi_O - \phi_W}{\Delta\xi_1}\right)\Delta\xi_2 - \cos\theta\left(\frac{\phi_{NE} - \phi_{SE}}{4\Delta\xi_2} - \frac{\phi_{NW} - \phi_{SW}}{4\Delta\xi_2}\right)\Delta\xi_2$$

$$-\cos\theta\left(\frac{\phi_{NE} - \phi_{NW}}{4\Delta\xi_1} - \frac{\phi_{SE} - \phi_{SW}}{4\Delta\xi_1}\right)\Delta\xi_1 + \left(\frac{\phi_N - \phi_O}{\Delta\xi_2} - \frac{\phi_O - \phi_S}{\Delta\xi_2}\right)\Delta\xi_1 = 0$$

The equation may be rearranged to write the nine-band form for the *interior* $(2 \leq i \leq N_C - 1, 2 \leq j \leq M_C - 1)$ cells

$$2\left(\frac{\Delta\xi_2}{\Delta\xi_1} + \frac{\Delta\xi_1}{\Delta\xi_2}\right)\phi_O - \frac{\Delta\xi_2}{\Delta\xi_1}\phi_E - \frac{\Delta\xi_2}{\Delta\xi_1}\phi_W - \frac{\Delta\xi_1}{\Delta\xi_2}\phi_N - \frac{\Delta\xi_1}{\Delta\xi_2}\phi_S + \frac{\cos\theta}{2}\phi_{NE}$$

$$-\frac{\cos\theta}{2}\phi_{SE} - \frac{\cos\theta}{2}\phi_{NW} + \frac{\cos\theta}{2}\phi_{SW} = 0$$

For the cells adjacent to the bottom wall, the following approximations may be used:

$$\left[\frac{\partial\phi}{\partial\xi_2}\right]_e = \frac{\frac{1}{4}(\phi_{NE} + \phi_N + \phi_E + \phi_O) - \phi_B}{\Delta\xi_2}, \left[\frac{\partial\phi}{\partial\xi_2}\right]_w = \frac{\frac{1}{4}(\phi_{NW} + \phi_N + \phi_W + \phi_O) - \phi_B}{\Delta\xi_2},$$

$$\left[\frac{\partial\phi}{\partial\xi_2}\right]_s = \frac{9\phi_O - \phi_N - 8\phi_B}{\Delta\xi_2},$$

$$\left[\frac{\partial\phi}{\partial\xi_1}\right]_n = \frac{\frac{1}{4}(\phi_{NE} + \phi_N + \phi_E + \phi_O) - \frac{1}{4}(\phi_{NW} + \phi_N + \phi_W + \phi_O)}{\Delta\xi_1}, \left[\frac{\partial\phi}{\partial\xi_1}\right]_s = 0$$

A second-order expression, Eq. (6.19), has been used to derive the derivative of ϕ with respect to ξ_2 at the southern face. Combining these approximations yields the following nodal equations:

Bottom wall $(2 \leq i \leq N_C - 1, j = 1)$:

$$\left(\frac{2\Delta\xi_2}{\Delta\xi_1} + \frac{4\Delta\xi_1}{\Delta\xi_2} \right)\phi_O - \frac{\Delta\xi_2}{\Delta\xi_1}\phi_E - \frac{\Delta\xi_2}{\Delta\xi_1}\phi_W - \frac{4}{3}\frac{\Delta\xi_1}{\Delta\xi_2}\phi_N + \frac{\cos\theta}{2}\phi_{NE} - \frac{\cos\theta}{2}\phi_{NW} = \frac{8}{3}\frac{\Delta\xi_1}{\Delta\xi_2}\phi_B$$

Likewise, the discrete equations at all other boundary cells may be derived, and are summarized below.

Top wall $(2 \leq i \leq N_C - 1, j = M_C)$:

$$\left(\frac{2\Delta\xi_2}{\Delta\xi_1} + \frac{4\Delta\xi_1}{\Delta\xi_2} \right)\phi_O - \frac{\Delta\xi_2}{\Delta\xi_1}\phi_E - \frac{\Delta\xi_2}{\Delta\xi_1}\phi_W - \frac{4}{3}\frac{\Delta\xi_1}{\Delta\xi_2}\phi_S + \frac{\cos\theta}{2}\phi_{SE} - \frac{\cos\theta}{2}\phi_{SW} = \frac{8}{3}\frac{\Delta\xi_1}{\Delta\xi_2}\phi_T$$

Left wall $(i = 1, 2 \leq j \leq M_C - 1)$:

$$\left(\frac{4\Delta\xi_2}{\Delta\xi_1} + \frac{2\Delta\xi_1}{\Delta\xi_2} \right)\phi_O - \frac{\Delta\xi_1}{\Delta\xi_2}\phi_N - \frac{\Delta\xi_1}{\Delta\xi_2}\phi_S - \frac{4}{3}\frac{\Delta\xi_2}{\Delta\xi_1}\phi_E + \frac{\cos\theta}{2}\phi_{NE} - \frac{\cos\theta}{2}\phi_{SE} = \frac{8}{3}\frac{\Delta\xi_2}{\Delta\xi_1}\phi_L$$

Right wall $(i = N_C, 2 \leq j \leq M_C - 1)$:

$$\left(\frac{4\Delta\xi_2}{\Delta\xi_1} + \frac{2\Delta\xi_1}{\Delta\xi_2} \right)\phi_O - \frac{\Delta\xi_1}{\Delta\xi_2}\phi_N - \frac{\Delta\xi_1}{\Delta\xi_2}\phi_S - \frac{4}{3}\frac{\Delta\xi_2}{\Delta\xi_1}\phi_W + \frac{\cos\theta}{2}\phi_{NW} - \frac{\cos\theta}{2}\phi_{SW} = \frac{8}{3}\frac{\Delta\xi_2}{\Delta\xi_1}\phi_R$$

Left bottom corner $(i = 1, j = 1)$:

$$\left(\frac{4\Delta\xi_2}{\Delta\xi_1} + \frac{4\Delta\xi_1}{\Delta\xi_2} + \frac{\cos\theta}{2} \right)\phi_O - \left(\frac{4}{3}\frac{\Delta\xi_1}{\Delta\xi_2} - \frac{\cos\theta}{2} \right)\phi_N - \left(\frac{4}{3}\frac{\Delta\xi_2}{\Delta\xi_1} - \frac{\cos\theta}{2} \right)\phi_E + \frac{\cos\theta}{2}\phi_{NE}$$

$$= \frac{8}{3}\frac{\Delta\xi_2}{\Delta\xi_1}\phi_L + \frac{8}{3}\frac{\Delta\xi_1}{\Delta\xi_2}\phi_B + (\phi_L + \phi_B)\cos\theta$$

Right bottom corner $(i = N_C, j = 1)$:

$$\left(\frac{4\Delta\xi_2}{\Delta\xi_1} + \frac{4\Delta\xi_1}{\Delta\xi_2} + \frac{\cos\theta}{2} \right)\phi_O - \left(\frac{4}{3}\frac{\Delta\xi_1}{\Delta\xi_2} - \frac{\cos\theta}{2} \right)\phi_N - \left(\frac{4}{3}\frac{\Delta\xi_2}{\Delta\xi_1} - \frac{\cos\theta}{2} \right)\phi_W + \frac{\cos\theta}{2}\phi_{NW}$$

$$= \frac{8}{3}\frac{\Delta\xi_2}{\Delta\xi_1}\phi_R + \frac{8}{3}\frac{\Delta\xi_1}{\Delta\xi_2}\phi_B + (\phi_R + \phi_B)\cos\theta$$

Left top corner $(i = 1, j = M_C)$:

$$\left(\frac{4\Delta\xi_2}{\Delta\xi_1} + \frac{4\Delta\xi_1}{\Delta\xi_2} + \frac{\cos\theta}{2}\right)\phi_O - \left(\frac{4}{3}\frac{\Delta\xi_1}{\Delta\xi_2} - \frac{\cos\theta}{2}\right)\phi_S - \left(\frac{4}{3}\frac{\Delta\xi_2}{\Delta\xi_1} - \frac{\cos\theta}{2}\right)\phi_E + \frac{\cos\theta}{2}\phi_{SE}$$

$$= \frac{8}{3}\frac{\Delta\xi_2}{\Delta\xi_1}\phi_L + \frac{8}{3}\frac{\Delta\xi_1}{\Delta\xi_2}\phi_T + (\phi_L + \phi_T)\cos\theta$$

Right top corner $(i = N_C, j = M_C)$:

$$\left(\frac{4\Delta\xi_2}{\Delta\xi_1} + \frac{4\Delta\xi_1}{\Delta\xi_2} + \frac{\cos\theta}{2}\right)\phi_O - \left(\frac{4}{3}\frac{\Delta\xi_1}{\Delta\xi_2} - \frac{\cos\theta}{2}\right)\phi_S - \left(\frac{4}{3}\frac{\Delta\xi_2}{\Delta\xi_1} - \frac{\cos\theta}{2}\right)\phi_W + \frac{\cos\theta}{2}\phi_{SW}$$

$$= \frac{8}{3}\frac{\Delta\xi_2}{\Delta\xi_1}\phi_R + \frac{8}{3}\frac{\Delta\xi_1}{\Delta\xi_2}\phi_T + (\phi_R + \phi_T)\cos\theta$$

The discrete equations were solved using the Gauss–Seidel method with an absolute convergence tolerance of 10^{-6} on a 40×40 mesh. A contour plot of the solution and the residual plot are shown in the figures below.

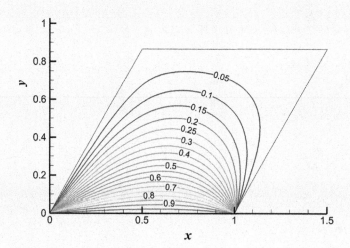

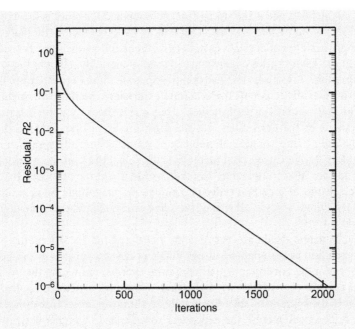

The solution is very close to the solution obtained using the FDM in Example 2.4. This example demonstrates the use of coordinate transformation along with the FVM for solving PDEs in complex geometries.

6.7 SUMMARY OF FDM AND FVM

This text covers two important methods for solving PDEs, namely the FDM and the FVM. Chapter 2 was dedicated to the FDM, while the present chapter is dedicated to the FVM. While the two methods have been compared and contrasted throughout the present chapter in bits and pieces, it is, perhaps, worthwhile to summarize their *pros* and *cons*, as well as to point out the nuances of the two methods at this juncture.

The FDM approximates and solves the governing PDE in its original form. Hence, the solution is referred to as the strong form solution. The differential operators in the governing equation are approximated by using interconnected nodes located within, and at the boundaries of, the computational domain. In contrast, in the FVM, the computational domain is first discretized into a set of control volumes or cells. The governing equation is first integrated over each of these control volumes and then solved. The solution, thus obtained, is not that of the original governing equation but an integral form thereof. Hence, the solution is known as the weak form solution. Since derivatives are ultimately approximated by algebraic equivalents in both methods, the fact that one is the strong form solution and the other is the weak form solution does not imply that that one is superior to the other simply for that reason.

The approximation of derivatives in both methods follow the same procedure, in which multiple Taylor series expansions are used to hierarchically eliminate higher order terms in the Taylor series expansions to generate higher order approximations to the derivatives. The use of Taylor series expansions implies fitting the nodal values with polynomials of high degree. This is not the only approach to obtain approximations to the derivatives, but the most prevalent one. Other approaches include the use of least square polynomials or splines. Since the approach used in approximating derivatives in both methods is the same, the truncation errors in the two methods are also comparable. In other words, the error produced by a finite difference scheme of a certain order is comparable, magnitude wise, to the error produced by a finite volume scheme of the same order. This has been demonstrated in Example 6.1.

The fundamental difference between the FDM and the FVM is in the way the boundary conditions are applied. In the FDM, nodes are located at the boundaries. In principle, the governing equation is valid everywhere within the open computational domain (open interval), and strictly speaking, it is to be applied at the boundaries of the computational domain only in a limiting sense. However, for most problems of practical interest, the dependent variable and its higher derivatives are continuous, and therefore, the governing equation can safely be (and should be) satisfied at the boundaries of the computational domain, as well. Unfortunately, in the FDM, if Dirichlet boundary conditions are applied, it is impossible to concurrently satisfy the governing equation at the boundaries without resulting in an overspecified problem. If Neumann or Robin boundary conditions are applied, it is possible to satisfy the governing equation at the boundary, as well. Concurrent satisfaction of the boundary condition (in the case of Neumann and Robin boundary conditions) and the governing equation at the boundary nodes in the FDM results in far superior accuracy than satisfying the boundary condition alone, as has been demonstrated in Example 2.2. In contrast with the FDM, in the FVM, no nodes exist on the boundary, and the boundary conditions are applied to cell faces. The fact that the governing equation is first integrated over all cells in the computational domain inherently implies that its applicability is extended to the extreme edges of the computational domain, i.e., in the true open interval. As mentioned earlier, this is exactly what is desired. This important difference between the two methods is schematically depicted in Fig. 6.10.

It is clear from Fig. 6.10 that in the FDM, the governing equation is only satisfied in part of the computational domain, while in the FVM, it is always satisfied in the entire computational domain no matter what the shape of the domain, how it is discretized, or what type the boundary conditions are. In the FDM, on the other hand, what fraction of the computational domain actually obeys the governing equation is dependent on the mesh and the type of boundary condition. This fundamental difference between the two methods manifests itself in differences in flux values computed by the two methods, especially at the boundaries. Since the fluxes of any physical

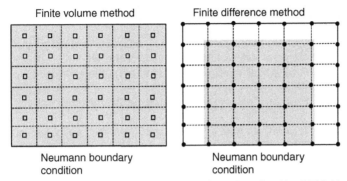

Finite volume method   Finite difference method

Neumann boundary   Neumann boundary
condition   condition

FIGURE 6.10 Schematic Representation of a Computational Domain with all Dirichlet Boundary Conditions, Except at the Bottom Boundary, Addressed Using the Finite Volume and FDMs

The gray shading denotes the region over which the governing PDE is satisfied in each of the two methods.

quantity (such as mass, energy, etc.) at the boundaries of the computational domain serve as the pillars of global conservation of that quantity, the two methods result in very different global conservation behavior. Based on our previous observation, in the case of the FDM, since the extent to which the governing equation's validity is extended to the edges of the computational domain depends on the mesh, it follows that global conservation in this method is mesh dependent. In contrast, in the FVM, since this extent is mesh independent, the global conservation properties of this method is also mesh independent. The FVM, as we have seen earlier, guarantees both local and global conservation. The aforementioned conservation properties, or lack thereof, of the two methods have been highlighted in the examples throughout this chapter.

The fact that the FVM is essentially an integral method offers one other notable advantage over the FDM. In cases where material properties exhibit discontinuities within the computational domain, the FVM can be applied with ease, while the FDM encounters problems. This is discussed in further detail in Section 7.3.2.

The chapter to follow paints a broader picture of the FVM by extending the method to control volumes of arbitrary shape. As we shall see in the next chapter, conservation laws usually result in divergence operators in PDEs. The FVM is ideally suited to equations containing divergence operators because of its inherent conservation properties, and outshines the FDM for such problems. For PDEs that are not borne out of conservation laws, the two methods are comparable. A final point worth noting is that although the FDM has been applied to unstructured grids and complex geometries, it is cumbersome to use and has almost faded to oblivion (the finite element method came as a replacement to the FDM to specifically address complex geometries). In contrast, the unstructured FVM has rapidly gained momentum in the past two decades, and is now the *de facto* method in many computational areas – most notably computational fluid dynamics, which has conservation laws at its core. The unstructured FVM is discussed in the chapter to follow.

REFERENCES

[1] Harlow FH, Welch JE. Numerical calculation of time-dependent viscous incompressible flow of fluid with a free surface. Phys Fluids 1965;8:2182–9.

[2] Whitaker S. Fundamental principles of heat transfer. Krieger Publishing Company: Malabar, FL; 1983.

[3] Ferziger JH, Perić M. Computational methods for fluid dynamics. 3rd ed. Springer: Berlin; 2002.

[4] Pletcher RH, Tannehill JC, Anderson D. Computational fluid mechanics and heat transfer. 3rd ed. CRC Press: Boca Raton, FL; 2011.

[5] Rhie CM, Chow WL. A Numerical study of the turbulent flow past an isolated airfoil with trailing edge separation. AIAA J 1983;21:1525–32.

[6] Miller TF, Schmidt FW. Use of a pressure weighted interpolation method for the solution of incompressible Navier-Stokes equations with non-staggered grid system. Numer Heat Transfer 1988;14:213–33.

[7] Majumdar S. Role of underrelaxation in momentum interpolation for calculation of flow with nonstaggered grids. Numer Heat Transfer 1988;13:125–32.

[8] Mazumder S. On the convergence of higher order upwind differencing schemes for tridiagonal iterative solution of the advection-diffusion equation. J Fluids Eng 2006;128(2):406–9.

[9] Leonard BP. A stable and accurate convective modelling procedure based on quadratic upstream interpolation. Comput Meth Appl Mech Eng 1979;19(1):59–98.

[10] Harten A, Engquist B, Osher S, Chakravarthy SR. Uniformly high order accurate essentially non-oscillatory schemes III. J Comput Phys 1987;71:231–303.

[11] Harten A. High resolution schemes for hyperbolic conservation laws. J Comput Phys 1983;49:357–93.

[12] Date AW. Introduction to computational fluid dynamics. Cambridge University Press: Cambridge, UK; 2005.

[13] Gaskell PH, Lau AKC. Curvature compensated convective transport: smart, a new boundedness preserving transport algorithm. Int J Numer Methods Fluids 1988;8:617–41.

[14] Alves MA, Oliviera PJ, Pinho FT. A convergent and universally bounded interpolation scheme for the treatment of advection. Int J Numer Methods Fluids 2003;41:47–75.

[15] Thompson JF, Warsi ZUA, Mastin CW. Numerical grid generation: foundations and applications. North-Holland: Amsterdam; 1985.

EXERCISES

6.1 Consider the following second-order linear ordinary differential equation and boundary conditions:

$$\frac{d^2\phi}{dx^2} = \exp(x), \qquad \phi(0) = 0, \phi(1) = 1.$$

Discretize the equation using the finite volume procedure and the central difference scheme. Use equal cell sizes with 10 cells. Solve the resulting discrete equations using TDMA. Also, derive the analytical solution. Plot the solution ($\phi(x)$vs. x) and also the error between the analytical and numerical solution.

Next, recompute the solution with 20 cells instead of 10. Compare the solutions between the two different meshes, and comment on your results.

6.2 Consider the following 1D diffusion equation and boundary conditions:

$$\frac{d}{dx}\left(\Gamma_\phi \frac{d\phi}{dx}\right) = S_\phi, \qquad \phi(0) = 0, \phi(1) = 1$$

where Γ_ϕ is the diffusion coefficient for ϕ. For the purposes of this exercise, you may assume $\Gamma_\phi = 1$. The source, S_ϕ, is given by $S_\phi = 2x - 1$.

a. Solve the equation analytically to determine the profile for ϕ. Also determine the flux at the left and right ends, where the flux is given by

$$J_\phi = -\Gamma_\phi \frac{d\varphi}{dx}$$

b. Discretize the equation using the finite volume technique. Use unequal grid spacing. The grid is to be generated using a power law, as explained below. If the length in the x direction is L, then

$\sum_{i=1}^{N_C} \Delta x_i = L$, where N_C is the number of cells, and Δx_i is the width of the ith cell. If we now assume that $\Delta x_{i+1} = s\Delta x_i$, where s is a stretching factor greater than 1, then we obtain the relation

$$\frac{1 - s^{N_C}}{1 - s} = \frac{L}{\Delta x_1}$$

Thus, if s and N_C are specified, one can obtain all the Δx_i. For this exercise, use $s = 1.02$. Solve the resulting discrete equations for various values of NC, starting from 10, using tridiagonal matrix inversion. Increase NC to 20, 40, and 80. Plot the error between the numerical and analytical solution on a single graph for all four mesh sizes. For each value of NC, also calculate the flux at the right and left ends.

c. Perform the exact exercise as in Part b, but now using the FDM with 11, 21, 41, and 81 nodes. Nodes are now located at faces of the cells. Tabulate your final results in three columns, which show exact flux, finite difference flux, and finite volume flux for various values of N_C. Comment on your results obtained in both Parts b and c.

6.3 Consider the steady-state 1D advection–diffusion equation subject to the boundary conditions shown below

$$\frac{d}{dx}\left(\rho u\phi - \Gamma_\phi \frac{d\phi}{dx}\right) = 0, \qquad \phi(0) = 0, \phi(1) = 1$$

where Γ_ϕ is the diffusion coefficient. For the purposes of this exercise, you may assume that Γ_ϕ and ρu are constants, such that the Peclet number is a global constant.

a. Write down the analytical solution to the above equation and boundary conditions.

b. Solve the same problem using the FVM and the following schemes: (i) first-order upwind, (ii) second-order upwind, (iii) QUICK, and (iv) exponential. For the first three schemes, use central differencing for the diffusion term. Use a uniform mesh. In each case, use 20, 40, 80, 160 cells for your calculations, and compare your results against the analytical solution. The calculations must be done for three different global (based on the total length) Peclet numbers: 0.1, 1, and 10. Note that schemes (ii) and (iii) will result in equation systems that are not tridiagonal. Use Gaussian elimination for solving the resulting set of equations.

c. Discuss your findings. Your results must be presented in a systematic and logical manner. For comparison with analytical solutions, it is best to plot the errors. Derivations of all finite volume equations, especially of boundary nodes must be presented in a neat manner for all four schemes.

6.4 Consider the 2D axisymmetric computational domain shown in the figure below.

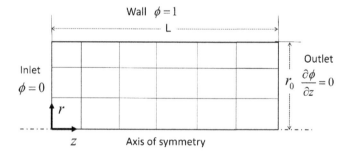

The objective is to solve the advection–diffusion in the computational domain shown above using the FVM with equal grid spacing in both the axial and radial directions. Use the velocity distribution used in Example 6.6 for the axial velocity, $u(r)$. Assume that the radial velocity is zero. The global Peclet number, defined as $Pe = \rho u_{avg} r_0 / \Gamma_\phi$, is equal to 10. Use the 1st UDS for the treatment of the advective flux and the central difference scheme for the treatment of the diffusive flux. The length of the domain is 0.5 and the radius is 0.1. Use a 200 (axial) $\times$ 40 (radial) grid for discretization, and the Stone's method for solution of the resultant linear system, with a convergence tolerance of 10^{-6}. Upon convergence, (a) make a contour plot of the solution, (b) compute the local flux (advective and diffusive) along each of the four outer boundaries, (c) integrate the flux over each of the four boundaries to compute the net incoming or outgoing flux through each boundary, and (d) comment on your findings.

6.5   Consider the following second-order linear PDE and boundary conditions (this is the same as Problem 3.4):

$$\frac{\partial^2 \phi}{\partial x^2} + \frac{\partial^2 \phi}{\partial y^2} = S_\phi = 50000 \cdot \exp[-50\{(1-x)^2 + y^2\}] \cdot [100\{(1-x)^2 + y^2\} - 2]$$

$$\phi(1, y) = 100(1 - y) + 500 \exp(-50 y^2)$$
$$\phi(0, y) = 500 \exp(-50\{1 + y^2\})$$
$$\phi(x, 0) = 100x + 500 \exp(-50(1 - x)^2)$$
$$\phi(x, 1) = 500 \exp(-50\{(1 - x)^2 + 1\})$$

The above system has an analytical solution, given by

$$\phi(x, y) = 500 \exp(-50\{(1 - x)^2 + y^2\}) + 100x(1 - y)$$

a.   Rerun the code that you developed for Problem 3.4, i.e., finite difference with Gauss–Seidel. Perform two separate calculations: (a) 41 × 41 nodes, and (b) 81 × 81 nodes. Once the solution converges (use same criterion as Problem 3.4), postprocess your results as follows:

- Derive a second-order accurate expression for $q = \partial\phi/\partial x$ or $= \partial\phi/\partial y$ for each boundary, where q is the flux.

- Integrate the flux over all nodes of a given boundary to compute a net flux on each of the four boundaries, i.e., $Q = \sum q \cdot A$, where A is the surface area (equal to Δx or Δy, depending on the surface).

- Compute the sum of the source term over the entire domain, i.e., $P = \sum_{i=1}^{N} \sum_{j=1}^{M} S_{i,j} V_{i,j}$, where $V_{i,j} = \Delta x \Delta y$ is the volume for interior nodes. For boundary nodes, use half or quarter (for corners) of the volume around the nodes.

b.   Develop a finite volume code for the above problem. Solve the final set of equations using Gauss–Seidel iterations with 40 and 80 cells in each direction (as in Part a). Use the same tolerance criterion as your finite difference code. Postprocess your converged solution in exactly the same fashion as outlined in Part a. Note that in this case, you have already derived flux expressions when you implemented boundary conditions, and the same expressions should be used to compute the fluxes. Also, it is not necessary to consider half and quarter volumes when computing the source term, because all cells have the same size in this case.

c.   Tabulate your results for both parts and for both mesh sizes as follows:

Method	Q(left)	Q (right)	Q (bottom)	Q (top)	Source, P	Imbalance, I
FDM						
FVM						
Analytical						

Recall that your overall flux balance over the entire domain gives:

$$Q(\text{right}) - Q(\text{left}) + Q(\text{top}) - Q(\text{bottom}) = \text{Source}, P$$

which implies, that the imbalance, I, is given by

$$I = Q(\text{right}) - Q(\text{left}) + Q(\text{top}) - Q(\text{bottom}) - P$$

Note: To calculate the fluxes for the analytical case, you have to differentiate (on paper) the analytical solution provided to you, set appropriate x and y values, and then integrate it (numerically) over each boundary. The source term value should be calculated in the same way, i.e., numerical integration from 0 to 1 in both x and y. For numerical integration, use any of the methods discussed in Chapter 8.

In addition to the table above, make two contour plots: (a) error in ϕ between FDM solution and analytical solution, and (b) error in ϕ between FVM solution and analytical solution. Make these two contour plots for both mesh sizes.

Discuss all your results and comment on what you have learnt from this exercise.

6.6   Repeat Example 6.7 but for a parallelogram of length 10 units. The height is to remain the same as in Example 6.7. Use 100 cells in the ξ_1 direction, and 10 cells in the ξ_2 direction. Use the Gauss–Seidel method for solving the resulting algebraic equations. Suggest one way to verify your solution.

Unstructured Finite Volume Method

7

The quest to find numerical solutions to partial differential equations (PDEs) in non-regular geometries led to coordinate transformations and the use of structured body-fitted meshes, as discussed in Chapters 1 and 2. For almost two decades (1970s and 1980s), structured body-fitted meshes remained the cutting-edge technology for numerical computations involving either the finite difference or the finite volume method. Additional geometrical complexities were dealt with by using multiple blocks or so-called block-structured mesh, in which each block (subset of the computational domain) is modeled using its own local grid line ordering and information is exchanged between blocks at their interfaces. Block-structured meshes are still widely used, especially in legacy codes in areas such as aeropropulsion and combustion. One of their biggest advantages is that they make possible the use of solvers based on line-wise sweeps. As discussed in Chapter 3, solution techniques such as the alternating direction implicit (ADI) method and Stone's method, which can only be used in the context of structured meshes, are computationally inexpensive and easy to implement, while still having reasonably good rates of convergence.

The drive toward unstructured finite volume technology in the early 1990s was prompted by two major perceived benefits. The first was the ability of unstructured meshes to model almost any shape – shapes far more complex than could ever be imagined with block-structured meshes. For example, meshing the internal and external passages of a shell-and-tube heat exchanger would be all but impossible with a block-structured mesh. The second perceived benefit was making full use of automatic unstructured mesh generation technology, which was already fairly mature by then, courtesy of the finite element method. Therefore, it made logical sense for researchers in this arena to adapt the finite volume method to unstructured mesh topology so that the inherent conservation properties of the finite volume method could be combined with the ability to treat extremely complex geometry to create a powerful modeling framework. In this chapter we extend the finite volume formulation developed in Chapter 6 to accommodate control volumes of arbitrary shape.

7.1 GAUSS DIVERGENCE THEOREM AND ITS PHYSICAL SIGNIFICANCE

As discussed in Chapter 6, the finite volume method is particularly suited to problems in which the governing equations represent conservation laws. In Chapter 6 the finite volume equation for a regular Cartesian control volume (or cell) was derived by

Numerical Methods for Partial Differential Equations: Finite Difference and Finite Volume Methods
http://dx.doi.org/10.1016/B978-0-12-849894-1.00007-X

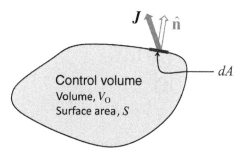

FIGURE 7.1 Schematic Representation of a Control Volume of Arbitrary Shape Showing a Differential Area on the Control Surface and the Directions of the Outward Pointing Unit Surface Normal Vector and the Flux Vector

integrating a generic elliptic PDE over the spatial bounds of the control volume [see derivation of Eq. (6.81)]. It is important to note, however, that closed-form analytical reduction of the multidimensional integrals in Eq. (6.81) was possible only because the cell in question had well-defined limits in both x and y-directions (i.e., w, e, s, and n). What if the cell was triangular or some other irregular shape? In such a scenario the procedure employed in Chapter 6 to derive the finite volume equation can no longer be used and a more general procedure is necessary. The divergence theorem enables us to do just that.

Let us consider a closed bounded region in space of volume V_O bounded by a surface of area S, as shown schematically in Fig. 7.1. Let $\mathbf{J}$ be a vector field that is continuous and has continuous first partial derivatives in this region. The divergence theorem, credited to Gauss,* states that

$$\int_{V_O} \nabla \cdot \mathbf{J} \, dV = \int_S \mathbf{J} \cdot \hat{\mathbf{n}} \, dA, \tag{7.1}$$

where $\hat{\mathbf{n}}$ is the outward-pointing unit surface normal to the differential area dA, as shown in Fig. 7.1. It is clear from Eq. (7.1) that the divergence theorem results in transformation of a volume integral to a surface integral (i.e., a reduction in dimensionality). It can also be similarly applied to a 2D control volume and would result in reduction from a surface to a line (or contour) integral.

In order to interpret the divergence theorem from a physical perspective, it is first necessary to assign a physical meaning to the vector field $\mathbf{J}$. Let us imagine that the vector $\mathbf{J}$ is a flux of some physical quantity. This physical quantity may be any fundamental quantity such as mass, momentum, energy, charge, or electric current. The flux of a quantity refers to the rate of transport of that quantity per unit time and per unit area. For example, mass flux would have units of kg/(m^2.s). Physically, flux is a vector because it has different components in different directions. For

---

*Johann Carl Friedrich Gauss** (1777–1855) was a German mathematician, known among his peers as the "prince of mathematicians." Among his numerous seminal contributions are the divergence theorem and his theory concerning the placement of nodes for numerical integration (Gaussian quadrature).

example, if a pipe is oriented at $45°$ to the ground, the components of mass flux in both the horizontal and vertical directions would be $1/\sqrt{2}$ times whatever the flux is along the pipe, such that the two components vectorially add up to reproduce total flux along the pipe. Since $\hat{n}$ is the outward-pointing unit surface normal, the quantity $\mathbf{J} \bullet \hat{n}$ represents the component of the flux normal to the surface of area dA. At some locations on the surface $\mathbf{J} \bullet \hat{n}$ may be outgoing or positive and at others $\mathbf{J} \bullet \hat{n}$ may be incoming or negative, depending on which way the vector $\mathbf{J}$ is pointing. For example, if $\mathbf{J}$ is the mass flux for all inlets $\mathbf{J} \bullet \hat{n}$ will be negative, while for all outlets it will be positive. When multiplied by the area and integrated over the entire bounding surface area, the quantity $\int_S \mathbf{J} \bullet \hat{n} \, dA$ represents the net imbalance – outgoing minus incoming – of the quantity at hand. Based on this physical understanding of the right-hand side of Eq. (7.1), it can now be safely concluded that the divergence of flux of a physical quantity in a given volume of space actually represents the net imbalance of that quantity within that volume. If the divergence is positive, then more goes out of the control volume than what comes in and vice versa.

In order to further understand the implications of the Gauss divergence theorem from a physical perspective, it is worth establishing a relationship between a conservation law and the theorem itself. To do so, let us consider a general conservation statement at steady state. The steady state assumption is not necessary, but is made at this point to keep our thought process relatively uncomplicated. Let us also assume that the conservation law being considered is for some physical quantity, denoted by ϕ, a scalar. Under these assumptions, the following general conservation statement may be written for a control volume of volume V_O:

$$\left[\begin{array}{c}\text{Rate at which} \phi \\ \text{comes into } V_O\end{array}\right] + \left[\begin{array}{c}\text{Rate at which} \phi \text{ is} \\ \text{generated within } V_O\end{array}\right] = \left[\begin{array}{c}\text{Rate at which} \phi \\ \text{goes out of } V_O\end{array}\right]. \tag{7.2}$$

Equation (7.2) may be rearranged to write

$$\left[\begin{array}{c}\text{Rate at which} \phi \\ \text{goes out of } V_O\end{array}\right] - \left[\begin{array}{c}\text{Rate at which} \phi \\ \text{comes into } V_O\end{array}\right] = \left[\begin{array}{c}\text{Rate at which} \phi \text{ is} \\ \text{generated within } V_O\end{array}\right]. \tag{7.3}$$

Based on the preceding discussion of the physical meaning of the divergence operator, we may write the left-hand side of Eq. (7.3) as

$$\int_{V_O} \nabla \bullet \mathbf{J}_\phi \, dV = \left[\begin{array}{c}\text{Rate at which} \phi \text{ is} \\ \text{generated within } V_O\end{array}\right], \tag{7.4}$$

where $\mathbf{J}_\phi$ is the flux of ϕ. If the rate of generation of ϕ per unit volume is denoted by S_ϕ, it follows then that

$$\int_{V_O} \nabla \bullet \mathbf{J}_\phi \, dV = \int_{V_O} S_\phi \, dV. \tag{7.5}$$

Equation (7.5) may be rearranged to yield:

$$\int_{V_O}\left[\nabla\bullet\mathbf{J}_\phi - S_\phi\right]dV = 0, \tag{7.6}$$

which implies that

$$\nabla\bullet\mathbf{J}_\phi = S_\phi, \tag{7.7}$$

since V_O could be arbitrary in size. If V_O is extremely small, then the quantity within square brackets in Eq. (7.6) may be assumed to be a constant over the volume and pulled outside the integral to produce Eq. (7.7). Essentially, it implies that Eq. (7.7) must be satisfied both locally and globally. The preceding derivation clearly shows that the divergence operator is inherently tied to conservation. In the physics or engineering literature, we often encounter scenarios where a quantity is referred to as *divergence free*. Such characterization of that quantity essentially implies that the quantity is conserved by virtue of the incoming amount exactly balancing the outgoing amount, and the divergence or net imbalance of its flux is zero.

At this juncture it is instructive to discuss the flux $\mathbf{J}_\phi$ from a physical standpoint. The flux or transported amount of the quantity ϕ may be caused by a variety of mechanisms. Transport mechanisms are deeply rooted in physics and a detailed discussion of the physical processes responsible for transport is beyond the scope of a numerical methods book. For further details, the reader may wish to consult a text on transport phenomena, such as the one by Bird et al. [1]. From a broad macroscopic viewpoint, the transport of a physical quantity (such as mass and energy) is caused by advection, diffusion, electromigration (drift), and other processes. The first two of these mechanisms are the most commonly prevalent. Advection refers to the transport of a quantity due to bulk motion of the medium, while diffusion (or conduction) refers to the transport of the quantity due to atomic or molecular interactions. From a macroscopic viewpoint, diffusion is generally expressed using the so-called *gradient diffusion hypothesis* in which, the diffusion flux is assumed to be proportional to the gradient of ϕ:

$$\mathbf{J}_\phi \propto \nabla\phi. \tag{7.8}$$

As a matter of fact, there are a myriad phenomenological laws (see Section 1.1 for explanation) in physics that are based on this gradient diffusion hypothesis. Some examples follow:

Heat conduction (Fourier law): $\mathbf{q}_c = -\kappa\nabla T$
Mass diffusion (Fick's law): $\mathbf{J}_d = -\rho D\nabla Y$
Current conduction (Ohm's law): $\mathbf{j}_c = -\sigma\nabla\phi$

It is clear that each of these laws is based on the gradient diffusion hypothesis. The Fourier law of heat conduction states that the conduction heat flux, $\mathbf{q}_c$, is proportional to the temperature gradient, ∇T, with a constant of proportionality κ, which is the so-called thermal conductivity. Fick's law of diffusion states that the mass diffusion flux, $\mathbf{J}_d$, is proportional to the mass fraction gradient, ∇Y, with a constant of proportionality ρD, which is the so-called dynamic diffusivity. Similarly, Ohm's law states that the conduction current flux (or current density), $\mathbf{j}_c$, is proportional to the electric potential gradient, $\nabla\phi$, with a constant of proportionality σ, which is the so-called electrical conductivity. The negative sign makes each of the laws conform to the second law of thermodynamics [1], which states that flux must always be from a higher potential to a lower potential – the potential being the driving force responsible for flux. In general, therefore, the diffusion (or conduction) flux of the vast majority of physical quantities may be written as

$$\mathbf{J}_\phi = -\Gamma\nabla\phi, \tag{7.9}$$

where Γ is the constant of proportionality and physically denotes a diffusion transport coefficient. Generally, this coefficient must be nonnegative in order for the law to conform to the second law of thermodynamics. Equation (7.9), when substituted into Eq. (7.7), yields

$$\nabla\bullet(\Gamma\nabla\phi) = -S_\phi. \tag{7.10}$$

Under the assumption of constant (space invariant) Γ, Eq. (7.10) further simplifies to

$$\nabla\bullet(\nabla\phi) = \nabla^2\phi = -S_\phi / \Gamma, \tag{7.11}$$

which is the familiar Poisson equation that was introduced in previous chapters. If the generation rate, S_ϕ, is zero, the resulting equation is a Laplace equation. The preceding discussion shows that canonical PDEs, such as the Laplace or Poisson equations, are nothing but manifestations of basic conservation laws under certain assumptions – pure diffusive transport and constant transport properties, to be precise. Consequently, these equations are often referred to as diffusion (or conduction) equations, and the Laplacian (∇^2) operator is often referred to as a diffusion operator in the physics or engineering literature. In Chapter 1, we already pointed out that the diffusion operator is elliptic.

As mentioned earlier, flux may include other mechanisms of transport such as advection and drift. In such cases, the resulting equation may be referred to as the advection–drift–diffusion equation. No matter how many transport mechanisms are included, the governing equation, assuming that it stems from a conservation law, can always be written in the form shown in Eq. (7.7). This is also referred to as the *conservative form* of the governing equations. In Section 7.2, this important finding is utilized to perform finite volume integration on an unstructured mesh.

7.2 DERIVATION OF FINITE VOLUME EQUATIONS ON AN UNSTRUCTURED MESH

We now return to the central issue of how to perform finite volume integration of the governing PDE at hand on a control volume of arbitrary convex shape. It has already been established (Section 7.1) that the governing equation may stem from a conservation law and may, therefore, be written in divergence or conservative form in the vast majority of cases. For the derivation in this section, the Poisson equation is once again considered as the governing equation so that the connection between what is derived here and what was derived in the preceding chapters is easily evident. Moreover, finite volume integration of diffusion operators (in contrast with advection) offers special challenges for an unstructured mesh because the diffusion flux is dependent on the gradient of the dependent variable itself. The final section of this chapter presents the finite volume integration of the advection–diffusion equation.

Rather than consider the Poisson equation in its traditional form (i.e., $\nabla^2 \phi = S_\phi$), we consider a more general form of the equation, namely

$$\nabla \bullet (\Gamma \nabla \phi) = S_\phi, \tag{7.12}$$

where the constant Γ, which physically represents a transport coefficient, has been reinserted. This is the more general form of the Poisson equation in the sense that it allows for nonconstant transport coefficients, as might be encountered in actual physical problems. Note that the quantity ϕ is a scalar. Consequently, the gradient $\nabla \phi$ is a vector. This makes the product $\Gamma \nabla \phi$ a vector. We will refer to this quantity as our flux vector, **J** (the minus sign can be reinserted at will to lend a physical meaning to flux). As before, the first step in a finite volume formulation is to integrate the governing equation over a generic control volume; in this case one of arbitrary shape. A schematic of such a control volume is shown in Fig. 7.2. Although the derivation to follow is applicable to any closed shape, in principle, a convex polyhedron (or polygon in 2D) is considered in Fig. 7.2 simply because most mesh generators generate meshes that are comprised of cells that are bounded by a finite number of discrete straight edges or planar faces rather than continuous curves or warped surfaces. As in Chapter 6, these control volumes will be referred to as cells and their planar bounding surfaces (or line segments

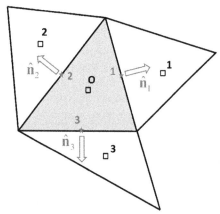

FIGURE 7.2 An Unstructured Stencil Showing a Control Volume, O, Being Surrounded by Three Neighboring Control Volumes

The Hollow Squares Denote the Geometric Centroids of the Cells (Cell Centers), While the Stars Denote the Geometric Centroids of the Bounding Faces (Face Centers).

in 2D) as faces. Similarly, the centroids of the cells and faces will be referred to as cell centers and face centers, respectively. Integration of Eq. (7.12) over the cell O yields

$$\int_{V_O} \nabla \cdot (\Gamma \nabla \phi) dV = \int_{V_O} S_\phi \, dV. \tag{7.13}$$

Since $\Gamma \nabla \phi$ is a vector, the Gauss divergence theorem can be applied directly to the left-hand side of Eq. (7.13) to yield

$$\int_S (\Gamma \nabla \phi) \cdot \hat{n} \, dA = \int_{V_O} S_\phi \, dV, \tag{7.14}$$

where S is the bounding surface area of the cell O. The volume integral on the right-hand side of Eq. (7.14) can be performed in one of two ways. The first option is to perform an exact integration. However, this is possible only if the function S_ϕ is known in explicit functional form and the control volume is of regular shape. Neither of these two criteria are generally satisfied. The second option is to apply the definition of the volume average [see Eq. (6.1)] such that

$$\int_S (\Gamma \nabla \phi) \cdot \hat{n} \, dA = \int_{V_O} S_\phi \, dV = \bar{S}_{\phi,O} V_O \approx S_{\phi,O} V_O, \tag{7.15}$$

where $\bar{S}_{\phi,O}$ denotes the volume-averaged value of the function S_ϕ. In the final step of Eq. (7.15), it has been assumed that the average value of S_ϕ over the cell is the same as its value at the geometric centroid of the cell O (as denoted by $S_{\phi O}$). This approximation becomes exact if the function S_ϕ is linear. If it is nonlinear, the approximation becomes exact only in the limiting case of the cell becoming vanishingly small (i.e., as the mesh is refined, the error in the approximation reduces). To illustrate this point, let us first consider the linear function $S_\phi(x) = x$. Therefore, $\int_0^1 S_\phi(x) dx = \int_0^1 x \, dx = 1/2$. Using the approximation, on the other hand, yields $\int_0^1 S_\phi(x) dx = S_\phi(1/2).1 = 1/2$ (i.e., the approximate and exact solutions are identical). Next, let us consider the nonlinear function $S_\phi(x) = \exp(x)$. In this case $\int_0^1 S_\phi(x) dx = \int_0^1 \exp(x) dx = e - 1 = 1.718$. Using the approximation, on the other hand, yields $\int_0^1 S_\phi(x) dx = S_\phi(1/2).1 = \exp(1/2).1 = 1.648$, which implies an error of 4.03%. Were the same integrations to be performed from 0 to 0.5, the error would become 1.03%. The exact magnitude of the error in this approximation will depend on the degree of nonlinearity of S_ϕ. It is, therefore, problem dependent. However, the important point is that the error behaves in a similar fashion as the truncation error, making this approximation consistent with the discretization process. To avoid confusion, we will refrain from using the approximation sign in the following equations since the discrete algebraic form is an approximation of the differential form to begin with.

Since the cell in question is bounded by a set of discrete planar faces, the surface integral on the left-hand side of Eq. (7.15) may be written as a discrete summation over faces to yield

$$
\int_S (\Gamma \nabla \phi) \bullet \hat{\mathbf{n}} \, dA = \sum_{f=1}^{N_{f,O}} \int_{S_f} (\Gamma \nabla \phi) \bullet \hat{\mathbf{n}} \, dA
$$
$$
= \sum_{f=1}^{N_{f,O}} \left[(\Gamma \nabla \phi)_f \bullet \hat{\mathbf{n}}_f \right] A_f = \sum_{f=1}^{N_{f,O}} \Gamma_f \left[(\nabla \phi)_f \bullet \hat{\mathbf{n}}_f \right] A_f = S_{\phi,O} V_O
$$
$$ (7.16) $$

where $N_{f,O}$ is the number of faces of cell O, and A_f their respective surface areas. In the second half of Eq. (7.16), the subscript f, used to qualify quantities such as Γ_f and $(\nabla \phi)_f$, denotes the average values of those quantities at the face. Once again, it is assumed that the average values at the face are equal to their values evaluated at the face centers. The final step is to express the quantities Γ_f and $(\nabla \phi)_f$ in terms of cell center values, since all dependent variables are stored only at cell centers in the finite volume formulation.

Prior to performing this final step for an irregular cell (as shown in Fig. 7.2), let us first reduce Eq. (7.16) for a perfectly Cartesian stencil such as the one considered in Fig. 6.6. Since the derivation in Chapter 6 using direct integration was done for the original Poisson equation [Eq. (7.12)] with $\Gamma=1$, we make the same assumption here as well. For the regular Cartesian cell shown in Fig. 6.6, Eq. (7.16) may be written as

$$
\left[(\nabla \phi)_e \bullet \hat{\mathbf{n}}_e \right] A_e + \left[(\nabla \phi)_w \bullet \hat{\mathbf{n}}_w \right] A_w + \left[(\nabla \phi)_n \bullet \hat{\mathbf{n}}_n \right] A_n + \left[(\nabla \phi)_s \bullet \hat{\mathbf{n}}_s \right] A_s = S_{\phi,O} V_O. \quad (7.17)
$$

Noting that $\hat{\mathbf{n}}_e = \hat{\mathbf{i}}, \hat{\mathbf{n}}_w = -\hat{\mathbf{i}}$, $A_e = A_w = \Delta y$, $\hat{\mathbf{n}}_n = \hat{\mathbf{j}}, \hat{\mathbf{n}}_s = -\hat{\mathbf{j}}$, $A_n = A_s = \Delta x$, and $V_O = \Delta x \, \Delta y$, Eq. (7.17) reduces to

$$
\left[(\nabla \phi)_e \bullet \hat{\mathbf{i}} \right] \Delta y - \left[(\nabla \phi)_w \bullet \hat{\mathbf{i}} \right] \Delta y + \left[(\nabla \phi)_n \bullet \hat{\mathbf{j}} \right] \Delta x - \left[(\nabla \phi)_s \bullet \hat{\mathbf{j}} \right] \Delta x = S_{\phi,O} \Delta x \, \Delta y, \quad (7.18)
$$

which, upon further reduction, yields

$$
\left[\frac{\partial \phi}{\partial x} \Big|_e \right] \Delta y - \left[\frac{\partial \phi}{\partial x} \Big|_y \right] \Delta y + \left[\frac{\partial \phi}{\partial y} \Big|_n \right] \Delta x - \left[\frac{\partial \phi}{\partial y} \Big|_s \right] \Delta x = S_{\phi,O} \Delta x \, \Delta y. \quad (7.19)
$$

Equation (7.19) is identical to Eq. (6.81) indicating that, irrespective of whether direct integration is performed or the Gauss divergence theorem is used, the resulting finite volume equation is the same. For a general unstructured stencil, expressing Γ_f and $(\nabla \phi)_f$ in terms of cell center values requires several additional pieces of information and extensive preprocessing of the mesh. This is discussed next.

7.3 PROCESSING AND STORAGE OF GEOMETRIC (MESH) INFORMATION

As discussed in Chapter 1, the connectivity between cells (i.e., which cell is connected to which other cells) in an unstructured mesh is not automatically known. Only the mesh generator is privy to such information, and this information must be communicated to the unstructured finite volume–based PDE solver code. There are several options that may be explored to perform this task. Many commercial mesh generators provide a library of application programming interfaces (APIs) that can be interfaced with either the mesh generator during runtime or be used as postprocessors for the mesh file generated by the mesh generator to extract relevant information pertaining to the mesh and geometry. For example, the MSC Nastran™ mesh generator offers such a capability. Another option is to write user-defined functions that can be attached to the solver during runtime to extract desired information. For example, such a strategy could be used with commercial computational fluid dynamics (CFD) codes such as Fluent™ or CFD-ACE+™. The advantage of this second approach is that even boundary conditions can be set using the software's graphical user interface and then extracted. This is an important consideration since boundary conditions are inherently tied to the geometry. The third option is to use an open-source unstructured mesh generator. Gmsh [2] and TetGen [3], for example, are two popular mesh generators in this category. Extraction of mesh-related information is one of the most tedious tasks in unstructured PDE solver development. However, it is not one that can be bypassed because development of an unstructured mesh generator from the ground up is far from trivial and is better left to experts in that discipline. Matlab™ supplies a toolbox that includes a simple unstructured mesh generator. This may be used for preliminary learning purposes.

Irrespective of the mesh generator used and the strategy adopted to extract the mesh and geometry information, the following data must be made available to the unstructured PDE solver.

Geometry-related information:
- Whether the geometry is 2D or 3D: *geom_type*
- Total number of cells: *ncells*
- Total number of faces: *nfaces*
- Total number of boundary faces: *nbfaces*
- Total number of vertices (or nodes): *nnodes*
- Cell center coordinates: *xc(ncells)*, *yc(ncells)*, *zc(ncells)* (only in 3D)
- Face center coordinates: *xf(nfaces)*, *yf(nfaces)*, *zf(nfaces)* (only in 3D)
- Vertex or nodal coordinates: *xv(nnodes)*, *yv(nnodes)*, *zv(nnodes)* (only in 3D)
- Surface normal: *sn(nfaces,2)* (in 2D) or *sn(nfaces,3)* (in 3D)
- Cell volumes: *vol(ncells)*
- Face areas: *areaf(nfaces)*

Connectivity information:
- Number of faces of given cell: *nface(ncells)*
- Number of vertices (or nodes) of given face: *nfnode(nfaces)*
- Number of vertices (or nodes) of given cell: *ncnode(ncells)*
- Link from cell to face: *link_cell_to_face(ncells,nface)*
- Link from face to cell: *link_face_to_cell(nfaces,2)*
- Link from face to vertex (or node): *link_face_to_node(nfaces,nfnode)*
- Link from cell to vertex (or node): *link_cell_to_node(ncells,ncnode)*

- Link from face to boundary face: *link_face_to_bface(nfaces)*
- Link from boundary face to face: *link_bface_to_face(nbfaces)*

7.3.1 PROCESSING CONNECTIVITY INFORMATION

Processing connectivity information essentially entails extraction and storage of certain integers in appropriate formats so that this information can be readily used later. This is best explained using examples. Let us consider the mesh depicted in Fig. 7.3. Although very coarse and 2D, it is adequate for the purpose at hand. Before processing connectivity information is discussed, it is important to point out the following:

- Each cell has a unique (global) number. The cells are denoted by numbers enclosed in square boxes.
- Each face has a unique (global) number. The faces are denoted by numbers that are not enclosed in any outline shape.
- Each vertex (node) has a unique (global) number. The nodes are denoted by numbers enclosed in circles.
- Each face also has a local number (local to the cell that it bounds). So, a triangular cell will have local face numbers 1, 2, and 3.
- Each cell also has a local number (local to the face that it straddles). So, each face will have two local cells on either side of it with local numbers 1 and 2.
- Each boundary face has a unique (global) number. The boundary faces are denoted by numbers enclosed in triangles.

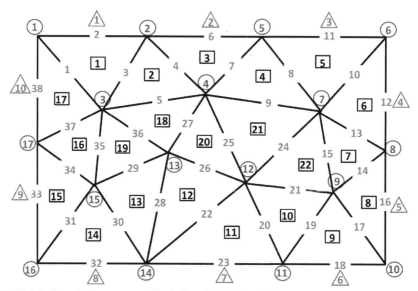

FIGURE 7.3 Sample Unstructured Mesh Showing Global Cell Numbers (Numbers Enclosed in Squares), Global Face Numbers (Unenclosed Numbers), Global Node Numbers (Numbers Enclosed in Circles), and Global Boundary Face Numbers (Numbers Enclosed in Triangles)

The spatial relationship between these local and/or global numbers (or indices) is termed *connectivity*. To illustrate how connectivity information is stored, let us first consider cell-to-face connectivity. This information is stored in a double-indexed integer array entitled *link_cell_to_face*. As an example, based on the mesh illustrated in Fig. 7.3, the following statements are true:

link_cell_to_face(20,1) = 25
link_cell_to_face(20,2) = 27
link_cell_to_face(20,3) = 26

The local face indices (1, 2, and 3 in this particular case) are assigned by the mesh generator and are quite arbitrary. The opposite relationship (i.e., face-to-cell connectivity) is similarly stored in another double-indexed integer array entitled *link_face_to_cell*. Once again, as an example:

link_face_to_cell(25,1) = 20
link_face_to_cell(25,2) = 21

The local cell indices (1 and 2) are also arbitrarily assigned by the mesh generator. However, we will see that this local numbering scheme is sometimes used to determine the direction of the surface normal despite it being arbitrary.

One question that needs to be addressed at this point is why both cell-to-face and face-to-cell connectivity information are needed. This question can be answered by considering the following two tasks: (1) compute the sum of normal fluxes (divergence) in each cell of the computational domain, and (2) compute the value of a scalar (such as temperature) at all the faces from cell center values. For the first task, let us assume that the normal outward-pointing flux at each face has already been computed and is readily available. The task simply involves summing them up. A code snippet to perform the first task is as follows:

CODE SNIPPET 7.1 DEMONSTRATION OF THE USE OF CELL-TO-FACE CONNECTIVITY

```
For icell = 1 : ncells  ! Loop over all cells
    divergence(icell) = 0
    For ifc = 1 : nface(icell)    ! Loop over number of faces of each cell
        gface = link_cell_to_face(icell,ifc) ! Making use of cell-to-face connectivity, i.e., getting
                                               global face index (gface) from global cell index (icell)
                                               and local face index (ifc)
        divergence(icell) = divergence(icell) + nflux(gface)*areaf(gface) ! Summation over normal
                                               fluxes assuming that the
                                               normal flux is already
                                               pointing outward
    End
End
```

It is clear from the above code snippet that performing the first task requires cell-to-face connectivity. Let us now consider a code snippet for the second task.

CODE SNIPPET 7.2 DEMONSTRATION OF THE USE OF FACE-TO-CELL CONNECTIVITY

```
For iface = 1 : nfaces  ! Loop over all faces
    ic1 = link_face_to_cell(iface,1) ! Get global index (ic1) of local Cell 1
    ic2 = link_face_to_cell(iface,2) ! Get global index (ic2) of local Cell 2
    face_value = interpolation(cell_value(ic1),cell_value(ic2)) ! Interpolation to get face value
                                                  from cell values
End
```

In this case, reverse connectivity (i.e., face-to-cell connectivity) is necessary. While one is derivable from the other, it is computationally wasteful since operations of the type cited above as examples are performed innumerable times within the overall solution procedure. Therefore, it is judicious to preprocess and store such information with the knowledge that integer storage is relatively less memory-intensive.

The storage of cell-to-vertex and face-to-vertex connectivity follows a procedure similar to that just described. As an example, using Fig. 7.3, we may write

link_face_to_node(25,1) = 4
link_face_to_node(25,2) = 12
and
link_cell_to_node(20,1) = 4
link_cell_to_node(20,2) = 13
link_cell_to_node(20,3) = 12

These vertex or nodal connectivity data are often used to compute interpolation functions, as will be discussed in the next subsection.

In an unstructured code, it is customary to tag the boundary faces and list them separately. This is advantageous from two viewpoints. First, not all information is relevant to interior faces; it need only be stored at boundary faces. For example, if a heat transfer computation is being performed with Newton cooling boundary conditions, the prescribed external convective heat transfer coefficient and ambient temperature need only be stored at the boundary faces. As a matter of fact, these quantities are not even relevant to interior faces. Since the total number of faces (*nfaces*) is generally much larger than the total number of boundary faces (*nbfaces*) (e.g., *nfaces* = 38 and *nbfaces* = 10 in the mesh shown in Fig. 7.3), it makes logical sense to store boundary condition information separately to save memory. In fact, as the mesh is refined, the fraction of boundary faces to total number of faces decreases. The second advantage concerns data access from computer memory. Without going into too much technical detail, it suffices to state here that data arrays are stored in computer memory as contiguous 1D blocks (or chunks). Data are fetched by setting a pointer at the beginning of the block along with the total size of the block to be fetched. Going back to the example just discussed, the convective

Total face based storage

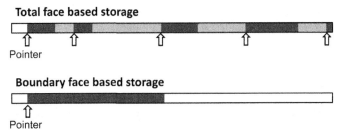

Pointer

Boundary face based storage

Pointer

FIGURE 7.4 Schematic Representation Showing Storage of Data in a Computer's Memory with Two Different Data Structures

The dark blocks indicate useful data, while the light gray blocks represent useless data.

heat transfer coefficient, when stored in a total face-based array would reside in computer memory in locations that are quite different than when stored in a boundary face–based array, as depicted schematically in Fig. 7.4. In the former case, these data can only be accessed by performing a logical IF check to determine whether the face in question is a boundary face or not and continually reset the memory pointer, while in the latter case, the entire chunk of data can be fetched in one step once that pointer is initially set.

In order to enable use of boundary face-based arrays, the connectivity between boundary face numbers and global face numbers must be preprocessed and stored. For example, based on the mesh illustrated in Fig. 7.3, we have

link_face_to_bface(33) = 9
link_bface_to_face(9) = 33

To highlight the advantages of storing boundary data in boundary face-based arrays as opposed to total face-based arrays, let us consider an example where we are interested in computing the area-weighted average of a certain scalar, ϕ, at the boundary of the entire computational domain. Mathematically, this is equivalent to computing $\int_S \phi \, dA \, / \int_S dA$. A code snippet for this task using a total face-based array for the values of ϕ at the boundary faces would resemble the following:

CODE SNIPPET 7.3 DEMONSTRATION OF THE USE OF A TOTAL FACE–BASED DATA STRUCTURE

```
sumphif = 0
sumarea = 0
For ifc = 1 : nfaces  ! Loop over all faces
    If (bface(ifc) == 1) Then  ! Filter boundary faces
        sumphif = sumphif + phif(ifc)*areaf(ifc) ! Note both area and phif are both stored using
                                    total face based data structure
        sumarea = sumarea + areaf(ifc)
    Endif
End
average = sumphif/sumarea
```

For the same task at hand, a code snippet using a boundary face–based array for the values of ϕ at the boundary faces would resemble the following:

CODE SNIPPET 7.4 DEMONSTRATION OF THE USE OF A BOUNDARY FACE–BASED DATA STRUCTURE AND BOUNDARY-FACE-TO-GLOBAL-FACE CONNECTIVITY

```
sumphif = 0
sumarea = 0
For ifbc = 1 : nbfaces  ! Loop over only boundary faces
    gfc = link_bface_to_face(ifbc)   ! Use connectivity array to get global face index (gfc) from
                                       boundary face index (ifbc)
        sumphif = sumphif + phif(ifbc)*areaf(gfc) ! Note that area uses total face based data structure
                                                    while  phif  uses  boundary  face  based  data
                                                    structure
    sumarea = sumarea + areaf(gfc)
End
average = sumphif/sumarea
```

As alluded to earlier, there are two shortcomings in the first approach. First, the loop is significantly longer and logical checks have to be performed to identify boundary faces from the full set of faces. Logical checks are computationally expensive when performed within the innermost loops of a program because of so-called branch mispredictions. Thus, even though the two approaches involve the same number of floating point operations, the first approach is likely to be significantly slower. Second, since the memory pointer has to shift several times in the first approach, this will further add to inefficiency.

7.3.2 INTERPOLATION

In the unstructured finite volume formulation, interpolations are necessary to determine face center and nodal values from cell center values. For example, Eq. (7.16) clearly shows that the face value of the transport coefficient, Γ_f, must be determined prior to solving the equation. Apart from a few special cases, it is customary to perform distance-weighted interpolation. This can be done by precalculating and storing interpolation functions as long as the mesh does not move or distort. First, let us consider the case of cell-to-face interpolation, as illustrated in Fig. 7.5.

There are several ways of performing distance-based interpolation. One is to perform inverse distance-weighted interpolation, as follows:

$$\phi_f = \frac{\phi_1 / d_1 + \phi_2 / d_2}{1 / d_1 + 1 / d_2},$$

(7.20)

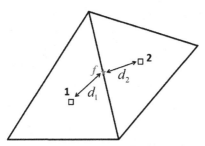

FIGURE 7.5 Unstructured Stencil Showing a Face Being Straddled by Two Cells

where d_1 and d_2 are distances between the face in question and cell centers 1 and 2, respectively. The error incurred by this type of interpolation is second order if the distances d_1 and d_2 are equal, and between first and second order if the distances are unequal [see derivation of Eq. (6.42)]. It is straightforward to show that if either $d_1 \rightarrow 0$ or $d_2 \rightarrow \infty$, $\phi_f \rightarrow \phi_1$. Conversely, if $d_1 \rightarrow \infty$ or $d_2 \rightarrow 0$, $\phi_f \rightarrow \phi_2$. Essentially, Eq. (7.20) indicates that the influence of a node on the face fades with increasing distance. While this makes logical sense, such an interpolation scheme is by no means the only choice at our disposal. Another alternative is to use inverse distance square–weighted interpolation, as follows:

$$\phi_f = \frac{\phi_1 / d_1^2 + \phi_2 / d_2^2}{1 / d_1^2 + 1 / d_2^2}. \tag{7.21}$$

This interpolation scheme also exhibits the same limiting behavior as the previous scheme. The only difference is that the influence of the node in this case fades more quickly with increasing distance. The idea behind inverse distance square-weighted interpolation stems from the laws of physics. For example, Coulomb's law states that the attractive force between two charged particles of opposite charge increases inversely as the square of the distance between them. Either of the two interpolation schemes is a valid option, although inverse distance-based interpolation is probably the more popular of the two choices, and is recommended. An alternative way of writing Eq. (7.20) is as follows:

$$\phi_f = \left(\frac{1 / d_1}{1 / d_1 + 1 / d_2} \right) \phi_1 + \left(1 - \frac{1 / d_1}{1 / d_1 + 1 / d_2} \right) \phi_2. \tag{7.22}$$

We now define a cell-to-face interpolation function

$$w_f = \frac{1 / d_1}{1 / d_1 + 1 / d_2}, \tag{7.23}$$

such that Eq. (7.22) can be rewritten as

$$\phi_f = w_f \phi_1 + \left(1 - w_f \right) \phi_2. \tag{7.24}$$

It is important to note that the function, w_f, can be precomputed and stored since it only requires geometric information (i.e., distances d_1 and d_2). Distances d_1 and d_2 can easily be computed using the coordinates of the cell centers and the face center. As mentioned earlier, these coordinates must be directly extracted from the mesh generator. The preprocessing and storage of cell-to-face interpolation functions and other interpolation functions, to be discussed shortly, save a tremendous amount of computational time. The savings can be quantified by computing a single face value using Eq. (7.22), which would require eleven multiplications and two square root operations if performed on the fly. In contrast, once the interpolation function has been computed and stored, the same computation would require only two multiplications using Eq. (7.24).

Cell-to-vertex interpolation functions may be computed using the same procedure. If a vertex (or node) is influenced by N cells (Fig. 7.6), then

$$w_{v,i} = \frac{1/d_i}{\sum\limits_{i=1}^{N} 1/d_i},$$

(7.25)

where the cell-to-vertex interpolation function, $w_{v,i}$, represents the contribution of the ith surrounding cell to the vertex, v. Once the interpolation function has been computed and stored, the vertex (or nodal) value of ϕ can be computed using

$$\phi_v = \frac{\phi_i/d_i}{\sum\limits_{i=1}^{N} 1/d_i} = \sum\limits_{i=1}^{N} w_{v,i}\,\phi_i$$

(7.26)

The cell-to-face or cell-to-vertex interpolation functions can only be computed using the connectivity data discussed in the preceding section. For example, a quantity at any face can only be computed as long as the indices of the two cells that share that face are known. This can be obtained from the *link_face_to_cell* array.

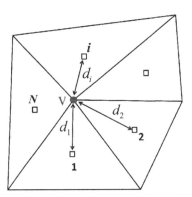

FIGURE 7.6 Unstructured Stencil Showing a Vertex (or Node) Surrounded by Cells

There is a special scenario in which the interpolation schemes, just discussed, violate the flux balance principle. This happens when there is a discontinuity in the transport property (or coefficient), Γ, at the interface between two control volumes. As discussed earlier, Γ is a transport property. Unlike the dependent variable ϕ, it need not be continuous. For example, at the interface between two materials, thermal conductivity may exhibit a jump. In fact, material property jump is one of the few scenarios where we encounter a true mathematical discontinuity in real life. To understand why the flux balance principle may be violated using a simple distance-based interpolation scheme, it is best to conduct a thought experiment on a simple heat conduction problem. Imagine two adjacent control volumes, one with a thermal conductivity of 100 W/m/K and the other with a thermal conductivity of 0 W/m/K, as shown in Fig. 7.7. Heat cannot flow across the interface since thermal conductivity on one side is zero, and any interpolation scheme we devise must enforce this physical constraint. Heat flux across the face is written as

$$J_f = -\kappa_f \left.\frac{\partial T}{\partial x}\right|_f . \tag{7.27}$$

Using Taylor series expansions and a central differencing scheme, it is straightforward to show that the expression on the right-hand side of Eq. (7.27) may be approximated as

$$J_f = -\kappa_f \left.\frac{\partial T}{\partial x}\right|_f \approx \kappa_f \frac{T_1 - T_2}{\delta_1 + \delta_2}. \tag{7.28}$$

Now, using distance-weighted interpolation (weighted arithmetic mean), thermal conductivity at the face may be written as [see Eq. (7.20)]

$$\kappa_f = \frac{\kappa_1/\delta_1 + \kappa_2/\delta_2}{1/\delta_1 + 1/\delta_2} = \frac{\kappa_1\delta_2 + \kappa_2\delta_1}{\delta_1 + \delta_2} \tag{7.29}$$

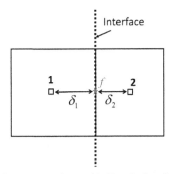

FIGURE 7.7 A Stencil Showing an Interface with Two Cells of Dissimilar Material Property (Transport Coefficient, Γ) on the Two Sides of the Interface

Setting $\kappa_2 = 0$ in Eq. (7.29) results in nonzero face conductivity and, subsequently, nonzero flux in Eq. (7.28). Therefore, the physical constraint of zero flux is violated. The central issue is the choice of interpolation scheme. Ideally, the interpolation scheme should have resulted in zero face thermal conductivity for this particular limiting case.

To derive an interpolation scheme that ensures flux balance across the face, we enforce continuity of the dependent variable (in this case, temperature) and its first derivative (heat flux). Thus, we may write

$$J_f = \kappa_1 \frac{T_1 - T_f}{\delta_1} = \kappa_2 \frac{T_f - T_2}{\delta_2}. \tag{7.30}$$

The last two expressions in Eq. (7.30) represent flux expressions at the interface approaching from the left and right sides. By setting them equal, we have enforced flux conservation across the interface. The continuity of temperature is implicit in Eq. (7.30) since the same interface temperature, T_f, has been used in both flux expressions. The final objective is to express flux in terms of the cell center values of the dependent variable. Since T_f is an unknown, this objective has not yet been met. This can be done by setting the fluxes in Eq. (7.30) equal to the flux expression in Eq. (7.28) such that

$$\kappa_f \frac{T_1 - T_2}{\delta_1 + \delta_2} = \kappa_1 \frac{T_1 - T_f}{\delta_1} = \kappa_2 \frac{T_f - T_2}{\delta_2} \tag{7.31}$$

Equation (7.31) represents two algebraic equations with two unknowns (i.e., T_f and κ_f). Solving, we obtain

$$T_f = \frac{\dfrac{\kappa_1 T_1}{\delta_1} + \dfrac{\kappa_2 T_2}{\delta_2}}{\dfrac{\kappa_1}{\delta_1} + \dfrac{\kappa_2}{\delta_2}}, \tag{7.32a}$$

and

$$\kappa_f = \frac{\kappa_1 \kappa_2}{\dfrac{\kappa_1 \delta_2 + \kappa_2 \delta_1}{\delta_1 + \delta_2}} \tag{7.32b}$$

The interface temperature, as given by Eq. (7.32a), is both inverse distance and thermal conductivity weighted, while the interface thermal conductivity as given by Eq. (7.32b) is an inverse distance-weighted harmonic mean (rather than the traditional inverse distance-weighted arithmetic mean) of the thermal conductivities of the two cell center values. Eq. (7.32b) would result in zero thermal conductivity (and zero flux) at the interface if either of the two materials has zero thermal conductivity.

In summary, when considering a scalar transported by diffusion across an interface, inverse distance-weighted interpolation may be used in general for both the scalar (dependent variable) and its transport coefficient if there is no discontinuity in the transport coefficient across the interface. In the event that a discontinuity of the transport coefficient does exist, both the value of the scalar at the interface and the value of the transport coefficient at the interface must be obtained using inverse distance-weighted harmonic interpolation, as given by Eq. (7.32).

7.3.3 CALCULATION OF VOLUME

Cell volume is one of the quantities that must be determined prior to solving the resulting finite volume equations, as is evident from Eq. (7.16). In most cases, cell volumes may be extracted directly from the mesh generator. However, there are instances where this may not be the case. Therefore, the procedure for calculation of the volume of an arbitrary polygon (in 2D) or arbitrary polyhedron (in 3D) is outlined in this section for the sake of completeness.

The basic building block in 2D is a triangle whereas in 3D it is a tetrahedron. Any other faceted shape (polygon or polyhedron) can be spilt into these two basic shapes. For example, a quadrilateral can be split into two triangles whereas a hexahedron can be split into five tetrahedrons. Therefore, this section focuses on calculating the volume of these two basic shapes.

Let us consider a vector, $\mathbf{q}$, such that $\mathbf{q} = x\,\hat{\mathbf{i}}$. Next, let us take the divergence of this vector field, such that

$$\nabla \bullet \mathbf{q} = \nabla \bullet (x\,\hat{\mathbf{i}}) = \left(\frac{\partial}{\partial x}\hat{\mathbf{i}} + \frac{\partial}{\partial y}\hat{\mathbf{j}} + \frac{\partial}{\partial z}\hat{\mathbf{k}} \right) \bullet (x\,\hat{\mathbf{i}}) = \frac{\partial}{\partial x}(x)\hat{\mathbf{i}} \bullet \hat{\mathbf{i}} = 1. \qquad (7.33)$$

Since the divergence of $\mathbf{q}$ is unity, we may write

$$V_O = \int_{V_O} dV = \int_{V_O} \nabla \bullet \mathbf{q}\, dV. \qquad (7.34)$$

Now, applying the Gauss divergence theorem, Eq. (7.1), we obtain

$$V_O = \int_{V_O} \nabla \bullet \mathbf{q}\, dV = \int_S \mathbf{q} \bullet \hat{\mathbf{n}}\, dA = \int_S (x\,\hat{\mathbf{i}}) \bullet \hat{\mathbf{n}}\, dA. \qquad (7.35)$$

Since the control volume in question is comprised of a set of discrete planar faces, each with a unique surface normal, the surface integral on the right-hand side of Eq. (7.35) may be split into a summation over the discrete faces to yield

$$V_O = \int_S (x\,\hat{\mathbf{i}}) \bullet \hat{\mathbf{n}}\, dA = \sum_{f=1}^{N_{f,O}} \int_{S_f} (x\,\hat{\mathbf{i}}) \bullet \hat{\mathbf{n}}_f\, dA = \sum_{f=1}^{N_{f,O}} \int_{S_f} x\, n_{x,f}\, dA, \qquad (7.36)$$

where $n_{x,f} = \hat{\mathbf{i}} \bullet \hat{\mathbf{n}}_f$ is the x-component of the surface normal at face f. Since the surface normal is unique on any discrete face and does not change with location on the face, we can further write

$$V_O = \sum_{f=1}^{N_{f,O}} n_{x,f} \int_{S_f} x\, dA. \qquad (7.37)$$

Since the coordinates of the geometrical centroid of any face (or face center) are given by

$$
x_f = \frac{1}{A_f} \int_{S_f} x \, dA
$$

$$
y_f = \frac{1}{A_f} \int_{S_f} y \, dA,
$$

$$
z_f = \frac{1}{A_f} \int_{S_f} z \, dA
$$

(7.38)

Eq. (7.37) may be further written as

$$
V_O = \sum_{f=1}^{N_{f,O}} n_{x,f} x_f A_f.
$$

(7.39)

Similar formulas may be derived by considering $\mathbf{q} = y \, \hat{\mathbf{j}}$ or $\mathbf{q} = z \, \hat{\mathbf{k}}$. In order to avoid the numerical precision errors that might occur when the cell is highly skewed, it is customary to take the average of the volumes obtained by alternative formulas; i.e., the volume can be computed as

$$
V_O = \frac{1}{2} \left(\sum_{f=1}^{N_{f,O}} n_{x,f} x_f A_f + \sum_{f=1}^{N_{f,O}} n_{y,f} y_f A_f \right) \text{in 2D}
$$

$$
V_O = \frac{1}{3} \left(\sum_{f=1}^{N_{f,O}} n_{x,f} x_f A_f + \sum_{f=1}^{N_{f,O}} n_{y,f} y_f A_f + \sum_{f=1}^{N_{f,O}} n_{z,f} z_f A_f \right) \text{in 3D}
$$

(7.40)

It has been assumed in Eq. (7.40) that 2D computations will be performed with the geometry on the x–y plane.

7.3.4 CALCULATION OF FACE AREAS AND SURFACE NORMALS

As in the case of cell volume, even though face areas and surface normals are usually extracted directly from the mesh generator, this section is included for the sake of completeness.

Calculation of the area of a face in 2D geometry essentially amounts to calculation of the length of a line segment. This can be done directly from the vertices of the face and needs no further discussion. The surface normal can be computed from the surface tangent by using vector relationships. Let us consider a 2D face (line segment) as shown in Fig. 7.8. First, the components of the unit tangent along the face are computed using

$$
t_{x,f} = \frac{x_2 - x_1}{\sqrt{(x_2 - x_1)^2 + (y_2 - y_1)^2}} = \frac{x_2 - x_1}{A_f}
$$

$$
t_{y,f} = \frac{y_2 - y_1}{\sqrt{(x_2 - x_1)^2 + (y_2 - y_1)^2}} = \frac{y_2 - y_1}{A_f}
$$

(7.41)

FIGURE 7.8 Schematic of a Face in 2D Showing the Two Vertices, the Unit Surface Tangent and the Unit Surface Normal

It is assumed once again that the 2D mesh is in the x–y plane. In order to obtain the unit normal from the unit tangent, we enforce the following two relationships:

$$\hat{\mathbf{n}}_f \cdot \hat{\mathbf{t}}_f = 0,$$
(7.42a)

and

$$\hat{\mathbf{n}}_f \times \hat{\mathbf{t}}_f = \hat{\mathbf{k}}$$
(7.42b)

The first of these two equations states that the normal and tangent are perpendicular to each other, while the second states that the cross product of these two vectors will result in a unit vector in the z-direction. This is true since the mesh is assumed to be on the x–y plane. Writing Eq. (7.42) explicitly in terms of the vector components results in

$$n_{x,f}t_{x,f} + n_{y,f}t_{y,f} = 0,$$
(7.43a)

$$n_{x,f}t_{y,f} - n_{y,f}t_{x,f} = 1.$$
(7.43b)

Solving, we obtain $n_{y,f} = -t_{x,f}$ and $n_{x,f} = t_{y,f}$, both of which are available from Eq. (7.41).

To compute the area of a face in 3D, as before, we first split the face into its basic triangular entities and then find the area of each triangle. Let us consider a generic triangle, as shown in Fig. 7.9. The area of this triangle may be written as

$$A_f = A_{123} = \frac{1}{2}\left|\mathbf{t}_1 \times \mathbf{t}_2\right| = \frac{1}{2}\left|\mathbf{n}\right|.$$
(7.44)

Note that the vectors used in Eq. (7.44) are not unit vectors (which are shown in Fig. 7.9), but vectors with their own respective lengths. The two tangents can easily be calculated from the vertex coordinates by subtracting the position vectors of the pair of nodes under consideration. Once the two tangents have been determined, their

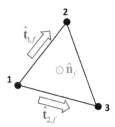

FIGURE 7.9 Schematic of the Triangular Face of a Cell in 3D Showing the Two Unit Surface Tangents and the Unit Surface Normal Pointing Out of the Plane of the Paper

cross product can be computed. The cross product represents a vector normal to the plane of the triangle, **n**, and is given by

$$\mathbf{n} = (t_{1y}t_{2z} - t_{1z}t_{2y})\hat{\mathbf{i}} + (t_{1z}t_{2x} - t_{1x}t_{2z})\hat{\mathbf{j}} + (t_{1x}t_{2y} - t_{1y}t_{2x})\hat{\mathbf{k}}. \tag{7.45}$$

One half of the magnitude of **n** is the area of the triangle [Eq. (7.44)], while the unit surface normal may be computed using

$$\hat{\mathbf{n}} = \frac{\mathbf{n}}{|\mathbf{n}|}. \tag{7.46}$$

While most mesh generators provide the surface normal, a critical issue is the direction in which it is pointing. For example, Fig. 7.2 shows that it is pointing out of cell O at face 1 but pointing into cell 1. Let us now consider a scenario in which the volume of cell 1 needs to be computed. Equation (7.39) enables us to perform this calculation, but it was derived with the assumption the all surface normals are pointing outward (the underlying premise of the Gauss divergence theorem). Therefore, we need to know which way (inward or outward) the surface normal at each face of cell 1 is pointing and flip its direction if necessary before applying Eq. (7.39).

It was mentioned earlier, when discussing face-to-cell connectivity, that although local *cell1* and *cell2* are assigned arbitrarily by the mesh generator, this assignment is often made use of by mesh generators to store the direction of the surface normal at the face. Most mesh generators store the surface normal such that it is pointing from local *cell1* to *cell2*. Thus, if *link_face_to_cell(25,1)* = *20* and *link_face_to_cell(25,2)* = *21* (see Fig. 7.3), then the surface normal at face 25 is stored such that it is pointing from cell 20 to cell 21. Once the direction of the surface normal at each face is known, the next task is to flip it on an as-needed basis. This is done by storing the direction of the surface normal from each cell's local viewpoint (i.e., locally in or out). A code snippet designed to perform this task is presented below, in which a double-indexed integer array, *snsign*, is used to store the direction of the surface normal from each cell's viewpoint.

CODE SNIPPET 7.5 DETERMINATION OF SURFACE NORMAL DIRECTION FROM THE VIEWPOINT OF EACH CELL

```
For icell = 1 : ncells  ! Loop over all cells
    For ifc = 1 : nface(icell)  ! Loop over number of faces of each cell
        gface = link_cell_to_face(icell,ifc)  ! Get global face index (gface) from global cell index
                                                (icell) and local face index (ifc)
        ic1 = link_face_to_cell(gface,1)  ! Global cell index of local cell1
        If (icell == ic1) Then  ! If cell1 is the same as the cell under consideration
            snsign(icell,ifc) = 1  ! Surface normal already pointing OUT
        Else
            snsign(icell,ifc) = -1  ! Surface normal pointing IN
        Endif
    End
End
```

Once the direction of the surface normal has been stored from each cell's local viewpoint, it can be readily used. Getting back to the example of computing the volume of a cell using Eq. (7.39), the following code snippet demonstrates how the *snsign* array may be used:

CODE SNIPPET 7.6 DEMONSTRATION OF THE USE OF THE LOCAL DIRECTION OF THE SURFACE NORMAL

```
For icell = 1 : ncells  ! Loop over all cells
    volume(icell) = 0
    For ifc = 1 : nface(icell)  ! Loop over number of faces of each cell
        gface = link_cell_to_face(icell,ifc)  ! Get global face index (gface)
        ! The next step is the summation of Eq. (7.39).
        volume(icell) = volume(icell) + sn(gface,1)*snsign(icell,ifc)*xf(gface)*areaf(gface)
    End
End
```

Note that the product $sn(gface,1)*snsign(icell,ifc)$ in Code snippet 7.6 produces the outward-pointing x-component of the surface normal, which is required for Eq. (7.39), as discussed earlier.

Our discussion on mesh and geometry processing and storage is complete at this juncture. We now return to discussion of the finite-volume equation [Eq. (7.16)] and express all quantities at the face in terms of cell center values to complete the formulation.

7.4 TREATMENT OF NORMAL AND TANGENTIAL FLUXES

Examination of Eq. (7.16) reveals that the central task at hand involves expressing the term $(\nabla\phi)_f \bullet \hat{\mathbf{n}}_f$ at each face of cell O in terms of the cell center values of neighboring cells. Henceforth, neighboring cells will simply be referred to as *neighbors*. In this section, we focus on just one generic face of cell O for our derivation. We start by deriving $(\nabla\phi)_f \bullet \hat{\mathbf{n}}_f$ for a 2D stencil and then derive it for a 3D stencil.

7.4.1 TWO-DIMENSIONAL GEOMETRY

In 2D geometry, the gradient of ϕ at face f is a vector and may be written as

$$(\nabla\phi)_f = \left.\frac{\partial\phi}{\partial x}\right|_f \hat{\mathbf{i}} + \left.\frac{\partial\phi}{\partial y}\right|_f \hat{\mathbf{j}} = \left[(\nabla\phi)_f \bullet \hat{\mathbf{i}}\right]\hat{\mathbf{i}} + \left[(\nabla\phi)_f \bullet \hat{\mathbf{j}}\right]\hat{\mathbf{j}}. \tag{7.47}$$

In Eq. (7.47), the vector $\nabla\phi$ has been decomposed into two mutually perpendicular directions: the Cartesian coordinate directions x and y. In principle, the vector $(\nabla\phi)_f$ could be decomposed into any two arbitrary directions that are mutually perpendicular, such as the normal and tangent to the face f, such that

$$(\nabla\phi)_f = \left[(\nabla\phi)_f \bullet \hat{\mathbf{n}}_f\right]\hat{\mathbf{n}}_f + \left[(\nabla\phi)_f \bullet \hat{\mathbf{t}}_f\right]\hat{\mathbf{t}}_f. \tag{7.48}$$

Next, we perform a dot product of Eq. (7.48) with the vector $\mathbf{l}_f$, which is the vector pointing from the cell center of O to the cell center of N, as illustrated in Fig. 7.10. The resulting equation is

$$(\nabla\phi)_f \bullet \mathbf{l}_f = \left[(\nabla\phi)_f \bullet \hat{\mathbf{n}}_f\right]\hat{\mathbf{n}}_f \bullet \mathbf{l}_f + \left[(\nabla\phi)_f \bullet \hat{\mathbf{t}}_f\right]\hat{\mathbf{t}}_f \bullet \mathbf{l}_f. \tag{7.49}$$

The quantity $\hat{\mathbf{n}}_f \bullet \mathbf{l}_f$ represents the distance between cell center O and cell center N in the direction normal to the face f. It is denoted by δ_f in Fig. 7.10. Despite not being mentioned in Section 7.3, this is another quantity that is precomputed and stored for each face so that it is readily available. Therefore, Eq. (7.49) may be rewritten as

$$(\nabla\phi)_f \bullet \hat{\mathbf{n}}_f = \frac{(\nabla\phi)_f \bullet \mathbf{l}_f}{\delta_f} - \frac{\left[(\nabla\phi)_f \bullet \hat{\mathbf{t}}_f\right]\hat{\mathbf{t}}_f \bullet \mathbf{l}_f}{\delta_f}, \tag{7.50}$$

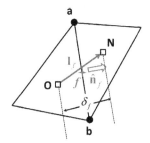

FIGURE 7.10 A 2D Unstructured Stencil Showing the Cell O and its Neighbor N

The face shared by the two faces is denoted by f.

where $\hat{\mathbf{t}}_f$ is the unit surface tangent pointing from b to a. As in the case of a regular Cartesian stencil (Chapter 6), we now perform two Taylor series expansions: one of O about f and the other of N about f. Of course, both Taylor series expansions will be 2D in this case and may be written as

$$
\phi_N = \phi_f + \left.\frac{\partial\phi}{\partial x}\right|_f (x_N - x_f) + \left.\frac{\partial\phi}{\partial y}\right|_f (y_N - y_f)
$$
$$
+ \frac{1}{2}\left.\frac{\partial^2\phi}{\partial x^2}\right|_f (x_N - x_f)^2 + \frac{1}{2}\left.\frac{\partial^2\phi}{\partial y^2}\right|_f (y_N - y_f)^2 + \frac{1}{2}\left.\frac{\partial^2\phi}{\partial x\partial y}\right|_f (x_N - x_f)(y_N - y_f) + \dots
$$

$$(7.51a)$$

$$
\phi_O = \phi_f + \left.\frac{\partial\phi}{\partial x}\right|_f (x_O - x_f) + \left.\frac{\partial\phi}{\partial y}\right|_f (y_O - y_f)
$$
$$
+ \frac{1}{2}\left.\frac{\partial^2\phi}{\partial x^2}\right|_f (x_O - x_f)^2 + \frac{1}{2}\left.\frac{\partial^2\phi}{\partial y^2}\right|_f (y_O - y_f)^2 + \frac{1}{2}\left.\frac{\partial^2\phi}{\partial x\partial y}\right|_f (x_O - x_f)(y_O - y_f) + \dots
$$

$$(7.51b)$$

Subtracting Eq. (7.51b) from Eq. (7.51a), we obtain

$$
\phi_N - \phi_O = \left.\frac{\partial\phi}{\partial x}\right|_f (x_N - x_O) + \left.\frac{\partial\phi}{\partial y}\right|_f (y_N - y_O)
$$
$$
+ \frac{1}{2}\left.\frac{\partial^2\phi}{\partial x^2}\right|_f \left[(x_N - x_f)^2 - (x_O - x_f)^2\right] + \frac{1}{2}\left.\frac{\partial^2\phi}{\partial y^2}\right|_f \left[(y_N - y_f)^2 - (y_O - y_f)^2\right].
$$
$$
+ \frac{1}{2}\left.\frac{\partial^2\phi}{\partial x\partial y}\right|_f \left[(x_N - x_f)(y_N - y_f) - (x_O - x_f)(y_O - y_f)\right] + \dots
$$

$$(7.52)$$

Noting that $\mathbf{l}_f = (x_N - x_O)\,\hat{\mathbf{i}} + (y_N - y_O)\,\hat{\mathbf{j}}$, Eq. (7.52) may be written as

$$
\phi_N - \phi_O \approx (\nabla\phi)_f \bullet \mathbf{l}_f,
$$

$$(7.53)$$

with the corresponding truncation error, ε, given by

$$
\varepsilon = \frac{1}{2}\left.\frac{\partial^2\phi}{\partial x^2}\right|_f \left[(x_N - x_f)^2 - (x_O - x_f)^2\right] + \frac{1}{2}\left.\frac{\partial^2\phi}{\partial y^2}\right|_f \left[(y_N - y_f)^2 - (y_O - y_f)^2\right]
$$
$$
+ \frac{1}{2}\left.\frac{\partial^2\phi}{\partial x\partial y}\right|_f \left[(x_N - x_f)(y_N - y_f) - (x_O - x_f)(y_O - y_f)\right] + \dots
$$

$$(7.54)$$

Substitution of Eq. (7.53) into Eq. (7.50) yields

$$
(\nabla\phi)_f \bullet \hat{\mathbf{n}}_f = \frac{\phi_N - \phi_O}{\delta_f} - \frac{\left[(\nabla\phi)_f \bullet \hat{\mathbf{t}}_f\right]\hat{\mathbf{t}}_f \bullet \mathbf{l}_f}{\delta_f} = \frac{\phi_N - \phi_O}{\delta_f} - \frac{J_{T,f}}{\delta_f}.
$$

$$(7.55)$$

Equation (7.55) essentially shows that flux normal to the face f has been decomposed into two components: one along the vector joining the two cell centers and another tangential to the face. If the vector $\mathbf{l}_f$ and the surface normal $\hat{\mathbf{n}}_f$ are coaligned, then $\hat{\mathbf{t}}_f \cdot \mathbf{l}_f$ is zero and the tangential flux vanishes. In such a scenario, we recover the simple central difference formula that was derived in Chapter 6 for a regular Cartesian stencil [Eq. (6.10)]. This observation does not imply that the tangential flux term – second term in Eq. (7.55) – will vanish only if we have a Cartesian mesh. The only criterion for this term to vanish is that vectors $\mathbf{l}_f$ and $\hat{\mathbf{n}}_f$ must be coaligned. This may happen, for example, in a triangular mesh in which all triangles are equilateral or in a tetrahedral mesh in which all tetrahedrons are of equal size, among many other scenarios.

Let us now focus our attention on calculating the tangential flux, $J_{T,f}$. Following the same procedure used to derive $(\nabla\phi)_f \cdot \mathbf{l}_f$ [Eq. (7.53)] it can easily be shown that $(\nabla\phi)_f \cdot \mathbf{t}_f \approx \phi_a - \phi_b$, where ϕ_a and ϕ_b are values of ϕ at the vertices a and b, respectively, as shown in Fig. 7.10. Therefore, it follows that

$$(\nabla\phi)_f \cdot \hat{\mathbf{t}}_f \approx \frac{\phi_a - \phi_b}{|\mathbf{t}_f|}, \tag{7.56}$$

where $|\mathbf{t}_f|$ is the length of the line segment ab (or area of face f in the case of a 2D mesh). The quantity $\hat{\mathbf{t}}_f \cdot \mathbf{l}_f$ can be computed directly using the coordinates of the vertices and cell centers, such that the final expression for tangential flux may be written as

$$J_{T,f} = \left[(\nabla\phi)_f \cdot \hat{\mathbf{t}}_f \right] \hat{\mathbf{t}}_f \cdot \mathbf{l}_f = \left[\frac{\phi_a - \phi_b}{A_f} \right] \frac{(x_a - x_b)(x_1 - x_O) + (y_a - y_b)(y_1 - y_O)}{A_f}. \tag{7.57}$$

Finally, substituting Eq. (7.55) and (7.57) into Eq. (7.16), we obtain

$$\sum_{f=1}^{N_{f,O}} \Gamma_f \left(\frac{\phi_{N(f)} - \phi_O}{\delta_f} - \left[\frac{\phi_{a(f)} - \phi_{b(f)}}{\delta_f A_f} \right] \hat{\mathbf{t}}_f \cdot \mathbf{l}_f \right) A_f = S_{\phi,O} V_O. \tag{7.58}$$

In Eq. (7.58), the value of ϕ at the neighboring cell N has been written as $\phi_{N(f)}$ rather than ϕ_N to indicate that this is the value at the cell that is a neighbor adjacent to the face f. Similarly, instead of denoting the value of ϕ at vertex a as ϕ_a, it has been denoted by $\phi_{a(f)}$ to imply that this is the vertex a of face f.

While the task of expressing the normal gradient at the face in terms of cell center values is now complete, it is only partly successful. Equation (7.58) still contains vertex values of ϕ that are unknown. There are two approaches to treating vertex values. The first is to express vertex values $\phi_{a(f)}$ and $\phi_{b(f)}$ in terms of neighboring cell center values using the cell-to-vertex interpolation functions derived in Section 7.3.2 [see Eq. (7.26)]. Upon substitution of these expressions into Eq. (7.58), an equation containing only cell center values would be obtained. Based on our discussions in Chapter 3, this is an *implicit* treatment of tangential flux because the resulting equation only has cell center values of ϕ, which, in fact, are unknowns. The second approach – an *explicit* approach – is one in which tangential flux is computed explicitly using previous iteration values and added to the source term, such that Eq. (7.58) becomes

$$\sum_{f=1}^{N_{f,O}} \Gamma_f \left(\frac{\phi_{N(f)} - \phi_O}{\delta_f} \right) A_f = S_{\phi,O} V_O + \sum_{f=1}^{N_{f,O}} \Gamma_f \left(\left[\frac{\phi_{a(f)}^* - \phi_{b(f)}^*}{\delta_f A_f} \right] \hat{\mathbf{t}}_f \bullet \mathbf{l}_f \right) A_f \quad (7.59)$$

where the superscript "*" once again denotes previous iteration values. Clearly, the explicit approach can be implemented only within the context of an iterative solver.

Both approaches have their respective *pros* and *cons*. The explicit approach is relatively straightforward to implement since the stencil of influence only includes immediate neighbors. In contrast, in the implicit approach, the stencil will extend beyond the immediate neighbors. The end result is that the coefficient matrix will be less sparse in the implicit approach (i.e., more memory will be needed to store it). The increased stencil of influence also implies that the treatment of boundary conditions will require additional caution. However, the implicit approach has the advantage that it will generally produce convergence that is superior to the explicit approach, in keeping with our findings in Chapter 3. The explicit approach is the more popular of the two approaches courtesy of the advantages cited earlier. An important point to note is that the degree of explicitness is dependent upon how skewed the mesh is. By skewness we mean how misaligned the vectors $\mathbf{l}_f$ and $\hat{\mathbf{n}}_f$ are. If the mesh is skewed, the tangential flux contribution [second term on the right-hand side of Eq. (7.59)] will be large compared with the normal flux contribution [left-hand side of Eq. (7.59)] resulting in a high degree of explicitness and poor convergence. Therefore, generating a mesh that has a relatively low degree of skewness is worth pursuing. Most commercial mesh generators provide statistics pertaining to skewness. For example, they might output a histogram of the angles in the mesh, along with the minimum and maximum angles. As discussed earlier, it is best to have angles concentrated around $60°$ for a triangular or tetrahedral mesh. Similarly, it is best to have angles concentrated around $90°$ for a mesh with quadrilaterals or hexahedrons. Poor convergence is often caused by cells that have extreme acute or obtuse angles. Therefore, it is imperative that attention be paid to "fixing" cells that contribute to the fringes of such a histogram of angles. Many mesh generators provide in-built tools that may be resorted to in an effort to improve skewness. For additional discussion on this topic the reader is referred to texts on unstructured mesh generation [4].

7.4.2 THREE-DIMENSIONAL GEOMETRY

Following the logic used to derive Eq. (7.48), the gradient of ϕ can be written in 3D using the normal unit vector and two unit tangents as

$$(\nabla \phi)_f = \left[(\nabla \phi)_f \bullet \hat{\mathbf{n}}_f \right] \hat{\mathbf{n}}_f + \left[(\nabla \phi)_f \bullet \hat{\mathbf{t}}_{1f} \right] \hat{\mathbf{t}}_{1f} + \left[(\nabla \phi)_f \bullet \hat{\mathbf{t}}_{2f} \right] \hat{\mathbf{t}}_{2f}, \quad (7.60)$$

where $\hat{\mathbf{t}}_{1f}$ and $\hat{\mathbf{t}}_{2f}$ are mutually orthogonal unit tangents on the face *f*. Unfortunately, it is not straightforward to compute unit tangents that are orthogonal to each other. As shown in Fig. 7.9, unit surface tangents are computed by connecting the vertices of a triangle. However, this does not guarantee that the two tangents will be orthogonal to each other. Therefore, Eq. (7.60) cannot be used in this case, and an alternative formulation must be developed.

Let us consider the vector identity given by Eq. (B.11). Using this identity the gradient of ϕ may be written as

$$(\nabla\phi)_f = \left[(\nabla\phi)_f \cdot \hat{\mathbf{n}}_f\right]\hat{\mathbf{n}}_f + \left[\hat{\mathbf{n}}_f \times (\nabla\phi)_f\right] \times \hat{\mathbf{n}}_f. \tag{7.61}$$

In this case the face under consideration is a triangle with the surface normal pointing outward from cell O toward cell N, as shown in Fig. 7.11(a). The triangular face is being shared by two tetrahedral cells: cell O and cell N. The triangular face projected on the plane of the paper (i.e., as viewed in the direction opposite to the surface normal) is shown in Fig. 7.11(b).

As in the 2D case, we perform a dot product of Eq. (7.61) with the vector $\mathbf{l}_f$, which is the vector pointing from the cell center of O to the cell center of N, as illustrated in Fig. 7.11(a). The resulting equation is

$$(\nabla\phi)_f \cdot \mathbf{l}_f = \left[(\nabla\phi)_f \cdot \hat{\mathbf{n}}_f\right]\hat{\mathbf{n}}_f \cdot \mathbf{l}_f + \left\{\left[\hat{\mathbf{n}}_f \times (\nabla\phi)_f\right] \times \hat{\mathbf{n}}_f\right\} \cdot \mathbf{l}_f. \tag{7.62}$$

Noting again that $\delta_f = \hat{\mathbf{n}}_f \cdot \mathbf{l}_f$ and $(\nabla\phi)_f \cdot \mathbf{l}_f \approx \phi_N - \phi_O$ (see Section 7.4.1), Eq. (7.62) may be rearranged and written to yield the component of the gradient vector normal to the face

$$(\nabla\phi)_f \cdot \hat{\mathbf{n}}_f = \frac{\phi_N - \phi_O}{\delta_f} - \frac{\left\{\left[\hat{\mathbf{n}}_f \times (\nabla\phi)_f\right] \times \hat{\mathbf{n}}_f\right\} \cdot \mathbf{l}_f}{\delta_f} = \frac{\phi_N - \phi_O}{\delta_f} - \frac{J_{T,f}}{\delta_f} \tag{7.63}$$

The tangential flux term in this case is the combined contribution of the two surface tangents mentioned earlier. Using Eq. (7.63) it can be written as

$$J_{T,f} = \left\{\left[\hat{\mathbf{n}}_f \times (\nabla\phi)_f\right] \times \hat{\mathbf{n}}_f\right\} \cdot \mathbf{l}_f. \tag{7.64}$$

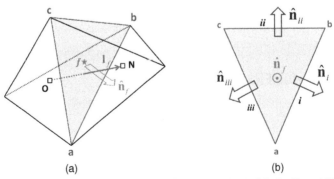

(a)   (b)

FIGURE 7.11 A 3D Stencil Showing (a) Two Adjacent Tetrahedral Cells (O and N) Sharing a Triangular Face, F, and (b) Projection of the Triangular Face on the Plane of the Paper

The next task is to express this tangential flux in terms of cell center values of ϕ. This can be done by considering a vector, $\mathbf{q}$, such that $\mathbf{q} = \phi\,\hat{\mathbf{i}}$. Next, let us take the divergence of this vector field such that

$$\nabla \cdot \mathbf{q} = \nabla \cdot (\phi\,\hat{\mathbf{i}}) = \left(\frac{\partial}{\partial x}\hat{\mathbf{i}} + \frac{\partial}{\partial y}\hat{\mathbf{j}} + \frac{\partial}{\partial z}\hat{\mathbf{k}} \right) \cdot (\phi\,\hat{\mathbf{i}}) = \frac{\partial\phi}{\partial x}. \tag{7.65}$$

Next, we integrate both sides of Eq. (7.65) over the area of the face f (triangle abc) to yield

$$\int_{A_f} \frac{\partial\phi}{\partial x}\,dA = \int_{A_f} \nabla \cdot \mathbf{q}\,dA. \tag{7.66}$$

Applying the definition of the average to the left-hand side of Eq. (7.66), it can be rewritten as

$$\left. \frac{\partial\phi}{\partial x} \right|_f A_f = \int_{A_f} \nabla \cdot \mathbf{q}\,dA, \tag{7.67}$$

where $\left. \frac{\partial\phi}{\partial x} \right|_f$ denotes the average value of the derivative at the face f. Next, we apply the Gauss divergence theorem (see Section 7.1) to the right-hand side of Eq. (7.67). However, instead of transforming a volume integral to a surface integral, this time we transform an area integral to a contour (or line) integral such that:

$$\left. \frac{\partial\phi}{\partial x} \right|_f = \frac{1}{A_f} \int_{A_f} \nabla \cdot \mathbf{q}\,dA = \frac{1}{A_f} \int_{P_f} \mathbf{q} \cdot \hat{\mathbf{n}}_e\,dl = \frac{1}{A_f} \int_{P_f} (\phi\,\hat{\mathbf{i}}) \cdot \hat{\mathbf{n}}_e\,dl, \tag{7.68}$$

where P_f denotes the perimeter of the triangular face f. In this case the perimeter of the triangular face is comprised of discrete edges. $\hat{\mathbf{n}}_e$ denotes a unit vector that is on the plane of the triangular face f and perpendicular to an edge e. Henceforth, it will be referred to as an edge normal. The three discrete edges are denoted by i, ii, and iii in Fig. 7.11(b), which also show the three edge normals. Since the perimeter of the triangle is comprised of three discrete edges, the contour integral on the right-hand side of Eq. (7.68) may be split into a discrete summation to yield

$$\left. \frac{\partial\phi}{\partial x} \right|_f = \frac{1}{A_f} \sum_{e=i}^{iii} \int_{L_e} (\phi\,\hat{\mathbf{i}}) \cdot \hat{\mathbf{n}}_e\,dl = \frac{1}{A_f} \sum_{e=i}^{iii} (\hat{\mathbf{i}} \cdot \hat{\mathbf{n}}_e) \int_{L_e} \phi\,dl, \tag{7.69}$$

where L_e is the length of edge e. Since the edge normal does not vary along the length of edge e, the final simplification step of Eq. (7.69) is made possible. Once again, using the definition of the average, Eq. (7.69) can be further simplified as

$$\left. \frac{\partial\phi}{\partial x} \right|_f = \frac{1}{A_f} \sum_{e=i}^{iii} (\hat{\mathbf{i}} \cdot \hat{\mathbf{n}}_e) \int_{L_e} \phi\,dl = \frac{1}{A_f} \sum_{e=i}^{iii} (\hat{\mathbf{i}} \cdot \hat{\mathbf{n}}_e) L_e \phi_e, \tag{7.70}$$

where ϕ_e denotes the average value of ϕ along the edge e.

The other two partial derivatives of ϕ at the face may also be derived using a similar procedure. The three partial derivatives may then be combined to yield the gradient of ϕ

$$
\begin{aligned}
(\nabla \phi)_f &= \left.\frac{\partial \phi}{\partial x}\right|_f \hat{\mathbf{i}} + \left.\frac{\partial \phi}{\partial y}\right|_f \hat{\mathbf{j}} + \left.\frac{\partial \phi}{\partial z}\right|_f \hat{\mathbf{k}} \\
&= \frac{1}{A_f} \sum_{e=i}^{iii} \left[(\hat{\mathbf{i}} \bullet \hat{\mathbf{n}}_e)\hat{\mathbf{i}} + (\hat{\mathbf{j}} \bullet \hat{\mathbf{n}}_e)\hat{\mathbf{j}} + (\hat{\mathbf{k}} \bullet \hat{\mathbf{n}}_e)\hat{\mathbf{k}} \right] L_e \phi_e = \frac{1}{A_f} \sum_{e=i}^{iii} [\hat{\mathbf{n}}_e] L_e \phi_e
\end{aligned}
\tag{7.71}
$$

Substituting Eq. (7.71) into Eq. (7.64), we obtain

$$
J_{T,f} = \left\{ \left[\hat{\mathbf{n}}_f \times \frac{1}{A_f} \sum_{e=i}^{iii} \hat{\mathbf{n}}_e L_e \phi_e \right] \times \hat{\mathbf{n}}_f \right\} \bullet \mathbf{l}_f = \left\{ \frac{1}{A_f} \sum_{e=i}^{iii} [(\hat{\mathbf{n}}_f \times \hat{\mathbf{n}}_e) \times \hat{\mathbf{n}}_f] L_e \phi_e \right\} \bullet \mathbf{l}_f.
\tag{7.72}
$$

Note that the cross product with $\hat{\mathbf{n}}_f$ in the second part of Eq. (7.72) has been inserted inside the summation since $\hat{\mathbf{n}}_f$ is edge independent. Next, we employ the vector identity given by Eq. (B.11). This identity expresses triple cross products in terms of dot products. Applying it to Eq. (7.72), we obtain

$$
J_{T,f} = \left\{ \frac{1}{A_f} \sum_{e=i}^{iii} [(\hat{\mathbf{n}}_f \bullet \hat{\mathbf{n}}_f)\hat{\mathbf{n}}_e - (\hat{\mathbf{n}}_f \bullet \hat{\mathbf{n}}_e)\hat{\mathbf{n}}_f] L_e \phi_e \right\} \bullet \mathbf{l}_f.
\tag{7.73}
$$

Since $\hat{\mathbf{n}}_e$ is a vector on the plane of the triangle, and $\hat{\mathbf{n}}_f$ is always perpendicular to the plane of the triangle, $\hat{\mathbf{n}}_f \bullet \hat{\mathbf{n}}_e = 0$. Moreover, noting that $\hat{\mathbf{n}}_f \bullet \hat{\mathbf{n}}_f = 1$, Eq. (7.73) can be simplified to

$$
J_{T,f} = \left\{ \frac{1}{A_f} \sum_{e=i}^{iii} [\hat{\mathbf{n}}_e] L_e \phi_e \right\} \bullet \mathbf{l}_f = \frac{1}{A_f} \sum_{e=i}^{iii} [\hat{\mathbf{n}}_e \bullet \mathbf{l}_f] L_e \phi_e.
\tag{7.74}
$$

As in the 2D case, if $\hat{\mathbf{n}}_f$ and $\mathbf{l}_f$ are coaligned, then $\hat{\mathbf{n}}_e \bullet \mathbf{l}_f = 0$ since $\hat{\mathbf{n}}_e$ and $\hat{\mathbf{n}}_f$, by definition, are perpendicular to each other. In such a scenario, the tangential flux will vanish, as is evident from Eq. (7.74). In a 3D mesh, such a scenario might occur either for a regular Cartesian hexahedral mesh (or bricks) or if the mesh is completely tetrahedral with equilateral triangles forming each tetrahedron (i.e., equal-sized tetrahedrons), among other scenarios.

A difficulty with using Eq. (7.74) to compute the tangential flux is that the edge normals, $\hat{\mathbf{n}}_e$, are not known. This problem can be circumvented by making use of the fact that the cross product of the unit tangent along the edge and the unit surface normal is actually the unit edge normal (i.e., $\hat{\mathbf{n}}_e = \hat{\mathbf{t}}_e \times \hat{\mathbf{n}}_f$), where the unit tangent along the edge is a unit vector joining the two vertices of the edge. Substitution of this new definition of the edge normal into Eq. (7.74) yields

$$
J_{T,f} = \frac{1}{A_f} \sum_{e=i}^{iii} [(\hat{\mathbf{t}}_e \times \hat{\mathbf{n}}_f) \bullet \mathbf{l}_f] L_e \phi_e.
\tag{7.75}
$$

Noting further that $\mathbf{t}_e = L_e \hat{\mathbf{t}}_e$ and using the vector identity given by Eq. (B.9), we obtain

$$J_{T,f} = \frac{1}{A_f} \sum_{e=i}^{iii} \phi_e \mathbf{t}_e \cdot [\hat{\mathbf{n}}_f \times \mathbf{l}_f] = \left(\frac{1}{A_f} \sum_{e=i}^{iii} \phi_e \mathbf{t}_e \right) \cdot [\hat{\mathbf{n}}_f \times \mathbf{l}_f]. \qquad (7.76)$$

Since the cross product $\hat{\mathbf{n}}_f \times \mathbf{l}_f$ is independent of the edges over which the summation is performed, it needs to be computed only once and only dotted with the resulting vector after the summation has been computed. This saves tremendous computational time compared with the scenario when Eq. (7.75) is used directly.

The final piece of the puzzle is to express the values of ϕ at the edges in terms of cell center or vertex (as in the 2D case) values. Recalling that ϕ_e is the average value of ϕ at the edge e, it can be approximated as the average of the two vertex values (i.e., $\phi_i = (\phi_a + \phi_b)/2$), and so on. Substituting these approximations into Eq. (7.76), we obtain

$$J_{T,f} = \left(\frac{1}{A_f} \sum_{e=i}^{iii} \frac{\phi_{a(e)} + \phi_{b(e)}}{2} \mathbf{t}_e \right) \cdot \left[\hat{\mathbf{n}}_f \times \mathbf{l}_f \right], \qquad (7.77)$$

where $\phi_{a(e)}$ and $\phi_{b(e)}$ denote the values of ϕ at the two vertices of edge e. Substitution of Eq. (7.77) into Eq. (7.63), followed by substitution of the resulting equation into Eq. (7.16), yields

$$\sum_{f=1}^{N_{f,O}} \Gamma_f \left(\frac{\phi_{N(f)} - \phi_O}{\delta_f} - \frac{1}{\delta_f} \left(\frac{1}{A_f} \sum_{e=1}^{N_{e(f)}} \frac{\phi_{a(e)} + \phi_{b(e)}}{2} \mathbf{t}_e \right) \cdot \left[\hat{\mathbf{n}}_f \times \mathbf{l}_f \right] \right) A_f = S_{\phi,O} V_O. \qquad (7.78)$$

In order to generalize the formulation, the summation of the three edges in the preceding equations has been replaced in Eq. (7.78) by a summation over $N_{e(f)}$ edges, which denotes the number of edges of face f. As in the 2D case, the final finite volume equation has some cell center values and some vertex values. As discussed in Section 7.4.1, either the implicit or the explicit approach may be employed to treat the term containing the vertex values.

Sections 7.4.1 and 7.4.2 outline the way in which finite volume equations for an unstructured stencil can be derived both in 2D and 3D. Section 7.5 presents the treatment of boundary conditions for an unstructured mesh. Assembly of the full set of discrete equations in matrix form is then discussed in Section 7.6.

7.5 BOUNDARY CONDITION TREATMENT

As in Chapters 2 and 6, the three canonical boundary condition types (i.e., Dirichlet, Neumann, and Robin) will be covered in this section. As discussed in Chapter 6, in the finite volume method, the boundary conditions are applied to the faces of control volumes that are adjacent to the boundary. This implies that when summation over faces in Eq. (7.16) is performed, boundary faces need to be treated specially. Prior to attempting to derive flux expressions at boundary faces using the three boundary condition types, it should be noted that it is customary to store the surface normal at boundary faces such that it is pointing out of the computational domain.

7.5.1 DIRICHLET BOUNDARY CONDITION

The treatment of a Dirichlet boundary condition in the context of a 2D mesh is presented in this section. The procedure presented can be applied to a 3D mesh without additional complexities. Let us consider the 2D stencil shown in Fig. 7.12.

The boundary face is denoted by B. The vector $\mathbf{l}$ for the boundary face, denoted by $\mathbf{l}_B$, points from the cell center O to the face center B, rather than to the cell center of the adjacent cell, which does not exist in this case. The boundary condition imposed on the face is the Dirichlet boundary condition, $\phi = \phi_B$.

The finite volume equation for cell O can be derived by starting with the general finite volume equation given by Eq. (7.16). Next, we split the loop over the faces into two loops: one over the interior faces and one over the boundary faces. Since cell O has only one boundary face in this particular case, Eq. (7.16) becomes

$$\sum_{\substack{f=1 \\ f \neq B}}^{N_{f,O}} \Gamma_f \left[(\nabla \phi)_f \cdot \hat{\mathbf{n}}_f \right] A_f + \Gamma_B \left[(\nabla \phi)_B \cdot \hat{\mathbf{n}}_B \right] A_B = S_{\phi,O} V_O. \tag{7.79}$$

The normal component of the gradient vector at the boundary can be expressed by following the same procedure as interior faces. First, we write the gradient vector in terms of the local normal and tangent vectors as

$$(\nabla \phi)_B = \left[(\nabla \phi)_B \cdot \hat{\mathbf{n}}_B \right] \hat{\mathbf{n}}_B + \left[(\nabla \phi)_B \cdot \hat{\mathbf{t}}_B \right] \hat{\mathbf{t}}_B. \tag{7.80}$$

Next, we perform a dot product of Eq. (7.80) with the vector $\mathbf{l}_B$ to yield

$$(\nabla \phi)_B \cdot \mathbf{l}_B = \left[(\nabla \phi)_B \cdot \hat{\mathbf{n}}_B \right] \hat{\mathbf{n}}_B \cdot \mathbf{l}_B + \left[(\nabla \phi)_B \cdot \hat{\mathbf{t}}_B \right] \hat{\mathbf{t}}_B \cdot \mathbf{l}_B. \tag{7.81}$$

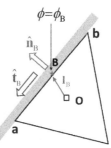

FIGURE 7.12 A 2D Unstructured Stencil Showing a Cell with One of its Faces Being a Boundary Face (B) with Dirichlet Boundary Condition Imposed on it

Using $\delta_B = \hat{\mathbf{n}}_B \cdot \mathbf{l}_B$ and noting that $(\nabla\phi)_B \cdot \mathbf{l}_B \approx \phi_B - \phi_O$, Eq. (7.81) may be rearranged to obtain an expression for the normal component of the gradient vector at the boundary:

$$(\nabla\phi)_B \cdot \hat{\mathbf{n}}_B = \frac{\phi_B - \phi_O}{\delta_B} - \frac{\left[(\nabla\phi)_B \cdot \hat{\mathbf{t}}_B\right]\hat{\mathbf{t}}_B \cdot \mathbf{l}_B}{\delta_B} = \frac{\phi_B - \phi_O}{\delta_B} - \frac{J_{T,B}}{\delta_B}, \qquad (7.82)$$

where tangential flux at the boundary is written as

$$J_{T,B} = \left[(\nabla\phi)_B \cdot \hat{\mathbf{t}}_B\right]\hat{\mathbf{t}}_B \cdot \mathbf{l}_B. \qquad (7.83)$$

Once again, if the surface normal at the boundary and the vector $\mathbf{l}$ are coaligned, then $\hat{\mathbf{t}}_B \cdot \mathbf{l}_B = 0$ and tangential flux vanishes. As in the case of interior faces, tangential flux can be computed by making use of vertex values of ϕ [cf. Eq. (7.56)], such that

$$J_{T,B} = \left[\frac{\phi_a - \phi_b}{A_B}\right]\hat{\mathbf{t}}_B \cdot \mathbf{l}_B, \qquad (7.84)$$

where A_B is the area (length in 2D) of the boundary face B. Substituting Eq. (7.82) and Eq. (7.84) (for flux at the boundary face) as well as Eq. (7.55) and Eq. (7.57) (for flux at the interior faces) into Eq. (7.79), we obtain

$$\sum_{\substack{f=1 \\ f\neq B}}^{N_{f,O}} \Gamma_f \left(\frac{\phi_{N(f)} - \phi_O}{\delta_f} - \left[\frac{\phi_{a(f)} - \phi_{b(f)}}{\delta_f A_f}\right]\hat{\mathbf{t}}_f \cdot \mathbf{l}_f\right) A_f$$
$$+ \Gamma_B \left(\frac{\phi_B - \phi_O}{\delta_B} - \frac{\phi_{a(B)} - \phi_{b(B)}}{\delta_B A_B}\hat{\mathbf{t}}_B \cdot \mathbf{l}_B\right) A_B = S_{\phi,O} V_O \qquad (7.85)$$

When assembling Eq. (7.85) in matrix form, the term containing ϕ_B must be transposed over to the right-hand side of the equation since it is a known quantity. These manipulations will be discussed in further detail in Section 7.6. Finally, a special case arises if the value of ϕ on the entire boundary patch is prescribed to be the same value. For example, the entire left boundary in Fig. 7.3 may have the same value of ϕ. In that case, the values of ϕ at the vertices a and b are identical and the tangential flux vanishes irrespective of the direction of the surface normal relative to the vector $\mathbf{l}$. Note that the approximation, $(\nabla\phi)_B \cdot \mathbf{l}_B \approx \phi_B - \phi_O$, is first order. Although, it is possible, in principle, to extend this to second order, the procedure is mathematically and algorithmically complicated for a general unstructured mesh. Therefore, the use of first-order boundary condition treatment in unstructured finite volume formulations of the diffusion operator has become routine.

7.5.2 NEUMANN BOUNDARY CONDITION

The value of ϕ is not prescribed at the boundary in this case. Rather, the normal gradient is prescribed, as shown in Fig. 7.13. The value of ϕ at the boundary (i.e., ϕ_B) is an unknown and can be determined only after the PDE has been solved.

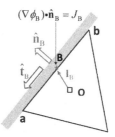

FIGURE 7.13 A 2D Unstructured Stencil Showing a Cell with One of its Faces Being a Boundary Face (B) with Neumann Boundary Condition Imposed on it

The finite volume equation for cell O can be derived using Eq. (7.79) as the starting point. In this case, the normal component of the flux is directly provided by the boundary condition, and is given by

$$(\nabla\phi_B)\bullet\hat{\mathbf{n}}_B = J_B. \tag{7.86}$$

Substituting Eq. (7.86) into Eq. (7.79) yields

$$\sum_{\substack{f=1 \\ f\neq B}}^{N_{f,O}}\Gamma_f\left[(\nabla\phi)_f\bullet\hat{\mathbf{n}}_f\right]A_f + \Gamma_B\left[J_B\right]A_B = S_{\phi,O}V_O. \tag{7.87}$$

The terms involving interior faces can be treated exactly as described in the preceding sections, but the steps involved are omitted here for the sake of brevity. Finally, the second term on the left-hand side of Eq. (7.87) must be transposed to the right-hand side prior to matrix assembly since it is a known quantity.

As mentioned earlier, the value of ϕ at the boundary (i.e., ϕ_B) in the case of the Neumann boundary condition is an unknown. However, this value is often of practical interest. For example, in a heat transfer calculation, the heat flux at the boundary may be specified (i.e., Neumann boundary condition) and the temperature at the boundary may be of interest. Obtaining the value of ϕ_B is a postprocessing step (i.e., one that needs to be performed once the PDE has been solved and the values of ϕ at all cell centers are already known). ϕ_B can be computed by utilizing the boundary portion of Eq. (7.85):

$$\frac{\phi_B - \phi_O}{\delta_B} - \frac{\phi_{a(B)} - \phi_{b(B)}}{\delta_B A_B}\hat{\mathbf{t}}_B\bullet\mathbf{l}_B = J_B. \tag{7.88}$$

If tangential flux is zero by virtue of $\hat{\mathbf{t}}_B\bullet\mathbf{l}_B = 0$, the value of ϕ_B can easily be calculated using the following explicit relationship:

$$\phi_B = \phi_O + \delta_B J_B. \tag{7.89}$$

If, however, the tangential flux is not zero, Eq. (7.88) can only be solved in an iterative manner as follows:

1. Obtain an initial estimate of ϕ_B using Eq. (7.89).
2. Compute the vertex values, $\phi_{a(B)}$ and $\phi_{b(B)}$, using face-to-vertex interpolation.

3. Recompute ϕ_B using Eq. (7.88).

4. Repeat 2–3 until the value of ϕ_B at each boundary face stops changing.

While this is a tedious and computationally involved process, the computational cost is not severe since it is performed as a one-off postprocessing step.

7.5.3 ROBIN BOUNDARY CONDITION

The Robin boundary condition is also known as the mixed boundary condition or a boundary condition of the third kind. As shown in Fig. 7.14, neither the value of ϕ nor the normal gradient is prescribed at the boundary in this case. Rather, a relationship between the two of the following form is specified

$$\alpha (\nabla \phi_B) \cdot \hat{\mathbf{n}}_B + \beta \phi_B = \gamma. \tag{7.90}$$

An expression for the normal component of the gradient vector can be derived by substituting ϕ_B from Eq. (7.90) into Eq. (7.82), resulting in

$$(\nabla \phi)_B \cdot \hat{\mathbf{n}}_B = \frac{\dfrac{\gamma}{\beta} - \dfrac{\alpha}{\beta}(\nabla \phi)_B \cdot \hat{\mathbf{n}}_B - \phi_O}{\delta_B} - \frac{\left[(\nabla \phi)_B \cdot \hat{\mathbf{t}}_B\right]\hat{\mathbf{t}}_B \cdot \mathbf{l}_B}{\delta_B}. \tag{7.91}$$

Rearranging, we obtain an expression for the normal component of the gradient of ϕ, written as

$$(\nabla \phi)_B \cdot \hat{\mathbf{n}}_B = \left(\frac{\gamma}{\beta} - \phi_O - \left[(\nabla \phi)_B \cdot \hat{\mathbf{t}}_B\right]\hat{\mathbf{t}}_B \cdot \mathbf{l}_B\right) \Big/ \left(\delta_B + \frac{\alpha}{\beta}\right). \tag{7.92}$$

Substituting Eq. (7.92) into Eq. (7.79), we obtain

$$\sum_{\substack{f=1 \\ f \neq B}}^{N_{f,O}} \Gamma_f \left[(\nabla \phi)_f \cdot \hat{\mathbf{n}}_f\right] A_f + \Gamma_B \left[\left(\frac{\gamma}{\beta} - \phi_O - \left[(\nabla \phi)_B \cdot \hat{\mathbf{t}}_B\right]\hat{\mathbf{t}}_B \cdot \mathbf{l}_B\right) \Big/ \left(\delta_B + \frac{\alpha}{\beta}\right)\right] A_B = S_{\phi,O} V_O. \tag{7.93}$$

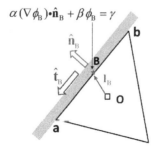

FIGURE 7.14 A 2D Unstructured Stencil Showing a Cell with One of its Faces Being a Boundary Face (B) with Robin Boundary Condition Imposed on it

Using the same procedure as that for interior faces for the treatment of the tangential flux at the boundary (as described in Section 7.5.1), Eq. (7.93) may be rewritten as

$$\sum_{\substack{f=1 \\ f \neq B}}^{N_{f,o}} \Gamma_f \left(\frac{\phi_{N(f)} - \phi_O}{\delta_f} - \left[\frac{\phi_{a(f)} - \phi_{b(f)}}{\delta_f A_f} \right] \hat{\mathbf{t}}_f \bullet \mathbf{l}_f \right) A_f$$

$$+ \Gamma_B \left[\left(\frac{\gamma}{\beta} - \phi_O - \left[\frac{\phi_{a(B)} - \phi_{b(B)}}{A_B} \right] \hat{\mathbf{t}}_B \bullet \mathbf{l}_B \right) \middle/ \left(\delta_B + \frac{\alpha}{\beta} \right) \right] A_B = S_{\phi,O} V_O$$

$$\text{(7.94)}$$

Once again, terms containing known quantities must first be transposed to the right-hand side of Eq. (7.94) prior to matrix assembly.

As in the case of Neumann boundary conditions, additional postprocessing is necessary in this particular case to extract the value of ϕ_B and the value of the flux at the boundary. Once again, if tangential flux is zero by virtue of $\hat{\mathbf{t}}_B \bullet \mathbf{l}_B$ being equal to zero, the procedure is straightforward and involves the following two steps:

1. Calculate the flux from Eq. (7.92) using $J_B = (\nabla \phi)_B \bullet \hat{\mathbf{n}}_B = \left(\frac{\gamma}{\beta} - \phi_O \right) \middle/ \left(\delta_B + \frac{\alpha}{\beta} \right)$.

2. Next, calculate the value of ϕ_B from Eq. (7.90) using $\phi_B = \frac{\gamma}{\beta} - \frac{\alpha}{\beta} J_B$.

In the event that tangential flux is not zero, an iterative procedure, as outlined below, must be used:

1. Obtain an initial estimate of J_B using $J_B = (\nabla \phi)_B \bullet \hat{\mathbf{n}}_B = \left(\frac{\gamma}{\beta} - \phi_O \right) \middle/ \left(\delta_B + \frac{\alpha}{\beta} \right)$.

2. Next, obtain an initial estimate of ϕ_B using $\phi_B = \frac{\gamma}{\beta} - \frac{\alpha}{\beta} J_B$.

3. Compute the vertex values, $\phi_{a(B)}$ and $\phi_{b(B)}$, using face-to-vertex interpolation.
4. Recompute J_B using Eq. (7.92).
5. Recompute ϕ_B using Eq. (7.90).
6. Repeat 3–5 until the values of ϕ_B and J_B at each boundary face stop changing.

Having derived finite volume equations for both interior cells and cells adjacent to the boundaries, the next step is to assemble them in a form amenable to numerical solution. This is discussed in Section 7.6.

7.6 ASSEMBLY AND SOLUTION OF DISCRETE EQUATIONS

The discrete finite volume equations that were derived in the preceding sections for 2D geometry will be used for the discussion in this section. These constitute Eq. (7.58) for interior cells and Eq. (7.85) for cells adjacent to boundaries with Dirichlet boundary conditions. In addition, the explicit approach for treatment of tangential fluxes (skew source) will be used to assemble the equations in matrix form since it is relatively straightforward and more commonly used.

For an interior cell O, Eq. (7.58) may be rewritten as:

$$A_{0,O}\phi_O + \sum_{j=1}^{N_{f,O}} A_{j,O}\phi_{j(O)} = -S_{\phi,O}V_O + S_{\text{skew},O} = Q_O. \tag{7.95}$$

where $A_{0,O}$ denotes the diagonal coefficient of the coefficient matrix $[A]$ corresponding to cell O, and $A_{j,O}$ denotes the off-diagonal elements in the same row. $\phi_{j(O)}$ denotes the jth neighbor of cell O, whose index can be obtained from the connectivity data discussed in Section 7.3. The quantities $A_{0,O}$ and $A_{j,O}$ are often referred to as link coefficients since they link one cell to another. As discussed in Chapter 3, the coefficient matrix is generally assembled in a way in which the diagonal is positive. Thus, the source term in Eq. (7.95) has a negative sign. The term, $S_{\text{skew},O}$, denotes the net so-called skew source contributed by the summation of tangential fluxes. Comparing Eq. (7.95) with Eq. (7.58) yields the following relationships:

$$A_{0,O} = \sum_{f=1}^{N_{f,O}} \frac{\Gamma_f A_f}{\delta_f}, \tag{7.96a}$$

$$A_{j,O} = -\frac{\Gamma_{j(O)}A_{j(O)}}{\delta_{j(O)}}, \tag{7.96b}$$

and

$$S_{\text{skew},O} = -\sum_{f=1}^{N_{f,O}} \Gamma_f \left(\frac{\phi_{a(f)}^* - \phi_{b(f)}^*}{\delta_f} \right) \hat{\mathbf{t}}_f \bullet \mathbf{l}_f. \tag{7.96c}$$

Generation of a computer program to compute these formulas requires the use of connectivity data. Understanding of this process can be facilitated by using the following code snippets to compute each of the above three quantities.

CODE SNIPPET 7.7 DEMONSTRATION OF CALCULATION OF LINK COEFFICIENTS

```
For icell = 1 : ncells   ! Loop over all cells
    Ao (icell) = 0
    Anb(icell,:) = 0
    For ifc = 1 : nface(icell)   ! Loop over number of faces of each cell
        gface = link_cell_to_face(icell,ifc) ! Get global face index (gface)
        If (bface(gface) == 1) Skip ! Skip over Boundary faces
        ic1 = link_face_to_cell(gface,1)  ! Global cell index of local cell1
        ic2 = link_face_to_cell(gface,2)  ! Global cell index of local cell2
        gammaf(gface) = wf(gface)*gamma(ic1)+(1-wf(gface))*gamma(ic2) ! Face value of Γ
                                  computed using cell-to-face interpolation function
        Ao(icell) = Ao(icell) + gammaf(gface)*areaf(gface)/deltaf(gface) ! Diagonal
        Anb(icell,ifc) = - gammaf(gface)*areaf(gface)/deltaf(gface) ! Off-Diagonal
    End
End
```

```
CODE SNIPPET 7.8  DEMONSTRATION OF CALCULATION OF SKEW
SOURCE AND RIGHT-HAND SIDE VECTOR Q

For icell = 1 : ncells   ! Loop over all cells
   skew(icell) = 0
   For ifc = 1 : nface(icell)    ! Loop over number of faces of each cell
      gface = link_cell_to_face(icell,ifc) ! Get global face index (gface)
      vb = link_face_to_vertex(gface,1) ! Get index of first vertex (b)
      va = link_face_to_vertex(gface,2) ! Get index of second vertex (a)
      ic1 = link_face_to_cell(gface,1) ! Global cell index of local cell1
      ic2 = link_face_to_cell(gface,2) ! Global cell index of local cell2
      Lf(1) = xc(ic2)-xc(ic1); Lf(2) = yc(ic2)-yc(ic1); ! Compute vector Lf
      tf(1) = xv(va)-xv(vb); tf(2) = yv(va)-yv(vb); ! Tangent
      utf(:) = tf(:)/areaf(ifc) ! Unit tangent
      utf_dot_Lf = utf(1)*Lf(1) + utf(2)*Lf(2) ! dot product
      ! In the step below, it is assumed that the vertices are in clockwise order. For anti-
         clockwise order, the sign must be positive
      skew(icell) = skew(icell) - gammaf(ifc)*utf_dot_Lf*(phiv(va)-phiv(vb))/deltaf(gface)
   End
   Q(icell) = skew(icell) – source(icell)*vol(icell)  ! Right hand side vector Q
End
```

Code snippet 7.8 actually computes the coefficients and sources for all cells including those adjacent to boundaries. Boundary face contributions are the only ones not included. In other words, the second term on the left-hand side of Eq. (7.85) has not been included. However, boundary face contributions may easily be included by running a loop over all boundary faces and then adding their contributions to the source term (right-hand side) of the adjacent cell, as shown in Code snippet 7.9.

```
CODE SNIPPET 7.9  DEMONSTRATION OF APPLICATION OF DIRICHLET
BOUNDARY CONDITION

For bface = 1 : nbfaces  ! Loop over all boundary faces
   gface = link_bface_to_face(bface)  ! Get global face index (gface)
   ic = link_face_to_cell(gface,1)  ! Cell index next to boundary
   Q(ic)  =  Q(ic)  +  gammaf(gface)*areaf(gface)*phib(bface)/delta(gface)  !Dirichlet
                                                            Boundary term
                                                            contribution
End
```

The code snippets shown above, coupled with the preceding discussion, clearly show the value of preprocessing and storing connectivity information in a format that can be readily accessed and used. Similarly, interpolation functions must be stored and available at all times. Once they are available, the computer program to execute assembly of the matrix can be both short and efficient.

Now that all details pertaining to generation and assembly of finite volume equations on an unstructured mesh have been presented, we can conclude by considering a flowchart (Fig. 7.15) of the solution algorithm such that certain subtleties can be elucidated and the modularity of the solution algorithm can be emphasized. It is designed for a steady-state calculation in which only a single elliptic PDE is solved on an unstructured mesh. Further, it is assumed that the approach used to treat tangential fluxes (skew sources) is the explicit method.

A new concept depicted by the solution algorithm in Fig. 7.15 is that of *inner iteration* versus *outer iteration*. What are they and why are they necessary? Linear algebraic equation solvers are often treated as blackboxes into which, aside from the

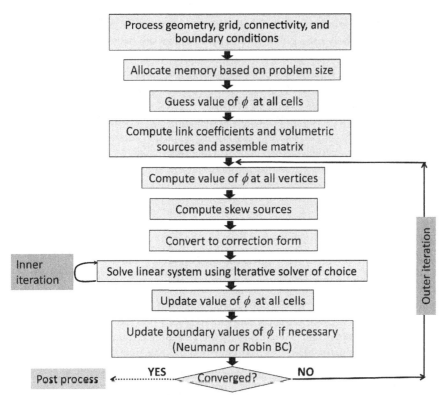

FIGURE 7.15 Flowchart of the Solution Algorithm for Solving an Elliptic PDE on an Unstructured Mesh Using the Explicit Approach to Treat Tangential Fluxes (Skew Sources)

coefficient matrix and the right hand–side vectors, only two other control parameters are fed. One of these is the maximum number of sweeps (or iterations), and the other is the residual convergence criterion, either on the absolute residual or on the normalized residual. These sweeps of the solver are referred to as inner iterations. During inner iterations, the coefficient matrix and right-hand vectors are inaccessible (or cannot be modified). If we imagine a scenario where the maximum number of iterations is set to 1, the algorithm depicted in Fig. 7.15 will essentially perform only the outer iteration loop, since there is actually no inner iteration to speak of. After every sweep of the linear equation solver, it will exit, the skew terms will be recomputed, and then the next sweep will be performed. While there is nothing conceptually wrong with this approach, it may be too conservative and computationally expensive, especially as convergence is approached. A different strategy may be conceived in which partial convergence of the equations is first attained by the linear equation solver, then the skew terms are updated, followed by partial convergence, and so on. In this approach the skew terms will be updated only at every outer iteration, enabling potential reduction in computational effort and time. Partial (as opposed to full) convergence is proposed because the skew terms will change at every outer iteration, and therefore, taking the equations to full convergence at every outer iteration using a large number of inner iterations would be wasteful. The idea of using nested inner and outer iteration loops is routinely used either to treat sources that are dependent on the dependent variable itself (i.e., the case when the source S_ϕ is a function of ϕ itself), or to couple two or more PDEs that are mathematically coupled but solved sequentially. This is discussed further in Section 8.4 in the context of nonlinear PDEs and in Section 8.5 in the context of coupled PDEs.

Finally, as shown in the solution algorithm (Fig. 7.15), an efficient code to solve a PDE can only be developed by, first, delineating steps that belong within and outside iteration loops and, second, writing code in a modular manner to avoid any redundancies. The procedures and equations presented in the preceding several sections are demonstrated in Example 7.1.

EXAMPLE 7.1

In this example we consider solution of a problem that was considered earlier in Example 2.4 in which the finite difference method in curvilinear coordinates was used. It was also considered in Example 6.7 in which the finite volume method in curvilinear coordinates was used. The objective is to find a solution to the Laplace equation on a rhombus shown in the figure below. The sides of the rhombus have lengths of 1 unit. Our aim here is to solve the problem using the unstructured finite volume method. The computational domain, the arrangement of the cells, and the boundary conditions are also shown in the figure below. The finite volume cells (mesh) used are the same as those used in Example 6.7, with the exception that the cells are arbitrarily numbered here and no coordinate transformation is invoked.

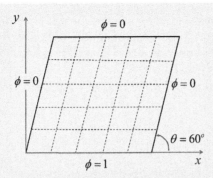

We developed a simple mesh generator to create the regular structured mesh sketched in the figure above. All cells, nodes, and faces were numbered using an arbitrary pattern, and connectivity information was generated and stored. Cell volumes, face areas, and face surface normals were computed as described in Sections 7.3.3 and 7.3.4. Cell-to-vertex interpolation functions were also developed, as described in Section 7.3.2. Simplifying Eq. (7.59) yields

$$\sum_{f=1}^{N_{f,O}} \left(\frac{\phi_{N(f)} - \phi_O}{\delta_f} \right) A_f = \sum_{f=1}^{N_{f,O}} \left(\left[\frac{\phi_{a(f)}^* - \phi_{b(f)}^*}{\delta_f} \right] \hat{\mathbf{t}}_f \cdot \mathbf{l}_f \right)$$

for interior cells. Comparing with the standard form shown in Eq. (7.95), we obtain

$$A_{o,O} = \sum_{f=1}^{N_{f,O}} \frac{A_f}{\delta_f}, A_{j,O} = -\frac{A_{f(j)}}{\delta_{f(j)}}, S_{\text{skew},O} = -\sum_{f=1}^{N_{f,O}} \left(\left[\frac{\phi_{a(f)}^* - \phi_{b(f)}^*}{\delta_f} \right] \hat{\mathbf{t}}_f \cdot \mathbf{l}_f \right)$$

which is similar to Eq. (7.96). Since the boundary conditions are of Dirichlet type with constant (uniform) values, Eq. (7.85) reduces to

$$\sum_{\substack{f=1 \\ f \neq B}}^{N_{f,O}} \left(\frac{\phi_{N(f)} - \phi_O}{\delta_f} - \left[\frac{\phi_{a(f)} - \phi_{b(f)}}{\delta_f A_f} \right] \hat{\mathbf{t}}_f \cdot \mathbf{l}_f \right) A_f + \left(\frac{\phi_B - \phi_O}{\delta_B} \right) A_B = 0,$$

for cells adjacent to boundaries. Comparing this equation with Eq. (7.95) yields

$$A_{o,O} = \sum_{f=1}^{N_{f,O}} \frac{A_f}{\delta_f}, A_{j,O} = -\frac{A_{f(j)}}{\delta_{f(j)}} \ \forall j \neq B, A_{j,O} = 0 \ \forall j = B,$$

$$S_{\text{skew},O} = -\sum_{\substack{f=1 \\ f \neq B}}^{N_{f,O}} \left(\left[\frac{\phi_{a(f)}^* - \phi_{b(f)}^*}{\delta_f} \right] \hat{\mathbf{t}}_f \cdot \mathbf{l}_f \right), S_O = \frac{A_B}{\delta_B} \phi_B$$

The resulting linear algebraic system was solved using Gauss–Seidel iterations with a tolerance of 10^{-6} on a mesh with 1600 cells. The figures below
show the converged solution as well as convergence behavior.

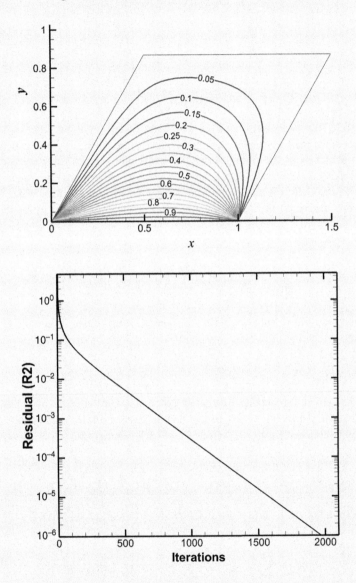

The solution is almost identical to that obtained in Example 6.7. There
are minor differences in the two solutions. For example, the boundary condition treatment in Example 6.7 was second order, whereas here it is first

order. Only one inner sweep of the Gauss–Seidel method was conducted for every outer iteration, and vertex values were immediately updated. Thus, each iteration shown in the convergence plot essentially corresponds to one Gauss–Seidel sweep. Convergence is reached in slightly less than 2000 iterations, whereas it was reached in just over 2000 iterations in Example 6.7.

7.7 FINITE VOLUME FORMULATION FOR ADVECTION–DIFFUSION EQUATION

As mentioned earlier, finite volume discretization of the elliptic diffusion term on an unstructured stencil poses the biggest challenge. The procedure for discretization of the diffusion term has already been presented in the preceding several sections. In this section, we extend the ideas already developed to the generalized unsteady advection–diffusion equation for the sake of completeness. The advection–diffusion equation is the cornerstone of transport phenomena and is widely used in a variety of applications in physics and engineering. Since the finite volume method is routinely used for solution of conservation equations that are of the generalized advection–diffusion form, this section is likely to serve as a quick reference for readers who may go on to pursue more applied subjects such as computational fluid dynamics. Finite volume formulation for the advection–diffusion equation on a Cartesian grid has already been discussed extensively in Chapter 6, along with several examples demonstrating the formulation.

The generalized unsteady advection–diffusion equation for a scalar ϕ, in its conservative form, is written as [1]

$$\frac{\partial}{\partial t}(\rho\phi) + \nabla \bullet (\rho \mathbf{U}\phi) = \nabla \bullet (\Gamma \nabla \phi) + S_\phi, \tag{7.97}$$

where ρ is the mass density of the medium. The first term of Eq. (7.97) physically represents storage of the scalar ϕ, and is often referred to simply as the transient term. In Chapter 5, this term was already introduced and discussed at length. The second term is the new advection term. Advection refers to transport of the scalar ϕ by bulk motion of the medium. Therefore, treatment of advection inherently implies prior knowledge of the flowfield. In Eq. (7.97), $\mathbf{U}$ represents the mass-averaged velocity vector of the fluid. In general, advective transport can be caused by any medium as long as it is moving, be it fluid or solid. For example, a moving conveyor belt, heated at one spot, will carry the heat with it. Therefore, strictly speaking, $\mathbf{U}$ represents the velocity of the medium. Nonetheless, $\mathbf{U}$ must be known prior to solution of Eq. (7.97). The quantity, $\rho\mathbf{U}$, represents the mass flux vector or mass flow rate per unit area. Note that $\mathbf{U}$ has units of m/s, while ρ has units of kg/m^3. Therefore, the product has units of kg/(m².s). In the absence of any motion of the medium (or static medium), Eq. (7.97) reduces to an unsteady diffusion equation – an equation that was discussed extensively in Chapter 5. Equation (7.97) may be rearranged and written in the following form

$$\frac{\partial}{\partial t}(\rho\phi) + \nabla \bullet (\rho\mathbf{U}\phi - \Gamma\nabla\phi) = S_\phi.$$

(7.98)

Defining the flux as

$$\mathbf{J}_\phi = \rho\mathbf{U}\phi - \Gamma\nabla\phi,$$

(7.99)

Eq. (7.98) may be written in compact form as

$$\frac{\partial}{\partial t}(\rho\phi) + \nabla \bullet \mathbf{J}_\phi = S_\phi.$$

(7.100)

It is clear that at steady state, Eq. (7.7) is recovered. The preceding discussion shows that the governing equation still represents a conservation equation. Only, instead of just a diffusive flux, we now have both diffusive and advective fluxes.

Next, we perform finite volume integration of Eq. (7.97) over a generic control volume, as shown in Fig. 7.2. This yields

$$\int_{V_O}\frac{\partial}{\partial t}(\rho\phi)dV + \int_{V_O}\nabla \bullet (\rho\mathbf{U}\phi)dV = \int_{V_O}\nabla \bullet (\Gamma\nabla\phi)dV + \int_{V_O}S_\phi\,dV.$$

(7.101)

The reduction of the right-hand side of Eq. (7.101) was already addressed in Sections 7.2–7.4 [cf. Eq. (7.13) and subsequent derivation]. Here, we will focus only on the left-hand side of Eq. (7.101). The transient term is simplified by applying the definition of the average, to yield

$$\int_{V_O}\frac{\partial}{\partial t}(\rho\phi)dV = \frac{\partial(\rho\phi)}{\partial t}\bigg|_O V_O.$$

(7.102)

Discretization of the time derivative can be accomplished using the methods and procedures outlined in Chapter 5 for parabolic PDEs. These include the forward Euler, the backward Euler, and the Crank–Nicolson method, and needs no additional discussion. In order to reduce the volume integral of the advective term, we apply the Gauss-divergence theorem [Section 7.1] to yield

$$\int_{V_O}\nabla \bullet (\rho\mathbf{U}\phi)dV = \int_S(\rho\mathbf{U}\phi) \bullet \hat{\mathbf{n}}\,dA,$$

(7.103)

where $\hat{\mathbf{n}}$ is the outward-pointing surface normal. As before, since the control volume in question is bounded by a set of discrete planar faces, the surface integral can be replaced by a summation over the faces to yield

$$\int_{V_O} \nabla \bullet (\rho \mathbf{U} \phi) dV = \sum_{f=1}^{N_{f,O}} (\rho \mathbf{U} \phi)_f \bullet \hat{\mathbf{n}}_f A_f = \sum_{f=1}^{N_{f,O}} (\rho \mathbf{U} \bullet \hat{\mathbf{n}} A)_f \phi_f = \sum_{f=1}^{N_{f,O}} G_f \phi_f. \quad (7.104)$$

The quantity $G_f = (\rho \mathbf{U} \bullet \hat{\mathbf{n}} A)_f$ represents the mass flow rate normal to the face f. Prior to solving the advection–diffusion equation at hand, this quantity must be available. Generally, computation of the advection–diffusion equation for a scalar is preceded by solution of the fluid flow equations, which provides the mass flux vector.

The final task is to express the face value of ϕ in terms of cell-center values of ϕ. As to how this is best done is a topic of intense research, debate, and discussion within the computational fluid dynamics community, and several methods for treatment of the advective flux have been provided in Section 6.3 for a Cartesian mesh. For further details, the reader is referred to texts on computational fluid dynamics [5,6]. In this section, we will discuss only the two most commonly used schemes to attain this objective: the central difference scheme and the upwind difference scheme, which have both been discussed in Chapter 6 for a structured mesh.

In the familiar central difference scheme, the value of ϕ_f is expressed in terms of the values of ϕ at the two cells adjacent to the face f using inverse distance-weighted interpolation, yielding

$$\int_{V_O} \nabla \bullet (\rho \mathbf{U} \phi) dV = \sum_{f=1}^{N_{f,O}} G_f [w_f \phi_O + (1 - w_f) \phi_{N(f)}], \quad (7.105)$$

wherein the cell-to-face interpolation function (see Section 7.3.2) has been used. It can be shown using Taylor series expansions that the error incurred in approximating ϕ_f using the interpolation formula utilized in Eq. (7.105) scales as the square of the distance between the cell face and the cell center, at least in the case where the two distances are equal. The second-order behavior of the truncation error in the central difference scheme makes it an attractive choice from a mathematical perspective. However, as discussed in considerable detail in Chapter 6, from a physical perspective, it is often argued that unlike diffusion, advection always transports information downstream, and therefore, only one of the two cells should influence the face value of ϕ depending on which direction the velocity vector is pointing. This argument has led to the so-called upwind difference scheme.

In the upwind difference scheme, the direction of the "wind" is considered to determine the value of ϕ_f. If the mass flux (or "wind" direction) is from cell O to cell N, then the value of ϕ_f is set to its value at the upwind or upstream cell, namely cell O, and vice versa. Mathematically, assuming that the surface normal is pointing from cell O to cell N, the upwind difference scheme may be written as

$$\phi_f = \begin{cases} \phi_O & \text{if } G_f > 0 \\ \phi_N & \text{if } G_f < 0 \end{cases}. \quad (7.106)$$

For computer programming, a convenient way to write the flux is to use so-called switching functions, such that the logic of the upwind difference scheme is automatically manifested:

$$G_f\phi_f = \frac{G_f + |G_f|}{2}\phi_O - \frac{G_f - |G_f|}{2}\phi_N.$$

(7.107)

Substitution of Eq. (7.107) into Eq. (7.104) yields an expression for the advective flux using the upwind difference scheme:

$$\int_{V_O} \nabla \bullet (\rho \mathbf{U}\phi)dV = \sum_{f=1}^{N_{f,O}} \frac{G_f + |G_f|}{2}\phi_O - \frac{G_f - |G_f|}{2}\phi_{N(f)}$$

(7.108)

Although the upwind difference scheme is, arguably, more realistic from a physical perspective, it has poor numerical accuracy. As discussed in Section 6.3, it results in numerical diffusion. This is because the resulting truncation error scales as the distance between the upstream cell-center and the face-center, rather than as the square of the distance, as in the central difference scheme. In other words, the scheme described by Eq. (7.106) is only first order accurate, and is often referred to in the literature as first-order upwind (FOU) scheme. Higher-order versions of this scheme are available, but they are sought with additional problems, as discussed in Section 6.3, and elsewhere [5,6].

In order to write the final discrete form of the advection–diffusion equation, we substitute equations (7.102) (for transient term), (7.108) (for advection term), and (7.55) (for diffusion term) into Eq. (7.101) to yield

$$\left.\frac{\partial(\rho\phi)}{\partial t}\right|_O V_O + \sum_{f=1}^{N_{f,O}} \frac{G_f + |G_f|}{2}\phi_O - \frac{G_f - |G_f|}{2}\phi_{N(f)}$$

$$= \sum_{f=1}^{N_{f,O}} \Gamma_f \left(\frac{\phi_{N(f)} - \phi_O}{\delta_f} - \frac{J_{T,f}}{\delta_f} \right) A_f + S_{\phi,O} V_O$$

(7.109)

The link coefficients for this equation may be written in the same manner as described in Section 7.6. Earlier, we had contended that the treatment of the elliptic diffusion term is the most challenging task in an unstructured finite volume formulation. Eq. (7.109) clearly shows that no tangential flux arises out of the advection term. The tangential flux is unique to the diffusion term because the diffusive flux depends on the gradient of ϕ, while the advective flux does not.

In summary, a complete finite volume formulation for an unstructured mesh has been presented in this chapter. Primary emphasis has been placed on the discretization of the elliptic diffusion operator because it leads to additional complexities involving tangential fluxes. The discrete equations were derived both for the diffusion (generalized Poisson) equation as well as for the generalized unsteady advection–diffusion equation. Finally, assembly of the discrete equations in matrix form and a solution algorithm has also been presented to complete the entire picture. Although the development of an unstructured finite volume code is more cumbersome than its structured counterpart, the payoff in terms of the geometric flexibility of the code is often worth the extra effort.

REFERENCES

[1] Bird RB, Stewart WE, Lightfoot EN. Transport phenomena. 2nd ed. New York: Wiley; 2001.

[2] Gmsh 2.8 Reference Manual. Available from: http://geuz.org/gmsh/doc/texinfo/gmsh.html.

[3] TetGen User Manual for Version 1.5.0. Available from: http://wias-berlin.de/software/tetgen/#Documentation.

[4] Liseikin VD. Grid generation methods. 2nd ed. Springer: Berlin; 2010.

[5] Ferziger J, Peric M. Computational methods for fluid dynamics. 2nd ed. Springer Verlag: Berlin; 1999.

[6] Date AW. Introduction to computational fluid dynamics. Cambridge University Press: Cambridge, UK; 2005.

EXERCISES

7.1 Using the Gauss-divergence theorem compute the following:

    a. Area of a triangle with coordinates $(1,2)$, $(3,5)$, and $(-1,1)$.

    b. Volume of a tetrahedron with coordinates $(1,1,1)$, $(1,4,3)$, $(-2,1,4)$, and $(4,-2,1)$.

    c. Outward-pointing unit surface normal vectors to all faces in both Parts (a) and (b).

7.2 Consider the stencil shown in the figure below. The values of a scalar ϕ at the cell-centers are also shown on the stencil. Using the Gauss-divergence theorem, calculate average value of $\nabla\phi$ at the cell O.

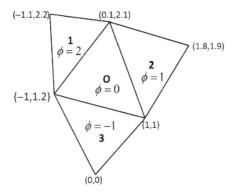

7.3 Consider the error expression given by Eq. (7.54) that resulted from expressing the gradient at the face in terms of cell-center values. Perform an analysis to establish the order of the error.

7.4 Section 7.5.1 presents the treatment of Dirichlet boundary conditions for a 2D stencil. Using the same procedure along with concepts presented in Section 7.4.2, derive an expression for the diffusion flux normal to a Dirichlet boundary for a 3D stencil in which the cell next to the boundary is a tetrahedron.

7.5 The equation to be solved is $\nabla^2 \phi = 0$. Consider the 2D stencil shown in the figure below. Derive the discrete equation for cell 0 using the finite volume formulation. Assume that all connectivity and geometric information is available to you already from the mesh generator. Your final equation must be an algebraic equation that has unknown terms (containing ϕ at the various cell-centers only) on the left-hand side of the equation, and known quantities (if any) on the right-hand side of the equation.

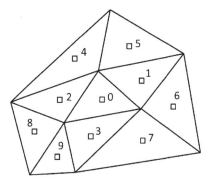

7.6 Consider the geometry, mesh and boundary conditions shown below. The symbols used around the various numbers in the figure are the same as in Fig. 7.3.

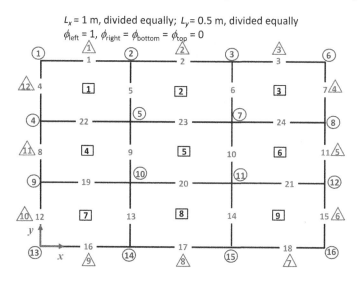

$L_x = 1$ m, divided equally; $L_y = 0.5$ m, divided equally
$\phi_{\text{left}} = 1$, $\phi_{\text{right}} = \phi_{\text{bottom}} = \phi_{\text{top}} = 0$

For this mesh, write a computer program to

a. Record all cell-center, face-center, and vertex coordinates.

b. Fill out all necessary connectivity arrays discussed in Section 7.3.

c. Calculate and record all volumes, face areas, and surface normal vectors.

d. Solve the Laplace equation using the finite volume method.

e. Compute the flux at the four boundaries. Is global flux conservation satisfied? Discuss your results.

7.7 Consider solution of the PDE $\nabla \cdot (\Gamma \nabla \phi) = 0$ in a rectangular domain, as shown in the figure below. The domain is 2 units in length (x direction) and 1 unit in height (y direction). As shown in the figure, it is comprised of two different materials with different values of Γ, the demarcation being along the $x = 1$ line. The unstructured finite volume method is to be used to solve the PDE, and the mesh is also shown in the figure. The boundary conditions are as follows: $\phi = 1$ at the left boundary, $\phi = 0$ at the right boundary, and zero fluxes (derivatives) on the top and bottom boundaries.

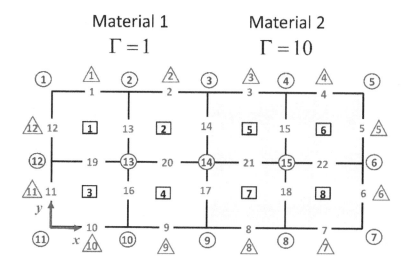

a. Solve the governing equation analytically subject to the boundary conditions described above. Note that since both top and bottom boundaries have zero flux boundary conditions, the governing equation essentially reduces to the 1D diffusion equation with two Dirichlet boundary conditions.

b. Write a computer program to solve the original 2D governing equation subject to the aforementioned four boundary conditions using the unstructured finite method. Compute the fluxes at the left and right boundaries, as well.

c. Compare the analytical solution with the numerical solution. Compare the analytically obtained fluxes with the numerically computed fluxes. Comment on your results from the point of view of accuracy of ϕ, the fluxes, as well as global conservation.

Miscellaneous Topics

The preceding six chapters of this text presented methods to solve various types of canonical partial differential equations (PDEs) using either the finite difference method or the finite volume method. In this chapter, we switch gears, and discuss three important miscellaneous topics: (a) interpolation, (b) numerical integration, and (c) finding the root(s) of single or coupled (simultaneous) nonlinear equation(s). These topics usually find their place at the beginning of most numerical analysis texts that cover differential equations. Here, the conscious choice was made to reverse the coverage sequence because two of the three aforementioned topics – interpolation and numerical integration – are often used for postprocessing results generated by PDE solvers. For example, one might desire to compute the net mass or heat flux at a certain boundary of the computational domain. This will require numerical integration. Similarly, one might wish to compute the value of the dependent variable at a point within the computational domain that is not co-located with one of the nodes, thereby necessitating interpolation. The third topic, namely root finding, is inherently related to solving discrete equations arising out of nonlinear PDEs. In fact, the Newton's method is the foundation of a class of very popular methods for solving nonlinear PDEs – the Newton–Krylov method [1]. Section 8.5, which is meant to serve as a preamble to advanced topics such as computational fluid dynamics or computational electromagnetics, introduces the reader to methods for solving coupled PDEs. We begin with interpolation.

8.1 INTERPOLATION

Interpolation is a ubiquitous mathematical operation. It is used in virtually all applications that involve any kind of calculation. It is so common that it is almost built into our intuitive senses. For example, if we see an hourly weather report, it is almost automatic that we think of the half-hourly state of the weather as being somewhere in between the hourly states. Essentially, our brain has involuntarily performed an interpolation.

Mathematically, the interpolation problem may be posed as follows: given a set (table) of data pairs, as shown in Table 8.1, what is the value of y at a location x_q, where $x_1 \leq x_q \leq x_n$, and $x_1 < x_2 < ... < x_{n-1} < x_n$? The locations, $x_1, x_2, ..., x_n$, are known as *nodes*. Location, x_q, at which the value of y is sought, is henceforth referred to as, the *query point*, and the corresponding value of y is denoted by y_q. The determination of y_q requires interpolation.

Table 8.1 Generic 1D Table with n Entries

x	x_1	x_2	...	x_{n-1}	x_n
y	y_1	y_2	...	y_{n-1}	y_n

Broadly speaking, interpolation methodologies that may be used to determine y_q fall under two categories. The first category is one in which values at the nodes, $y_1, y_2, ..., y_n$, must be rigorously satisfied by the interpolation formula. The second category is one in which values at the nodes have uncertainties or noise, as would occur in the case of experimental data or data generated by stochastic calculations. Therefore, it is not worthwhile to try and satisfy them precisely because the presence of noise, for example, would cause the data to have spurious high-frequency oscillations. Instead, an interpolation procedure is sought that provides a fit to the data that still embodies the low-frequency variations, but eliminates the high-frequency noise. This has given rise to least-square interpolation. Since this text only covers deterministic methods, and the main objective of this chapter is to facilitate postprocessing of data generated by deterministic PDE solvers, least-square interpolation is omitted. The reader is referred to standard texts on numerical methods [2, 3] for a discussion on least-square interpolation.

Another perspective from which interpolation may be classified is local (or piecewise) versus global interpolation. The most routinely encountered kind of interpolation is linear interpolation, in which piecewise lines are used to interpolate between any two adjacent nodes. Instead of lines, piecewise fits with higher degree polynomials (curves) may be used, resulting in spline interpolation. Global interpolation, on the other hand, involves fitting the entire dataset into a single polynomial of high degree, and is commonly referred to as *polynomial interpolation*. In the two sections to follow, both types of interpolations have been discussed.

8.1.1 POLYNOMIAL AND LAGRANGE INTERPOLATION

Polynomial interpolation involves fitting a polynomial of degree $n-1$ to a dataset comprised of n nodes, as shown in Table 8.1. Mathematically, such a polynomial may be written as

$$p(x) = \sum_{k=0}^{n-1} a_k x^k, \tag{8.1}$$

where $a_0, a_1, ..., a_{n-1}$ are the unknown polynomial coefficients. These coefficients are determined by enforcing that these polynomials satisfy the nodal values at all nodes:

$$p_i = p(x_i) = \sum_{k=0}^{n-1} a_k x_i^k = y_i \quad \forall i = 1, 2, ..., n. \tag{8.2}$$

Expanding Eq. (8.2) and writing it in matrix form yields

$$\begin{bmatrix} 1 & x_1 & x_1^2 & \cdots & x_1^{n-1} \\ 1 & x_2 & x_2^2 & \cdots & x_2^{n-1} \\ \vdots & \vdots & \vdots & \vdots & \vdots \\ \vdots & \vdots & \vdots & \vdots & \vdots \\ \vdots & \vdots & \vdots & \vdots & \vdots \\ 1 & x_n & x_n^2 & \cdots & x_n^{n-1} \end{bmatrix} \begin{bmatrix} a_0 \\ a_1 \\ \vdots \\ \vdots \\ \vdots \\ a_{n-1} \end{bmatrix} = \begin{bmatrix} y_1 \\ y_2 \\ \vdots \\ \vdots \\ \vdots \\ y_n \end{bmatrix}. \tag{8.3}$$

Equation (8.3) represents a linear system of equations, whose solution will yield the set of coefficients, $a_0, a_1, ..., a_{n-1}$. Once the coefficients have been determined, the queried value may be determined using

$$y(x_q) = y_q = \sum_{k=0}^{n-1} a_k x_q^k. \tag{8.4}$$

Although the preceding method using power polynomials is conceptually straightforward, from a computational standpoint, it is cumbersome and time consuming. Determination of the elements of the matrix shown on the left-hand side of Eq. (8.3) requires one multiplication (if recursive multiplication, i.e., $x^i = x \cdot x^{i-1}$, is used) for each element from the second column onwards. Thus, determination of the coefficient matrix requires a total of $n(n-2)$ multiplications. Furthermore, solution of the linear system, shown in Eq. (8.3), using Gaussian elimination requires approximately $n^3/3 + n^2/2$ long operations, as discussed in Section 3.1.1. Thus, the total number of long operations required to determine the interpolating polynomial, is approximately $n^3/3 + 3n^2/2$. Clearly, the cubic scaling of the number of long operations with the number of data points (or nodes) is undesirable, and renders this method difficult to use if the number of nodes is large.

The problem may be circumvented by using polynomials of a special form, referred to as Lagrange* polynomials. Lagrange polynomials are constructed out of Lagrange basis functions, and are written as

$$L_i(x) = \prod_{\substack{j=1 \\ j \neq i}}^{n} \frac{x - x_j}{x_i - x_j}. \tag{8.5}$$

These basis functions have special properties, which become evident by using $x = x_j$, which yields $L_i(x_j) = 0$, and $x = x_i$, which yields $L_i(x_i) = 1$. A $(n-1)$-th degree polynomial can be written in terms of the Lagrange basis functions as

$$p(x) = \sum_{i=1}^{n} y_i L_i(x). \tag{8.6}$$

***Joseph-Louis Lagrange** (1736–1813) was an Italian mathematician and astronomer. He is one of the most influential mathematicians of all time. Lagrange is famously known for inventing variational calculus, and for the development of methods to solve equations with constraints using Lagrange multipliers. He also transformed Newtonian mechanics into a new method of analysis known as Lagrangian mechanics, which became the foundation of mathematical physics in the nineteenth century. He was employed at the Prussian Academy of Sciences in Berlin for over twenty years, and later moved to Paris, where he eventually died.

Using the aforementioned special properties of Lagrange basis functions, it is easy to show from Eq. (8.6) that $p(x_i) = y_i$, i.e., the condition that a polynomial must satisfy the nodal values, is obeyed automatically. Finding the interpolating polynomial boils down to finding the Lagrange basis functions, which is easily done using Eq. (8.5). An example is considered next to demonstrate how Lagrange interpolating polynomials are determined.

EXAMPLE 8.1

The objective is to determine the Lagrange interpolating polynomial for the following data:

x	$1/4$	$1/3$	1
y	-1	2	7

Using Eq. (8.5), we obtain the following relations

$$L_1(x) = \prod_{\substack{j=1 \\ j\neq 1}}^{3} \frac{x-x_j}{x_1-x_j} = \left(\frac{x-\frac{1}{3}}{\frac{1}{4}-\frac{1}{3}}\right)\left(\frac{x-1}{\frac{1}{4}-1}\right) = 16\left(x-\tfrac{1}{3}\right)(x-1).$$

$$L_2(x) = \prod_{\substack{j=1 \\ j\neq 2}}^{3} \frac{x-x_j}{x_1-x_j} = \left(\frac{x-\frac{1}{4}}{\frac{1}{3}-\frac{1}{4}}\right)\left(\frac{x-1}{\frac{1}{3}-1}\right) = -18\left(x-\tfrac{1}{4}\right)(x-1)$$

$$L_3(x) = \prod_{\substack{j=1 \\ j\neq 3}}^{3} \frac{x-x_j}{x_3-x_j} = \left(\frac{x-\frac{1}{4}}{1-\frac{1}{4}}\right)\left(\frac{x-\frac{1}{3}}{1-\frac{1}{3}}\right) = 2\left(x-\tfrac{1}{4}\right)(x-\tfrac{1}{3})$$

Therefore, the Lagrange interpolating polynomial, using Eq. (8.6), may be written as

$$p(x) = \sum_{i=1}^{3} y_i L_i(x) = y_1 L_1(x) + y_2 L_2(x) + y_3 L_3(x)$$
$$= -16\left(x-\tfrac{1}{3}\right)(x-1) - 36\left(x-\tfrac{1}{4}\right)(x-1) + 14\left(x-\tfrac{1}{4}\right)(x-\tfrac{1}{3})$$

In order to compare the computational cost of this method with interpolation using power polynomials, let us now estimate the number of long operations required to determine the interpolating polynomial in this method. Computation of each $L_i(x)$ requires $(n-1)$ divisions and $(n-2)$ multiplications, as evident from the steps in Example 8.1. Since n such basis functions have to be determined, the total number of long operations is $n(n-1)(n-2)$, implying that the number of long operations in this method, also scales as n^3. However, the number of long operations may be significantly reduced by rewriting Eq. (8.5) as

$$L_i(x) = \prod_{\substack{j=1 \\ j \neq i}}^{n} \frac{x - x_j}{x_i - x_j} = \frac{\prod_{\substack{j=1 \\ j \neq i}}^{n} x - x_j}{\prod_{\substack{j=1 \\ j \neq i}}^{n} x_i - x_j} = \frac{\frac{1}{x - x_i} \prod_{j=1}^{n} x - x_j}{\prod_{\substack{j=1 \\ j \neq i}}^{n} x_i - x_j}. \tag{8.7}$$

The quantity, $\prod_{j=1}^{n} x - x_j$, is independent of i, and needs to be computed only once. The denominator of the rightmost expression of Eq. (8.7) requires $(n-2)$ multiplications. These many multiplications must be performed to compute each $L_i(x)$. On the other hand, the computation of the numerator requires only one division for each i. Thus, the total number of long operations is roughly $n(n-1)$, implying significant savings. Further savings can be achieved by computing the denominator using recursive multiplication.

To summarize, Lagrange interpolating polynomials are an efficient and elegant means of interpolation within large datasets. Determining them does not require any matrix inversion unlike the case of power polynomials. Finally, although not shown here, polynomials constructed upon Lagrange basis sets, can be extended to multiple dimensions with relative ease, as compared to other polynomial forms.

One of the advantages of using high-degree polynomials for interpolation is that the continuity of derivatives up to the $(n-1)$th derivative is guaranteed. However, computations for most practical applications do not require continuity of derivatives beyond a few orders. Therefore, using a polynomial of very high degree is not always necessary. On the downside, larger the degree of the polynomial, larger is the number of maxima or minima or wiggles. These wiggles can make a solution in the space between nodes unrealistic or even unphysical. Figure 8.1 shows a sixth degree interpolating polynomial, which is used to fit seven data points that represents

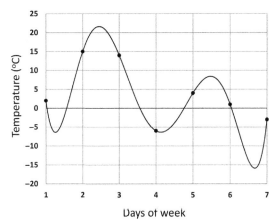

FIGURE 8.1 Interpolating Polynomial (Solid Curve) and Raw Data (Solid Circles) for a Sample Dataset

temperatures (in °C) on the seven days of a week. The data points are also shown in the same figure with solid circles. The sixth degree interpolating polynomial has five maxima or minima. On account of its wiggles, values predicted by the interpolating polynomial are often unrealistic. For example, day one and day two both have positive temperatures. Therefore, it is unrealistic to have a large negative temperature somewhere in between those two data points. The predicted value is not bounded by two end values. In contrast, variation between days three and four is monotonic and bounded. In many scientific computations, unbounded predictions are unacceptable, since they represent violation of physical laws. For example, if the x-axis in Fig. 8.1 represented spatial location instead of days, then by the second law of thermodynamics, it is impossible to get a temperature, which is beyond the bounds of these two end points for a pure heat conduction scenario without any heat sources or sinks.

Monotonicity and boundedness are both very important properties for the solution methodologies aimed toward solving governing equations that stem from physical conservation laws. As discussed in Section 6.3, in the computational fluid dynamics arena, for example, considerable research has been directed toward the development of higher order schemes for advection that preserve these two properties [4–6].

8.1.2 SPLINE INTERPOLATION

Interpolation using piecewise polynomials – referred to as spline interpolation – was developed precisely to address monotonicity and boundedness issues associated with the use of a high-degree polynomial interpolation. Linear interpolation, the most commonly used interpolation method, is in fact, a form of spline interpolation, in which lines are used to represent the spline segments. Despite its simplicity (some may even call it crude!), it owes its popularity to the fact that the interpolated value is always bounded between two end values, making it reliable for use in any situation. The downside of linear interpolation, of course, is that it only enforces continuity of the function (interpolated value) itself. Higher derivatives, which are often sought in scientific computations, are discontinuous, rendering it somewhat limiting in scope.

As mentioned earlier, a spline is comprised of piecewise polynomials. The locations (independent variable values) at which these polynomials switch from one to the next are referred to as *knots* of the spline. In general, the knots and the nodes may not be co-located. Mathematically, a spline may be written as

$$S(x) = \begin{cases} s_1(x) & \eta_1 \le x \le \eta_2 \\ s_2(x) & \eta_2 \le x \le \eta_3 \\ \vdots & \\ s_{n-1}(x) & \eta_{n-1} \le x \le \eta_n \end{cases}, \tag{8.8}$$

where the knots are denoted by $\eta_1, \eta_2, ..., \eta_n$, and $s_1, s_2, ..., s_{n-1}$ represent the piecewise polynomials used in each segment. In general, there are no rules to abide by when it comes to the placement of the knots. However, it is customary to make the two ends co-located with the two end knots, i.e., $\eta_1 = x_1$, and $\eta_n = x_n$. Also, for the sake of convenience, in most cases, the interior knots are co-located with the interior nodes. However, there are some exceptions to this practice. For example, in Subbotin quadratic splines, the interior knots ($\eta_2, ..., \eta_{n-1}$) are placed such that $\eta_i = 2x_i - \eta_{i+1}$. Starting with $i = n-1$, the position of the other knots may be computed using $\eta_{n-1} = 2x_{n-1} - \eta_n = 2x_{n-1} - x_n, \eta_{n-2} = 2x_{n-2} - \eta_{n-1}, ..., \eta_2 = 2x_2 - \eta_3$.

Irrespective of where the knots are placed, interpolation using splines must satisfy certain conditions to ensure that the resulting interpolant is "smooth." In the discussion to follow, we will assume that all the knots are co-located with the nodes, i.e., $\eta_i = x_i$ for all i. In order to understand the aforementioned conditions, first, let us consider the scenario where the piecewise polynomials are lines, i.e., $s_i(x) = a_i x + b_i$. Since each line segment has two unknown constants, and there are a total of $n-1$ intervals [Eq. (8.8)] a total of $2(n-1)$ unknowns need to be determined. Each line segment must satisfy two nodal values (continuity of the function is automatically enforced this way), resulting in $2(n-1)$ equations.

If the piecewise polynomials are of degree 2 (quadratic), then, a total of $3(n-1)$ unknown constants need to be determined. Continuity conditions for the function, once again, result in $2(n-1)$ equations. In addition, the continuity of the first derivative at all interior knots results in $n-2$ additional equations. Thus, the total number of equations is $2(n-1) + n - 2 = 3n - 4$, which falls one short of the number of unknowns. Usually, the final equation is determined by setting the derivative at either of the two end knots to some value (typically, zero). It is due to the flexibility in setting this value (and also whether it is set at the first or last knot) that quadratic spline fits are not unique. It is worth noting that setting the derivatives at either end essentially amounts to setting the local slope of the spline function. Thus, depending on what value the slope is set to and at which of the two knots it is set, the results can actually look noticeably different.

For splines using cubic polynomials, the total number of undetermined constants is $4(n-1)$. The continuity of the function provides $2(n-1)$ equations. Continuity of the first and second derivatives at the interior knots provide an additional $2(n-2)$ equations, taking the number of equations to $4n - 6$. The two remaining equations required to make the number of equations equal to the number of unknowns are customarily derived by setting the second derivative equal to zero at the two end knots. Thus, as opposed to quadratic splines, cubic splines are unique since there is no ambiguity with regard to which end knot the final condition is set at. Furthermore, since the two flexible conditions are applied to the second derivative, rather than the first (as in the case of quadratic splines), the flexible conditions have minimal impact on the shape of the spline curve, at least from a purely visual standpoint. Such a cubic spline is also known as a *natural cubic spline*. Since cubic splines are, by far, the most popular of all splines that use piecewise polynomials, the mathematical procedure for deriving them are discussed next. The same procedure may, just as easily, be extended to splines of other degrees. We consider the case when the nodes and knots are co-located, i.e., $\eta_i = x_i$ for all i. For a cubic spline, the second derivative of each spline segment is a line. Therefore, one may write

$$s_i''(x) = s_i''(\eta_i) \frac{\eta_{i+1} - x}{\eta_{i+1} - \eta_i} + s_i''(\eta_{i+1}) \frac{x - \eta_i}{\eta_{i+1} - \eta_i}, \tag{8.9}$$

where the primes used in the superscript are in keeping with standard notations used in calculus. Integrating Eq. (8.9) twice, we obtain

$$s_i(x) = \frac{s_i''(\eta_i)}{6} \frac{(\eta_{i+1} - x)^3}{\eta_{i+1} - \eta_i} + \frac{s_i''(\eta_{i+1})}{6} \frac{(x - \eta_i)^3}{\eta_{i+1} - \eta_i} + c_i x + d_i, \tag{8.10}$$

where c_i and d_i are integration constants. Equation (8.10) may be rewritten by using a different pair of integration constants:

$$s_i(x) = \frac{s_i''(\eta_i)}{6} \frac{(\eta_{i+1} - x)^3}{\eta_{i+1} - \eta_i} + \frac{s_i''(\eta_{i+1})}{6} \frac{(x - \eta_i)^3}{\eta_{i+1} - \eta_i} + C_i(x - \eta_i) + D_i(\eta_{i+1} - x). \tag{8.11}$$

Further, in order to determine the unknown constants C_i and D_i, we apply the condition that the spline must satisfy the prescribed value at the nodes (same as knots in this derivation), i.e., $s_i(\eta_i) = y_i$, and $s_i(\eta_{i+1}) = y_{i+1}$. This yields

$$s_i(\eta_i) = \frac{s_i''(\eta_i)}{6}(\eta_{i+1} - \eta_i)^2 + D_i(\eta_{i+1} - \eta_i), \tag{8.12a}$$

$$s_i(\eta_{i+1}) = \frac{s_i''(\eta_{i+1})}{6}(\eta_{i+1} - \eta_i)^2 + C_i(\eta_{i+1} - \eta_i). \tag{8.12b}$$

Solving Eq. (8.12) for C_i and D_i, and substituting the result into Eq. (8.11), we obtain

$$\begin{aligned} s_i(x) = & \frac{s_i''(\eta_i)}{6} \frac{(\eta_{i+1} - x)^3}{\eta_{i+1} - \eta_i} + \frac{s_i''(\eta_{i+1})}{6} \frac{(x - \eta_i)^3}{\eta_{i+1} - \eta_i} \\ & + \left(\frac{y_{i+1}}{\eta_{i+1} - \eta_i} - \frac{s_i''(\eta_{i+1})(\eta_{i+1} - \eta_i)}{6} \right)(x - \eta_i) \cdot \\ & + \left(\frac{y_i}{\eta_{i+1} - \eta_i} - \frac{s_i''(\eta_i)(\eta_{i+1} - \eta_i)}{6} \right)(\eta_{i+1} - x) \end{aligned} \tag{8.13}$$

One of the conditions that have not been used thus far is the continuity of the first derivative at the interior knots. To apply this condition, we first differentiate Eq. (8.13) to yield

$$\begin{aligned} s_i'(x) = & -\frac{s_i''(\eta_i)}{2} \frac{(\eta_{i+1} - x)^2}{\eta_{i+1} - \eta_i} + \frac{s_i''(\eta_{i+1})}{2} \frac{(x - \eta_i)^2}{\eta_{i+1} - \eta_i} \\ & + \left(\frac{y_{i+1}}{\eta_{i+1} - \eta_i} - \frac{s_i''(\eta_{i+1})(\eta_{i+1} - \eta_i)}{6} \right) - \left(\frac{y_i}{\eta_{i+1} - \eta_i} - \frac{s_i''(\eta_i)(\eta_{i+1} - \eta_i)}{6} \right) \cdot \end{aligned} \tag{8.14}$$

Substitution of $x = \eta_i$ in Eq. (8.14), followed by simplification, yields

$$s_i'(\eta_i) = -\left[\frac{s_i''(\eta_i)}{3} + \frac{s_i''(\eta_{i+1})}{6} \right](\eta_{i+1} - \eta_i) + \left(\frac{y_{i+1} - y_i}{\eta_{i+1} - \eta_i} \right). \tag{8.15}$$

By analogy with Eq. (8.14), one may write

$$\begin{aligned} s_{i-1}'(x) = & -\frac{s_{i-1}''(\eta_{i-1})}{2} \frac{(\eta_i - x)^2}{\eta_i - \eta_{i-1}} + \frac{s_{i-1}''(\eta_i)}{2} \frac{(x - \eta_{i-1})^2}{\eta_i - \eta_{i-1}} \\ & + \left(\frac{y_i}{\eta_i - \eta_{i-1}} - \frac{s_{i-1}''(\eta_i)(\eta_i - \eta_{i-1})}{6} \right) - \left(\frac{y_{i-1}}{\eta_i - \eta_{i-1}} - \frac{s_{i-1}''(\eta_{i-1})(\eta_i - \eta_{i-1})}{6} \right) \cdot \end{aligned} \tag{8.16}$$

Substituting $x = \eta_i$ in Eq. (8.16), and simplifying, yields

$$s'_{i-1}(\eta_i) = \left[\frac{s''_{i-1}(\eta_i)}{3} + \frac{s''_{i-1}(\eta_{i-1})}{6} \right](\eta_i - \eta_{i-1}) + \left(\frac{y_i - y_{i-1}}{\eta_i - \eta_{i-1}} \right). \tag{8.17}$$

Setting the right sides of Eqs (8.15) and (8.17) equal (continuity of first derivative), we obtain

$$-\left[\frac{s''_i(\eta_i)}{3} + \frac{s''_i(\eta_{i+1})}{6} \right](\eta_{i+1} - \eta_i) + \left(\frac{y_{i+1} - y_i}{\eta_{i+1} - \eta_i} \right)$$
$$= \left[\frac{s''_{i-1}(\eta_i)}{3} + \frac{s''_{i-1}(\eta_{i-1})}{6} \right](\eta_i - \eta_{i-1}) + \left(\frac{y_i - y_{i-1}}{\eta_i - \eta_{i-1}} \right). \tag{8.18}$$

Noting also that by continuity of the second derivative, we have $s''_i(\eta_i) = s''_{i-1}(\eta_i)$, and by making use of this in Eq. (8.18), we obtain

$$-\left[\frac{s''_i(\eta_i)}{3} + \frac{s''_i(\eta_{i+1})}{6} \right](\eta_{i+1} - \eta_i) + \left(\frac{y_{i+1} - y_i}{\eta_{i+1} - \eta_i} \right)$$
$$= \left[\frac{s''_i(\eta_i)}{3} + \frac{s''_{i-1}(\eta_{i-1})}{6} \right](\eta_i - \eta_{i-1}) + \left(\frac{y_i - y_{i-1}}{\eta_i - \eta_{i-1}} \right). \tag{8.19}$$

Next, to write Eq. (8.19) in compact form, we make use of the following notations

$$\delta_i = \eta_{i+1} - \eta_i, \delta_{i-1} = \eta_i - \eta_{i-1},$$
$$u_i = s''_i(\eta_i) = S''(\eta_i), u_{i-1} = s''_{i-1}(\eta_{i-1}) = S''(\eta_{i-1}), u_{i+1} = s''_i(\eta_{i+1}) = S''(\eta_{i+1}), \tag{8.20}$$
$$Y_i = y_{i+1} - y_i, Y_{i-1} = y_i - y_{i-1}.$$

Substitution of Eq. (8.20) into Eq. (8.19) yields

$$\delta_i u_{i+1} + 2(\delta_i + \delta_{i-1})u_i + \delta_{i-1}u_{i-1} = 6\left(\frac{Y_i}{\delta_i} - \frac{Y_{i-1}}{\delta_{i-1}} \right) \quad \forall i = 2,3,...,n-1. \tag{8.21}$$

Finally, the two end conditions for the second derivative yield

$$u_1 = s''_1(\eta_1) = 0,$$
$$u_n = s''_{n-1}(\eta_n) = 0. \tag{8.22}$$

Equations (8.21) and (8.22) constitute a tridiagonal system of equations that may be easily solved using the tridiagonal matrix algorithm (TDMA) discussed in Section 3.1.2, to yield $u_1, u_2, ..., u_n$. Using the notations provided by Eq. (8.20), the final spline function, given by Eq. (8.13), may be written as

$$s_i(x) = \frac{u_i}{6}\frac{(\eta_{i+1} - x)^3}{\delta_i} + \frac{u_{i+1}}{6}\frac{(x - \eta_i)^3}{\delta_i} + \left(\frac{y_{i+1}}{\delta_i} - \frac{u_{i+1}\delta_i}{6} \right)(x - \eta_i) + \left(\frac{y_i}{\delta_i} - \frac{u_i\delta_i}{6} \right)(\eta_{i+1} - x),$$
$$\tag{8.23}$$

where $\eta_i \leq x \leq \eta_{i+1}$. The complete step-by-step algorithm for obtaining a cubic spline fit to the dataset given by Table 8.1 is described next.

ALGORITHM: CUBIC SPLINE FIT (INTERPOLATION) TO 1D DATASET

Step 1: Set the knots of the spline to be the same as nodes: $\eta_i = x_i$.
Step 2: Compute the elements of the tridiagonal matrix using Eqs (8.20), (8.21), and (8.22).
Step 3: Solve the tridiagonal system using TDMA to determine $u_1, u_2, ..., u_n$.
Step 4: Use Eq. (8.23) to obtain an interpolated value at a given query point.

An example is considered next to demonstrate interpolation using cubic splines.

EXAMPLE 8.2

In this example we consider the same dataset that was used to generate Fig. 8.1, and perform a cubic spline fit to the data. The raw data, in tabular form, is as follows:

x	1	2	3	4	5	6	7
y	2	15	14	−6	4	1	−3

After execution of Step 3 of the above algorithm, the unknown coefficients in the spline fit are available. These coefficients, which represent the second derivatives at the nodes, are given by: $u_1 = 0$, $u_2 = -10.4692307692308$, $u_3 = -42.1230769230769$, $u_4 = 64.9615384615385$, $u_5 = -37.7230769230769$, $u_6 = 7.93076923076923$, and $u_7 = 0$. The spline fit obtained using these coefficients and Eq. (8.23) is shown in this figure. For comparison, the polynomial fit shown in Fig. 8.1 is also plotted.

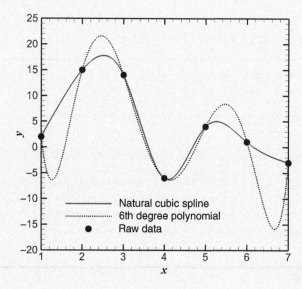

It is clear from the given figure that the natural cubic spline yields a much smoother fit to the data than the polynomial interpolant. All unnecessary

wiggles are completely eliminated, and the spline follows the raw data closely. The raw data has two maxima and one minimum, and so does the spline.
The fact that the number of extrema is also reproduced accurately is particularly attractive in scenarios where not only is the value of the function sought,
but also derivatives of the function are of interest.

As illustrated in Example 8.2, one of the greatest properties of the natural cubic
spline is its smoothness. To assess its smoothness, let us consider a twice-differentiable function $f(x)$, which is approximated by the natural cubic spline $S(x)$ in the interval
$[a, b]$. The error between the exact function and the approximating spline may be
written as $e(x) = f(x) - S(x)$. It follows then that $e''(x) = f''(x) - S''(x)$, since both
$f(x)$ and $S(x)$ are twice differentiable. The quantity, $e''(x)$, represents the local error in
curvature between the function and the approximating spline. It is a measure of how
smooth the spline is compared with the function it is being used to approximate. Since
$e''(x)$ may be positive or negative, we will assess the square of this quantity, rather
than the quantity itself (similar to computing the L^2Norm discussed in Chapter 3).
Furthermore, to assess the total error over the entire interval $[a, b]$, we will examine
$[e''(x)]^2$ over the interval. Rearranging the above equation, and integrating, we obtain

$$\int_a^b [f''(x)]^2\,dx = \int_a^b [e''(x) + S''(x)]^2\,dx$$

$$= \int_a^b [e''(x)]^2\,dx + \int_a^b [S''(x)]^2\,dx + \int_a^b [2e''(x)S''(x)]dx. \tag{8.24}$$

Using integration by parts, the last term in Eq. (8.24) may be written as

$$2\int_a^b [e''(x)S''(x)]dx = 2[S''(x)e'(x)]_{x=a}^{x=b} - 2\int_a^b [e'(x)S'''(x)]dx. \tag{8.25}$$

For a natural cubic spline, the second derivative of the spline at the two ends is
zero, i.e., $S''(a) = S''(b) = 0$. Also, the third derivative of the spline is a constant (in
each segment of the spline) since the spline is a cubic spline. Deriving the third derivative from Eq. (8.9), and substituting it into Eq. (8.25), we obtain

$$2\int_a^b [e''(x)S''(x)]dx = -2\sum_{i=1}^n \left[\frac{s_i''(\eta_{i+1}) - s_i''(\eta_i)}{\eta_{i+1} - \eta_i} \right] \int_{\eta_i}^{\eta_{i+1}} [e'(x)]dx$$

$$= -2\sum_{i=1}^n \left[\frac{s_i''(\eta_{i+1}) - s_i''(\eta_i)}{\eta_{i+1} - \eta_i} \right] [e(\eta_{i+1}) - e(\eta_i)] = 0 \tag{8.26}$$

where the final expression is equal to zero because at each knot, the error (difference
between the spline and the function) is zero, by definition. Substitution of the result
of Eq. (8.26) into Eq. (8.24) results in

$$\int_a^b [f''(x)]^2\,dx = \int_a^b [e''(x)]^2\,dx + \int_a^b [S''(x)]^2\,dx. \tag{8.27}$$

Since $[e''(x)]^2$ is always positive, the first integral on the right side of Eq. (8.27) is also always positive. Hence, it follows that

$$\int_a^b [f''(x)]^2\, dx \ge \int_a^b [S''(x)]^2\, dx, \tag{8.28}$$

which implies that the curvature of the spline, in a least-squared sense, over the entire interval of fit, is smaller than the original function that is being fitted. In summary, natural cubic splines are desirable over all other kinds of polynomial splines because of their smoothness. Another kind of spline that is widely used is the B-spline. B-splines are also known as basis splines, and are particularly suited to interpolation within multidimensional datasets.

The objective of Section 8.1 was to provide a brief overview of interpolation using either polynomials or splines. Clearly, the few methods discussed here constitute the proverbial "tip of the iceberg." However, it is sufficient information to enable postprocessing of data generated by PDE solvers, which is the central focus of this text. Readers interested in an in-depth study of the subject of interpolation are referred to advanced texts on the subject [2, 3, 7]. Next, we switch our attention to numerical integration.

8.2 NUMERICAL INTEGRATION

Closed-form analytical integration is not possible for the vast majority of functions encountered in scientific and engineering analysis. To complicate matters further, in many instances, the function is not available in analytical form, but is available only as a set (table) of discrete values, similar to what was considered in the preceding section. For example, if the PDE solver computes the unknown ϕ at discrete nodes, and an average value of ϕ is sought at one of the boundaries, one has no alternative but to resort to numerical integration.

8.2.1 INTEGRATION USING SPLINES

We begin our discussion by considering a continuous function $f(x)$ in the closed interval $[a, b]$. We assume that the function is only known at a set of prescribed locations, $a = x_1, x_2, ..., x_n = b$, called nodes. The scenario is illustrated in Fig. 8.2. For the sake of this discussion, the actual function from which the nodal values were generated is also shown in the figure. The objective is to determine the integral of the function, namely

$$I = \int_a^b f(x)\, dx, \tag{8.29}$$

which, geometrically, represents the area underneath the curve. Since the assumption is that closed-form analytical integration of the function is not possible, or that the

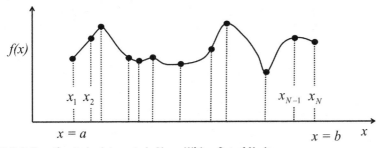

FIGURE 8.2 Function to be Integrated, Along With a Set of Nodes

function itself is not available in analytical form, the objective is really to find an approximation to I that makes use of only the prescribed nodal values (Fig. 8.2).

The fundamental philosophy behind most numerical integration procedures is to fit the discrete data using polynomials or splines, and subsequently employ the fitted entity for integration. For example, if the discrete dataset is fitted by a zeroth degree spline, either left-continuous or right-continuous, as illustrated in Fig. 8.3, the result is the lower bound integral or the upper bound integral, mathematically written as

$$I_L = \sum_{i=1}^{n-1} (x_{i+1} - x_i) \, \mathrm{MIN}(f(x_i), f(x_{i+1})), \tag{8.30a}$$

$$I_U = \sum_{i=1}^{n-1} (x_{i+1} - x_i) \, \mathrm{MAX}(f(x_i), f(x_{i+1})). \tag{8.30b}$$

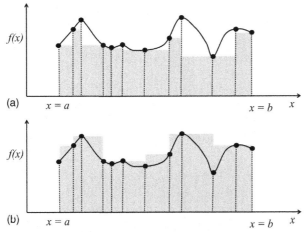

FIGURE 8.3 Geometric Representation of Numerical Integration

(a) Lower bound integral, and (b) upper bound integral.

Both I_L and I_U are approximations to I. If a general spline fit of arbitrary degree is used instead, and it is assumed that the knots are co-located with the nodes, then the approximate form of the integral becomes

$$I \approx \sum_{i=1}^{n-1} \int_{x_i}^{x_{i+1}} S(x)\,dx. \tag{8.31}$$

If the spline is of first degree (piecewise lines), then Eq. (8.31) may be written as

$$I \approx \sum_{i=1}^{n-1} \int_{x_i}^{x_{i+1}} \left[f(x_i)\tfrac{x_{i+1}-x}{x_{i+1}-x_i} + f(x_{i+1})\tfrac{x-x_i}{x_{i+1}-x_i} \right]dx. \tag{8.32}$$

Performing the inner definite integration analytically, and substituting the result into Eq. (8.32) yields

$$I \approx \sum_{i=1}^{n-1} \frac{1}{2}(x_{i+1}-x_i)\left[f(x_i)+f(x_{i+1})\right] = I_T, \tag{8.33}$$

which is the integration formula for the well-known *trapezoidal rule*. The trapezoidal rule derives its name from the fact that the area under the function is approximated using the sum of a series of trapezoids, as illustrated in Fig. 8.4. It is clear from the geometric interpretation shown in Fig. 8.4 that if the function has a local minimum or maximum in the interval (x_i, x_{i+1}), or if the function is monotonic but strongly non-linear in (x_i, x_{i+1}), then the accuracy of the trapezoidal rule will be poor.

In the case when the function $f(x)$ is known analytically, but its closed-form ana-lytical integration is not possible, an arbitrary number of nodes (or quadrature points) can be utilized to obtain an accurate answer. Typically, equally spaced quadrature points are used. In such a scenario, one important question that is often raised is the minimum number of nodes (or quadrature points) that must be used to obtain an answer with a desired level of accuracy. In the case of the trapezoidal rule, this ques-tion can be answered by analyzing its error. If $f(x)$ is a twice-differentiable function, then the difference (error) between the exact (analytical) integration result and the trapezoidal formula is written as [2]

$$E_T = I - I_T = -\frac{1}{12}(b-a)h^2 f''(\xi), \tag{8.34}$$

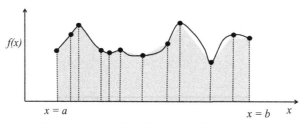

FIGURE 8.4 **Geometric Representation of the Trapezoidal Rule**

where I and I_T are given by Eqs (8.29) and (8.33), respectively, and $h = x_{i+1} - x_i$ is the nodal spacing, chosen to be uniform. Eq. (8.34) is similar to the error term obtained for the central difference formula in Chapter 2 [Eq. (2.10)]. In Eq. (8.34), ξ is a value of x that lies in the open interval (a, b). Using the error expression given by Eq. (8.34), it is possible to compute the value of h, and subsequently, the number of quadrature points that must be used for a prescribed error. This is demonstrated next through an example.

EXAMPLE 8.3

The objective is to determine the number of quadrature points that will be necessary to compute the integral $I = \int_0^1 \exp(-x^2)\,dx$ accurately up to at least four decimal places using the trapezoidal rule. Also, we seek to estimate, without actually computing the integral, the maximum percentage error in the solution if the answer is accurate up to four decimal places.

To answer these two questions, we start with the error expression for the trapezoidal rule given by Eq. (8.34). The maximum error will occur when $|f''(\xi)|$ is maximum. Noting that $f(x) = \exp(-x^2)$, and differentiating twice, we obtain $f''(x) = (4x^2 - 2)\exp(-x^2)$. Thus, $f''(0) = -2$, and $f''(1) = 2\exp(-1)$, and the maximum value of $|f''(\xi)| = 2$. Now setting the error to 10^{-4} (four decimal places), from Eq. (8.34), we obtain $\dfrac{1}{12}(1) \cdot h^2 \cdot 2 \leq 10^{-4}$, which yields

$h \leq 0.0245$. This yields $\dfrac{1}{n-1} \leq 0.0245$, or $n \geq 41$. Hence, at least 41 quadrature points are necessary to guarantee a solution that is accurate up to the fourth decimal place.

In order to estimate maximum percentage error in the solution, we need to find an estimate for the minimum value of the integral. Since the function is a monotonically decreasing function in $[0,1]$, the lowest value the integral can take is $\exp(-1^2) \cdot (1-0) = e^{-1} = 0.3678$. Therefore, the maximum possible percentage error that would be incurred if the solution is accurate up to four decimal places is $10^{-4} \times 100 / 0.3678 = 0.027\%$.

Returning to our discussion on using splines for numerical integration, we have established that using first degree splines results in the trapezoidal rule. Next, we will develop the framework for using natural cubic splines since, as we have proven in the preceding section, such splines have excellent smoothness properties. To do so, we substitute Eq. (8.23) into Eq. (8.31), and perform the inner integration analytically. After simplification, this yields

$$I \approx -\frac{1}{24}\sum_{i=1}^{n-1}(x_{i+1}-x_i)^3(u_{i+1}+u_i)+\frac{1}{2}\sum_{i=1}^{n-1}(x_{i+1}-x_i)[f(x_{i+1})+f(x_i)], \quad (8.35)$$

where u_i are defined in Eq. (8.20), and are to be determined by solving the tridiagonal system given by Eqs (8.21) and (8.22). Next, we explore the accuracy of the spline-based numerical integration methods by conducting an example calculation.

EXAMPLE 8.4

In this example, we explore the accuracy and efficiency of integration of $I = \int_{-2}^{2} x\exp(x)\,dx$ using two spline-based methods: (a) first degree spline (trapezoidal rule) and (b) natural cubic spline. The analytical solution to the integral may be obtained using integration by parts, and yields $I = e^2 + 3e^{-2} = 7.7950619$.

For numerical integration, we will employ the following nodes (quadrature points): $[-2,-1,0,1,2]$, such that the spacing between the nodes is uniform, and is denoted by $h = 1$. For the selected set of nodes, the trapezoidal rule [Eq. (8.33)] yields

$$I \approx \frac{h}{2}[f(a)+f(b)]+h\sum_{i=2}^{4}f(a+(i-1)h)=\frac{1}{2}[f(-2)+f(2)]+f(-1)+f(0)+f(1)$$

$$=\frac{1}{2}[-0.2706705+14.778112]-0.3678794+0+2.7182818 = 9.6041232$$

The percentage error in the solution is 23.2%.

The solution using natural cubic splines is somewhat more involved because it entails solution of a system of five equations [Eqs (8.21) and (8.22)] to determine the u_i. This yields $u_1 = 0$, $u_2 = 0.74102826$, $u_3 = -0.17358313$, $u_4 = 14.05571859$, and $u_5 = 0$. Substitution into Eq. (8.35) yields

$$I \approx -\frac{h^3}{24}\sum_{i=1}^{n-1}(u_{i+1}+u_i)+\frac{h}{2}\sum_{i=1}^{n-1}[f(x_{i+1})+f(x_i)]$$

$$= -\frac{1}{24}[u_5+u_4+u_4+u_3+u_3+u_2+u_2+u_1]$$

$$+\frac{1}{2}[f(x_5)+f(x_4)+f(x_4)+f(x_3)+f(x_3)+f(x_2)+f(x_2)+f(x_1)]$$

$$= -\frac{1}{24}[u_5+2(u_4+u_3+u_2)+u_1]+\frac{1}{2}[f(x_5)+2(f(x_4)+f(x_3)+f(x_2))+f(x_1)]$$

$$= -\frac{1}{24}[0+2(14.05571859-0.17358313+0.74102826)+0]$$

$$+\frac{1}{2}[14.778112+2(2.7182818+0-0.3678794)-0.2706705]= 8.3855262$$

The resulting percentage error in the solution is 7.6%. As expected, the natural cubic spline yields a more accurate result. However, this increased accuracy comes at a significantly increased computational cost. While the trapezoidal rule requires just five functional evaluations, the cubic spline requires not only five functional evaluations, but also determination of the values of u_i. As discussed in Section 3.1.2, the tridiagonal solver requires n long operations in addition to operations required to compute the coefficients of the tridiagonal matrix and the right side vector. These *pros* and *cons* of the two methods relative to each other must be considered when choosing which method to use.

8.2.2 INTEGRATION USING LAGRANGE POLYNOMIALS

Another option to perform numerical integration of a function is to fit a polynomial to the discrete dataset, and then integrate the subsequent polynomial function. For this purpose, it is convenient to use Lagrange polynomials rather than power polynomials. We begin by writing the exact integral in approximate form as

$$I = \int_a^b f(x)\,dx \approx \int_a^b p(x)\,dx, \tag{8.36}$$

where, as in Section 8.1, $p(x)$ is the fitting polynomial. Substituting Eq. (8.6) into Eq. (8.36) and interchanging the summation with the integration, we obtain

$$I \approx \sum_{i=1}^n f(x_i) \int_a^b L_i(x)\,dx = \sum_{i=1}^n w_i f(x_i), \tag{8.37}$$

where w_i are known as *weights*. In this particular case, it is clear from Eq. (8.37) that the weights are given by

$$w_i = \int_a^b L_i(x)\,dx. \tag{8.38}$$

It is worth pointing out that the weights given by Eq. (8.38) are independent of the function being integrated. Therefore, once computed, they may be used for any function. Next, we consider an example that demonstrates the computation of the weights and their use in the context of polynomial integration.

EXAMPLE 8.5

In this example, we compute $I = \int_{-2}^{2} x\exp(x)\,dx$ using the polynomial integration technique. This is the same problem considered in Example 8.4, and therefore, for comparison, we employ the same quadrature points, namely $[-2,-1,0,1,2]$. Using Eq. (8.38), in conjunction with Eq. (8.5), we obtain

$$L_1(x)=\prod_{\substack{j=1 \\ j\neq 1}}^{5}\frac{x-x_j}{x_1-x_j}=\left(\frac{x+1}{-2+1}\right)\left(\frac{x-0}{-2-0}\right)\left(\frac{x-1}{-2-1}\right)\left(\frac{x-2}{-2-2}\right)=\frac{1}{24}(x+1)x(x-1)(x-2).$$

Therefore,

$$w_1=\int_{-2}^{2}L_1(x)\,dx=\frac{14}{45}.$$ Using a similar procedure, we obtain $w_2=\frac{64}{45}$,

$$w_3=\frac{8}{15},\ w_4=\frac{64}{45},\ w_5=\frac{14}{45}.$$ Therefore, the integral, using Eq. (8.37), becomes

$$\begin{aligned}I &\approx w_1 f(x_1)+w_2 f(x_2)+w_3 f(x_3)+w_4 f(x_4)+w_5 f(x_5)\\&=\frac{14}{45}\times-0.2706705+\frac{64}{45}\times-0.3678794+0+\frac{64}{45}\times2.7182818+\frac{14}{45}\times14.778112\\&=7.8562208\end{aligned}$$

The percentage error in the solution is only 0.78%, which is significantly lower than even using cubic splines (Example 8.4). This is probably because a fourth-degree polynomial was used in this case, and the function was monotonically increasing in the entire interval making the polynomial fit almost perfect.

The two methods of numerical integration discussed this far, namely spline- and polynomial-based integration, can be applied to both closed-form analytical functions as well as discrete datasets (function provided as a table of values). In the former case, the nodes or quadrature points can be arbitrarily chosen, while in the latter case the quadrature points are usually made to co-locate with the nodes in the discrete data table. In the next section, we discuss Gaussian quadrature formulas, in which the location of the nodes is not arbitrary but based on strict mathematical principles.

8.2.3 GAUSSIAN QUADRATURE

Although the polynomial- and spline-based integration methods offer the advantage that the location of the nodes can be flexible, concurrently, they have the disadvantage that if the nodes are shifted, the weights have to be rederived or recomputed. Furthermore, the accuracy of the answer is dependent upon the location of the nodes. This leads to the following critical question: what is the optimum location of the nodes? The answer to this question was provided by Gauss, in his groundbreaking work on numerical integration, which is outlined here.

Let $q(x)$ be a polynomial of degree n, such that

$$\int_{a}^{b}q(x)x^k=0 \quad \forall k=0,1,2,...,n-1. \tag{8.39}$$

Also, let $x_1, x_2, ..., x_n$ be the n roots of the polynomial $q(x)$. Gauss' theorem states that the quadrature formula, given by Eq. (8.37), will yield exact answers for integration of any function that is a polynomial of degree less than or equal to $2n - 1$ provided the quadrature points are placed on the roots of $q(x)$. The left-hand side of Eq. (8.39), for various values of k, represent moments of $q(x)$. $k = 0$ results in the zeroth moment, $k = 1$ results in the first moment, and so on.

The implications of Gauss' theorem are best divulged by considering an example. Let us consider a scenario in which we want to compute the integral of an arbitrary function in the interval $[-1, 1]$ using just three nodes. In this case, we will not choose the location of the quadrature points (or nodes), but will use Gauss' theorem to determine their locations. Since $n = 3$ in this particular example, $2n - 1 = 5$. Thus, if the polynomial $q(x)$ is chosen such that Eq. (8.39) is satisfied, then placing the three quadrature points at the roots of $q(x)$ will yield an exact answer for the integration of any function that can be expressed as a polynomial of degree 5 or less. This means that the integration of $f(x) = x^5 - 2x^4 + 3$, for example, will be exact with just 3 nodes – a remarkable result! In practice, all functions can be written in polynomial form using Taylor series expansions as long as the function is differentiable multiple times. For example, $f(x) = x\exp(x)$ may be written as $f(x) = x + x^2 + x^3/2! + ... + x^m/(m-1)! + ...$ For such functions, Gaussian nodes will not produce an exact answer, but an answer whose accuracy will depend on how large the higher degree terms are. Since integral of the function is the sum of the integral of each of the polynomial terms, and since the integrals up to the 5th degree are exact with just 3 Gaussian nodes in this particular example, the resulting error will only depend on how erroneous the integration of the higher degree terms are.

The critical unanswered question at this juncture is how the function $q(x)$ and its roots are to be determined. For the particular example under consideration, $q(x)$ must be a polynomial of degree 3. Therefore, we may write

$$q(x) = c_0 + c_1 x + c_2 x^2 + c_3 x^3, \tag{8.40}$$

where c_0 through c_3 are undetermined constants. To determine these constants, we enforce Eq. (8.39) for $k = 0$, 1, and 2. This yields

$$\int_{-1}^{1} q(x)\,dx = \int_{-1}^{1} (c_0 + c_1 x + c_2 x^2 + c_3 x^3)\,dx = 2c_0 + \frac{2}{3}c_2 = 0, \tag{8.41a}$$

$$\int_{-1}^{1} q(x)x\,dx = \int_{-1}^{1} (c_0 x + c_1 x^2 + c_2 x^3 + c_3 x^4)\,dx = \frac{2}{3}c_1 + \frac{2}{5}c_3 = 0, \tag{8.41b}$$

$$\int_{1}^{1} q(x)x^2\,dx = \int_{-1}^{1} (c_0 x^2 + c_1 x^3 + c_2 x^4 + c_3 x^5)\,dx = \frac{2}{3}c_0 + \frac{2}{5}c_2 = 0. \tag{8.41c}$$

Combining Eqs (8.41a) and (8.41c) yields $c_0 = c_2 = 0$. The remaining equation, Eq. (8.41b), has two unknowns, and can, therefore, be satisfied by an infinite number of combinations. One obvious choice is to use $c_1 = -(3/2)$ and $c_3 = 5/2$. With this choice, the polynomial becomes

$$q(x) = -\frac{3}{2}x + \frac{5}{2}x^3, \tag{8.42}$$

which, in fact, is a Legendre polynomial of degree 3. The roots of $q(x)$ are 0 and $\pm\sqrt{3/5}$, which become the 3 Gaussian nodes for this particular case. Since the choice of the constants c_1 and c_3 resulted in $q(x)$ being a Legendre polynomial, this type of quadrature scheme is often referred to as the *Gauss–Legendre quadrature scheme*. Although somewhat tedious, it is straightforward to derive $q(x)$ for higher values of n, and show that with the appropriate choice of coefficients, we do always arrive at the Legendre polynomial of the nth degree. Hence, in practice, one can blindly use the roots of the Legendre polynomials as the nodes in a Gauss–Legendre quadrature scheme without having to go through the derivation.

Having derived the quadrature node locations, the next task is to derive the weights. Gauss' theorem stipulates that the result of integration of a polynomial of degree less than or equal to $2n - 1$ will be exact if n quadrature points are used. If $f(x)$ is a polynomial of degree $2n - 1$ or less, and it is divided by a polynomial $q(x)$, which is of degree n, then it follows that the quotient will be a polynomial of degree at most $n - 1$, and the remainder will be a polynomial of degree less than or equal to $n - 1$. For example, if $f(x) = x^5 - 2x^4 + 3$ is divided by $q(x) = -(3/2)x + (5/2)x^3$, then the result is the quotient $v(x) = (2/5)x^2 - (4/5)x + (6/25)$, and the remainder is $r(x) = -(6/5)x^2 + (9/25)x + 3$. In general, therefore, we may write

$$I = \int_a^b f(x)\,dx = \int_a^b [q(x)v(x) + r(x)]\,dx. \tag{8.43}$$

Since $v(x)$ is a polynomial of degree $n - 1$, it follows that it is a linear combination of terms containing $x^0, x^1, ..., x^{n-1}$. In the example above, it is a linear combination of x^0, x^1, and x^2 since $n = 3$. Therefore, the first integral on the right-hand side of Eq. (8.43) becomes a linear combination of $\int_a^b q(x)x^0\,dx, \int_a^b q(x)x^1\,dx, ..., \int_a^b q(x)x^{n-1}\,dx$, which are all equal to zero by Gauss' theorem [Eq. (8.39)]. Therefore, Eq. (8.43) reduces to

$$I = \int_a^b f(x)\,dx = \int_a^b r(x)\,dx. \tag{8.44}$$

Since $r(x)$ is also a polynomial of degree $n - 1$ or less, its integration, according to Gauss' theorem, must yield exact answers. Noting further that $r(x)$ is a linear combination of $x^0, x^1, ..., x^{n-1}$, the integration of each of these terms must be exact. Reverting back to our example problem for which we already derived the quadrature node locations, ($n = 3$ case with interval $[-1, 1]$), based on the above argument, we may, therefore, write

$$\int_{-1}^{1} 1 \cdot dx = 2 = w_1 \cdot 1 + w_2 \cdot 1 + w_3 \cdot 1, \tag{8.45a}$$

$$\int_{-1}^{1} x \, dx = 0 = -w_1 \cdot \sqrt{\tfrac{3}{5}} + w_2 \cdot 0 + w_3 \cdot \sqrt{\tfrac{3}{5}}, \tag{8.45b}$$

$$\int_{-1}^{1} x^2 \, dx = \frac{2}{3} = w_1 \cdot \tfrac{3}{5} + w_2 \cdot 0 + w_3 \cdot \tfrac{3}{5}. \tag{8.45c}$$

Solution of Eq. (8.45) yields $w_1 = 5/9, w_2 = 8/9$, and $w_3 = 5/9$, which are the three required weights.

In summary, the three-point Gauss–Legendre quadrature formula to integrate any function in the interval $[-1, 1]$ is

$$I = \int_{-1}^{1} f(x) \, dx = \sum_{i=1}^{3} w_i f(x_i) = \frac{5}{9} f\left(-\sqrt{\frac{3}{5}}\right) + \frac{8}{9} f(0) + \frac{5}{9} f\left(\sqrt{\frac{3}{5}}\right). \tag{8.46}$$

This formula has the remarkable feature that it is exact for the integration of any function that can be represented by a polynomial of degree 5 or less.

Although the Gaussian nodes and weights were derived for an integral in the interval $[-1, 1]$, the same nodes and weights can be applied to integrals in other intervals by a simple change of variables. To illustrate this, let us consider computation of the following integral using Gaussian quadrature:

$$I = \int_{a}^{b} f(z) \, dz. \tag{8.47}$$

Let us now use the substitution $z = [(b-a)/2]x + (b+a)/2$. It follows, then, that $dz = [(b-a)/2]dx$. Substitution into Eq. (8.47), followed by use of Eq. (8.46), yields

$$I = \int_{a}^{b} f(z) \, dz = \frac{b-a}{2} \int_{-1}^{1} f(\tfrac{b-a}{2}x + \tfrac{b+a}{2}) \, dx \approx \frac{b-a}{2} \sum_{i=1}^{n} w_i f(\tfrac{b-a}{2}x_i + \tfrac{b+a}{2}). \tag{8.48}$$

Following the procedure used to derive Eq. (8.46), higher order quadrature schemes may also be derived. These quadrature nodes and weights have

Table 8.2 Quadrature Nodes and Weights for Gauss–Legendre Quadrature Schemes of Various Orders. Eq. (8.48) is to be Used for the Computation of the Integral

Orders of Quadrature (n)	Quadrature Nodes ($\pm x_i$)	Quadrature Weights (w_i)
2	0.577350269189626	1.000000000000000
3	0.000000000000000	0.888888888888889
	0.774596669241483	0.555555555555556
5	0.000000000000000	0.568888888888889
	0.538469310105683	0.478628670499366
	0.906179845938664	0.236926885056189
10	0.148874338981631	0.295524224714753
	0.433395394129247	0.269266719309994
	0.679409568299024	0.219086362515983
	0.865063366688985	0.149451349150581
	0.973906528517172	0.066671344308688
20	0.076526521133497	0.152753387130725
	0.227785851141645	0.149172986472603
	0.373706088715419	0.142096109318382
	0.510867001950827	0.131688638449176
	0.636053680726515	0.118194531961518
	0.746331906460150	0.101930119817240
	0.839116971822218	0.083276741576704
	0.912234428251325	0.062672048334109
	0.963971927277913	0.040601429800386
	0.993128599185094	0.017614007139152

already been tabulated in available texts and handbooks. The most reliable source for such data is the invaluable handbook by Abramowitz and Stegun [8]. For convenience, Table 8.2 lists the quadrature nodes and weights necessary for a few selected Gaussian quadrature schemes up to $n = 20$. It should be noted that the quadrature nodes are symmetric about $x = 0$, as was shown in the preceding derivation. Therefore, it is sufficient to tabulate only half of the nodal values and weights. Also, the order in which the nodes and corresponding weights are used is not important, since the quadrature formula, Eq. (8.48), is additive.

One important advantage of Gaussian quadrature formulas is that the end points are not used in the integration. In many computations, end points have discontinuities, and therefore, using the functional values at the end points is not desirable. For example, if the Poisson equation is solved on a rectangular domain with four Dirichlet boundary conditions with four different values, the four corners have discontinuities in the dependent variable, ϕ. Therefore, if one were to compute the average value of ϕ at each of the four boundaries, it is desirable to avoid using values at the four corners.

Next, a code snippet that demonstrates use of the data in Table 8.2 for integration of an arbitrary function in the interval $[a, b]$, is provided.

CODE SNIPPET 8. 1: INTEGRATION USING 10-POINT GAUSSIAN QUADRATURE

```
! Read in (load) the 5 nodes and weights from Table 8.2 for n = 10.
xi(1) = 0.148874338981631; wi(1) = 0.295524224714753
xi(2) = 0.433395394129247; wi(2) = 0.269266719309994
xi(3) = 0.679409568299024; wi(3) = 0.219086362515983
xi(4) = 0.865063366688985; wi(4) = 0.149451349150581
xi(5) = 0.973906528517172; wi(5) = 0.066671344308688
For i = 1:5  ! Generate nodes and weights of the other half
  xi(5+i) = -xi(i) ! Nodes get negative value
  wi(5+i) = wi(i) ! Weight remains the same
End
! Now, compute integral. a and b are limits of the integral (inputs)
ngauss = 10
c1 = (b+a)/2
c2 = (b-a)/2
sum_gauss = 0
For i = 1:ngauss
  zi = c1 + c2*xi(i)
  sum_gauss = sum_gauss + wi(i)*function(zi)  ! The functional values need to be evaluated (inputs)
End
integral = 2*c2*sum_gauss
```

Although, in theory, Gaussian quadrature is expected to yield very accurate results for relatively low-order quadrature schemes, it remains to be seen if the accuracy is comparable to the other methods of numerical integration discussed earlier. To shed light on this issue, next, we compute the same integral considered in the preceding several examples.

EXAMPLE 8.6

In this example, the integral $I = \int_{-2}^{2} x \exp(x)\, dx$ is computed using the Gauss–Legendre quadrature formula. This is the same problem considered in the previous two examples in this chapter. Since comparing the accuracy of this method is one of the goals, we employ the same number of quadrature points as in the previous examples, namely 5.

Transforming x_i to z_i (here, $z_i - 2x_i$), and using Table 8.2 (for $n = 5$), we get $z_1 = 0, z_2 = 1.0769386, z_3 = 1.8123596, z_4 = -1.0769386$, and $z_5 = -1.8123596$. Also, $w_2 = w_4$, and $w_3 = w_5$. Now using Eq. (8.48), we obtain

$$I \approx 2[w_1 f(z_1) + w_2 f(z_2) + w_3 f(z_3) + w_4 f(z_4) + w_5 f(z_5)]$$
$$= 2[0 + 0.4786287 \times 3.1615455 + 0.2369268 \times 11.100\ 49 +$$
$$0.4786287 \times -0.3668448 + 0.2369268 \times -0.2959011]$$
$$= 7.7950413$$

The exact analytical solution is 7.7950619, thereby yielding a percentage error of $2.6 \times 10^{-4}\%$, which is several orders of magnitude superior to any of the other methods considered thus far. In fact, using a three-point Gaussian formula yields a result of 7.737793 (0.73% error). A summary of the results obtained with various methods considered in the preceding examples is tabulated below.

Methods	Results $(= I = \int_{-2}^{2} x\exp(x)\,dx)$	Percentage Error (with analytical)
Analytical (exact)	7.7950619	—
Trapezoidal rule (five nodes)	9.6041232	23.2
Cubic spline (five nodes)	8.3855262	7.6
Lagrange polynomial (five nodes)	7.8562208	0.78
Gauss–Legendre (five nodes)	7.7950413	2.6×10^{-4}
Gauss–Legendre (three nodes)	7.737793	0.73

Clearly, the Gauss–Legendre quadrature outshines all other methods by a considerable margin.

Thus far, in this section, the example considered to compare and contrast various methods of numerical integration is one in which the function, $f(x) = x\exp(x)$, is monotonically increasing in the interval $[-2, 2]$. Although this example sheds considerable light on the *pros* and *cons* of the various methods, the real test of the methods lies in the case where the function is oscillatory. We will now consider such an example.

EXAMPLE 8.7

In this example, the integral $I = \int_{-2}^{2} (2x - 1)\cos x\,dx$ is computed using the four methods discussed in this section. As in the previous examples, we employ 5 quadrature points for integration. For the trapezoidal rule, cubic spline, and Lagrange polynomial, the nodes are chosen to be located at $[-2, -1, 0, 1, 2]$. The results are tabulated here.

Methods	Results $(= I = \int_{-2}^{2} (2x - 1)\cos(x)\,dx)$	Percentage Errors (with analytical)
Analytical (exact)	−1.8185949	—
Trapezoidal rule (five nodes)	−1.6644578	8.5

Methods	Results $(= I = \int_{-2}^{2} (2x - 1)\cos(x)\, dx)$	Percentage Errors (with analytical)
Cubic spline (five nodes)	−1.8365756	1
Lagrange polynomial (five nodes)	−1.8112574	0.4
Gauss–Legendre (five nodes)	−1.8185964	8.4×10^{-5}

As in Example 8.6, in this case too, the Gauss–Legendre quadrature outshines the other methods by a considerable margin. For this problem, the accuracy of the cubic spline method is much closer to that of the Lagrange polynomial method compared with what was found for Example 8.6.

Since the Gauss–Legendre quadrature scheme far outshines the other schemes for numerical integration, it would, obviously, be the method of choice for numerical integration in most cases, at least among the methods presented here. One big disadvantage of this method, however, is that the nodes are not flexible. This implies that the functional values must be available at the Gauss–Legendre nodes. While this is not a problem if a closed-form analytical function is being integrated, it does become a hindrance if the function is available only in discrete tabular form. In such a case, the values of the function at the quadrature points must be first determined using interpolation. As demonstrated in Section 8.1, in general, natural cubic spline interpolation is the method of choice for this purpose. While polynomial interpolation may sometimes yield excellent results, it is not always desirable because of the large number of wiggles that are often exhibited by high-degree polynomials. To summarize, for functions represented by discrete tabular data, it is recommended that spline interpolation, in conjunction with Gauss–Legendre quadrature, be used.

The methods discussed in this section constitute a small subset of the methods used to perform numerical integration. Other popular methods include the Simpson's rule, the Newton–Cotes quadrature, the Gauss–Lobatto quadrature, and the Gaussian quadrature of moments. In general, the latter two, derived from Gauss' theorem, yield significantly more accurate results than the former two. The use of these schemes is as straightforward as the ones described here, and the reader is referred to Abramowitz and Stegun [8] for details.

8.3 NEWTON–RAPHSON METHOD FOR NONLINEAR EQUATIONS

The Newton–Raphson method, commonly known as the Newton's method, is one of the most prevalent methods for computation of roots of nonlinear equations. The history of this method can be traced back to the early independent works of both Isaac Newton and Joseph Raphson, but it was Thomas Simpson who first proposed this method as the present-day iterative method.

8.3.1 SINGLE NONLINEAR EQUATION (SCALAR CASE)

We begin by considering an arbitrary smooth function, $f(x)$, whose roots are sought. In general, if the function is nonlinear, it is difficult to find its roots. The crudest method to find the root (or roots) may be a simple guess-and-check approach. Let us start with a guess for the root, which we denote by $x^{(0)}$. Henceforth, we will refer to it as the initial guess. Assuming that the function $f(x)$ is infinitely differentiable, it may be expanded using a Taylor series to write

$$f(x) = f\left(x^{(0)}\right) + \left(x - x^{(0)}\right)f'\left(x^{(0)}\right) + \frac{\left(x - x^{(0)}\right)^2}{2!}f''\left(x^{(0)}\right) + \dots \qquad (8.49)$$

To determine the roots, we may now use the right-hand side of Eq. (8.49), and set it equal to zero, yielding

$$f\left(x^{(0)}\right) + \left(x - x^{(0)}\right)f'\left(x^{(0)}\right) + \frac{\left(x - x^{(0)}\right)^2}{2!}f''\left(x^{(0)}\right) + \dots = 0. \qquad (8.50)$$

Unfortunately, Eq. (8.50) is a nonlinear polynomial equation in x, whose solution is, arguably, as difficult to obtain as solving the original nonlinear equation. As an approximation, the higher-order terms may be discarded to yield

$$f\left(x^{(0)}\right) + \left(x - x^{(0)}\right)f'\left(x^{(0)}\right) \approx 0. \qquad (8.51)$$

Rearrangement of Eq. (8.51) yields an approximate solution to $f(x) = 0$:

$$x \approx x^{(0)} - \frac{f\left(x^{(0)}\right)}{f'\left(x^{(0)}\right)}. \qquad (8.52)$$

The solution given by Eq. (8.52) is approximate because the higher-order terms of the Taylor series were discarded when arriving at Eq. (8.51) from Eq. (8.50). The higher-order terms may indeed be identically zero or negligible in two scenarios. The first scenario is if all the higher derivatives starting from the second derivative are zero. This implies that the function must have been a linear function to begin with. Hence, Newton's method, as it should, defaults to yielding the exact solution for the root of a linear function in one update irrespective of the initial guess. The second scenario is one in which the initial guess is exceedingly close to the root. If $\left(x - x^{(0)}\right) \rightarrow 0$, it implies that the higher order terms containing powers of $\left(x - x^{(0)}\right)$ will vanish even faster. In Newton's method, the update formula given by Eq. (8.52) is used repeatedly (iteratively) to bring the initial guess (at each iteration) closer to the root, such that, Eq. (8.51) is eventually valid in an exact sense. The overall idea is analogous to approximating a curve by piecewise lines.

Newton's method has a clear and simple geometric interpretation. Starting from an initial guess, $x^{(0)}$, a tangent to the function is drawn at $x^{(0)}$, i.e., $f'\left(x^{(0)}\right)$ is constructed. Wherever this tangent intersects the x-axis is the new predicted root, as mathematically expressed by Eq. (8.52). This new predicted root may now be used

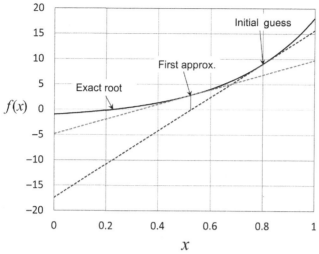

FIGURE 8.5 Geometric Interpretation of the Newton–Raphson Method

The solid curve is the function, $f(x) = exp(3x) - 2$, whose root is being sought. The dotted lines represent tangents at the initial guess (0.8) and at the first approximation (0.527). The exact root is 0.231.

as the new initial guess, and the process may be repeated. The process of successive approximation is illustrated graphically in Fig. 8.5.

Based on the procedure just discussed, the stepwise algorithm of the Newton's method for computing roots of a nonlinear equation is presented next.

ALGORITHM: NEWTON'S METHOD FOR FINDING ROOTS OF A NONLINEAR EQUATION

Step 1: Start with a guess for the root: $x = x^{(0)}$.

Step 2: Differentiate the function analytically to find its derivative.

Step 3: Determine the next approximation to the root using $x^{(n+1)} \approx x^{(n)} - \dfrac{f\left(x^{(n)}\right)}{f'\left(x^{(n)}\right)}$.

Step 4: Update the solution: $x^{(n+1)} = x^{(n)} + \omega(x^{(n+1)} - x^{(n)})$.

Step 5: Repeat Steps 3–4 until convergence.

As in Chapter 3, convergence is defined as the state when the residual decreases to a prescribed tolerance, and the residual is defined as the L^2Norm of the equation being solved. In this case, since the objective is to solve $f(x) = 0$, we define the residual as $R2 = \sqrt{[f(x)]^2}$.

One of the shortcomings of the Newton's method is that, unlike what is shown in the example in Fig. 8.5, the root may not always be approached monotonically, and convergence may not always be smooth. In some cases, the approach direction may switch after a few iterations, and the residual may increase before decreasing again, and this may even occur multiple times and finally lead to divergence. One of the strategies to mitigate such oscillations in the convergence is to use *linear relaxation*. Step 4 of the algorithm presented above shows the use of a *linear relaxation factor*, ω. For nonlinear equations, it is customary to use $0 \le \omega \le 1$. Noting that $x^{(n+1)} - x^{(n)}$ is the change in the solution between successive iterations, instead of allowing the full predicted change to take effect, one can choose to update the solution by a fraction of the predicted change. This idea leads to the formula shown in step 4, and is referred to as *under-relaxation*. If $\omega = 1$, the old solution, $x^{(n)}$, simply gets replaced by the new one, $x^{(n+1)}$. The use of a linear relaxation (or damping) factor is quite common in iterative solution procedures. In fact, a linear relaxation may also be used in the context of all the iterative solvers discussed in Chapter 3 (consider, in particular, Exercise problem 3.3). Typically, for linear problems of the kind discussed in Chapter 3, it is not necessary to employ linear relaxation to stabilize convergence. However, a linear relaxation factor greater than unity is sometimes used to accelerate convergence in classical iterative methods for solving linear systems. In such a case (i.e., $\omega > 1$), the relaxation is referred to as *successive over-relaxation* (SOR). With the advent of powerful solvers such as the Krylov subspace solvers and the multigrid method, the popularity of SOR has faded over the past several decades. For nonlinear problems, however, linear relaxation continues to be used, and in many cases, is imperative for stable convergence. Next, we consider an example problem that exhibits oscillatory convergence, and demonstrates the role of the linear relaxation factor in mitigating such behavior.

EXAMPLE 8.8

In this example, we seek the solution to the nonlinear equation $f(x) = \dfrac{1}{1 + \exp(-x)} - \dfrac{1}{3} = 0$. The exact solution is -0.693147. To apply Newton's method, we start with an initial guess of 2. The roots and residual values at successive iterations are shown in the table here. Two cases – one without relaxation, and the other with an under-relaxation factor of 0.5 – are tabulated.

| Iterations (n) | Without relaxation ($\omega = 1$) | | With under-relaxation ($\omega = 0.5$) | |
	$x^{(n)}$	$R2^{(n)}$	$x^{(n)}$	$R2^{(n)}$
0	2.00000	0.547464	2.00000	0.547464
1	−3.21426	0.294700	−0.60713	0.019381
2	4.720566	0.657835	−0.64957	0.009752
3	−70.431	0.333333	−0.67121	0.004893
4	1.29×10^{30}	0.666667	−0.68214	0.002451

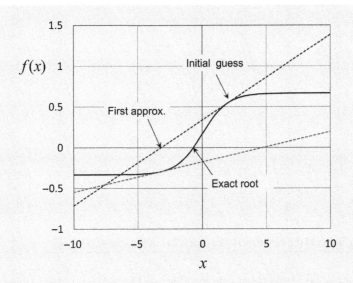

The function is shown in this plot above, along with graphical interpretation of the first two iterations. As is evident from the graph, starting from the initial guess of 2, the first predicted root overshoots the exact root (-0.693) and attains a value of -3.214. In the next iteration, the predicted root changes sign again. This causes the residual to oscillate. If instead, an under-relaxation factor of 0.5 is used, the residual decreases smoothly, and the root is approached monotonically from one direction, as is clear from the table.

In general, Newton's method becomes unstable when an initial guess or a predicted root is at a location where the derivative of the function is close to zero. Other than using relaxation, another strategy to control fluctuations in the solution is to enforce a constraint on the intermediate predictions of the root. While this may not be always feasible, the physical description of the problem may open the door to construct such constraints. For example, if the nonlinear equation is being solved for mole fractions of a chemical species, the physically realizable value of the root is between 0 and 1, which may be used as a constraint.

Since a nonlinear equation, by definition, has many roots, it is not always easy to determine the desired root. It is generally believed that if an initial guess sufficiently close to the desired root is used, the algorithm will yield the desired root. Many numerical methods texts advocate using a different algorithm, such as the bisection method, to narrow down the root prior to using it as an initial guess in the Newton's method. However, even in such a scenario, there is no guarantee that the algorithm will yield the desired root. For example, let us consider solution of the equation $f(x) = x\cos(2x) = 0$. For ease of understanding, a graph of the function is shown in Fig. 8.6. Let us now assume that our goal is to find the root between 0 and 1. The exact value of this root is

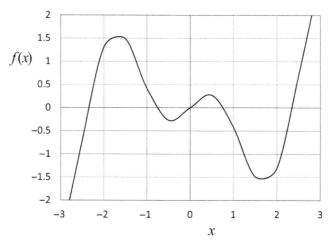

FIGURE 8.6 Graph of the Function $f(x) = x\cos(2x)$ in the interval $[-3, 3]$

0.785. If 0.5 is used as the initial guess, Newton's method converges to 0.785. If 0.3 is used as the initial guess, Newton's method converges to 0, which is the root to the immediate left of the desired root. If, however, 0.4 is used as the initial guess, Newton's method converges to -0.803, which is neither of the two roots closest to the initial guess, but one root further to the left. Once again, this unpredictability is related to the fact that the slope of the function at 0.4 is small, as is evident from Fig. 8.6.

To mitigate the difficulty associated with vanishingly small values of the local derivative in Newton's method, the secant method was developed. In the secant method, the derivative in Newton's method is replaced by its finite difference approximation, yielding

$$x^{(n+1)} \approx x^{(n)} - \frac{f\left(x^{(n)}\right)}{f'\left(x^{(n)}\right)} \approx x^{(n)} - \frac{f\left(x^{(n)}\right)}{\left[f\left(x^{(n)}\right) - f\left(x^{(n-1)}\right)\right]/\left[x^{(n)} - x^{(n-1)}\right]}. \quad (8.53)$$

The iterations of the secant method must commence with two initial guesses, $x^{(-1)}$ and $x^{(0)}$, so that the derivative can be computed in the first iteration. In subsequent iterations, two values are available, and can be readily used. If the secant method is used for the solution of the aforementioned equation, $f(x) = x\cos(2x) = 0$ (Fig. 8.6), significantly different behavior than Newton's method is exhibited. For example, if the two initial guesses are -0.1 and 0.5, the secant method converges to 0. If the two initial guesses are 0.5 and 1, it converges to 0.785, and if the two initial guesses are 2 and 3, the method converges to 2.356. In other words, the two initial guesses may be used to envelop the root and guide the convergence to the desired root. Another important advantage of the secant method is that it requires only one functional evaluation per iteration. In contrast, Newton's method requires two functional evaluations – the function and its derivative. Computing the derivative in the secant method requires only one division no matter how complicated the function may be.

Despite the inadequacies of the Newton–Raphson method, it continues to be one of the most popular methods for finding roots of nonlinear equations. It owes is popularity to its high rate of convergence. Let the error between the exact solution, $x^{(E)}$, and the solution after the $(n+1)$th iteration be written as $\varepsilon^{(n+1)} = x^{(n+1)} - x^{(E)}$. Using the update formula for Newton's method, we may write

$$\varepsilon^{(n+1)} = x^{(n+1)} - x^{(E)} = x^{(n)} - \frac{f(x^{(n)})}{f'(x^{(n)})} - x^{(E)} = \varepsilon^{(n)} - \frac{f(x^{(n)})}{f'(x^{(n)})} = \frac{\varepsilon^{(n)} f'(x^{(n)}) - f(x^{(n)})}{f'(x^{(n)})}.$$

(8.54)

The second expression onward in Eq. (8.54) is only valid if $x^{(n)}$ is sufficiently close to $x^{(E)}$ such that the higher order terms can be justifiably neglected. Using a Taylor series expansion, we may also write

$$f(x^{(E)}) = f(x^{(n)} - \varepsilon^{(n)})$$
$$= f(x^{(n)}) - \varepsilon^{(n)} f'(x^{(n)}) + \frac{1}{2}\left[\varepsilon^{(n)}\right]^2 f''(x^{(n)}) - \frac{1}{6}\left[\varepsilon^{(n)}\right]^3 f'''(x^{(n)}) + \dots$$

(8.55)

By Taylor's theorem, there exists a value of x, denoted by $\xi^{(n)}$, in the interval $(x^{(n)}, x^{(n)} - \varepsilon^{(n)})$, for which

$$f(x^{(n)}) - \varepsilon^{(n)} f'(x^{(n)}) + \frac{1}{2}\left[\varepsilon^{(n)}\right]^2 f''(x^{(n)}) - \frac{1}{6}\left[\varepsilon^{(n)}\right]^3 f'''(x^{(n)}) + \dots$$
$$= f(x^{(n)}) - \varepsilon^{(n)} f'(x^{(n)}) + \frac{1}{2}\left[\varepsilon^{(n)}\right]^2 f''(\xi^{(n)})$$

(8.56)

Also, by the definition of the exact solution, we have $f(x^{(E)}) = 0$. Substitution of Eq. (8.56) and this definition into Eq. (8.55) yields

$$f(x^{(n)}) - \varepsilon^{(n)} f'(x^{(n)}) + \frac{1}{2}\left[\varepsilon^{(n)}\right]^2 f''(\xi^{(n)}) = 0,$$

(8.57)

which, upon substitution into the final expression of Eq. (8.54), yields

$$\varepsilon^{(n+1)} = \frac{1}{2}\left[\varepsilon^{(n)}\right]^2 \frac{f''(\xi^{(n)})}{f'(x^{(n)})}.$$

(8.58)

Equation (8.58) shows that if the initial guess for the root is sufficiently close to the exact root, in successive iterations, the error scales in a quadratic fashion. Hence, Newton's method is said to have quadratic convergence. Sufficient closeness ensures that the local derivatives, appearing on the right-hand side of Eq. (8.58), do not change significantly from iteration to iteration. It is due to the violation of this particular criterion that Newton's method does not always converge. Unfortunately, the degree of closeness cannot be quantified *a priori* without plotting the function first. Using a similar analysis, it can be shown that the secant method also converges in a superlinear fashion but not as fast as Newton's method. Its error in successive iterations scales with a power equal to 1.62 [2] instead of 2.

8.3.2 SYSTEM OF NONLINEAR EQUATIONS (VECTOR CASE)

In this section, Newton's method for a single nonlinear equation is generalized for a system of nonlinear algebraic equations. If a system of N nonlinear equations with N unknowns is considered, then each of the equations can be represented by a separate multidimensional function, and then set equal to zero, yielding

$$\begin{aligned}
f_1(\phi_1, \phi_2, ..., \phi_N) &= 0 \\
f_2(\phi_1, \phi_2, ..., \phi_N) &= 0 \\
&\vdots \\
f_N(\phi_1, \phi_2, ..., \phi_N) &= 0
\end{aligned} \quad (8.59)$$

The unknowns, $\phi_1, \phi_2, ..., \phi_N$, will henceforth be denoted by $[\phi]$. As in the scalar case, the solution process must start with a guess. Let this initial guess be denoted by $[\phi^{(0)}]$. Expanding each of the functions in Eq. (8.59) using a Taylor series, we may write

$$f_1(\phi_1, \phi_2, ..., \phi_N) = f_1(\phi_1^{(0)}, \phi_2^{(0)}, ..., \phi_N^{(0)}) + \left.\frac{\partial f_1}{\partial \phi_1}\right|^{(0)} \Delta\phi_1$$

$$+ \left.\frac{\partial f_1}{\partial \phi_2}\right|^{(0)} \Delta\phi_2 + ... + \left.\frac{\partial f_1}{\partial \phi_N}\right|^{(0)} \Delta\phi_N + \cdots = 0$$

$$f_2(\phi_1, \phi_2, ..., \phi_N) = f_2(\phi_1^{(0)}, \phi_2^{(0)}, ..., \phi_N^{(0)}) + \left.\frac{\partial f_2}{\partial \phi_1}\right|^{(0)} \Delta\phi_1$$

$$+ \left.\frac{\partial f_2}{\partial \phi_2}\right|^{(0)} \Delta\phi_2 + ... + \left.\frac{\partial f_2}{\partial \phi_N}\right|^{(0)} \Delta\phi_N + \cdots = 0 \quad , \quad (8.60)$$

$$\vdots$$

$$f_N(\phi_1, \phi_2, ..., \phi_N) = f_N(\phi_1^{(0)}, \phi_2^{(0)}, ..., \phi_N^{(0)}) + \left.\frac{\partial f_N}{\partial \phi_1}\right|^{(0)} \Delta\phi_1$$

$$+ \left.\frac{\partial f_N}{\partial \phi_2}\right|^{(0)} \Delta\phi_2 + ... + \left.\frac{\partial f_N}{\partial \phi_N}\right|^{(0)} \Delta\phi_N + \cdots = 0$$

where $\Delta\phi_i = \phi_i - \phi_i^{(0)}$ represents the change in the solution (roots) from one iteration to the next. Equation (8.60) may be written in matrix form as

$$\begin{bmatrix}
\left.\frac{\partial f_1}{\partial \phi_1}\right|^{(0)} & \left.\frac{\partial f_1}{\partial \phi_2}\right|^{(0)} & \cdots & \left.\frac{\partial f_1}{\partial \phi_N}\right|^{(0)} \\
\left.\frac{\partial f_2}{\partial \phi_1}\right|^{(0)} & \left.\frac{\partial f_2}{\partial \phi_2}\right|^{(0)} & \cdots & \left.\frac{\partial f_2}{\partial \phi_N}\right|^{(0)} \\
\vdots & \vdots & \cdots & \vdots \\
\left.\frac{\partial f_N}{\partial \phi_1}\right|^{(0)} & \left.\frac{\partial f_N}{\partial \phi_2}\right|^{(0)} & \cdots & \left.\frac{\partial f_N}{\partial \phi_N}\right|^{(0)}
\end{bmatrix}
\begin{bmatrix}
\Delta\phi_1 \\
\Delta\phi_2 \\
\vdots \\
\Delta\phi_N
\end{bmatrix}
=
\begin{bmatrix}
-f_1(\phi_1^{(0)}, \phi_2^{(0)}, ..., \phi_N^{(0)}) \\
-f_2(\phi_1^{(0)}, \phi_2^{(0)}, ..., \phi_N^{(0)}) \\
\vdots \\
-f_N(\phi_1^{(0)}, \phi_2^{(0)}, ..., \phi_N^{(0)})
\end{bmatrix} \cdot \quad (8.61)$$

The square matrix on the left-hand side of Eq. (8.61), containing the partial derivatives, is known as the *Jacobian matrix*, or *Jacobian*. Denoting it by $[J]$, we can write the general update formula for the Newton's method as follows:

$$[J]^{(n)}[\Delta\phi]^{(n)} = -[f]^{(n)}, \tag{8.62}$$

where $[\Delta\phi]^{(n)} = [\phi]^{(n+1)} - [\phi]^{(n)}$. The superscripts within parenthesis, as before, denote iteration number. Equation (8.62) represents a system of linear equations that must be solved at each iteration of the Newton's method. Therefore, the same issues (memory, efficiency) discussed in Chapter 3 for solution of linear systems may be encountered if the number of dependent variables (equal to N in this case) is large. In particular, in many applications of Newton's method, the Jacobian matrix may be fairly full. The algorithm for applying Newton's method to a system of nonlinear equations is presented next.

ALGORITHM: NEWTON'S METHOD FOR FINDING THE ROOTS OF A SYSTEM OF NONLINEAR EQUATIONS

Step 1: Start with a guess for the roots: $[\phi] = [\phi]^{(0)}$.

Step 2: Differentiate the functions analytically to find all partial derivatives necessary to compute the elements of the Jacobian matrix.

Step 3: Determine the next approximation to the root by solving $[J]^{(n)}[\Delta\phi]^{(n)} = -[f]^{(n)}$. If an iterative solver is used for this purpose, iterations and convergence checks must be implemented within this step.

Step 4: Update the solution: $[\phi]^{(n+1)} = [\phi]^{(n)} + \omega\left([\phi]^{(n+1)} - [\phi]^{(n)}\right)$.

Step 5: Repeat Steps 3–4 until convergence.

In this case, the residual is the $L^2 Norm$ of all the functions, i.e., $R2 = \sqrt{\sum_{i=1}^{N}(f_i)^2}$.

One of the most time-consuming steps in the Newton's method for a system of nonlinear equations is the computation of the partial derivatives (Step 2) and the Jacobian matrix. The scalar case, described in the preceding section, required computation of 1 function and 1 derivative per iteration. In the present vector case, however, N functions and N^2 partial derivatives must be evaluated, thereby destroying the linear scale-up of the workload with increasing number of dependent variables. In addition to the fact that the solution of the matrix system in Step 3 may be cumbersome, computing the Jacobian itself may be a bottleneck if the number of dependent variables is large. To alleviate this problem, it is customary to employ numerical difference formulas for computation of the partial derivatives, which essentially reduces the Newton's method to the secant method. An example is considered next to demonstrate the Newton's method for a system of nonlinear equations.

EXAMPLE 8.9

In this example, we seek the solution to the system of nonlinear equations as follows:

$$x^2 y - \exp[(1-x)y] = 1$$
$$x^3 y + y^2 x = 6$$

Newton's method is to be used to accomplish this task.

The exact solution to the system is $x = 1$ and $y = 2$.

To implement the Newton's method, we first rewrite the equations with the notations used above:

$$f_1 = \phi_1^2 \phi_2 - \exp[(1 - \phi_1)\phi_2] - 1$$
$$f_2 = \phi_1^3 \phi_2 + \phi_2^2 \phi_1 - 6$$

Differentiating each function with respect to each independent variable, we obtain the Jacobian matrix, written as

$$\begin{bmatrix} \dfrac{\partial f_1}{\partial \phi_1} & \dfrac{\partial f_1}{\partial \phi_2} \\ \dfrac{\partial f_2}{\partial \phi_1} & \dfrac{\partial f_2}{\partial \phi_2} \end{bmatrix} = \begin{bmatrix} 2\phi_1\phi_2 + \phi_2\exp[(1-\phi_1)\phi_2] & \phi_1^2 - (1-\phi_1)\exp[(1-\phi_1)\phi_2] \\ 3\phi_1^2\phi_2 + \phi_2^2 & \phi_1^3 + 2\phi_1\phi_2 \end{bmatrix}.$$

Next, we exercise the algorithm of the Newton's method with initial guesses of $\phi_1^{(0)} = \phi_2^{(0)} = 0.5$. The table below shows the behavior of the algorithm when no relaxation ($\omega = 1$) is used. As shown in the table, the solution diverges quickly.

Iterations	Residual, R2	ϕ_1	ϕ_2
1	5.1094733	4.2839944	6.0160056
2	192.0555591	88.3012160	−235.9159759

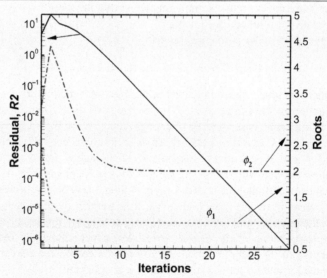

If under-relaxation with $\omega = 0.5$ is applied instead, the solution converges to the desired tolerance of 10^{-6} in 29 iterations, as shown by the above figure. As in the case of a single nonlinear equation, this example demonstrates the instability of the Newton's method, and the efficacy of using an under-relaxation factor in the solution of strongly nonlinear equations. Although the use of under-relaxation disrupts the quadratic convergence (as is evident from the above residual plot) of the nonrelaxed Newton's method, it is often unavoidable.

The Newton's method for a nonlinear system is used in a variety of applications. Examples include the solution of the equations of chemical kinetics, equations in control algorithms, and optimization algorithms using Lagrange multipliers, among others. In most cases, by careful tweaking of the under-relaxation factor, using an "educated" initial guess, and placing constraints on the roots during intermediate iterations, one can arrive at a solution within a reasonable number of iterations. Next, we discuss the application of the Newton's method to solution of PDEs.

8.4 APPLICATION OF THE NEWTON–RAPHSON METHOD TO SOLVING NONLINEAR PDEs

In Section 3.3, the solution of PDEs with nonlinear source terms was discussed. The strategy used therein was to linearize the source using a Taylor series expansion in an effort to make the system of linearized equations diagonally dominant. Many PDEs encountered in practical scientific and engineering computations, however, have nonlinearity in the left-hand side of the equation either due to the presence of non-linear dependence of the dependent variable within the differential operators, or due to the material properties being a function of the dependent variable. The Navier–Stokes equation, which governs fluid flow, is a classic example of the former case. The left-hand side of the x component of this equation, for example, contains terms such as $\frac{\partial}{\partial x}(\rho uu)$, where u is the x component of the velocity vector, and is the dependent variable in the equation. Clearly, such terms make the Navier–Stokes equations nonlinear. Heat conduction with temperature-dependent thermal conductivity is an example of the latter type of nonlinearity. The steady-state heat conduction equation, for example, may be written as $\nabla \cdot (\kappa \nabla T) = 0$, where κ is the thermal conductivity and T is temperature. If κ is, in addition, a function of temperature, as is the case for most materials encountered in everyday life, the governing equation for heat conduction becomes nonlinear. The Newton's method, being a general-purpose method for solving nonlinear algebraic equations, can be applied directly to the algebraic equations arising from the discretization of such nonlinear PDEs. The following example demonstrates how this may be accomplished.

EXAMPLE 8.10

The objective is to apply the Newton's method to solving the following non-linear differential equation:

$$\frac{\partial}{\partial x}(\phi^2) + \frac{\partial}{\partial y}(\phi^2) = \frac{\partial}{\partial x}\left((1+\phi)\frac{\partial \phi}{\partial x}\right) + \frac{\partial}{\partial y}\left((1+\phi)\frac{\partial \phi}{\partial y}\right),$$

with the following boundary conditions: $\phi(x,0) = \phi(0,y) = 0$, and $\phi(x,1) = \phi(1,y) = 1$. The above PDE is similar in form to the classical advection-diffusion equation, in which the non linearity directly appears in the left-hand side advection term, and the nonlinearity in the right-hand side appears due to the dependence of material property on the dependent variable (consider the term $\nabla \cdot (\Gamma \nabla \phi)$ with Γ being a linear function of ϕ).

For purposes of discretization of the governing PDE, we will use the finite difference method outlined in Chapter 2, although the finite volume procedure, outlined in Chapter 6, may just as easily be used. Since the computational domain is a square, finite difference approximations can be derived easily using a regular Cartesian grid with grid spacings $\Delta x = 1/(N-1)$, and $\Delta y = 1/(M-1)$, where N and M are the number of nodes in the x and y directions, respectively. Here, we will use $N = M = 101$. For ease of discretization, we first rewrite the governing equation as

$$\frac{\partial}{\partial x}\left(\phi^2\right) + \frac{\partial}{\partial y}\left(\phi^2\right) = \frac{\partial}{\partial x}\left((1+\phi)\frac{\partial \phi}{\partial x}\right) + \frac{\partial}{\partial y}\left((1+\phi)\frac{\partial \phi}{\partial y}\right) = \frac{\partial^2 \phi}{\partial x^2} + \frac{\partial^2 \phi}{\partial y^2} + \frac{\partial}{\partial x}\left(\phi\frac{\partial \phi}{\partial x}\right) + \frac{\partial}{\partial y}\left(\phi\frac{\partial \phi}{\partial y}\right)$$

$$= \frac{\partial^2 \phi}{\partial x^2} + \frac{\partial^2 \phi}{\partial y^2} + \frac{\partial}{\partial x}\left(\frac{1}{2}\frac{\partial (\phi^2)}{\partial x}\right) + \frac{\partial}{\partial y}\left(\frac{1}{2}\frac{\partial (\phi^2)}{\partial y}\right) = \frac{\partial^2 \phi}{\partial x^2} + \frac{\partial^2 \phi}{\partial y^2} + \frac{1}{2}\left[\frac{\partial^2 (\phi^2)}{\partial x^2} + \frac{\partial^2 (\phi^2)}{\partial y^2}\right].$$

Now, applying the central difference approximation, for the interior nodes $(2 \leq i \leq N-1, 2 \leq j \leq M-1)$, we obtain

$$\frac{\phi_{i+1,j}^2 - \phi_{i-1,j}^2}{2\Delta x} + \frac{\phi_{i,j+1}^2 - \phi_{i,j-1}^2}{2\Delta y} = \left[\left(\frac{\phi_{i+1,j} - 2\phi_{i,j} + \phi_{i-1,j}}{(\Delta x)^2}\right) + \left(\frac{\phi_{i,j+1} - 2\phi_{i,j} + \phi_{i,j-1}}{(\Delta y)^2}\right)\right]$$
$$+ \frac{1}{2}\left[\left(\frac{\phi_{i+1,j}^2 - 2\phi_{i,j}^2 + \phi_{i-1,j}^2}{(\Delta x)^2}\right) + \left(\frac{\phi_{i,j+1}^2 - 2\phi_{i,j}^2 + \phi_{i,j-1}^2}{(\Delta y)^2}\right)\right].$$

Next, we rewrite the discrete equation in terms of the unique global index $k = (j-1)N + i$ (Section 2.5) to yield

$$\frac{\phi_{k+1}^2 - \phi_{k-1}^2}{2\Delta x} + \frac{\phi_{k+N}^2 - \phi_{k-N}^2}{2\Delta y} = \left[\left(\frac{\phi_{k+1} - 2\phi_k + \phi_{k-1}}{(\Delta x)^2}\right) + \left(\frac{\phi_{k+N} - 2\phi_k + \phi_{k-N}}{(\Delta y)^2}\right)\right]$$
$$+ \frac{1}{2}\left[\left(\frac{\phi_{k+1}^2 - 2\phi_k^2 + \phi_{k-1}^2}{(\Delta x)^2}\right) + \left(\frac{\phi_{k+N}^2 - 2\phi_k^2 + \phi_{k-N}^2}{(\Delta y)^2}\right)\right].$$

In order to use the Newton's method, we define its functions:

$$f_k = \frac{\phi_{k+1}^2 - \phi_{k-1}^2}{2\Delta x} + \frac{\phi_{k+N}^2 - \phi_{k-N}^2}{2\Delta y} - \left[\left(\frac{\phi_{k+1} - 2\phi_k + \phi_{k-1}}{(\Delta x)^2}\right) + \left(\frac{\phi_{k+N} - 2\phi_k + \phi_{k-N}}{(\Delta y)^2}\right)\right]$$
$$- \frac{1}{2}\left[\left(\frac{\phi_{k+1}^2 - 2\phi_k^2 + \phi_{k-1}^2}{(\Delta x)^2}\right) + \left(\frac{\phi_{k+N}^2 - 2\phi_k^2 + \phi_{k-N}^2}{(\Delta y)^2}\right)\right] \quad \forall k \in \text{interior}.$$

$$f_k = \phi_k - \phi_{boundary} \quad \forall k \in \text{boundary}$$

In each of the above equations, there are at most five dependent variables. Therefore, there will only be five nonzero partial derivatives, as follows:

$$\frac{\partial f_k}{\partial \phi_{k+1}} = \frac{\phi_{k+1}}{\Delta x} - \frac{1+\phi_{k+1}}{(\Delta x)^2}, \frac{\partial f_k}{\partial \phi_{k-1}} = -\frac{\phi_{k-1}}{\Delta x} - \frac{1+\phi_{k-1}}{(\Delta x)^2}, \frac{\partial f_k}{\partial \phi_{k+N}} = \frac{\phi_{k+N}}{\Delta y} - \frac{1+\phi_{k+N}}{(\Delta y)^2}, \frac{\partial f_k}{\partial \phi_{k-N}} = -\frac{\phi_{k-N}}{\Delta y} - \frac{1+\phi_{k-N}}{(\Delta y)^2},$$

$$\frac{\partial f_k}{\partial \phi_k} = \frac{2(1+\phi_k)}{(\Delta x)^2} + \frac{2(1+\phi_k)}{(\Delta y)^2} \quad \forall k \in \text{interior}$$

$$\frac{\partial f_k}{\partial \phi_k} = 1 \quad \forall k \in \text{boundary}$$

The above equations can be used to set up the matrix given by Eq. (8.61). The resulting matrix is banded with five diagonals, identical in structure to the one shown in Chapter 3 [Eq. (3.39)]. Since the matrix is expected to be fairly large (10^4 unknowns in this case), we cannot employ Gaussian elimination, and will instead use an iterative solver. In general, any of the iterative solvers discussed in Chapters 3 and 4 may be used for this purpose. Here, the Stone's strongly implicit method (Section 3.2.5) is used to solve the system of equations. The iterative solution of the update equation constitutes Step 3 of the Newton's method algorithm discussed above. This iterative solution of the update equation [Eq. (8.61)] must be performed for each iteration of the Newton's method. An absolute tolerance of 10^{-6} is used for convergence of both the Stone's method (inner iterations in step 3), as well as the convergence of the Newton's method (outer iteration). An initial guess of $\phi = 0.5$, which is the average of the two prescribed boundary values, is used for all the interior nodes to start Newton's iteration.

Newton (outer) iteration number	Number of inner iterations in Stone's method
1	819
2	933
3	690
4	158
5	2
6	1

The table shows the number of iterations required by the Stone's method to bring the residuals down to 10^{-6} for each Newton iteration. Initially, the number of iterations is large because the starting residuals are large. With subsequent Newton iterations, the linearized system becomes closer and closer to the nonlinear system (the higher order terms in the Taylor series expansion become negligible as $\Delta \phi \to 0$), and the starting residuals become smaller and smaller. Thus, the linear solver requires very few iterations to converge to the desired tolerance. The following figures show the convergence behavior of Newton's (outer) iterations, as well as the final solution.

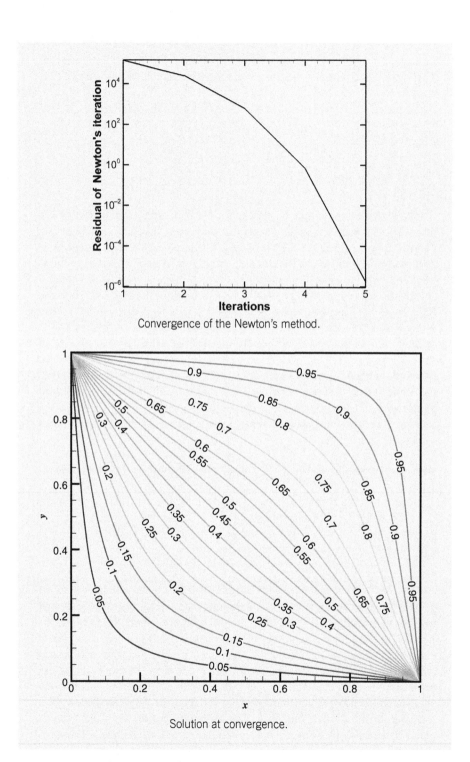

Convergence of the Newton's method.

Solution at convergence.

> As shown in the above figure, the convergence of the Newton's method exhibits superlinear behavior. Although the convergence behavior appears more or less quadratic, it is difficult to ascertain if it is exactly so. On account of the fact that the update equation is not solved exactly, quadratic convergence is often difficult to achieve in such scenarios especially if a tight tolerance is not used for the inner iterations. The solution with a 101×101 mesh, as expected, is smooth. This example demonstrates that the Newton's method can be effectively combined with the linear algebraic equation solvers discussed in Chapters 3 and 4 to solve nonlinear PDEs with reasonable computational efficiency.

Over the past two decades, a wide variety of methods that are specifically geared toward the solution of nonlinear PDEs have been developed. These methods aim to deviate from the aforedemonstrated conventional scheme of solving a linear system of equations (the linearized update equation) repeatedly within an outer iteration designed to address the nonlinearity in the problem. While the conventional scheme allows us to use mature technologies, such as linear algebraic equation solvers and the Newton's method, it can sometimes be prohibitive from a computational efficiency standpoint. Arguably, the two most powerful methods geared specifically to address nonlinear PDEs are as follows: (a) nonlinear multigrid methods, and (b) Newton–Krylov methods.

The standard nonlinear multigrid method is often referred to as the full approximation scheme [9]. In this method, instead of solving the linear correction equations on hierarchical mesh levels, nonlinear correction equations are solved. Of course, this requires a nonlinear equation solver, and either the Newton's method or the secant method is typically used. One of the biggest challenges in this method is the derivation of the nonlinear correction equations, especially if this method were to be used on unstructured meshes (algebraic multigrid). Often, one has to resort to correction equations that are local and not global, thereby adversely affecting convergence. Nonetheless, the method enjoys all the advantages of the linear multigrid method, namely low memory requirement, and rapid damping of the errors when the problem size is large. Nonlinear multigrid methods have already witnessed a fair amount of success in the computational fluid dynamics arena [10,11], and continue to be developed further.

In Newton–Krylov methods [1], the Newton iteration, designed to address the nonlinearity is the outer iteration, and a preconditioned Krylov subspace–based solver is designed to solve the linear update equation using inner iterations. In principle, it is no different to that demonstrated in Example 8.10, with the exception that the linear algebraic equation solver would be a preconditioned Krylov subspace–based solver, such as CGS or GMRES, instead of the Stone's method. Since the inner iterations usually are stopped at a relatively loose (large) tolerance level, the convergence does not always strictly follow a quadratic pattern, as was discussed in Example 8.10. Ultimately, the balance between the tolerance level used for the inner iterations and the convergence of the outer iterations is determined by some numerical experimentation for the problem at hand – the time to convergence being the most important metric.

One notable disadvantage of the Newton–Krylov method is that repeated calculation and storage of the Jacobian is expensive. Another disadvantage is that since the Jacobian matrix changes from one outer iteration to the next, it also has to be preconditioned separately in each outer iteration, further increasing the computational cost. In order to alleviate these two issues, the Jacobian-free Newton–Krylov method was developed [12,13]. As the name suggests, in this method, the Jacobian is not explicitly computed and stored. Instead, the outer Newton iteration directly uses the Krylov subspace vectors to construct an equation that yields the change vector. The method has gained tremendous popularity in recent times, especially in the context of finite element–based solvers for the advection-diffusion-reaction equations of reacting flow. An informative review of this method and the Newton–Krylov method, in general, may be found in Knoll and Keys [14].

8.5 SOLUTION OF COUPLED PDES

Until this juncture, this text has been dedicated to the solution of a single PDE – both linear and nonlinear. Many problems in science and engineering, however, require solution of a set of coupled PDEs. The reason for this is deeply rooted in the physics of how matter behaves. The conservation laws that govern how matter behaves work in unison. Very often, violation of one law leads to the violation of another. For example, if mass is not conserved, then energy conservation will be automatically violated because mass carries energy with it. In other situations, the conservation law may apply to a vector field, as in the case of linear or angular momentum conservation, which immediately results in more than one governing PDE.

Based on the preceding discussion, one might form the impression that all this effort directed toward solution of a single PDE may have been wasted. In fact, it is quite the contrary, as will become apparent from the discussion to follow. There are two vastly different approaches to the solution of a set of coupled PDEs: (a) the segregated solution approach, and (b) the coupled solution approach. They are schematically depicted in Fig. 8.7.

In the segregated solution approach, each PDE is solved sequentially and in isolation. This means that if the PDE has ϕ_1 as its dependent variable, then, during its solution, only ϕ_1 is allowed to change, and all other ϕ_i are frozen. Of course, the discretized equations arising out of each PDE must be solved using an iterative linear algebraic equation solver, and this would require iterations. These iterations are referred to as *inner iterations*. In order to address the coupling between the various PDEs, the entire sequence of the solution is enclosed within a bigger iteration loop, referred to as the *outer iteration*. This type of iterative coupling between the PDEs is known as *Picard iteration*. The inner iterations in this algorithm are generally continued until each PDE has attained partial convergence – typically, about two to three orders of magnitude. It is not worthwhile reducing the residuals to very tight tolerances because the outer iteration has to be repeated anyway.

In the coupled solution approach, all PDEs are discretized and solved in a coupled manner. Essentially, a matrix containing link coefficients of all the PDEs is

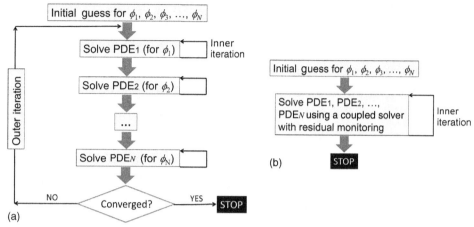

FIGURE 8.7 Schematic Representation of Two Different Approaches Used to Solve a Set of Coupled Linear PDEs

(a) Segregated solution approach and (b) coupled solution approach.

assembled in an appropriate format, and the resulting system is solved using an iterative solver. Convergence of this global system of equations is monitored. In this case, no outer iterations are necessary as long as the system of equations is linear.

The aforementioned two approaches have their respective *pros* and *cons*. The biggest advantage of the segregated solution approach is its modularity. Solution of a single PDE, which has been the focus of this text thus far, is considered a mature area in numerical analysis. In the segregated solution approach, this knowledge and the tools borne out of it can be readily used. The modularity offered by the segregated solution approach also enables easy extension of the code. For example, an isothermal fluid flow code can be extended with relative ease to include heat transfer. Performing the same task in a code that employs the coupled solution approach is not as straightforward. It would involve reformatting the data structure of the code to a significant degree, and would require the developer to delve deep into the code to make the necessary modifications. Furthermore, once developed, in the coupled solution approach, it would not be easy to deactivate the heat transfer calculations and operate the code in isothermal mode. In contrast, in the segregated solution approach, activating or deactivating any one or more of the PDEs is almost trivial. This attribute of the segregated solution approach makes it an attractive choice for multiphysics codes, wherein additional physics (PDEs) can be added or removed without hampering the functionality of the rest of the code.

Low memory usage is yet another advantage of the segregated solution approach. The same memory stack can be reused for the storage of the link coefficients and sources of the various PDEs since the solution process is sequential. Therefore, irrespective of the number of PDEs being solved, the memory requirement for the solver remains almost unchanged. In contrast, for a coupled solution, the amount of memory required could easily reach prohibitive limits. Let us consider the case of a reactive flow calculation, in which 20 species mass conservation equations (i.e., 20 PDEs) need to be solved in addition to the fluid flow and energy equations. If the mesh is comprised of even as low as 10^5 cells, the memory required to store the coefficient matrix for these 20 species conservation equations is $10^5 \times 6 \times 20^2 \times 8 = 1.92$ GB. In this estimate, the number 6 represents a rough estimate of the number of neighbors of each cell (equal to the number of nonzero elements in each row), 20^2 represents the size of the coupling matrix (explained later) of the 20 PDEs, and 8 represents the number of bytes required to store a number in double precision. Clearly, storage of such a matrix would be quite difficult even for a high-end processor. In contrast, the segregated solution approach would require storage of a matrix of size $10^5 \times 6 \times 8 = 4.8$ MB.

The solution flowchart depicted in Fig. 8.7b for the coupled solution approach is one that applies only to a system of linear PDEs. If the system of PDEs was nonlinear, an outer loop, similar to the Newton's iteration loop would be needed to address the nonlinearity of the system, and the coupled solver would then solve the update equation (which is a linear equation) of the Newton's method, albeit in coupled form. In other words, the advantage of the original coupled solution approach of not having to perform outer iterations would be sacrificed. In contrast, in the segregated solution approach, since an outer iteration loop is already in place, it would serve the dual purpose of coupling the PDEs and addressing the nonlinearity of the system. Admittedly, a nonlinear system may require more outer iterations than a linear system. Nonetheless, the overall algorithm would remain unchanged due to the introduction of nonlinearity in the system of equations.

Despite all of the above shortcomings, the popularity of the coupled solution approach has increased tremendously over the past two decades primarily due to one major advantage over the segregated solution approach: superior convergence. The popularity has been stimulated, in large part, by the dramatic increase in the available runtime memory of modern-day computers. The coupled solution approach has demonstrated superior performance over the segregated solution approach in application areas such as solution of the Navier–Stokes equation [4, 5] for fluid flow, Maxwell's equation for electromagnetics [15], the radiative transfer equation [16–18], the species conservation equation [19,20], and the Boltzmann transport equation [21], among others. In many cases, it has enabled attainment of a converged solution, when the segregated solution approach has completely failed.

Coupled solution of PDEs requires restructuring of the coefficient matrix. Through a simple example, we elucidate how two coupled PDEs may be manipulated to generate a coefficient matrix wherein the coupling between the two dependent variables is implicit. Let us consider two coupled Poisson equations of the following form:

$$\nabla^2 \phi = \psi, \tag{8.63a}$$

$$\nabla^2 \psi = \phi. \tag{8.63b}$$

If discretized on a 2D uniform orthogonal Cartesian mesh using the finite difference method, they yield

$$\left(\frac{2}{(\Delta x)^2} + \frac{2}{(\Delta y)^2} \right) \phi_O - \frac{1}{(\Delta x)^2} \phi_W - \frac{1}{(\Delta x)^2} \phi_E - \frac{1}{(\Delta y)^2} \phi_S - \frac{1}{(\Delta y)^2} \phi_N = -\psi_O, \tag{8.64a}$$

$$\left(\frac{2}{(\Delta x)^2} + \frac{2}{(\Delta y)^2}\right)\psi_O - \frac{1}{(\Delta x)^2}\psi_W - \frac{1}{(\Delta x)^2}\psi_E - \frac{1}{(\Delta y)^2}\psi_S - \frac{1}{(\Delta y)^2}\psi_N = -\phi_O. \quad (8.64b)$$

In the segregated solution approach, Eq. (8.64a) will be solved first by holding ψ constant. This will be followed by the solution of Eq. (8.64b), wherein ϕ will be held constant. The two sets of equations will be solved repeatedly within an outer iteration until convergence.

To use the coupled solution approach, instead, we first rewrite Eq. (8.64) in its fully implicit form:

$$\left(\frac{2}{(\Delta x)^2} + \frac{2}{(\Delta y)^2}\right)\phi_O + \psi_O - \frac{1}{(\Delta x)^2}\phi_W - \frac{1}{(\Delta x)^2}\phi_E - \frac{1}{(\Delta y)^2}\phi_S - \frac{1}{(\Delta y)^2}\phi_N = 0, \quad (8.65a)$$

$$\left(\frac{2}{(\Delta x)^2} + \frac{2}{(\Delta y)^2}\right)\psi_O + \phi_O - \frac{1}{(\Delta x)^2}\psi_W - \frac{1}{(\Delta x)^2}\psi_E - \frac{1}{(\Delta y)^2}\psi_S - \frac{1}{(\Delta y)^2}\psi_N = 0. \quad (8.65b)$$

Equation (8.65) may be also written in matrix form as follows:

$$
\begin{bmatrix} \frac{2}{(\Delta x)^2} + \frac{2}{(\Delta x)^2} & 1 \\ 1 & \frac{2}{(\Delta x)^2} + \frac{2}{(\Delta x)^2} \end{bmatrix}
\begin{bmatrix} \phi_O \\ \psi_O \end{bmatrix}
+
\begin{bmatrix} -\frac{1}{(\Delta x)^2} & 0 \\ 0 & -\frac{1}{(\Delta x)^2} \end{bmatrix}
\begin{bmatrix} \phi_W \\ \psi_W \end{bmatrix}
$$

$$
+
\begin{bmatrix} -\frac{1}{(\Delta x)^2} & 0 \\ 0 & -\frac{1}{(\Delta x)^2} \end{bmatrix}
\begin{bmatrix} \phi_E \\ \psi_E \end{bmatrix}
$$

$$
+
\begin{bmatrix} -\frac{1}{(\Delta y)^2} & 0 \\ 0 & -\frac{1}{(\Delta y)^2} \end{bmatrix}
\begin{bmatrix} \phi_S \\ \psi_S \end{bmatrix}
$$

$$
+
\begin{bmatrix} -\frac{1}{(\Delta y)^2} & 0 \\ 0 & -\frac{1}{(\Delta y)^2} \end{bmatrix}
\begin{bmatrix} \phi_N \\ \psi_N \end{bmatrix}
=
\begin{bmatrix} 0 \\ 0 \end{bmatrix}
$$

$$(8.66)$$

Equation (8.66) is the familiar five-band form of the discrete equation, with the notable exception that it is in vector form. It is worth noting that all unknowns have been treated in an implicit manner in this rearranged form, and the system of equations written in this form is often referred to as a *block implicit* system of equations. Each position in the coefficient matrix, instead of having a single number (scalar), now has a 2×2 matrix, and each unknown is represented by a 2×1 vector or column matrix. The 2×2 matrices are known as *coupling matrices*. In general, if N PDEs are being solved in a coupled manner, the coupling matrices will be of size $N \times N$, thereby explaining why 20^2 was used in an earlier example to estimate the size of the coefficient matrix arising from the coupled solution of 20 species conservation equations.

In general, a set of N coupled PDEs, when discretized, may be written in the following form:

$$[A_O][\phi_O] + \sum_{nb}[A_{nb}][\phi_{nb}] = [Q_O] \quad \forall O \in 1,2,...K, \tag{8.67}$$

where, as in preceding chapters, K is the total number of nodes or cells in the computational domain. $[A_O]$ is the self-coupling matrix and is of size $N \times N$, while $[\phi_O]$ is a $N \times 1$ column matrix comprised of the dependent variables at node or cell O. In the preceding 2×2 example, $[A_O] = \begin{bmatrix} \frac{2}{(\Delta x)^2} + \frac{2}{(\Delta x)^2} & 1 \\ 1 & \frac{2}{(\Delta x)^2} + \frac{2}{(\Delta x)^2} \end{bmatrix}$, while

$[\phi_O] = [\phi_O \quad \psi_O]^T$. The summation in Eq. (8.67) is over all the neighboring nodes or cells of node or cell O. $[A_{nb}]$ represents the coupling matrices of the neighbors or off-diagonal elements, while $[\phi_{nb}]$ is a column matrix comprised of the dependent variables at node or cell nb. Similarly, $[Q_O]$ is a $N \times 1$ column matrix comprised of the right-hand side of the individual PDEs.

As far as the solution of the equations is concerned, many of the classical iterative solvers discussed in Chapter 3 can be reformulated to write in vector form. For example, if the Gauss–Seidel method is used to solve Eq. (8.66) using a bottom-to-top and left-to-right sweeping pattern, following Chapter 3 [Eq. (3.32)], the iteration equation may be written as

$$\begin{bmatrix} \frac{2}{(\Delta x)^2} + \frac{2}{(\Delta x)^2} & 1 \\ 1 & \frac{2}{(\Delta x)^2} + \frac{2}{(\Delta x)^2} \end{bmatrix} \begin{bmatrix} \phi_O \\ \psi_O \end{bmatrix}^{(n+1)} = -\begin{bmatrix} -\frac{1}{(\Delta x)^2} & 0 \\ 0 & -\frac{1}{(\Delta x)^2} \end{bmatrix} \begin{bmatrix} \phi_W \\ \psi_W \end{bmatrix}^{(n+1)}$$

$$-\begin{bmatrix} -\frac{1}{(\Delta x)^2} & 0 \\ 0 & -\frac{1}{(\Delta x)^2} \end{bmatrix} \begin{bmatrix} \phi_E \\ \psi_E \end{bmatrix}^{(n)}$$

$$-\begin{bmatrix} -\frac{1}{(\Delta y)^2} & 0 \\ 0 & -\frac{1}{(\Delta y)^2} \end{bmatrix} \begin{bmatrix} \phi_S \\ \psi_S \end{bmatrix}^{(n+1)} \cdot \tag{8.68}$$

$$-\begin{bmatrix} -\frac{1}{(\Delta y)^2} & 0 \\ 0 & -\frac{1}{(\Delta y)^2} \end{bmatrix} \begin{bmatrix} \phi_N \\ \psi_N \end{bmatrix}^{(n)}$$

Like in Chapter 3 [Eq. (3.32)], Eq. (8.68) is an explicit iteration equation. However, it is in vector form, and its execution requires solution of a 2×2 system of equations. In the general case, following Eq. (8.67), the iteration equation of the Gauss–Seidel method will be a $N \times N$ system of equations. Since the number of coupled PDEs that are encountered for physical problems are generally small (typically a few tens, at most), it is customary to solve this linear system using Gaussian elimination, especially since the coupling matrices may be fairly full (nonsparse). Such as method is known as the *block Gauss–Seidel* method.

The multigrid method was identified in Chapter 4 as one of the most powerful methods for efficient solution of linear systems. Since the Gauss–Seidel method is often the smoother of choice for multigrid methods, the vector form of the Gauss–Seidel method (block Gauss–Seidel), just discussed, can be easily adopted for the coupled multigrid solution of the discrete equations resulting from a coupled set of PDEs without loss of generality or efficiency [22,23]. As a matter of fact, the vector form of the algebraic multigrid method [24] has been used extensively, and is often considered the default solver in many commercial computational fluid dynamics codes.

For structured mesh computations, the alternating direction implicit (ADI) method, described in Chapter 3, may, just as easily, be adopted to vector form. For example, Eq. (8.66) may be written in the following *block tridiagonal form* for row- and column-wise sweeps, respectively:

$$
\begin{bmatrix} \frac{2}{(\Delta x)^2}+\frac{2}{(\Delta x)^2} & 1 \\ 1 & \frac{2}{(\Delta x)^2}+\frac{2}{(\Delta x)^2} \end{bmatrix} \begin{bmatrix} \phi_O \\ \psi_O \end{bmatrix}^{(n+1)} + \begin{bmatrix} -\frac{1}{(\Delta x)^2} & 0 \\ 0 & -\frac{1}{(\Delta x)^2} \end{bmatrix} \begin{bmatrix} \phi_W \\ \psi_W \end{bmatrix}^{(n+1)}
$$
$$
+ \begin{bmatrix} -\frac{1}{(\Delta x)^2} & 0 \\ 0 & -\frac{1}{(\Delta x)^2} \end{bmatrix} \begin{bmatrix} \phi_E \\ \psi_E \end{bmatrix}^{(n+1)}
$$
$$
= -\begin{bmatrix} -\frac{1}{(\Delta y)^2} & 0 \\ 0 & -\frac{1}{(\Delta y)^2} \end{bmatrix} \begin{bmatrix} \phi_S \\ \psi_S \end{bmatrix}^{(n+1)} \quad (8.69a)
$$
$$
- \begin{bmatrix} -\frac{1}{(\Delta y)^2} & 0 \\ 0 & -\frac{1}{(\Delta y)^2} \end{bmatrix} \begin{bmatrix} \phi_N \\ \psi_N \end{bmatrix}^{(n)}
$$

$$
\begin{bmatrix} \frac{2}{(\Delta x)^2}+\frac{2}{(\Delta x)^2} & 1 \\ 1 & \frac{2}{(\Delta x)^2}+\frac{2}{(\Delta x)^2} \end{bmatrix} \begin{bmatrix} \phi_O \\ \psi_O \end{bmatrix}^{(n+1)} + \begin{bmatrix} -\frac{1}{(\Delta y)^2} & 0 \\ 0 & -\frac{1}{(\Delta y)^2} \end{bmatrix} \begin{bmatrix} \phi_S \\ \psi_S \end{bmatrix}^{(n+1)}
$$
$$
+ \begin{bmatrix} -\frac{1}{(\Delta y)^2} & 0 \\ 0 & -\frac{1}{(\Delta y)^2} \end{bmatrix} \begin{bmatrix} \phi_N \\ \psi_N \end{bmatrix}^{(n+1)}
$$
$$
= -\begin{bmatrix} -\frac{1}{(\Delta x)^2} & 0 \\ 0 & -\frac{1}{(\Delta x)^2} \end{bmatrix} \begin{bmatrix} \phi_W \\ \psi_W \end{bmatrix}^{(n+1)} \quad (8.69b)
$$
$$
- \begin{bmatrix} -\frac{1}{(\Delta x)^2} & 0 \\ 0 & -\frac{1}{(\Delta x)^2} \end{bmatrix} \begin{bmatrix} \phi_E \\ \psi_E \end{bmatrix}^{(n)}
$$

Block tridiagonal solvers, required to solve Eq. (8.69), are relatively easy to develop using the Thomas algorithm outlined in Section 3.1.2, except that scalar multiplications must be replaced by matrix–vector multiplications, and division operations must be replaced by inverse operations, followed by matrix–vector multiplications. However, inverse matrices are not determined explicitly for computational efficiency reasons, but are replaced by the solution of a set of algebraic equations, as discussed in Chapter 4 [see discussion immediately preceding Eq. (4.68)]. Next, an example that demonstrates the coupled solution of a set of linear PDEs is presented.

EXAMPLE 8.11

In this example, we aim to solve three coupled linear PDEs of the following general form:

$$\nabla \cdot \left(\sum_{j=1}^{3} \Gamma_{ij} \nabla \phi_j \right) = 0 \quad \forall i = 1, 2, 3, \text{ where } \Gamma_{ij} \text{ is a tensor. These governing equa-}$$

tions are formulated after the equations of multicomponent diffusion that are used to model diffusion of chemical species in chemical engineering applications. In such equations, Γ_{ij} would represent the multicomponent diffusion coefficient, while ϕ_j would represent the species concentration. For simplicity, we will use the following constant values for Γ_{ij}: $\Gamma_{ij} = \begin{bmatrix} 1 & 0.4 & -0.2 \\ 0.4 & 1 & -0.3 \\ -0.2 & -0.3 & 1 \end{bmatrix}$.

The computational domain under consideration is a 2D unit square with the following boundary conditions:

Left: $\phi_1(0, y) = 1$, and $\phi_2(0, y) = \phi_3(0, y) = 0$.
Right: $\phi_3(1, y) = 1$, and $\phi_1(1, y) = \phi_2(1, y) = 0$.
Bottom: $\phi_2(x, 0) = 1$, and $\phi_1(x, 0) = \phi_3(x, 0) = 0$.
Top: $\phi_3(x, 1) = 1$, and $\phi_1(x, 1) = \phi_2(x, 1) = 0$.

In 2D Cartesian coordinates, the governing equation may be rewritten as

$$\Gamma_{11} \left(\frac{\partial^2 \phi_1}{\partial x^2} + \frac{\partial^2 \phi_1}{\partial y^2} \right) + \Gamma_{12} \left(\frac{\partial^2 \phi_2}{\partial x^2} + \frac{\partial^2 \phi_2}{\partial y^2} \right) + \Gamma_{13} \left(\frac{\partial^2 \phi_3}{\partial x^2} + \frac{\partial^2 \phi_3}{\partial y^2} \right) = 0$$

$$\Gamma_{21} \left(\frac{\partial^2 \phi_1}{\partial x^2} + \frac{\partial^2 \phi_1}{\partial y^2} \right) + \Gamma_{22} \left(\frac{\partial^2 \phi_2}{\partial x^2} + \frac{\partial^2 \phi_2}{\partial y^2} \right) + \Gamma_{23} \left(\frac{\partial^2 \phi_3}{\partial x^2} + \frac{\partial^2 \phi_3}{\partial y^2} \right) = 0 \quad .$$

$$\Gamma_{31} \left(\frac{\partial^2 \phi_1}{\partial x^2} + \frac{\partial^2 \phi_1}{\partial y^2} \right) + \Gamma_{32} \left(\frac{\partial^2 \phi_2}{\partial x^2} + \frac{\partial^2 \phi_2}{\partial y^2} \right) + \Gamma_{33} \left(\frac{\partial^2 \phi_3}{\partial x^2} + \frac{\partial^2 \phi_3}{\partial y^2} \right) = 0$$

Discretizing using the finite difference procedure yields

$$\sum_{k=1}^{3} \Gamma_{1k} \left(\frac{\phi_{k,i+1,j} - 2\phi_{k,i,j} + \phi_{k,i-1,j}}{(\Delta x)^2} + \frac{\phi_{k,i,j+1} - 2\phi_{k,i,j} + \phi_{k,i,j-1}}{(\Delta y)^2} \right) = 0$$

$$\sum_{k=1}^{3} \Gamma_{2k} \left(\frac{\phi_{k,i+1,j} - 2\phi_{k,i,j} + \phi_{k,i-1,j}}{(\Delta x)^2} + \frac{\phi_{k,i,j+1} - 2\phi_{k,i,j} + \phi_{k,i,j-1}}{(\Delta y)^2} \right) = 0 \quad .$$

$$\sum_{k=1}^{3} \Gamma_{3k} \left(\frac{\phi_{k,i+1,j} - 2\phi_{k,i,j} + \phi_{k,i-1,j}}{(\Delta x)^2} + \frac{\phi_{k,i,j+1} - 2\phi_{k,i,j} + \phi_{k,i,j-1}}{(\Delta y)^2} \right) = 0$$

In this example, both the segregated and the coupled solution approach will be explored, and the two approaches will be compared and contrasted.

In the coupled solution approach, the above equations are solved in a coupled manner using the block ADI method. Following the procedure described to derive Eq. (8.69), block tridiagonal matrices are set up for all three equations together, and row-wise and column-wise sweeps (inner iterations) are performed to sweep through the computational domain. An absolute tolerance of 10^{-6} was used for convergence.

In the segregated solution approach, the above equations are written such that only the diagonal term is treated implicitly, while the others are treated explicitly. This yields the following three semi-implicit equations:

$$\Gamma_{11}\left(\frac{\phi_{1,i+1,j}-2\phi_{1,i,j}+\phi_{1,i-1,j}}{(\Delta x)^2}+\frac{\phi_{1,i,j+1}-2\phi_{1,i,j}+\phi_{1,i,j-1}}{(\Delta y)^2}\right)$$

$$=-\sum_{k=2}^{3}\Gamma_{1k}\left(\frac{\phi_{k,i+1,j}-2\phi_{k,i,j}+\phi_{k,i-1,j}}{(\Delta x)^2}+\frac{\phi_{k,i,j+1}-2\phi_{k,i,j}+\phi_{k,i,j-1}}{(\Delta y)^2}\right)$$

$$\Gamma_{22}\left(\frac{\phi_{2,i+1,j}-2\phi_{2,i,j}+\phi_{2,i-1,j}}{(\Delta x)^2}+\frac{\phi_{2,i,j+1}-2\phi_{2,i,j}+\phi_{2,i,j-1}}{(\Delta y)^2}\right)$$

$$=-\sum_{\substack{k=1\\k\neq2}}^{3}\Gamma_{2k}\left(\frac{\phi_{k,i+1,j}-2\phi_{k,i,j}+\phi_{k,i-1,j}}{(\Delta x)^2}+\frac{\phi_{k,i,j+1}-2\phi_{k,i,j}+\phi_{k,i,j-1}}{(\Delta y)^2}\right)$$

$$\Gamma_{33}\left(\frac{\phi_{3,i+1,j}-2\phi_{3,i,j}+\phi_{3,i-1,j}}{(\Delta x)^2}+\frac{\phi_{3,i,j+1}-2\phi_{3,i,j}+\phi_{3,i,j-1}}{(\Delta y)^2}\right)$$

$$=-\sum_{k=1}^{2}\Gamma_{3k}\left(\frac{\phi_{k,i+1,j}-2\phi_{k,i,j}+\phi_{k,i-1,j}}{(\Delta x)^2}+\frac{\phi_{k,i,j+1}-2\phi_{k,i,j}+\phi_{k,i,j-1}}{(\Delta y)^2}\right)$$

Here, the above equations were solved using the ADI method. Two different strategies were employed in the segregated solution approach. The first strategy was to use only one inner iteration (sweep) both row- and column-wise. This implies that in a single outer iteration, the computational domain is swept through twice (one row-wise, one column-wise). In the second strategy, a residual criterion was set for the inner iterations, and an absolute tolerance of 10^{-3} was used for the inner iterations. In this case, the number of inner iterations varied from outer iteration to outer iteration. For both strategies, an absolute tolerance of 10^{-6} was used for the outer iterations. The mesh size used was 101×101.

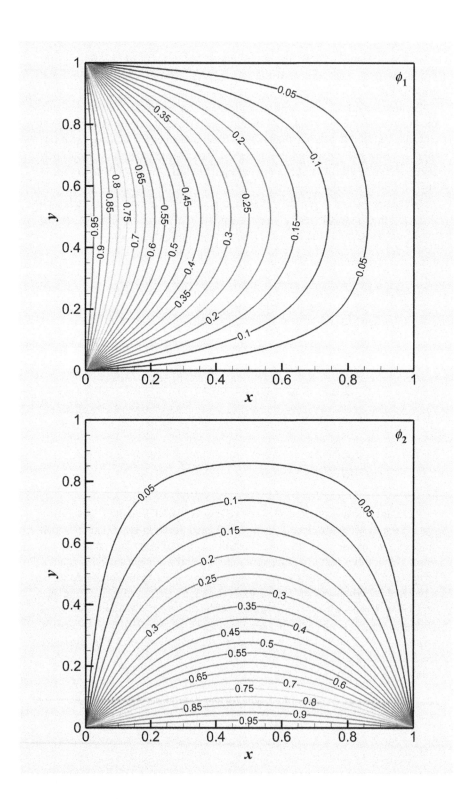

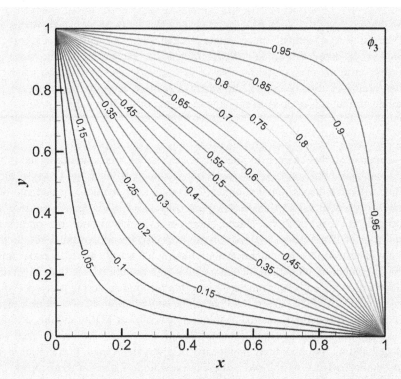

The solution (contour plots) obtained for the three dependent variables is shown and in the figure here. It is identical for all methods used, as it should be. The table shows the number of iterations and the CPU time taken for the computations on a 2.2 GHz Intel core i7 processor. The number of iterations listed is the number of outer iterations in the case of the segregated solver, and the number of inner iterations for the coupled solver (there is only one iteration loop in the coupled solver).

Solution method	Number of iterations	CPU time (s)
Segregated solution with strategy 1	7454	26.1
Segregated solution with strategy 2	2882	33.7
Coupled solution	5116	27.5

Some interesting trends are shown by the data in the table above. First, for the segregated solution approach, when inner iterations are used (strategy 2), the number of outer iterations goes down significantly. However, the decrease in number of iterations is not reflected in the CPU time, which does not decrease. This is because each outer iteration is now significantly more expensive. When the coupled solver is used, again, the number of iterations decreases when compared with strategy 1 (this strategy and the coupled solver are similar; one uses ADI equation by equation, while the other uses block ADI), but the CPU time does not decrease accordingly. Again, this indicates that one sweep of the block ADI is more expensive the two segregated sweeps of the scalar ADI method. This example elucidates the *pros* and *cons* of segregated versus coupled solution of PDEs. In general, coupled solution provides benefits in terms of the total number of iterations. However, that may often come at an increased computational cost. The balance between these two conflicting attributes is quite problem dependent, and difficult to judge *a priori*.

To summarize, this chapter on selected miscellaneous topics was designed to tie a few loose ends. First, no text on numerical solution of PDEs can be considered complete without a discussion of nonlinear PDEs and their solution. Second, the vast majority of practical applications that require solution of PDEs require solution of not just one, but a set of PDEs. Hence, some coverage, albeit brief, on solution of a coupled set of PDEs was warranted. The chapter also introduced the reader to the paradigm of nested inner and outer iterations. Finally, a few truly miscellaneous topics such as interpolation, numerical integration, and root finding of nonlinear algebraic equations was included because these topics often materialize during postprocessing of data generated by PDE solvers. Of course, the solution of a set of nonlinear algebraic equations is inherently related to solution of nonlinear PDEs, and this connection has also been established in this chapter. The all-important Newton's method was introduced and its role in solving nonlinear PDEs was delineated. While a lot more remains untouched, it is envisioned that the introduction of some of these topics – in particular, nonlinear PDEs and coupled PDEs – will make the reader comfortable in embarking upon the task of understanding and developing advanced codes that are geared toward specific applications of the aforementioned topics.

REFERENCES

[1] Kelly CT. *Solving Nonlinear Equations with Newton's Method*, SIAM Series in Fundamentals of Algorithms, (2003), ISBN 978-0-89871-546-0.

[2] Ward WE, Kincaid DR. Numerical Mathematics and Computing. 7th ed. Cengage Learning; 2012.

[3] Hoffman JD, Frankel S. Numerical Methods for Engineers and Scientists. 2nd ed. New York: Marcel Decker Inc; 2001.

[4] Ferziger JH, Perić M. Computational Methods for Fluid Dynamics. 3rd ed. Springer: Berlin; 2002.

[5] Pletcher RH, Tannehill JC, Anderson D. Computational Fluid Mechanics and Heat Transfer. 3rd ed. CRC Press: Boca Raton, FL; 2011.

[6] Harten A, Engquist B, Osher S, Chakravarthy S. Uniformly high order essentially nonoscillatory schemes, III. J Comput Phys 1987;71:231–303.

[7] de Boor C. A Practical Guide to Splines. Revised ed. In: Marsden JE, Sirovich L, editors. Applied Mathematical Sciences 2001: Springer: Berlin; 27.

[8] Abramowitz M, Stegun IA, editors. Handbook of Mathematical Functions. New York: Dover Publications; 1965.

[9] Wesseling P. An Introduction to Multigrid Methods. R.T. Edwards Inc: Philadelphia, PA; 1992.

[10] Drikakis D, Iliev OP, Vassileva DP. A nonlinear multigrid method for three-dimensional incompressible Navier-Stokes equations. J Comput Phys 1998;146:310–21.

[11] Mavriplis DJ. An assessment of linear versus non-linear multigrid methods for unstructured mesh solvers. J Comput Phys 2002;175:302–25.

[12] Brown PN, Saad Y. Hybrid Krylov methods for nonlinear systems of equations. SIAM J Sci Stat Comput 1990;11:450–81.

[13] Chan TF, Jackson KR. Nonlinearly preconditioned Krylov subspace methods for discrete Newton algorithms. SIAM J Sci Stat Comput 1984;5:533–42.

[14] Knoll DA, Keyes DE. Jacobian-free Newton-Krylov methods: a survey of approaches and applications. J Comput Phys 2004;193:357–97.

[15] Thompson RJ, Wilson A, Moeller T, Merkle CL, A Strong Conservative Implicit Riemann Solver for Coupled Navier-Stokes and Full Maxwell Equations, Proceedings of the Fifty First AIAA Aerospace Sciences Meeting, Grapevine, Texas, Paper Number 2013–0210, 2013.

[16] Mathur SR, Murthy JY. Acceleration of anisotropic scattering computations using coupled ordinates method (COMET). J Heat Transfer 2001;123(3):607–12.

[17] Mazumder S. A new numerical procedure for coupling radiation in participating media with other modes of heat transfer. J Heat Transfer 2005;127(9):1037–45.

[18] Ravishankar M, Mazumder S, Kumar A. Finite-volume formulation and solution of the P_3 equations of radiative transfer on unstructured meshes. J Heat Transfer 2010;132(2). Article number 023402 (1–11).

[19] Mazumder S. Critical assessment of the stability and convergence of the equations of multi-component diffusion. J Comput Phys 2006;212(2):383–92.

[20] Kumar A, Mazumder S. Coupled solution of the species conservation equations using unstructured finite-volume method. Int J Numer Meth Fluids 2010;64(4):409–42.

[21] Loy JM, Mathur SR, Murthy JY. A coupled ordinates method for convergence acceleration of the phonon boltzmann transport equation. J Heat Transfer 2015;137(1). Article number 012402 (1–9).

[22] Vanka SP. Block-implicit multigrid solution of Navier-Stokes equations in primitive variables. J Comput Phys 1986;65:138–58.

[23] de Lemos MJS. A block-implicit numerical procedure for simulation of buoyant swirling flows in a model furnace. Int J Numer Meth Fluids 2003;43:281–99.

[24] Raw MJ, A coupled algebraic multigrid method for the 3D Navier–Stokes equations. In: Hackbusch W, Wittum G, editors. Fast solvers for flow problems. Proceedings of the tenth GAMM-Seminar. Kiel, Germany, Notes Numer. Fluid Mech., vol. 49. Wiesbaden: Vieweg; 1995. pp. 204–215.

EXERCISES

8.1 Generate a data table of 10 equally spaced nodes by using the function $f(x) = 1/x$ in the interval $[1,10]$. Then, obtain a fit to the data (write computer programs if necessary) using the following three methods:

   a. Power polynomial of degree 9.

   b. Lagrange polynomial.

   c. Natural cubic spline.

Compare the accuracy of the three methods by making a plot of the three curve-fits and the original function on the same graph. Comment on your findings.

8.2 Which of the three methods considered in Problem 1 would be most unsuitable for extrapolation (rather than interpolation), and why? Hint: Find the value of the function at $x = 0.5$ and $x = 10.5$ using the three curve-fits that you have obtained.

8.3 Compute the integral $I = \int\limits_{1}^{2} \dfrac{x}{(e^x - 1)^2}\, dx$ using the following three methods.

  a. Trapezoidal rule.

  b. Polynomial integration.

  c. Cubic spline integration.

  d. Gauss–Legendre integration.

In each case, use five quadrature points. The exact analytical solution is $I = -\dfrac{1}{2}\left[\dfrac{e^x + 1}{e^x - 1}\right]_{1}^{2}$. Comment on the accuracy of each of the four methods.

8.4 Discuss what problems, if any, might be encountered if the lower limit of the integral in Problem 3 is changed from 1 to 0. How will you modify each of the above methods, if at all, to compute $I = \int\limits_{0}^{2} \dfrac{x}{(e^x - 1)^2}\, dx$? Which method is best suited for this computation and why?

8.5 Find the first 10 positive roots of the transcendental equation $x\tan x = 1$ using Newton's method. You may want to first plot the function to help devise a solution strategy.

8.6 Using the backward Euler method (Chapter 5) to discretize the time derivative, find the solution, $\phi(t)$ and $\psi(t)$, in the range $t \in [0, 1]$ to the following system of ordinary differential equations using Newton's method:

$$\frac{d\psi}{dt} = 10\psi^2 \exp[-1/\phi]$$
$$\frac{d\phi}{dt} = -\phi + 1 + \psi \exp[-1/\phi]$$

The initial conditions are $\phi(0) = 1$ and $\psi(0) = 1$. Use a time step size of 0.01.

8.7 Consider the following second-order nonlinear ordinary differential equation and boundary conditions:

$$\frac{d^2\phi}{dx^2} = \exp(C\phi), \quad \phi(0) = 0, \phi(1) = 1, \text{ where } C \text{ is a constant.}$$

Discretize the equation using the central difference scheme. Use equal node spacing with 41 nodes. Obtain the solution to the resulting algebraic equations using Newton's method for $C = 2$ and $C = 4$. Set the tolerance for the residual to 10^{-10}. Plot the solution ($\phi(x)$ versus x). Also, plot the residuals ($R2$ versus number of iterations) on a semilog plot.

8.8   Consider the following set of coupled PDEs to be solved on a unit square with a uniform Cartesian mesh and the finite difference method:

$$\frac{\partial}{\partial x}(\phi^2) + \frac{\partial}{\partial y}(\phi\psi) = \frac{\partial^2 \phi}{\partial x^2} + \frac{\partial^2 \phi}{\partial y^2}$$

$$\frac{\partial}{\partial x}(\phi\psi) + \frac{\partial}{\partial y}(\psi^2) = \frac{\partial^2 \psi}{\partial x^2} + \frac{\partial^2 \psi}{\partial y^2} \quad ,$$

The boundary conditions are $\phi(0, y) = \psi(0, y) = 1$, with a value of zero on all other boundaries for both dependent variables. Obtain the solution to the above system on an 81×81 mesh using:

a.   The segregated solution approach with the Gauss–Seidel method for inner iterations.

b.   The coupled solution approach with the block Gauss–Seidel method.

In both the cases, use a convergence tolerance of 10^{-6}. The central difference scheme is to be used for discretization of all terms. Use inertial damping or under-relaxation if deemed necessary to attain convergence. Comment on your results, especially from the perspectives of computational efficiency, ease of implementation, and memory usage of the two methods.

Useful Relationships in Matrix Algebra

In this appendix, useful definitions, terminology, and relationships in matrix algebra that have been utilized in this text are summarized.

A matrix, $[A]$, of size $N \times M$ has N rows and M columns, and may be written as

$$[A] = \begin{bmatrix} a_{1,1} & a_{1,2} & \cdots & a_{1,j} & \cdots & a_{1,M} \\ a_{2,1} & a_{2,2} & \cdots & a_{2,j} & \cdots & a_{2,M} \\ \vdots & \vdots & \vdots & \vdots & \vdots & \vdots \\ a_{N,1} & a_{N,2} & \cdots & \cdots & \cdots & a_{N,M} \end{bmatrix}, \qquad (A.1)$$

where $a_{i,j}$ are the *elements* of the matrix. The first subscript, i, denotes the row number, while the second subscript, j, denotes the column number. If the number of rows and columns are the same, i.e., $N = M$, the matrix is known as a *square matrix*. If the number of rows of the matrix is equal to 1 ($N = 1$), the resulting matrix is known as a *row matrix*, and is written as

$$[A] = \begin{bmatrix} a_1 & a_1 & \cdots & a_i & \cdots & a_M \end{bmatrix}. \qquad (A.2)$$

For simplicity, the elements of a row matrix may be denoted by a single subscript, and written as a_i. Conversely, if the number of columns of the matrix is equal to 1 ($M = 1$), the resulting matrix is known as a column matrix, and is written as

$$[A] = \begin{bmatrix} a_1 \\ a_2 \\ \vdots \\ a_j \\ \vdots \\ a_N \end{bmatrix}. \qquad (A.3)$$

As in the case of a row matrix, the elements of a column matrix may also be denoted by a single index. A column matrix is also sometimes referred to as a *vector*. It is a vector in hyperdimensional space. It is useful to liken such hyperdimensional vectors to vectors encountered in Euclidean geometry except that hyperdimensional vectors have more than three components and therefore cannot be represented graphically. Likewise, a matrix is often referred to as a *tensor* of order 2 (see Appendix C for details).

Numerical Methods for Partial Differential Equations: Finite Difference and Finite Volume Methods
http://dx.doi.org/10.1016/B978-0-12-849894-1.00009-3

The *transpose* of a matrix is formed by interchanging its row elements with its column elements. Thus, the transpose of a $N \times M$ matrix is a $M \times N$ matrix. It follows then that the transpose of a row matrix is a column matrix and vice versa. Mathematically, if $[B] = [A]^T$, then the elements of matrix $[B]$ are given by $b_{i,j} = a_{j,i}$.

Matrix–matrix multiplication involves multiplying the row elements of the first matrix with the column elements of the second. This necessitates that the number of columns of the first matrix must be the same as the number of rows of the second matrix. Thus, a $N \times M$ matrix can be multiplied with a $M \times K$ matrix, since the number of columns of the first matrix and the number of rows in the second matrix are both equal to M. The reverse is not possible, i.e., multiplication of a $M \times K$ matrix with a $N \times M$ matrix is invalid. The preceding rule for multiplication also implies that matrix multiplication, in general, is not commutative, i.e., $[A][B] \neq [B][A]$. If two compatible matrices, $[A]$ and $[B]$ are multiplied such that $[C] = [A][B]$, then the elements of the matrix $[C]$ may be computed using

$$c_{i,k} = \sum_{j=1}^{M} a_{i,j} b_{j,k},$$ (A.4)

where the matrix $[A]$ is of size $N \times M$ and the matrix $[B]$ is of size $M \times K$. The product matrix $[C]$ will be of size $N \times K$.

A common operation used in computational linear algebra is the *matrix–vector multiplication*. It is no different than a matrix–matrix multiplication, except that in this case, the second matrix is a column matrix or vector of size $M \times 1$. Therefore, for simplicity of notation, the second subscript is usually dropped, and the elements of the product matrix are written as

$$c_i = \sum_{j=1}^{M} a_{i,j} b_j.$$ (A.5)

In this case, the product matrix $[C]$ will be of size $N \times 1$, i.e., it is also a vector. To summarize, matrix–vector multiplication results in a vector, while matrix–matrix multiplication results in a tensor.

A special case arises when a row matrix (or transpose of a vector) is multiplied with another vector. The result is a scalar. Such a product is known as the *dot product* or *scalar product* or *inner product* and is similar to the dot product between two vectors in geometry, which also yields a scalar. The dot product is a commutative operation. Mathematically, the dot product is computed using

$$c = \sum_{i=1}^{N} a_i b_i,$$ (A.6)

where again for the sake of compatibility, the row matrix $[A]$ and the column matrix $[B]$ must have the same length. When a column matrix (or vector) is multiplied with its own transpose, the resulting dot product is the square of the so-called L^2Norm or *Euclidean norm*. It represents the square of the length of the vector. Thus,

$$|A| = \sqrt{[A]^T[A]} = \sqrt{\sum_{i=1}^{N} a_i^2}. \tag{A.7}$$

As discussed extensively throughout this text, the Euclidean norm of the residual vector is conventionally used to monitor convergence.

A square matrix $[A]$ is said to be *symmetric* if $[A]^T = [A]$. This implies that the diagonal of the matrix constitutes a mirror about which the elements may be reflected, and such reflections will not alter the matrix. A matrix $[A]$ is said to be *positive definite* if for any nonzero vector $[X]$, the product $[X]^T[A][X] > 0$. Conversely, a matrix $[A]$ is said to be *negative definite* if $[X]^T[A][X] < 0$.

As in the case of geometry, if the dot product of two vectors $[X]$ and $[Y]$ is equal to zero, i.e., $[X]^T[Y] = 0$, then the two vectors $[X]$ and $[Y]$ are said to be *orthogonal*. On the other hand, if the dot product is zero with respect to another matrix $[A]$ such that $[X]^T[A][Y] = 0$, then the two vectors $[X]$ and $[Y]$ are said to be *A-orthogonal*.

The *minor*, $M_{i,j}$, of a square matrix $[A]$, sometimes referred to as the *(i,j)-minor*, is determined by computing the determinant of the matrix that remains after striking out the *i*th row and *j*th column of $[A]$. Thus, for a 3×3 $[A]$ matrix written as

$$[A] = \begin{bmatrix} a_{1,1} & a_{1,2} & a_{1,3} \\ a_{2,1} & a_{2,2} & a_{2,3} \\ a_{3,1} & a_{3,2} & a_{3,3} \end{bmatrix}, \text{ the minors may be written as } M_{1,1} = \det\begin{bmatrix} a_{2,2} & a_{2,3} \\ a_{3,2} & a_{3,3} \end{bmatrix},$$

$$M_{1,2} = \det\begin{bmatrix} a_{2,1} & a_{2,3} \\ a_{3,1} & a_{3,3} \end{bmatrix}, \quad M_{1,3} = \det\begin{bmatrix} a_{2,1} & a_{2,2} \\ a_{3,1} & a_{3,2} \end{bmatrix}, \text{ and so on. The } \textit{cofactors} \text{ of}$$

matrix $[A]$ may be written in terms of its minors as

$$C_{i,j} = (-1)^{i+j} M_{i,j}. \tag{A.8}$$

The *determinant* of a matrix may be expressed in terms of its cofactors using the so-called Laplace's formula to yield

$$\det[A] = A_{i,1}C_{i,1} + A_{i,2}C_{i,2} + \ldots + A_{i,N}C_{i,N} = \sum_{j=1}^{N} A_{i,j}C_{i,j}, \tag{A.9a}$$

$$\det[A] = A_{1,j}C_{1,j} + A_{2,j}C_{2,j} + \ldots + A_{N,j}C_{N,j} = \sum_{i=1}^{N} A_{i,j}C_{i,j}. \tag{A.9b}$$

Either formula, given by Eq. (A.9), may be used to compute the determinant. Equation (A.9a) uses a row-wise (fixed *i*) scan, while Eq. (A.9) uses a column-wise (fixed *j*) scan.

Using *Cramer's rule*, the *inverse* of a matrix may be expressed in terms of its cofactors and determinant, as follows:

$$[A]^{-1} = \frac{[C(A)]^T}{\det[A]}, \tag{A.10}$$

where the symbol $[C(A)]$ has been used to denote the cofactor matrix of $[A]$, whose elements are given by Eq. (A.8).

Listed below are a few additional useful matrix identities that have been employed throughout the text:

$$\left[[A]^T\right]^T = [A], \tag{A.11}$$

$$\left[[A][B]\right]^T = [B]^T[A]^T, \tag{A.12}$$

$$\left[[A][B]\right]^{-1} = [B]^{-1}[A]^{-1}, \tag{A.13}$$

$$\left[[A]^T\right]^{-1} = \left[[A]^{-1}\right]^T, \tag{A.14}$$

$$\det\left[[A][B]\right] = \det[A] \cdot \det[B], \tag{A.15}$$

$$\det\left[[A]^{-1}\right] = \frac{1}{\det[A]}. \tag{A.16}$$

Useful Relationships in Vector Calculus

B

In this Appendix, symbols and notations commonly encountered in vector calculus have been expanded and elaborated upon to convey their precise meanings. The most commonly used coordinate systems, namely, Cartesian, cylindrical, and spherical coordinates have been utilized for expansions.

The *del* operator is the most commonly used operator in vector calculus. It is denoted by the Greek symbol nabla (∇). Its mathematical representation requires a coordinate system. For the Cartesian, cylindrical, and spherical coordinate systems, it is written as

$$\text{Cartesian}(x,y,z): \nabla \equiv \hat{\mathbf{i}}\frac{\partial}{\partial x} + \hat{\mathbf{j}}\frac{\partial}{\partial y} + \hat{\mathbf{k}}\frac{\partial}{\partial z}, \tag{B.1a}$$

$$\text{Cylindrical}(r,\theta,z): \nabla \equiv \hat{\mathbf{r}}\frac{\partial}{\partial r} + \hat{\boldsymbol{\theta}}\frac{1}{r}\frac{\partial}{\partial \theta} + \hat{\mathbf{z}}\frac{\partial}{\partial z}, \tag{B.1b}$$

$$\text{Spherical}(r,\theta,\psi): \nabla \equiv \hat{\mathbf{r}}\frac{\partial}{\partial r} + \hat{\boldsymbol{\theta}}\frac{1}{r}\frac{\partial}{\partial \theta} + \hat{\boldsymbol{\psi}}\frac{1}{r\sin\theta}\frac{\partial}{\partial \psi}. \tag{B.1c}$$

It is evident from Eq. (B.1) that the *del* operator is a vector operator. When operating on a scalar, it yields the *gradient* of the scalar. Thus, the gradient of a scalar, ϕ, in the above three coordinate systems may be written as

$$\text{Cartesian}(x,y,z): \nabla\phi \equiv \hat{\mathbf{i}}\frac{\partial\phi}{\partial x} + \hat{\mathbf{j}}\frac{\partial\phi}{\partial y} + \hat{\mathbf{k}}\frac{\partial\phi}{\partial z}, \tag{B.2a}$$

$$\text{Cylindrical}(r,\theta,z): \nabla\phi \equiv \hat{\mathbf{r}}\frac{\partial\phi}{\partial r} + \hat{\boldsymbol{\theta}}\frac{1}{r}\frac{\partial\phi}{\partial \theta} + \hat{\mathbf{z}}\frac{\partial\phi}{\partial z}, \tag{B.2b}$$

$$\text{Spherical}(r,\theta,\psi): \nabla\phi \equiv \hat{\mathbf{r}}\frac{\partial\phi}{\partial r} + \hat{\boldsymbol{\theta}}\frac{1}{r}\frac{\partial\phi}{\partial \theta} + \hat{\boldsymbol{\psi}}\frac{1}{r\sin\theta}\frac{\partial\phi}{\partial \psi}. \tag{B.2c}$$

Clearly, the gradient of a scalar is a vector, with its three components as shown in Eq. (B.2). When operating on a vector, $\mathbf{q}$, there are two possibilities: either to perform a dot product with the vector or to perform a cross product with the vector. The dot product yields the *divergence* of the vector, and is written as

$$\text{Cartesian}(x,y,z): \nabla \bullet \mathbf{q} = \frac{\partial q_x}{\partial x} + \frac{\partial q_y}{\partial y} + \frac{\partial q_z}{\partial z}, \tag{B.3a}$$

$$\text{Cylindrical}(r,\theta,z): \nabla \bullet \mathbf{q} = \frac{1}{r}\frac{\partial(rq_r)}{\partial r} + \frac{1}{r}\frac{\partial q_\theta}{\partial \theta} + \frac{\partial q_z}{\partial z}, \tag{B.3b}$$

$$\text{Spherical}(r,\theta,\psi): \nabla \bullet \mathbf{q} = \frac{1}{r^2}\frac{\partial(r^2 q_r)}{\partial r} + \frac{1}{r\sin\theta}\frac{\partial(\sin\theta\, q_\theta)}{\partial \theta} + \frac{1}{r\sin\theta}\frac{\partial q_\psi}{\partial \psi}. \tag{B.3c}$$

In contrast, the cross product yields the *curl* of the vector, and is written as

$$\text{Cartesian}(x,y,z): \nabla \times \mathbf{q} = \left(\frac{\partial q_z}{\partial y} - \frac{\partial q_y}{\partial z}\right)\hat{\mathbf{i}} + \left(\frac{\partial q_x}{\partial z} - \frac{\partial q_z}{\partial x}\right)\hat{\mathbf{j}} + \left(\frac{\partial q_y}{\partial x} - \frac{\partial q_x}{\partial y}\right)\hat{\mathbf{k}}, \tag{B.4a}$$

$$\text{Cylindrical}(r,\theta,z): \nabla \times \mathbf{q} = \left(\frac{1}{r}\frac{\partial q_z}{\partial \theta} - \frac{\partial q_\theta}{\partial z}\right)\hat{\mathbf{r}} + \left(\frac{\partial q_r}{\partial z} - \frac{\partial q_z}{\partial r}\right)\hat{\boldsymbol{\theta}} + \frac{1}{r}\left(\frac{\partial(rq_\theta)}{\partial r} - \frac{\partial q_r}{\partial \theta}\right)\hat{\mathbf{z}}, \tag{B.4b}$$

$$\text{Spherical}(r,\theta,\psi): \quad \begin{aligned} \nabla \times \mathbf{q} = &\frac{1}{r\sin\theta}\left(\frac{\partial(\sin\theta\, q_\psi)}{\partial \theta} - \frac{\partial q_\theta}{\partial \psi}\right)\hat{\mathbf{r}} \\ &+ \frac{1}{r}\left(\frac{1}{\sin\theta}\frac{\partial q_r}{\partial \psi} - \frac{\partial(rq_\psi)}{\partial r}\right)\hat{\boldsymbol{\theta}} + \frac{1}{r}\left(\frac{\partial(rq_\theta)}{\partial r} - \frac{\partial q_r}{\partial \theta}\right)\hat{\boldsymbol{\psi}} \end{aligned} \tag{B.4c}$$

The *scalar Laplacian* may be readily derived by substituting $\mathbf{q} = \nabla\phi$ in Eq. (B.3) to yield

$$\text{Cartesian}(x,y,z): \nabla^2\phi = \nabla \bullet (\nabla\phi) = \frac{\partial^2\phi}{\partial x^2} + \frac{\partial^2\phi}{\partial y^2} + \frac{\partial^2\phi}{\partial z^2}, \tag{B.5a}$$

$$\text{Cylindrical}(r,\theta,z): \nabla^2\phi = \nabla \bullet \nabla\phi = \frac{1}{r}\frac{\partial}{\partial r}\left(r\frac{\partial\phi}{\partial r}\right) + \frac{1}{r^2}\frac{\partial^2\phi}{\partial\theta^2} + \frac{\partial^2\phi}{\partial z^2}, \tag{B.5b}$$

$$\text{Spherical}(r,\theta,\psi): \quad \begin{aligned} \nabla^2\phi = \nabla \bullet \nabla\phi = &\frac{1}{r^2}\frac{\partial}{\partial r}\left(r^2\frac{\partial\phi}{\partial r}\right) + \frac{1}{r^2\sin\theta}\frac{\partial}{\partial\theta}\left(\sin\theta\frac{\partial\phi}{\partial\theta}\right) \\ &+ \frac{1}{r^2\sin^2\theta}\frac{\partial^2\phi}{\partial\psi^2} \end{aligned} \tag{B.5c}$$

Next, we present an identity that has been utilized in the text. For the sake of brevity, it is only presented in the Cartesian coordinate system, although it is just as easily applicable to the other two coordinate systems. Performing a dot product of Eq. (B.2a) with the three Cartesian unit vectors, it is straightforward to show that

$$\frac{\partial \phi}{\partial x} = (\nabla \phi) \cdot \hat{\mathbf{i}}, \frac{\partial \phi}{\partial y} = (\nabla \phi) \cdot \hat{\mathbf{j}}, \frac{\partial \phi}{\partial z} = (\nabla \phi) \cdot \hat{\mathbf{k}}, \tag{B.6}$$

which upon substituting back into Eq. (B.2a), yields

$$\nabla \phi \equiv \left[(\nabla \phi) \cdot \hat{\mathbf{i}} \right] \hat{\mathbf{i}} + \left[(\nabla \phi) \cdot \hat{\mathbf{j}} \right] \hat{\mathbf{j}} + \left[(\nabla \phi) \cdot \hat{\mathbf{k}} \right] \hat{\mathbf{k}}. \tag{B.7}$$

Equation (B.7) is valid for any coordinate system in which the three unit vectors are mutually orthogonal.

We end this Appendix by presenting a few additional vector identities that have been used throughout this text. For any three vectors denoted by **a**, **b**, and **c**, the following identities hold:

$$\mathbf{a} \cdot (\mathbf{b} \cdot \mathbf{c}) = (\mathbf{a} \cdot \mathbf{b}) \cdot \mathbf{c}, \tag{B.8}$$

$$\mathbf{a} \cdot (\mathbf{b} \times \mathbf{c}) = \mathbf{b} \cdot (\mathbf{c} \times \mathbf{a}) = \mathbf{c} \cdot (\mathbf{a} \times \mathbf{b}), \tag{B.9}$$

$$\mathbf{a} \times (\mathbf{b} \times \mathbf{c}) = (\mathbf{a} \cdot \mathbf{c})\mathbf{b} - (\mathbf{a} \cdot \mathbf{b})\mathbf{c}, \tag{B.10}$$

$$(\mathbf{a} \times \mathbf{b}) \times \mathbf{c} = (\mathbf{a} \cdot \mathbf{c})\mathbf{b} - (\mathbf{b} \cdot \mathbf{c})\mathbf{a}. \tag{B.11}$$

Tensor Notations and Useful Relationships

This appendix provides a brief overview of Cartesian tensor notations with the intention to facilitate the derivations and discussions provided in Sections 2.8 and 6.6 of this text. Additional derivations that clarify some of the equations presented in Section 2.8 are also presented in the second-half of this appendix.

The purpose of using Cartesian tensor notations is to make algebraic expressions compact and convenient to write. The fundamental rule in using Cartesian tensor notations is that if a single term contains a repeated index (subscript), it represents a summation over all three Cartesian components. Thus, $a_i b_i c_j$ is a compact way of writing $\sum_{i=1}^{3} a_i b_i c_j$. Here, since the index i is repeated, the summation is implied. Thus, in expanded form, $a_i b_i c_j = a_1 b_1 c_j + a_2 b_2 c_j + a_3 b_3 c_j$. Similarly, the term $a_i b_{ij} c_j$, when expanded, will result in nine terms, since both indices are repeated in this expression. Thus,

$$
\begin{aligned}
a_i b_{ij} c_j &= a_i b_{i1} c_1 + a_i b_{i2} c_2 + a_i b_{i3} c_3 \\
&= (a_1 b_{11} c_1 + a_2 b_{21} c_1 + a_3 b_{31} c_1) + (a_1 b_{12} c_2 + a_2 b_{22} c_2 + a_3 b_{32} c_2) \\
&\quad + (a_1 b_{13} c_3 + a_2 b_{23} c_3 + a_3 b_{33} c_3).
\end{aligned}
\tag{C.1}
$$

Since summation is commutative, the order of summation does not alter the result.

Using Cartesian tensor notations, the divergence of the vector $\mathbf{q}$ may be written as

$$
\nabla \bullet \mathbf{q} = \frac{\partial q_i}{\partial x_i}
\tag{C.2}
$$

Similarly, the scalar Laplacian may be written as

$$
\nabla^2 \phi = \frac{\partial}{\partial x_i} \left(\frac{\partial \phi}{\partial x_i} \right).
\tag{C.3}
$$

It is worth noting that since the repeated index is summed over, it is essentially the so-called dummy index. In other words, it can be arbitrarily switched to a different symbol. For example, the term $a_i b_{ij} c_j$ could just as easily have been written as $a_k b_{kj} c_j$ without altering the result. By the same token, since the repeated index has a special meaning, one has to be careful that it is switched in all occurrences within the term, and not just in a few locations.

In the discussion to follow, the derivation of Eqs (2.90) and (2.91) is presented, since these equations make use of Cartesian tensor notations, on account of which their derivation had to be deferred until the present appendix. For simplicity, we will use a 2D framework for our derivations. We begin with the 2D rendition of the Jacobian of forward transformation, which, following Eq. (2.88), may be written as

$$[J] = \begin{bmatrix} \dfrac{\partial x_1}{\partial \xi_1} & \dfrac{\partial x_1}{\partial \xi_2} \\ \dfrac{\partial x_2}{\partial \xi_1} & \dfrac{\partial x_2}{\partial \xi_2} \end{bmatrix}. \tag{C.4}$$

The inverse of $[J]$ may be determined readily using Cramer's rule to yield

$$[J]^{-1} = \frac{1}{J} \begin{bmatrix} \dfrac{\partial x_2}{\partial \xi_2} & -\dfrac{\partial x_1}{\partial \xi_2} \\ -\dfrac{\partial x_2}{\partial \xi_1} & \dfrac{\partial x_1}{\partial \xi_1} \end{bmatrix}, \tag{C.5}$$

where $J = \det[J]$. The cofactor matrix of $[J]$ may be written as

$$[\beta] = \begin{bmatrix} \dfrac{\partial x_2}{\partial \xi_2} & -\dfrac{\partial x_2}{\partial \xi_1} \\ -\dfrac{\partial x_1}{\partial \xi_2} & \dfrac{\partial x_1}{\partial \xi_1} \end{bmatrix}. \tag{C.6}$$

By comparing Eq. (C.5) with Eq. (C.6), it is clear that $[J]^{-1} = \dfrac{1}{J}[\beta]^T$, which is essentially Eq. (A.10). Equation (C.5) may be combined with Eq. (C.6) and the 2D rendition of Eq. (2.88) to write

$$\begin{bmatrix} \dfrac{\partial \xi_1}{\partial x_1} & \dfrac{\partial \xi_1}{\partial x_2} \\ \dfrac{\partial \xi_2}{\partial x_1} & \dfrac{\partial \xi_2}{\partial x_2} \end{bmatrix} = [J]^{-1} = \frac{1}{J} \begin{bmatrix} \beta_{11} & \beta_{21} \\ \beta_{12} & \beta_{22} \end{bmatrix}. \tag{C.7}$$

Next, we consider the 2D rendition of Eq. (2.89). Substitution of Eq. (C.7) into it yields

$$\begin{bmatrix} \dfrac{\partial \phi}{\partial x_1} \\ \dfrac{\partial \phi}{\partial x_2} \end{bmatrix} = \frac{1}{J} \begin{bmatrix} \beta_{11} & \beta_{12} \\ \beta_{21} & \beta_{22} \end{bmatrix} \begin{bmatrix} \dfrac{\partial \phi}{\partial \xi_1} \\ \dfrac{\partial \phi}{\partial \xi_2} \end{bmatrix} = \frac{1}{J} \begin{bmatrix} \beta_{11}\dfrac{\partial \phi}{\partial \xi_1} + \beta_{12}\dfrac{\partial \phi}{\partial \xi_2} \\ \beta_{21}\dfrac{\partial \phi}{\partial \xi_1} + \beta_{22}\dfrac{\partial \phi}{\partial \xi_2} \end{bmatrix}. \tag{C.8}$$

Using mathematical induction to extend to 3D, we obtain

$$\frac{\partial \phi}{\partial x_i} = \frac{1}{J}\left[\beta_{i1}\frac{\partial \phi}{\partial \xi_1} + \beta_{i2}\frac{\partial \phi}{\partial \xi_2} + \beta_{i3}\frac{\partial \phi}{\partial \xi_3}\right], \tag{C.9}$$

which, when written using Cartesian tensor notations, yields Eq. (2.90).

Next, we present the derivation of Eq. (2.91) from Eq. (2.90). In order to do so, we consider a single element of Eq. (C.9) [same as Eq. (2.90)] in 2D. For example,

$$\frac{\partial \phi}{\partial x_1} = \frac{1}{J}\left[\beta_{11}\frac{\partial \phi}{\partial \xi_1} + \beta_{12}\frac{\partial \phi}{\partial \xi_2}\right], \tag{C.10}$$

which, using the product rule of derivatives, may also be written as

$$\frac{\partial \phi}{\partial x_1} = \frac{1}{J}\left[\frac{\partial}{\partial \xi_1}(\beta_{11}\phi) - \phi\frac{\partial}{\partial \xi_1}(\beta_{11}) + \frac{\partial}{\partial \xi_2}(\beta_{12}\phi) - \phi\frac{\partial}{\partial \xi_2}(\beta_{12})\right]. \tag{C.11}$$

Substituting Eq. (C.6) for β_{11} and β_{12} into the second and fourth terms of the right-hand side of Eq. (C.11), we obtain

$$\frac{\partial \phi}{\partial x_1} = \frac{1}{J}\left[\frac{\partial}{\partial \xi_1}(\beta_{11}\phi) - \phi\frac{\partial^2 x_2}{\partial \xi_1 \partial \xi_2} + \frac{\partial}{\partial \xi_2}(\beta_{12}\phi) + \phi\frac{\partial^2 x_2}{\partial \xi_1 \partial \xi_2}\right] = \frac{1}{J}\left[\frac{\partial}{\partial \xi_1}(\beta_{11}\phi) + \frac{\partial}{\partial \xi_2}(\beta_{12}\phi)\right].$$
$$\tag{C.12}$$

Applying mathematical induction to extend Eq. (C.12) to 3D, and using Cartesian tensor notations, yields Eq. (2.91). The above derivations also highlight the simultaneous use of matrix and tensor notations.

Index

Printed in the United States
By Bookmasters